Style (cont.)
  Clichés 71
  Comparison 84
  Conciseness 90
  Contractions 101
  Defining Terms 116
  Direct Address 126
  Double Negatives 153
  Emphasis 167
  English, Varieties of 173
  Euphemisms 180
  Expletives 182
  Figures of Speech 189
  Gobbledygook 233
  Idioms 248
  Intensifiers 264
  Jargon 285
  Nominalizations 346
  Pace 367
  Parallel Structure 370
  Point of View 385
  Positive Writing 386
  Repetition 453
  Rhetorical Questions 490
  Sentence Variety 505
  Subordination 516
  Technical Writing Style 521
  Telegraphic Style 522
  Tone 532
  "You" Viewpoint 575
Word Choice 568
  Abstract / Concrete Words 6
  Antonyms 33
  Buzzwords 57
  Connotation / Denotation 96
  Dictionaries 123
  Foreign Words in English 194
  Functional Shift 224
  Malapropisms 315
  Synonyms 518
  Thesaurus 529
  Vague Words 546

Usage
See page 545.

Sentences and Paragraphs
Sentence Construction 498
  Appositives 39
  Clauses 70
  Complements 88
  Expletives 182
  Modifiers 334
  Objects 356
  Phrases 380
  Restrictive and Nonrestrictive
    Elements 470
  Syntax 518
Sentence Faults 503
  Comma Splice 77
  Dangling Modifiers 113
  Garbled Sentences 225
  Mixed Constructions 334
  Run-on Sentences 490

Grammar 234
  Agreement 23
  Case 62
  English as a Second Language 169
  Gender 226
  Mood 336
  Number (Grammar) 351
  Person 376
  Possessive Case 387
  Pronoun Reference 405
  Tense 523
  Voice 557
Parts of Speech 374
  Adjectives 12
  Adverbs 19
  Articles 39
  Conjunctions 95
  Functional Shift 224
  Interjections 264
  Nouns 349
  Prepositions 390
  Pronouns 405
  Verbals 546
  Verbs 549

Punctuation and Mechanics
Mechanics
  Abbreviations 2
  Acronyms and Initialisms 11
  Ampersands 33
  Capitalization 59
  Compound Words 90
  Contractions 101
  Dates 115
  Italics 283
  Numbers 352
  Prefixes 389
  Proofreaders' Marks 410
  Proofreading 411
  Spelling 512
  Suffixes 517
Punctuation 434
  Apostrophes 34
  Brackets 52
  Colons 75
  Commas 77
  Dashes 114
  Ellipses 161
  Exclamation Marks 180
  Hyphens 245
  Parentheses 373
  Periods 375
  Question Marks 436
  Quotation Marks 443
  Semicolons 497
  Slashes 508

Sentence Fragments 304
Paragraphs 36
Coherence 71
Transitions 537
Essay 362

Handbook of
Technical Writing

About the Authors

Gerald J. Alred is Professor of English at the University of Wisconsin–Milwaukee, where he teaches courses in the Graduate Professional Writing Program. He is author of numerous scholarly articles and several standard bibliographies on business and technical communication. He is Associate Editor of the *Journal of Business Communication* and a recipient of the prestigious Jay R. Gould Award for "profound scholarly and textbook contributions to the teaching of business and technical writing."

Charles T. Brusaw was a faculty member at NCR Corporation's Management College, where he developed and taught courses in professional writing, editing, and presentation skills for the corporation worldwide. Previously, he worked in advertising, technical writing, and public relations. He has been a communications consultant, an invited speaker at academic conferences, and a teacher of business writing at Sinclair Community College.

Walter E. Oliu served as Chief of the Publishing Services Branch at the U.S. Nuclear Regulatory Commission, where he managed the agency's printing, graphics, editing, and publishing programs. He also developed the public-access standards for and managed daily operations of the agency's public Web site. He has taught at Miami University of Ohio, Slippery Rock State University, and as an adjunct faculty member at Montgomery College and George Mason University.

Ninth Edition

Handbook of Technical Writing

Gerald J. Alred

Charles T. Brusaw

Walter E. Oliu

Bedford/St. Martin's   Boston ◆ New York

FOR BEDFORD/ST. MARTIN'S
Developmental Editors: Amy Hurd Gershman, Rachel Goldberg
Editorial Assistant: Marisa Feinstein
Production Supervisor: Andrew Ensor
Senior Marketing Manager: Karita dos Santos
Project Management: Books By Design, Inc.
Text Design: Books By Design, Inc.
Cover Design: Billy Boardman
Cover Art: Technical Abstract © Stockbyte
Composition: Pine Tree Composition, Inc.
Printing and Binding: RR Donnelley & Sons Company

President: Joan E. Feinberg
Editorial Director: Denise B. Wydra
Editor in Chief: Karen S. Henry
Director of Marketing: Karen R. Soeltz
Director of Editing, Design, and Production: Marcia Cohen
Assistant Director of Editing, Design, and Production: Elise S. Kaiser
Manager, Publishing Services: Emily Berleth

Library of Congress Control Number: 2007943428

Copyright © 2009, 2006, 2003 by Bedford/St. Martin's

All rights reserved. No part of this book may be reproduced, stored in a retrieval system, or transmitted in any form or by any means, electronic, mechanical, photocopying, recording, or otherwise, except as may be expressly permitted by the applicable copyright statutes or in writing by the Publisher.

Manufactured in the United States of America.

4 3 2 1
d c b a

For information, write: Bedford/St. Martin's, 75 Arlington Street, Boston, MA 02116   (617-399-4000)

ISBN-10: 1-4576-1061-2
ISBN-13: 978-1-4576-1061-5

ACKNOWLEDGMENTS

Figure B–1. Corporate Blog. "GM FastLane Blog." *http://fastlane.gmblogs.com.* Gm.com. Reprinted with the permission of General Motors Corporation.
Figure D–3. Dictionary Entry for "regard." From *The American Heritage Dictionary of the English Language,* Fourth Edition. Copyright © 2006 by Houghton Mifflin Company. Reprinted by permission. All rights reserved.

Acknowledgments and copyrights are continued at the back of the book on page 577, which constitutes an extension of the copyright page.

Contents

Contents by Topic	*inside front cover*
Preface	ix
Five Steps to Successful Writing	xv
Checklist of the Writing Process	xxiii
***Handbook of Technical Writing*: Alphabetical Entries**	1–576
Index	579
Commonly Misused Words and Phrases	627
Model Documents and Figures by Topic	*inside back cover*

Preface

Like previous editions, the ninth edition of the *Handbook of Technical Writing* is a comprehensive, easy-access guide to all aspects of technical communication in the classroom and on the job. It places writing in a real-world context with quick reference to hundreds of business writing topics and scores of model documents and visuals. Meeting the needs of today's writers, the ninth edition includes expanded coverage of audience and context and reflects the impact that e-mail and other technology have had on workplace communication. This comprehensive reference tool is accompanied by a robust Web site that works together with the text to offer expanded resources online.

Helpful Features

The ESL Tips boxes throughout the book offer special advice for multilingual writers. In addition, the Contents by Topic on the inside front cover includes a list of entries—ESL Trouble Spots—that may be of particular interest to nonnative speakers of English.

Digital Tips and Web Links boxes direct readers to specific, related resources on the companion Web site. The Digital Tips in the book suggest ways to use technology to simplify complex writing tasks, such as incorporating track changes and creating styles and templates. Expanded Digital Tips on the Web site offer step-by-step instructions for completing each task. Web Links in the book point students to related resources on the companion site, such as model documents, tutorials, and links to hundreds of useful, related Web sites.

Ethics Notes throughout the text highlight the ethical concerns of today's technical writers and offer advice for dealing with these concerns. A thorough discussion of copyright and plagiarism clarifies what plagiarism is in the digital age and highlights the ethical aspects of using and documenting sources appropriately.

New to This Edition

As mentioned above, our focus in revising the *Handbook* for this edition has been to address the impact that technology has had on workplace communication. We have updated our coverage of correspondence and other entries throughout the book to show that there is often more than one appropriate medium for a particular message. A report, for example, can be sent as a hard copy, an e-mail attachment, or an e-mail itself. To address this issue, we have expanded our rhetorical advice on analyzing context and audience and have added new information on instant messaging, blogs, and other means by which today's writers communicate. We have also thoroughly updated coverage of grammar, usage, and style, and have made the following additional improvements:

- **Expanded coverage of the latest types of writing for the Web** discusses FAQs and blogs as forms of collaborative writing and promotion. A new entry on content management suggests how writers can use this technology to electronically access, share, and revise a wide variety of digital forms.

- **New information on environmental-impact statements** reflects current environmental policy and ethics. Covering the scope, language, and organization of these statements, the new entry features a link to the Environmental Protection Agency Web site and a full-length example.

- **A new entry on repurposing** explains how writers can use content for multiple purposes and audiences by adapting it for different contexts and mediums.

- **Detailed job-search entries** discuss social-networking Web sites such as MySpace and Facebook and their relationship to current job-search issues.

- **Updated coverage of research and documentation** helps students find, use, and integrate sources effectively in their writing. Real-world documentation models and a visual guide to citing sources make this challenging topic more accessible.

- **Updated Digital Tips** throughout the book focus on using technology to assist with a variety of writing tasks, such as using wikis for collaborative documents and conducting meetings from remote locations.

- **New and updated sample documents and visuals** reflect the prominence of e-mail in the workplace. Other updated visuals include charts, graphs, drawings, tables, internationally recognized symbols, illustrated descriptions and instructions, brochure and newsletter pages, presentation slides, and more.

- **An updated companion Web site at** *bedfordstmartins.com/ alredtech* helps instructors take advantage of the *Handbook*'s potential as a text for face-to-face, online, or hybrid classes by offering lesson plans, handouts, teaching tips, and assignment ideas. For students, the Web site includes additional sample documents, useful tutorials, expanded Digital Tips, and links to hundreds of useful Web sites keyed to the *Handbook*'s main entries.

How to Use This Book

The *Handbook of Technical Writing* is made up of alphabetically organized entries with color tabs. Within each entry, underlined cross-references such as "**formal reports**" link readers to related entries that contain further information. Many entries present advice and guidelines in the form of convenient Writer's Checklists.

The *Handbook*'s alphabetical organization enables readers to find specific topics quickly and easily; however, readers with general questions will discover several alternate ways to find information in the book and on its companion Web site at *bedfordstmartins.com/alredtech*.

- **Contents by Topic.** The complete Contents by Topic on the inside front cover groups the alphabetical entries into topic categories. This topical key can help a writer focusing on a specific task or problem browse all related entries; it is also useful for instructors who want to correlate the *Handbook* with standard textbooks or their own course materials.

- **Commonly Misused Words and Phrases.** The list of Commonly Misused Words and Phrases on pages 627–28 extends the Contents by Topic by listing all the usage entries, which appear in *italics* throughout the book.

- **Model Documents and Figures by Topic.** The topically organized list of model documents and figures on the inside back cover makes it easier to browse the book's most commonly referenced sample documents and visuals to find specific examples of technical communication genres.

- **Checklist of the Writing Process.** The checklist on pages xxiii–xxiv helps readers reference key entries in a sequence useful for planning and carrying out a writing project.

- **Comprehensive Index.** The Index lists all the topics covered in the book, including subtopics within the main entries in the alphabetical arrangement.

Acknowledgments

For their invaluable comments and suggestions for this edition of *Handbook of Technical Writing*, we thank the following reviewers who responded to our questionnaire: Dana Anderson, Indiana University, Bloomington; Daniel Ding, Ferris State University; Daniel Fitzstephens, University of Colorado; Karen Griggs, Indiana University–Purdue University, Fort Wayne; Lila M. Harper, Central Washington University; Douglas Jerolimov, University of Virginia; John F. Lee, University of Texas at San Antonio; Joseph P. McCallus, Columbus State University; Barbara J. McCleary, University of Hartford; Laura Osborne, Stephen

F. Austin State University; Suzanne Kesler Rumsey, Indiana University–Purdue University, Fort Wayne; Michael Stephans, Bloomsburg University of Pennsylvania; Babette Wald, California State University, Dominguez Hills; Paul Walker, Northern Arizona University; and Thomas L. Warren, Oklahoma State University.

For their helpful reviews of the model documents, we thank Patricia C. Click, University of Virginia; Barbara D'Angelo, Arizona State University; Karen Gookin, Central Washington University; Dale Jacobson, University of North Dakota; Nancy Nygaard, University of Wisconsin–Milwaukee; and Linda Van Buskirk, Cornell University.

We owe special thanks to Michelle M. Schoenecker, University of Wisconsin–Milwaukee, for her outstanding contribution to the ninth edition, especially for her work on the entries "blogs," "FAQs," and "repurposing." Michelle's workplace experience and her graduate studies in professional writing were invaluable—her keen analysis and cheerful perspective brought fresh energy to the project.

We are indebted to Kenneth J. Cook, President, Ken Cook Co., for his ongoing support of this and earlier editions. For this edition, we thank especially Melissa Marney, Marketing Coordinator, and Wendy Ballard, Technical Writer, both of Ken Cook Co.

We thank Dave Clark, University of Wisconsin–Milwaukee, and Matthias Jonas, Manpower Inc., for developing the entry "content management." Thanks as well go to Stuart Selber, Pennsylvania State University, for his review and advice for the entry "repurposing." We appreciate the help of Gail M. Boviall, Department of Mathematical Sciences at the University of Wisconsin–Milwaukee, for her advice on the entry "mathematical equations."

We appreciate Rebekka Andersen and Richard Hay, University of Wisconsin–Milwaukee, for their continuing work on "documenting sources." Thanks especially to Sara Eaton Gaunt for her excellent work in updating this complicated section. Thanks also go to Erik Thelen, Marquette University, and Mohan Limaye, Boise State University, for their helpful advice and counsel. In addition, we are very much indebted to the many reviewers and contributors not named here who helped us shape the first eight editions.

We wish to thank Bedford/St. Martin's for supporting this book, especially Joan Feinberg, President, and Karen Henry, Editor in Chief. We are grateful to Emily Berleth, Manager of Publishing Services at Bedford/St. Martin's, and Herb Nolan of Books By Design for their patience and expert guidance. Finally, we wish to thank Amy Hurd Gershman and Rachel Goldberg, our developmental editors at Bedford/St. Martin's, whose professionalism and collegiality helped produce an outstanding edition.

We offer heartfelt thanks to Barbara Brusaw for her patience and time spent preparing the manuscript for the first five editions. We also gratefully acknowledge the ongoing contributions of many students and instructors at the University of Wisconsin–Milwaukee. Finally, special thanks go to Janice Alred for her many hours of substantive assistance and for continuing to hold everything together.

G. J. A.
C. T. B.
W. E. O.

Five Steps to Successful Writing

Successful writing on the job is not the product of inspiration, nor is it merely the spoken word converted to print; it is the result of knowing how to structure information using both text and design to achieve an intended purpose for a clearly defined audience. The best way to ensure that your writing will succeed—whether it is in the form of a memo, a résumé, a proposal, or a Web page—is to approach writing using the following steps:

1. Preparation
2. Research
3. Organization
4. Writing
5. Revision

You will very likely need to follow those steps consciously—even self-consciously—at first. The same is true the first time you use new software, interview a candidate for a job, or chair a committee meeting. With practice, the steps become nearly automatic. That is not to suggest that writing becomes easy. It does not. However, the easiest and most efficient way to write effectively is to do it systematically.

As you master the five steps, keep in mind that they are interrelated and often overlap. For example, your readers' needs and your purpose, which you determine in step 1, will affect decisions you make in subsequent steps. You may also need to retrace steps. When you conduct research, for example, you may realize that you need to revise your initial understanding of the document's purpose and audience. Similarly, when you begin to organize, you may discover the need to return to the research step to gather more information.

The time required for each step varies with different writing tasks. When writing an informal memo, for example, you might follow the first three steps (preparation, research, and organization) by simply listing the points in the order you want to cover them. In such situations, you gather and organize information in your mind as you consider your purpose and audience. For a formal report, the first three steps require well-organized research, careful note-taking, and detailed outlining. For a routine e-mail message to a coworker, the first four steps merge as you type the information onto the screen. In short, the five steps expand, contract, and at times must be repeated to fit the complexity or context of the writing task.

Dividing the writing process into steps is especially useful for collaborative writing, in which you typically divide work among team members, keep track of a project, and save time by not duplicating effort. For details on collaborating with others and using electronic tools to help you manage the process, see **collaborative writing**.*

Preparation

Writing, like most professional tasks, requires solid **preparation**. In fact, adequate preparation is as important as writing a draft. In preparation for writing, your goal is to accomplish the following four major tasks:

- Establish your primary purpose.
- Assess your audience (or readers) and the context.
- Determine the scope of your coverage.
- Select the appropriate medium.

Establishing Your Purpose. To establish your primary **purpose** simply ask yourself what you want your readers to know, to believe, or to be able to do after they have finished reading what you have written. Be precise. Often a writer states a purpose so broadly that it is almost useless. A purpose such as "to report on possible locations for a new research facility" is too general. However, "to compare the relative advantages of Paris, Singapore, and San Francisco as possible locations for a new research facility so that top management can choose the best location" is a purpose statement that can guide you throughout the writing process. In addition to your primary purpose, consider possible secondary purposes for your document. For example, a secondary purpose of the research-facilities report might be to make corporate executive readers aware of the staffing needs of the new facility so that they can ensure its smooth operation regardless of the location selected.

Assessing Your Audience and Context. The next task is to assess your **audience**. Again, be precise and ask key questions. Who exactly is your reader? Do you have multiple readers? Who needs to see or to use the document? What are your readers' needs in relation to your subject? What are their attitudes about the subject? (Skeptical? Supportive? Anxious? Bored?) What do your readers already know about the subject? Should you define basic terminology, or will such definitions merely bore, or even impede, your readers? Are you communicating with international readers and therefore dealing with issues inherent in **global communication**?

---

*In this discussion, as elsewhere throughout this book, words and phrases underlined and set in an alternate typeface refer to specific alphabetical entries.

For the research-facilities report, the readers are described as "top management." Who is included in that category? Will one of the people evaluating the report be the Human Resources Manager? If so, that person likely would be interested in the availability of qualified professionals as well as in the presence of training, housing, and perhaps even recreational facilities available to potential employees in each city. The Purchasing Manager would be concerned about available sources for materials needed by the facility. The Marketing Manager would give priority to the facility's proximity to the primary markets for its products and services and the transportation options that are available. The Chief Financial Officer would want to know about land and building costs and about each country's tax structure. The Chief Executive Officer would be interested in all this information and perhaps more. As with this example, many workplace documents have audiences composed of multiple readers. You can accommodate their needs through one of a number of approaches described in the entry **audience**.

In addition to knowing the needs and interests of your readers, learn as much as you can about the **context**. Simply put, context is the environment or circumstances in which writers produce documents and within which readers interpret their meanings. Everything is written in a context, as illustrated in many entries and examples throughout this book. To determine the effect of context on the research-facilities report, you might ask both specific and general questions about the situation and about your readers' backgrounds: Is this the company's first new facility, or has the company chosen locations for new facilities before? Have the readers visited all three cities? Have they already seen other reports on the three cities? What is the corporate culture in which your readers work, and what are its key values? What specific factors,

---

ESL TIPS for Considering Audiences

In the United States, **conciseness**, **coherence**, and **clarity** characterize good writing. Make sure readers can follow your writing, and say only what is necessary to communicate your message. Of course, no writing style is inherently better than another, but to be a successful writer in any language, you must understand the cultural values that underlie the language in which you are writing. See also **awkwardness**, **copyright**, **global communication**, **English as a second language**, and **plagiarism**.

Throughout this book we have included ESL Tips boxes like this one with information that may be particularly helpful to nonnative speakers of English. See the Contents by Topic on the inside front cover for listings of ESL Tips and ESL Trouble Spots, entries that may be of particular help to ESL writers.

such as competition, finance, and regulation, are recognized as important within the organization?

Determining the Scope. Determining your purpose and assessing your readers and context will help you decide what to include and what not to include in your writing. Those decisions establish the **scope** of your writing project. If you do not clearly define the scope, you will spend needless hours on research because you will not be sure what kind of information you need or even how much. Given the purpose and audience established for the report on facility locations, the scope would include such information as land and building costs, available labor force, cultural issues, transportation options, and proximity to suppliers. However, it probably would not include the early history of the cities being considered or their climate and geological features, unless those aspects were directly related to your particular business.

Selecting the Medium. Finally, you need to determine the most appropriate medium for communicating your message. Professionals on the job face a wide array of options—from **e-mail**, fax, voice mail, videoconferencing, and Web sites to more traditional means like **letters**, **memos**, **reports**, telephone calls, and face-to-face **meetings**.

The most important considerations in selecting the appropriate medium are the audience and the purpose of the communication. For example, if you need to collaborate with someone to solve a problem or if you need to establish rapport with someone, written exchanges could be far less efficient than a phone call or a face-to-face meeting. However, if you need precise wording or you need to provide a record of a complex message, communicate in writing. If you need to make information that is frequently revised accessible to employees at a large company, the best choice might be to place the information on the company's intranet site. If reviewers need to submit their written comments about a proposal, you can either send them paper copies of the proposal that can be faxed or scanned, or you can send them the word-processing file to insert their comments electronically. The comparative advantages and primary characteristics of the most typical means of communication are discussed in **selecting the medium**.

Research

The only way to be sure that you can write about a complex subject is to thoroughly understand it. To do that, you must conduct adequate **research**, whether that means conducting an extensive investigation for a major proposal—through interviewing, library and Internet research, and careful **note-taking**—or simply checking a company Web site and jotting down points before you send an e-mail to a colleague.

Methods of Research. Researchers frequently distinguish between primary and secondary research, depending on the types of sources consulted and the method of gathering information. *Primary research* refers to the gathering of raw data compiled from interviews, direct observation, surveys, experiments, **questionnaires**, and audio and video recordings, for example. In fact, direct observation and hands-on experience are the only ways to obtain certain kinds of information, such as the behavior of people and animals, certain natural phenomena, mechanical processes, and the operation of systems and equipment. *Secondary research* refers to gathering information that has been analyzed, assessed, evaluated, compiled, or otherwise organized into accessible form. Such forms or sources include books, articles, reports, Web documents, e-mail discussions, and brochures. Use the methods most appropriate to your needs, recognizing that some projects will require several types of research and that collaborative projects may require those research tasks to be distributed among team members.

Sources of Information. As you conduct research, numerous sources of information are available to you, including the following:

- Your own knowledge and that of your colleagues
- The knowledge of people outside your workplace, gathered through **interviewing for information**
- Internet sources, including Web sites, directories, archives, and discussion groups
- Library resources, including databases and indexes of articles as well as books and reference works
- Printed and electronic sources in the workplace, such as various **correspondence**, reports, and Web intranet documents

Consider all sources of information when you begin your research and use those that are appropriate and useful. The amount of research you will need to do depends on the scope of your project. See also **documenting sources**.

Organization

Without organization, the material gathered during your research will be incoherent to your readers. To organize information effectively, you need to determine the best way to structure your ideas; that is, you must choose a primary **method of development**.

Methods of Development. An appropriate method of development is the writer's tool for keeping information under control and the readers' means of following the writer's presentation. As you analyze the information you have gathered, choose the method that best suits your subject, your readers' needs, and your purpose. For example, if you were writing instructions for assembling office equipment, you would naturally present the steps of the process in the order readers should perform them: the **sequential method of development**. If you were writing about the history of an organization, your account would most naturally go from the beginning to the present: the **chronological method of development.** If your subject naturally lends itself to a certain method of development, use it—do not attempt to impose another method on it.

Often you will need to combine methods of development. For example, a persuasive brochure for a charitable organization might combine a specific-to-general method of development with a **cause-and-effect method of development**. That is, you could begin with persuasive case histories of individual people in need and then move to general information about the positive effects of donations on recipients.

Outlining. Once you have chosen a method of development, you are ready to prepare an outline. **Outlining** breaks large or complex subjects into manageable parts. It also enables you to emphasize key points by placing them in the positions of greatest importance. By structuring your thinking at an early stage, a well-developed outline ensures that your document will be complete and logically organized, allowing you to focus exclusively on writing when you begin the rough draft. An outline can be especially helpful for maintaining a collaborative-writing team's focus throughout a large project. However, even a short letter or memo needs the logic and structure that an outline provides, whether the outline exists in your mind or on-screen or on paper.

At this point, you must begin to consider **layout and design** elements that will be helpful to your readers and appropriate to your subject and purpose. For example, if **visuals** such as photographs or tables will be useful, this is a good time to think about where they may be positioned to be most effective and if they need to be prepared by someone else while you are writing and revising the draft. The outline can also suggest where **headings**, **lists**, and other special design features may be useful.

Writing

When you have established your purpose, your readers' needs, and your scope and have completed your research and your outline, you will be well prepared to write a first draft. Expand your outline into **paragraphs**, without worrying about **grammar**, refinements of language **usage**, or **punctuation**. Writing and revising are different activities; refinements come with **revision**.

Write the rough draft, concentrating entirely on converting your outline into sentences and paragraphs. You might try writing as though you were explaining your subject to a reader sitting across from you. Do not worry about a good opening. Just start. Do not be concerned in the rough draft about exact word choice unless it comes quickly and easily—concentrate instead on ideas.

Even with good preparation, writing the draft remains a chore for many writers. The most effective way to get started and keep going is to use your outline as a map for your first draft. Do not wait for inspiration—you need to treat writing a draft as you would any on-the-job task. The entry **writing a draft** describes tactics used by experienced writers—discover which ones are best suited to you and your task.

Consider writing an **introduction** last because then you will know more precisely what is in the body of the draft. Your opening should announce the subject and give readers essential background information, such as the document's primary purpose. For longer documents, an introduction should serve as a frame into which readers can fit the detailed information that follows.

Finally, you will need to write a **conclusion** that ties the main ideas together and emphatically makes a final significant point. The final point may be to recommend a course of action, make a prediction or a judgment, or merely summarize your main points—the way you conclude depends on the purpose of your writing and your readers' needs.

Revision

The clearer finished writing seems to the reader, the more effort the writer has likely put into its **revision**. If you have followed the steps of the writing process to this point, you will have a rough draft that needs to be revised. Revising, however, requires a different frame of mind than does writing the draft. During revision, be eager to find and correct faults and be honest. Be hard on yourself for the benefit of your readers. Read and evaluate the draft as if you were a reader seeing it for the first time.

Check your draft for accuracy, completeness, and effectiveness in achieving your purpose and meeting your readers' needs and expectations. Trim extraneous information: Your writing should give readers exactly what they need, but it should not burden them with unnecessary information or sidetrack them into loosely related subjects.

Do not try to revise for everything at once. Read your rough draft several times, each time looking for and correcting a different set of problems or errors. Concentrate first on larger issues, such as **unity** and **coherence**; save mechanical corrections, like **spelling** and **punctuation**, for later **proofreading**. See also **ethics in writing**.

Finally, for important documents, consider having others review your writing and make suggestions for improvement. For collaborative writing, of course, team members must review each other's work on segments of the document as well as the final master draft. Use the Checklist of the Writing Process on pages xxiii–xxiv to guide you not only as you revise but also throughout the writing process.

WEB LINK | **Style Guides and Standards**

Organizations and professional associations often follow such guides as *The Chicago Manual of Style, MLA Style Manual and Guide to Scholarly Publishing,* and *United States Government Printing Office Style Manual* to ensure consistency in their publications on issues of usage, format, and documentation. Because advice in such guides often varies, some organizations set their own standards for documents. Where such standards or specific style guides are recommended or required by regulations or policies, you should follow those style guidelines. For a selected list of style guides and standards, see *bcs.bedfordstmartins.com/alredtech* and select *Links for Handbook Entries.*

Checklist of the Writing Process

This checklist arranges key entries of the *Handbook of Technical Writing* according to the sequence presented in Five Steps to Successful Writing, which begins on page xv. This checklist is useful both for following the steps and for diagnosing writing problems.

Preparation 389

- ✔ Establish your **purpose** 435
- ✔ Identify your **audience** or **readers** 42, 448
- ✔ Consider the **context** 98
- ✔ Determine your **scope** of coverage 493
- ✔ **Select the medium** 494

Research 459

- ✔ **Brainstorm** to determine what you already know 53
- ✔ Conduct **research** 459
- ✔ Take notes (**note-taking**) 347
- ✔ **Interview for information** 270
- ✔ Create and use **questionnaires** 437
- ✔ Avoid **plagiarism** 383
- ✔ **Document sources** 129

Organization 361

- ✔ Choose the best **methods of development** 329
- ✔ **Outline** your notes and ideas 362
- ✔ Develop and integrate **visuals** 552
- ✔ Consider **layout and design** 295

Writing a Draft 569

- ✔ Select an appropriate **point of view** 385
- ✔ Adopt an appropriate **style** and **tone** 513, 532
- ✔ Use effective **sentence construction** 498
- ✔ Construct effective **paragraphs** 367
- ✔ Use **quotations** and **paraphrasing** 445, 372
- ✔ Write an **introduction** 276
- ✔ Write a **conclusion** 93
- ✔ Choose a **title** 529

Revision 490

- ✔ Check for **unity** and **coherence** 543, 71
  - **conciseness** 90
  - **pace** 367
  - **transition** 537
- ✔ Check for **sentence variety** 505
  - **emphasis** 167
  - **parallel structure** 370
  - **subordination** 516
- ✔ Check for **clarity** 68
  - **ambiguity** 32
  - **awkwardness** 44

- logic errors 312
- positive writing 386
- voice 557
- ✔ Check for **ethics in writing** 177
  - biased language 46
  - copyright 101
  - plagiarism 383
- ✔ Check for appropriate **word choice** 568
  - abstract / concrete words 6
  - affectation and jargon 22, 285
  - clichés 71
  - connotation / denotation 96
  - defining terms 116
- ✔ Eliminate problems with grammar 234
  - agreement 23
  - case 62
  - modifiers 334
  - pronoun reference 405
  - sentence faults 503
- ✔ Review mechanics and **punctuation** 434
  - abbreviations 2
  - capitalization 59
  - contractions 101
  - dates 115
  - italics 283
  - numbers 352
  - proofreading 411
  - spelling 512

Handbook of
Technical Writing

a / an

A and *an* are indefinite **articles** because the **noun** designated by the article is not a specific person, place, or thing but is one of a group.

- ▶ She installed *a* program.
  [not a specific program but an unnamed program]

Use *a* before words or abbreviations beginning with a consonant or consonant sound, including *y* or *w*.

- ▶ *A* manual is available online.
- ▶ It was *a* historic event for the Institute.
  [*Historic* begins with the consonant *h*.]
- ▶ We received *a* DNA sample.
- ▶ The year's activities are summarized in *a* one-page report.
  [*One* begins with the consonant sound "wuh."]

Use *an* before words or abbreviations beginning with a vowel or a consonant with a vowel sound.

- ▶ The report is *an* overview of the year's activities.
- ▶ The applicant arrived *an* hour early.
  [*Hour* begins with a silent *h*.]
- ▶ She bought *an* SLR digital camera.
  [*SLR* begins with a vowel sound "ess."]

Do not use unnecessary indefinite articles in a sentence.

- ▶ Fill with *a* half *a* pint of fluid.
  [Choose one article and eliminate the other.]

See also **adjectives**.

a lot

A lot is often incorrectly written as one word (*alot*). The phrase *a lot* is informal and normally should not be used in technical writing. Use *many* or *numerous* for estimates or give a specific number or amount.

▶ We received a̶ l̶o̶t̶ o̶f̶ *many* e-mails supporting the new policy.

abbreviations

DIRECTORY
Using Abbreviations   2
Writer's Checklist: Using Abbreviations   3
Forming Abbreviations   3
  Names of Organizations   3
  Measurements   4
  Personal Names and Titles   4
  Common Scholarly Abbreviations   4

Abbreviations are shortened versions of words or combinations of the first letters of words (Corp./*Corp*oration, URL/*U*niform *R*esource *L*ocator). Abbreviations, if used appropriately, can be convenient for both the reader and the writer. Like symbols, they can be important space savers in technical writing.

Abbreviations that are formed by combining the initial letter of each word in a multiword term are called *initialisms*. Initialisms are pronounced as separate letters (AC or ac/*a*lternating *c*urrent). Abbreviations that combine the first letter or letters of several words—and can be pronounced—are called *acronyms* (PIN/*p*ersonal *i*dentification *n*umber, laser/*l*ight *a*mplification by *s*timulated *e*mission of *r*adiation).

Using Abbreviations

In business, industry, and government, specialists and those working together on particular projects often use abbreviations. The most important consideration in the use of abbreviations is whether they will be understood by your **audience**. The same abbreviation, for example, can have two different meanings (NEA stands for both the National Education Association and the National Endowment for the Arts). Like **jargon**, shortened forms are easily understood within a group of specialists; outside the group, however, shortened forms might be incomprehensible. In fact, abbreviations can be easily overused, either as an **affectation** or in a misguided attempt to make writing concise, even with **instant messaging** where abbreviations are often used. Remember that **memos**, **e-mail**, or **reports** addressed to specific people may be read by other people—you must consider those secondary audiences as well. A good rule to follow is "when in doubt, spell it out."

> **WRITER'S CHECKLIST** Using Abbreviations
>
> - Except for commonly used abbreviations (U.S., a.m.), spell out a term to be abbreviated the first time it is used, followed by the abbreviation in parentheses. Thereafter, the abbreviation may be used alone.
> - In long documents, repeat the full term in parentheses after the abbreviation at regular intervals to remind readers of the abbreviation's meaning, as in "Remember to submit the CAR (Capital Appropriations Request) by. . . ."
> - Do not add an additional period at the end of a sentence that ends with an abbreviation. (The official name of the company is DataBase, Inc.)
> - For abbreviations specific to your profession or discipline, use a style guide recommended by your professional organization or company. (A list of style guides appears at the end of **documenting sources**.)
> - Write acronyms in capital letters without periods. The only exceptions are acronyms that have become accepted as common nouns, which are written in lowercase letters, such as *scuba* (*s*elf-*c*ontained *u*nderwater *b*reathing *a*pparatus).
> - Generally, use periods for lowercase initialisms (a.k.a., e.d.p., p.m.) but not for uppercase ones (GDP, IRA, UFO). Exceptions include geographic names (U.S., U.K., E.U.) and the traditional expression of academic degrees (B.A., M.B.A., Ph.D.).
> - Form the plural of an acronym or initialism by adding a lowercase *s*. Do not use an **apostrophe** (CARs, DVDs).
> - Do not follow an abbreviation with a word that repeats the final term in the abbreviation (HIV transmission *not* HIV virus transmission).
> - Do not make up your own abbreviations; they will confuse readers.

Forming Abbreviations

Names of Organizations.   A company may include in its name a term such as *Brothers*, *Incorporated*, *Corporation*, *Company*, or *Limited Liability Company*. If the term is abbreviated in the official company name that appears on letterhead stationery or on its Web site, use the abbreviated form: *Bros.*, *Inc.*, *Corp.*, *Co.*, or *LLC*. If the term is not abbreviated in the official name, spell it out in writing, except with addresses, footnotes, **bibliographies**, and **lists** where abbreviations may be used. Likewise, use an ampersand (&) only if it appears in the official company name. For names of divisions within organizations, terms such as *Department* and *Division* should be abbreviated only when space is limited (*Dept.* and *Div.*).

Measurements.   Except for abbreviations that may be confused with words (*in.* for *inch* and *gal.* for *gallon*), abbreviations of measurement do not require periods (*yd* for *yard* and *qt* for *quart*). Abbreviations of units of measure are identical in the singular and plural: 1 *cm* and 15 *cm* (*not* 15 *cms*). Some abbreviations can be used in combination with other symbols (°F for *degrees Fahrenheit* and *ft²* for *square feet*).

The following list includes abbreviations for the basic units of the International System of Units (SI), the metric system. This system not only is used in science but also is used in international commerce and trade.

MEASUREMENT	UNIT	ABBREVIATION
length	meter	m
mass	kilogram	kg
time	second	s
electric current	ampere	A
thermodynamic temperature	kelvin	K
amount of substance	mole	mol
luminous intensity	candela	cd

For additional definitions and background, see the National Institute of Standards and Technology Web site at *http://physics.nist.gov/cuu/Units/units.html*. For information on abbreviating dates and time, see **numbers**.

Personal Names and Titles.   Personal names generally should not be abbreviated: Thomas (*not* Thos.) and William (*not* Wm.). An academic, civil, religious, or military title should be spelled out and in lowercase when it does not precede a name. ("The *captain* wanted to check the orders.") When they precede names, some titles are customarily abbreviated (Dr. Smith, Mr. Mills, Ms. Katz). See also **Ms./Miss/Mrs.**

An abbreviation of a title may follow the name; however, be certain that it does not duplicate a title before the name (Angeline Martinez, Ph.D. *or* Dr. Angeline Martinez). When addressing **correspondence** and including names in other documents, you normally should spell out titles (The Honorable Mary J. Holt; Professor Charles Matlin). Traditionally, periods are used with academic degrees, although they are sometimes omitted (M.A./MA, M.B.A./MBA, Ph.D./PhD).

Common Scholarly Abbreviations.   The following is a partial list of abbreviations commonly used in reference books and for documenting sources in research papers and reports. Other than in formal scholarly work, generally avoid such abbreviations.

anon.	anonymous
bibliog.	bibliography, bibliographer, bibliographic
ca., c.	*circa*, "about" (used with approximate dates: ca. 1756)
cf.	*confer*, "compare"
chap.	chapter
diss.	dissertation
ed., eds.	edited by, editor(s), edition(s)
e.g.	*exempli gratia*, "for example" (see <u>e.g./i.e.</u>)
esp.	especially
et al.	*et alii*, "and others"
etc.	*et cetera*, "and so forth" (see <u>etc.</u>)
ff.	and the following page(s) or line(s)
GPO	Government Printing Office, Washington, D.C.
i.e.	*id est*, "that is"
MS, MSS	manuscript, manuscripts
n., nn.	note, notes (used immediately after page number: 56n., 56n.3, 56nn.3–5)
N.B., n.b.	*nota bene*, "take notice, mark well"
n.d.	no date (of publication)
n.p.	no place (of publication); no publisher; no page
p., pp.	page, pages
proc.	proceedings
pub.	published by, publisher, publication
rev.	revised by, revised, revision; review, reviewed by (Spell out "review" where "rev." might be ambiguous.)
rpt.	reprinted by, reprint
sec., secs.	section, sections
sic	so, thus; inserted in <u>brackets</u> ([*sic*]) after a misspelled or wrongly used word in quotations
supp., suppl.	supplement
trans.	translated by, translator, translation
UP	University Press (used in MLA style, as in "Oxford UP")
viz.	*videlicet*, "namely"
vol., vols.	volume, volumes
vs., v.	*versus*, "against" (*v.* preferred in titles of legal cases)

> **WEB LINK** Using Abbreviations
>
> For links to Web sites specifying standard abbreviations and acronyms, including U.S. Postal Service abbreviations, see *bedfordstmartins.com/alredtech* and select *Links for Handbook Entries*.

above

Avoid using *above* to refer to a preceding passage or **visual** because its reference is vague and often an **affectation**. The same is true of *aforesaid* and *aforementioned*. (See also **former/latter**.) To refer to something previously mentioned, repeat the **noun** or **pronoun**, or construct your **paragraph** so that your reference is obvious.

▶ Please complete and submit ~~the above~~ *your travel voucher* by March 1.

absolutely

Absolutely means "definitely," "entirely," "completely," or "unquestionably." Avoid it as a redundant **intensifier** to mean "very" or "much."

▶ We are ~~absolutely~~ certain we can meet the deadline.

abstract / concrete words

Abstract words refer to general ideas, qualities, conditions, acts, or relationships—intangible things that cannot be detected by the five senses (sight, hearing, touch, taste, and smell), such as *learning, leadership,* and *technology*. *Concrete* words identify things that can be perceived by the five senses, such as *diploma, manager,* and *keyboard*.

Abstract words must frequently be further defined or described.

▶ The accident investigation team needs freedom *to interview the maintenance technicians*.

Abstract words are best used with concrete words to help make intangible concepts specific and vivid.

▶ *Transportation* [abstract] was limited to *buses* [concrete] and *commuter trains* [concrete].

The example in Figure A–1 represents seven levels of abstraction from most abstract to most concrete; the appropriate level depends on your **purpose** in writing and on the **context** in which you are using the word. See also **word choice**.

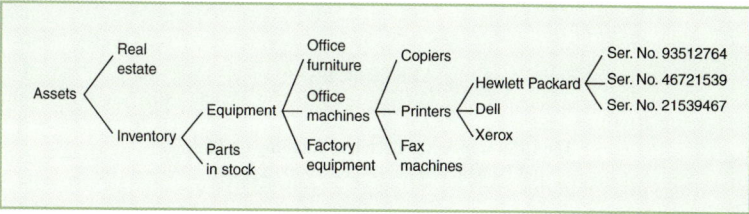

FIGURE A–1. Abstract-to-Concrete Words

abstracts

An abstract summarizes and highlights the major points of a **formal report**, **trade journal article**, dissertation, or other long work. Its primary purpose is to enable readers to decide whether to read the work in full. For a discussion of how summaries differ from abstracts, see **executive summaries**.

Although abstracts, typically 200 to 250 words long, are published with the longer works they condense, they can also be published separately in periodical indexes and by abstracting services (see **research**). For this reason, an abstract must be readable apart from the original document.

Types of Abstracts

Depending on the kind of information they contain, abstracts are often classified as descriptive or informative. A *descriptive abstract* summarizes the **purpose**, **scope**, and methods used to arrive at the reported findings. It is a slightly expanded **table of contents** in sentence and paragraph form. A descriptive abstract need not be longer than several sentences. An *informative abstract* is an expanded version of the descriptive abstract. In addition to information about the purpose, scope, and research methods used, the informative abstract summarizes any results, **conclusions**, and recommendations. The informative abstract retains the **tone** and essential scope of the original work, omitting its details. The first four sentences of the abstract shown in Figure A–2 alone would be descriptive; with the addition of the sentences that detail the conclusions of the report, the abstract becomes informative.

The type of abstract you should write depends on your **audience** and the organization or publication for which you are writing. Informative abstracts work best for wide audiences that need to know conclusions and recommendations; descriptive abstracts work best for compilations, such as proceedings and progress reports, that do not contain conclusions or recommendations.

8 abstracts

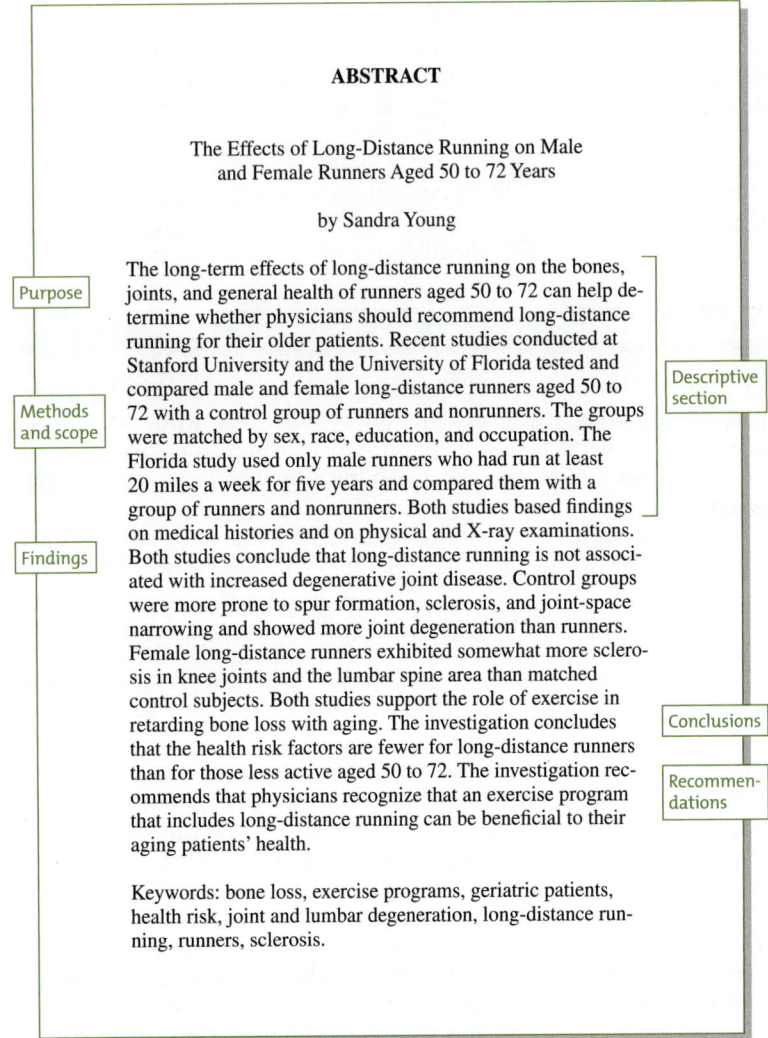

FIGURE A–2. Informative Abstract (for an Article)

Writing Strategies

Write the abstract *after* finishing the report or document. Otherwise, the abstract may not accurately reflect the longer work. Begin with a topic sentence that announces the subject and scope of your original document. Then, using the major and minor headings of your outline or table of contents to distinguish primary ideas from secondary ones, decide what material is relevant to your abstract. (See **outlining**.) Write with **clarity** and **conciseness**, eliminating unnecessary words and ideas. Do not, however, become so terse that you omit articles (*a*, *an*, *the*) and important transitional words and phrases (*however*, *therefore*, *but*, *next*). Write complete sentences, but avoid stringing together a group of short sentences end to end; instead, combine ideas by using **subordination** and **parallel structure**. Spell out all but the most common **abbreviations**. In a report, an abstract follows the title page and is numbered page iii.

accept / except

Accept is a **verb** meaning "consent to," "agree to take," or "admit willingly." ("I *accept* the responsibility.") *Except* is normally used as a preposition meaning "other than" or "excluding." ("We agreed on everything *except* the schedule.")

acceptance / refusal letters (for employment)

When you decide to accept a job offer, you can notify your new employer by telephone or in a meeting—but to make your decision official, you need to send an acceptance in writing. What you include in your message and whether you send a **letter** or an **e-mail** depends on your previous conversations with your new employer. See also **correspondence**. Figure A–3 shows an example of a job acceptance letter written by a graduating student with substantial experience (see his **résumé** in Figure R–7 on page 475).

Note that in the first paragraph of Figure A–3, the writer identifies the job he is accepting and the salary he has been offered—doing so can avoid any misunderstandings about the job or the salary. In the second paragraph, the writer details his plans for relocating and reporting for work. Even if the writer discussed these arrangements during earlier conversations, he needs to confirm them, officially, in this written message. The writer concludes with a brief but enthusiastic statement that he looks forward to working for the new employer.

When you decide to reject a job offer, send a written job refusal to make that decision official, even if you have already notified the employer during a meeting or on the phone. Writing to an employer is an important goodwill gesture. Be especially tactful and courteous—the employer you are refusing has spent time and effort interviewing you and may have counted on your accepting the job. Remember, you may

acceptance / refusal letters

> Dear Ms. Castro:
>
> I am pleased to accept your offer of $44,500 per year as a Graphic Designer with the Natural History Museum.
>
> After graduation, I plan to leave Pittsburgh on Tuesday, June 2. I should be able to find living accommodations and be ready to report for work on Monday, June 15. If you need to reach me prior to this date, please call me at 412-555-1212 (cell) or contact me by e-mail at jgoodman@aol.com.
>
> I look forward to joining the marketing team and working with the excellent support staff I met during the interview.
>
> Sincerely,
>
> *Joshua Goodman*
>
> Joshua Goodman

FIGURE A–3. Acceptance Letter (for Employment)

apply for another job at that company in the future. In Figure A–4, an example of a job refusal, the applicant mentions something positive about his contact with the employer and refers to the specific job offered. He indicates his serious consideration of the offer, provides a log-

> Dear Mr. Vallone:
>
> I enjoyed talking with you about your opening for a manager of aerospace production at your Rockford facility, and I seriously considered your generous offer.
>
> After giving the offer careful thought, however, I have decided to accept a management position with a research-and-development firm. I feel that the job I have chosen is better suited to my long-term goals.
>
> I appreciate your consideration and the time you spent with me. I wish you success in filling the position.
>
> Sincerely,
>
> *Robert Mandillo*
>
> Robert Mandillo

FIGURE A–4. Refusal Letter (for Employment)

ical reason for the refusal, and concludes on a pleasant note. (See his résumé in Figure R–8 on page 476) For further advice on handling refusals and negative messages generally, see **refusal letters**.

accuracy / precision

Accuracy means error-free and correct. *Precision*, mathematically, refers to the "degree of refinement with which a measurement is made." Thus, a measurement carried to five decimal places (5.28371) is more precise than one carried to two decimal places (5.28), but it is not necessarily more accurate.

acknowledgment letters

One way to build goodwill with colleagues and clients is to send an acknowledgment, letting them know that something they sent arrived and expressing thanks. It is usually a short, polite note. The example shown in Figure A–5 could be sent as a **letter** or an **e-mail**. See also **correspondence**.

Dear Mr. Evans:

I received your comprehensive report today. When I finish studying it in detail, I'll send you our cost estimate for the installation of the ECAT 1240 Pet Scanner.

Thank you for preparing such a thorough analysis.

Regards,

Roger Vonblatz

FIGURE A–5. Acknowledgment

acronyms and initialisms

Acronyms are formed by combining the first letter or letters of several words; they are pronounced as words and written without periods (ANSI/*American National Standards Institute*). Initialisms are formed by combining the initial letter of each word in a multiword term; they are pronounced as separate letters (STC/Society for Technical Communication). See **abbreviations** for the use of acronyms and initialisms.

activate / actuate

Both *activate* and *actuate* mean "make active," although *actuate* is usually applied only to mechanical processes.

- ▶ The relay *actuates* the trip hammer. [mechanical process]
- ▶ The electrolyte *activates* the battery. [chemical process]
- ▶ The governor *activated* the National Guard. [legal process]

active voice (*see* voice)

ad hoc

Ad hoc is Latin for "for this" or "for this particular occasion." An ad hoc committee is one set up temporarily to consider a particular issue. The term has been fully assimilated into English and thus does not have to be italicized. See also **foreign words in English**.

adapt / adept / adopt

Adapt is a verb meaning "adjust to a new situation." *Adept* is an adjective meaning "highly skilled." *Adopt* is a verb meaning "take or use as one's own."

- ▶ The company will *adopt* a policy of finding engineers who are *adept* managers and who can *adapt* to new situations.

adjectives

DIRECTORY
Limiting Adjectives 13
Comparison of Adjectives 14
Placement of Adjectives 15
Use of Adjectives 16

adjectives

An adjective is any word that modifies a **noun** or **pronoun**. *Descriptive adjectives* identify a quality of a noun or pronoun. *Limiting adjectives* impose boundaries on the noun or pronoun.

- ▶ *hot* surface [descriptive]
- ▶ *three* phone lines [limiting]

Limiting Adjectives

Limiting adjectives include the following categories:

- Articles (*a*, *an*, *the*)
- Demonstrative adjectives (*this*, *that*, *these*, *those*)
- Possessive adjectives (*my*, *your*, *his*, *her*, *its*, *our*, *their*)
- Numeral adjectives (*two*, *first*)
- Indefinite adjectives (*all*, *none*, *some*, *any*)

Articles. Articles (*a*, *an*, *the*) are traditionally classified as adjectives because they modify nouns by either limiting them or making them more specific. See also **articles** and **English as a second language**.

Demonstrative Adjectives. A demonstrative adjective points to the thing it modifies, specifying the object's position in space or time. *This* and *these* specify a closer position; *that* and *those* specify a more remote position.

- ▶ *This* report is more current than *that* report, which Human Resources distributed last month.
- ▶ *These* sales figures are more recent than *those* reported last week.

Demonstrative adjectives often cause problems when they modify the nouns *kind*, *type*, and *sort*. Demonstrative adjectives used with those nouns should agree with them in number.

- ▶ *this* kind, *these* kinds; *that* type, *those* types

Confusion often develops when the preposition *of* is added (*this kind of*, *these kinds of*) and the object of the preposition does not conform in number to the demonstrative adjective and its noun. See also **agreement** and **prepositions**.

- ▶ *This* **kind of** human resources ~~policies are~~ *policy is* standard.
- ▶ *These* **kinds of** human resources ~~policy is~~ *policies are* standard.

Avoid using demonstrative adjectives with words like *kind*, *type*, and *sort* because doing so can easily lead to vagueness. Instead, be more specific. See also **kind of / sort of**.

Possessive Adjectives. Because possessive adjectives (*my*, *your*, *his*, *her*, *its*, *our*, *their*) directly modify nouns, they function as adjectives, even though they are pronoun forms (*my* idea, *her* plans, *their* projects). See also **functional shift**.

Numeral Adjectives. Numeral adjectives identify quantity, degree, or place in a sequence. They always modify count nouns. Numeral adjectives are divided into two subclasses: cardinal and ordinal. A *cardinal adjective* expresses an exact quantity (*one* pencil, *two* computers); an *ordinal adjective* expresses degree or sequence (*first* quarter, *second* edition).

In most writing, an ordinal adjective should be spelled out if it is a single word (*tenth*) and written in figures if it is more than one word (*312th*). Ordinal numbers can also function as adverbs. (John arrived *first*.) See also **first / firstly** and **numbers**.

Indefinite Adjectives. Indefinite adjectives do not designate anything specific about the nouns they modify (*some* CD-ROMs, *all* designers). The articles *a* and *an* are included among the indefinite adjectives (*a* chair, *an* application).

Comparison of Adjectives

Most adjectives in the positive form show the comparative form with the suffix *-er* for two items and the superlative form with the suffix *-est* for three or more items.

- ▶ The first report is *long*. [positive form]
- ▶ The second report is *longer*. [comparative form]
- ▶ The third report is *longest*. [superlative form]

Many two-syllable adjectives and most three-syllable adjectives are preceded by the word *more* or *most* to form the comparative or the superlative.

- ▶ The new media center is *more* impressive than the old one. It is the *most* impressive in the county.

A few adjectives have irregular forms of comparison (*much*, *more*, *most*; *little*, *less*, *least*).

Some adjectives (*round*, *unique*, *exact*, *accurate*), often called *absolute words*, are not logically subject to comparison.

> **ESL TIPS** for Using Adjectives
>
> Do not add *-s* or *-es* to an adjective to make it plural.
>
> ▶ the *long* trip
> ▶ the *long* trips
>
> Capitalize adjectives of origin (city, state, nation, continent).
>
> ▶ the *Venetian* canals
> ▶ the *Texas* longhorn steer
> ▶ the *French* government
> ▶ the *African* deserts
>
> In English, verbs of feeling (for example, *bore, interest, surprise*) have two adjectival forms: the present participle (*-ing*) and the past participle (*-ed*). Use the present participle to describe what causes the feeling. Use the past participle to describe the person who experiences the feeling.
>
> ▶ We heard the *surprising* election results.
> [The *election results* cause the feeling.]
>
> ▶ Only the losing candidate was *surprised* by the election results.
> [The *candidate* experienced the feeling of surprise.]
>
> Adjectives follow nouns in English in only two cases: when the adjective functions as a subjective complement
>
> ▶ That project is not *finished*.
>
> and when an adjective phrase or clause modifies the noun.
>
> ▶ The project *that was suspended temporarily* . . .
>
> In all other cases, adjectives are placed before the noun.
>
> When there are multiple adjectives, it is often difficult to know the right order. The guidelines illustrated in the following example would apply in most circumstances, but there are exceptions. (Normally do not use a phrase with so many stacked **modifiers**.) See also **articles**.
>
> ▶ The six extra-large rectangular brown cardboard take-out containers
>   determiner | number | comment | size | shape | color | material | qualifier | noun

Placement of Adjectives

When limiting and descriptive adjectives appear together, the limiting adjectives precede the descriptive adjectives, with the articles usually in the first position.

adjustment letters

▶ *The ten yellow* taxis were sold at auction.
[article (*The*), limiting adjective (*ten*), descriptive adjective (*yellow*)]

Within a sentence, adjectives may appear before the nouns they modify (the attributive position) or after the nouns they modify (the predicative position).

▶ *The small* jobs are given priority. [attributive position]

▶ The exposure is *brief*. [predicative position]

Use of Adjectives

Nouns often function as adjectives to clarify the meaning of other nouns.

▶ The *accident* report prompted a *product* redesign.

When adjectives modifying the same noun can be reversed and still make sense or when they can be separated by *and* or *or*, they should be separated by commas.

▶ The division seeks *bright*, *energetic*, *creative* designers.

Notice that there is no comma after *creative*. Never use a comma between a final adjective and the noun it modifies. When an adjective modifies a phrase, no comma is required.

▶ We need an *updated Web page design*.
[*Updated* modifies the phrase *Web page design*.]

Writers sometimes string together a series of nouns used as adjectives to form a unit modifier, thereby creating stacked (jammed) **modifiers**, which can confuse readers. See also **word choice**.

adjustment letters

An adjustment **letter** or **e-mail** is written in response to a **complaint** and tells a customer or client what your organization intends to do about the complaint. Although sent in response to a problem, an adjustment letter actually provides an excellent opportunity to build goodwill for your organization. An effective adjustment letter, such as the examples shown in Figures A–6 and A–7, can not only repair any damage done but also restore the customer's confidence in your company.

   No matter how unreasonable the complaint, the **tone** of your response should be positive and respectful. Avoid emphasizing the problem, but do take responsibility for it when appropriate. Focus your

adjustment letters

INTERNET SERVICES CORPORATION
10876 Crispen Way
Chicago, Illinois 60601

May 11, 2009

Mr. Jason Brandon
4319 Anglewood Street
Tacoma, WA 98402

Dear Mr. Brandon:

We are sorry that your experience with our customer support help line did not go smoothly. We are eager to restore your confidence in our ability to provide dependable, high-quality service. Your next three months of Internet access will be complimentary as our sincere apology for your unpleasant experience.

Providing dependable service is what is expected of us, and when our staff doesn't provide quality service, it is easy to understand our customers' disappointment. I truly wish we had performed better in our guidance for setup and log-on procedures and that your experience had been a positive one. To prevent similar problems in the future, we plan to use your letter in training sessions with customer support personnel.

We appreciate your taking the time to write us. It helps to receive comments such as yours, and we conscientiously follow through to be sure that proper procedures are being met.

Yours truly,

Inez Carlson

Inez Carlson, Vice President
Customer Support Services

www.isc.com

FIGURE A–6. Adjustment Letter (When Company Is at Fault)

> Dear Mr. Sanchez:
>
> Enclosed is your Addison Laptop Computer, which you shipped to us on August 31.
>
> Our technical staff reports that the laptop was damaged by exposure to high levels of humidity. You stated in your letter that you often use your laptop on a covered patio. Doing so in a high-humidity environment, as is typical in Louisiana, can result in damage to the internal circuitry of your computer—as described on page 32 of your Addison Owner's Manual.
>
> We have replaced the damaged circuitry and thoroughly tested your laptop. To ensure that a repetition of your recent experience does not occur, we recommend you avoid leaving your laptop exposed to high humidity for extended periods.
>
> If you should find that the problem recurs, please call us at 800-555-0990. We will be glad to work with you to find a solution.
>
> Sincerely,
>
> Customer Service Department

FIGURE A–7. Partial Adjustment (Accompanying a Product)

response on what you are doing to correct the problem. Settle such matters quickly and courteously, and lean toward giving the customer or client the benefit of the doubt at a reasonable cost to your organization. See also **refusal letters**.

Full Adjustments

Before granting an adjustment to a claim for which your company is at fault, first determine what happened and what you can do to satisfy the customer. Be certain that you are familiar with your company's adjustment policy—and be careful with **word choice**:

▶ We have just received your letter of May 7 about our ~~defective~~ gas grill.

Saying something is "defective" could be ruled in a court of law as an admission that the product is in fact defective. When you are in doubt, seek legal advice.

Grant adjustments graciously: A settlement made grudgingly will do more harm than good. Not only must you be gracious, but you must also acknowledge the error in such a way that the customer will not lose confidence in your company. Emphasize early what the reader will consider good news.

- Enclosed is a replacement for the damaged part.
- Yes, you were incorrectly billed for the delivery.
- Please accept our apologies for the error in your account.

If an explanation will help restore your reader's confidence, explain what caused the problem. You might point out any steps you may be taking to prevent a recurrence of the problem. Explain that customer feedback helps your firm keep the quality of its product or service high. Close pleasantly, looking forward, not back. Avoid recalling the problem in your closing (do not write, "Again, we apologize . . .").

The adjustment letter in Figure A–6, for example, begins by accepting responsibility and offers an apology for the customer's inconvenience (note the use of the pronouns *we* and *our*). The second paragraph expresses a desire to restore goodwill and describes specifically how the writer intends to make the adjustment. The third paragraph expresses appreciation to the customer for calling attention to the problem and assures him that his complaint has been taken seriously.

Partial Adjustments

You may sometimes need to grant a partial adjustment—even if a claim is not really justified—to regain the lost goodwill of a customer or client. If, for example, a customer incorrectly uses a product or service, you may need to help that person better understand the correct use of that product or service. In such a circumstance, remember that your customer or client believes that his or her claim is justified. Therefore, you should give the explanation before granting the claim—otherwise, your reader may never get to the explanation. If your explanation establishes customer responsibility, do so tactfully. Figure A–7 is an example of a partial adjustment letter. See also **correspondence**.

adverbs

An adverb modifies the action or condition expressed by a **verb**.

- The wrecking ball hit the side of the building *hard*.
  [The adverb tells *how* the wrecking ball hit the building.]

An adverb also can modify an **adjective**, another adverb, or a **clause**.

- The brochure design used *remarkably* bright colors.
  [*Remarkably* modifies the adjective *bright*.]

▶ The redesigned brake pad lasted *much* longer.
[*Much* modifies the adverb *longer*.]

▶ *Surprisingly*, the engine failed.
[*Surprisingly* modifies the clause *the engine failed*.]

Types of Adverbs

A simple adverb can answer one of the following questions:

Where? (adverb of place)

▶ Move the display *forward* slightly.

When? or *How often?* (adverb of time)

▶ Replace the thermostat *immediately*.

▶ I worked overtime *twice* this week.

How? (adverb of manner)

▶ Add the solvent *cautiously*.

How much? (adverb of degree)

▶ The *nearly* completed report was sent to the director.

An interrogative adverb can ask a question (*Where? When? Why? How?*):

▶ *How* many hours did you work last week?

▶ *Why* was the hard drive reformatted?

A conjunctive adverb can modify the clause that it introduces as well as join two independent clauses with a **semicolon**. The most common conjunctive adverbs are *however, nevertheless, moreover, therefore, further, then, consequently, besides, accordingly, also,* and *thus*.

▶ I rarely work on weekends; *however*, this weekend will be an exception.

In this example, note that a semicolon precedes and a comma follows *however*. The conjunctive adverb (*however*) introduces the independent clause (*this weekend will be an exception*) and indicates its relationship to the preceding independent clause (*I rarely work on weekends*). See also **transition**.

Comparison of Adverbs

Most one-syllable adverbs show comparison with the suffixes *-er* and *-est*.

- This copier is *fast*. [positive form]
- This copier is *faster* than the old one. [comparative form]
- This copier is the *fastest* of the three tested. [superlative form]

Most adverbs with two or more syllables end in *-ly*, and most adverbs ending in *-ly* are compared by inserting the comparative *more* or *less* or the superlative *most* or *least* in front of them.

- The patient recovered *more quickly* than the staff expected.
- *Most surprisingly*, the engine failed during the final test phase.

A few irregular adverbs require a change in form to indicate comparison (*well, better, best; badly, worse, worst; far, farther, farthest*).

- The training program functions *well*.
- Our training program functions *better* than most others in the industry.
- Many consider our training program the *best* in the industry.

Placement of Adverbs

An adverb usually should be placed in front of the verb it modifies.

- The pilot *methodically* performed the preflight check.

An adverb may, however, follow the verb (or the verb and its object) that it modifies.

- The system failed *unexpectedly*.
- They replaced the hard drive *quickly*.

An adverb may be placed between a helping verb and a main verb.

- In this temperature range, the pressure will *quickly* drop.

Adverbs such as *only, nearly, almost, just,* and *hardly* should be placed immediately before the words they limit. See also **modifiers** and **only**.

affect / effect

Affect is a **verb** that means "influence." ("The decision could *affect* the company's stock value.") *Effect* can function as a **noun** that means "result" ("The decision had a positive *effect*") or as a verb that means "bring about" or "cause." However, avoid *effect* as a verb when you can replace it with a less formal word, such as *make* or *produce*.

▶ The new technician will ~~effect~~ *make* several changes to network security.

affectation

Affectation is the use of language that is more formal, technical, or showy than necessary to communicate information to the reader. Affectation is a widespread writing problem in the workplace because many people feel that affectation lends a degree of authority to their writing. In fact, affectation can alienate customers, clients, and colleagues.

Affected writing forces readers to work harder to understand the writer's meaning. It typically contains inappropriate abstract, highly technical, or foreign words and is often liberally sprinkled with trendy **buzzwords**.

◆ **ETHICS NOTE** **Jargon** and **euphemisms** can become affectation, especially if their purpose is to hide relevant facts or give a false impression of competence. See **ethics in writing**. ✦

Writers easily slip into affectation through the use of long variants—words created by adding prefixes and suffixes to simpler words (*orientate* for *orient*; *utilization* for *use*). Unnecessarily formal words (such as *penultimate* for *next to last*), created words using *-ese* (such as *managementese*), and outdated words (such as *aforesaid*) can produce affectation. (See also **above**.) Elegant variation—attempting to avoid repeating a word within a paragraph by substituting a pretentious synonym—is also a form of affectation. Either repeat the term or use a pronoun.

▶ The use of modules in the assembly process has increased production~~. Modular utilization~~ *, and it* has ~~also~~ cut costs.

Another type of affectation is **gobbledygook**, which is wordy, roundabout writing with many legal- and technical-sounding terms (such as *wherein* and *morphing*).

| **WEB LINK** | **Affected Writing Revised** |

For an example of regulations that are revised to eliminate various forms of affectation, see *bedfordstmartins.com/alredtech* and select *Model Documents Gallery*.

Understanding the possible reasons for affectation is the first step toward avoiding it. The following are some causes of affectation.

- *Impression.* Some writers use pretentious language in an attempt to impress the reader with fancy words instead of evidence and logic.
- *Insecurity.* Writers who are insecure about their facts, conclusions, or arguments may try to hide behind a smoke screen of pretentious words.
- *Imitation.* Perhaps unconsciously, some writers imitate the poor writing they see around them.
- *Intimidation.* A few writers, consciously or unconsciously, try to intimidate or overwhelm their readers with words, often to protect themselves from criticism.
- *Initiation.* Writers who are new to a field often feel that one way to prove their professional expertise is to use as much technical terminology and jargon as possible.
- *Imprecision.* Writers who are having trouble being precise sometimes find that an easy solution is to use a vague, trendy, or pretentious word.

See also **clichés**, **conciseness**, and **nominalizations**.

affinity

Affinity refers to the attraction of two persons or things to each other. *Affinity* should not be used to mean "ability" or "aptitude."

▶ She has an ~~affinity~~ *aptitude* for problem solving.

agreement

DIRECTORY
Subject-Verb Agreement   24
Compound Subjects   26
Pronoun-Antecedent Agreement   27

Grammatical agreement is the correspondence in form between different elements of a sentence to indicate **number**, **person**, **gender**, and **case**.

agreement

A subject and its **verb** must agree in number.

- The *design is* acceptable.
  [The singular subject, *design*, requires the singular verb, *is*.]
- The new *products are* going into production soon.
  [The plural subject, *products*, requires the plural verb, *are*.]

A subject and its verb must agree in person.

- *I am* the designer.
  [The first-person singular subject, *I*, requires the first-person singular verb, *am*.]
- *They are* the designers.
  [The third-person plural subject, *they*, requires the third-person plural verb, *are*.]

A **pronoun** and its antecedent must agree in person, number, gender, and case.

- The *employees* report that *they* are more efficient in the new facility.
  [The third-person plural subject, *employees*, requires the third-person plural pronoun, *they*.]
- *Kaye McGuire* will meet with the staff on Friday, when *she* will assign duties.
  [The third-person singular subject, *Kaye McGuire*, requires *she*, the third-person feminine pronoun, in the subjective case.]

See also **sentence construction**.

Subject-Verb Agreement

Subject-verb agreement is not affected by intervening **phrases** and **clauses**.

- *One* in twenty hard drives we receive from our suppliers *is* faulty.
  [The verb, *is*, must agree in number with the subject, *one*, not *hard drives* or *suppliers*.]

The same is true when **nouns** fall between a subject and its verb.

- Only *one* of the emergency lights *was* functioning.
  [The subject of the verb is *one*, not *lights*.]
- *Each* of the switches *controls* a separate circuit.
  [The subject of the verb is *each*, not *switches*.]

Note that *one* and *each* are normally singular.

Indefinite pronouns such as *some*, *none*, *all*, *more*, and *most* may be singular or plural, depending on whether they are used with a mass noun ("*Most* of the oil *has* been used") or with a count noun ("*Most* of the drivers *know* why they are here"). Mass nouns are singular, and count nouns are plural. Other words, such as *type*, *part*, *series*, and *portion*, take singular verbs even when they precede a phrase containing a plural noun.

▶ A *series* of meetings *was* held to improve the product design.

▶ A large *portion* of most annual reports *is* devoted to promoting the corporate image.

Modifying phrases can obscure a simple subject.

▶ The *advice* of two engineers, one lawyer, and three executives *was* obtained prior to making a commitment.
[The subject of the verb *was* is *advice*.]

Inverted word order can cause problems with agreement.

▶ From this work *have come* several important *improvements*.
[The subject of the verb is *improvements*, not *work*.]

The number of a subjective **complement** does not affect the number of the verb—the verb must always agree with the subject.

▶ The *topic* of his report *is* employee benefits.
[The subject of the sentence is *topic*, not *benefits*.]

A subject that expresses measurement, weight, mass, or total often takes a singular verb even when the subject word is plural in form. Such subjects are treated as a unit.

▶ *Four weeks is* the normal duration of the training program.

A verb following the relative pronoun *who* or *that* agrees in number with the noun to which the pronoun refers (its antecedent).

▶ Steel is one of those *industries* that *are* most affected by global competition. [*That* refers to *industries*.]

▶ She is one of those *employees* who *are* rarely absent.
[*Who* refers to *employees*.]

The word *number* sometimes causes confusion. When used to mean a specific number, it is singular.

▶ *The number* of committee members *was* six.

When used to mean an approximate number, it is plural.

▶ *A number* of people *were* waiting for the announcement.

Relative pronouns (*who*, *which*, *that*) may take either singular or plural verbs, depending on whether the antecedent is singular or plural. See also **who / whom**.

▶ He is a manager *who seeks* the views of others.

▶ He is one of those managers *who seek* the views of others.

Some abstract nouns are singular in meaning but plural in form: *mathematics*, *news*, *physics*, and *economics*.

▶ *Mathematics is* essential in computer science.

Some words, such as the plural *jeans* and *scissors*, cause special problems.

▶ The *scissors were* ordered last week.
[The subject is the plural *scissors*.]

▶ *A pair* of scissors *is* on order.
[The subject is the singular *pair*.]

A book with a plural title requires a singular verb.

▶ *Basic Medical Laboratory Techniques* is an essential resource.

A collective noun (*committee*, *faculty*, *class*, *jury*) used as a subject takes a singular verb when the group is thought of as a unit and a plural verb when the individuals in the group are thought of separately.

▶ The *committee is* unanimous in its decision.

▶ The *committee are* returning to their offices.

A clearer way to emphasize the individuals would be to use a phrase.

▶ The *committee members are* returning to their offices.

Compound Subjects

A compound subject is composed of two or more elements joined by a **conjunction** such as *and*, *or*, *nor*, *either . . . or*, or *neither . . . nor*. Usually, when the elements are connected by *and*, the subject is plural and requires a plural verb.

▶ *Professional writing and translation are* prerequisites for this position.

One exception occurs when the elements connected by *and* form a unit or refer to the same thing. In that case, the subject is regarded as singular and takes a singular verb.

- *Bacon and eggs is* a high-cholesterol meal.
- Our greatest *challenge and business opportunity* is the Internet.

A compound subject with a singular element and a plural element joined by *or* or *nor* requires that the verb agree with the closer element.

- Neither the director nor the *project assistants were* available.
- Neither the project assistants nor the *director was* available.

If *each* or *every* modifies the elements of a compound subject, use the singular verb.

- *Each* manager and supervisor *has* a production goal to meet.
- *Every* manager and supervisor *has* a production goal to meet.

Pronoun-Antecedent Agreement

Every pronoun must have an antecedent—a noun to which it refers. See also **pronoun reference**.

- When *employees* are hired, *they* must review the policy manual. [The pronoun *they* refers to the antecedent *employees*.]

Gender. A pronoun must agree in gender with its antecedent.

- *Mr. Swivet* in the accounting department acknowledges *his* share of responsibility for the misunderstanding, just as *Ms. Barkley* in the research division must acknowledge *hers*.

Traditionally, a masculine, singular pronoun was used to agree with such indefinite antecedents as *anyone* and *person* ("*Each* may stay or go as *he* chooses"). Because such usage ignores or excludes women, use alternatives when they are available. One solution is to use the plural. Another is to use both feminine and masculine pronouns, although that combination is clumsy when used too often.

- ~~Every employee~~ *All employees* must sign ~~his~~ *their* time ~~card~~ *cards*.
- Every employee must sign his *or her* time card.

Do not attempt to avoid expressing gender by resorting to a plural pronoun when the antecedent is singular. You may be able to avoid the pronoun entirely.

> Every employee must sign ~~their~~ *a* time card.

Avoid gender-related stereotypes in general references, as in "the nurse . . . *she*" or "the doctor . . . *he*." What if the nurse is male or the doctor female? See also **biased language**.

Number. A pronoun must agree with its antecedent in number. Many problems of agreement are caused by expressions that are not clear in number.

> Although the typical engine runs well in moderate temperatures, ~~they~~ *it* often ~~stall~~ *stalls* in extreme cold.

Use singular pronouns with the antecedents *everybody* and *everyone* unless to do so would be illogical because the meaning is obviously plural. See also **everybody / everyone**.

> *Everyone* pulled *his or her* share of the load.

> *Everyone* thought my plan should be revised, and I really couldn't blame *them*.

Collective nouns may use a singular or plural pronoun, depending on the meaning.

> The *committee* agreed to the recommendations only after *it* had deliberated for days.
> [*committee* thought of as collective singular]

> The *committee* quit for the day and went to *their* respective homes.
> [*committee* thought of as plural]

Demonstrative **adjectives** sometimes cause problems with agreement of number. *This* and *that* are used with singular nouns, and *these* and *those* are used with plural nouns. Demonstrative adjectives often cause problems when they modify the nouns *kind*, *type*, and *sort*. When used with those nouns, demonstrative adjectives should agree with them in number.

> *this* kind, *these* kinds; *that* type, *those* types

Confusion often develops when the **preposition** *of* is added (*this kind of*, *these kinds of*) and the object of the preposition does not agree in number with the demonstrative adjective and its noun.

▶ This kind of retirement ~~plans are~~ *plan is* best.

Avoid that error by remembering to make the demonstrative adjective, the noun, and the object of the preposition—all three—agree in number. The agreement makes the sentence not only correct but also more precise. Using demonstrative adjectives with words like *kind*, *type*, and *sort* can easily lead to vagueness. See **kind of / sort of**.

Compound Antecedents. A compound antecedent joined by *or* or *nor* is singular if both elements are singular and plural if both elements are plural.

▶ Neither the *engineer* nor the *technician* could do *his* job until *he* understood the new concept.

▶ Neither the *executives* nor the *directors* were pleased at the performance of *their* company.

When one of the antecedents connected by *or* or *nor* is singular and the other plural, the pronoun agrees with the closer antecedent.

▶ Either the *computer* or the *printers* should have *their* serial numbers registered.

▶ Either the *printers* or the *computer* should have *its* serial number registered.

A compound antecedent with its elements joined by *and* requires a plural pronoun.

▶ *Seon Ju and Juanita* took *their* layout drawings with them.

If both elements refer to the same person, however, use the singular pronoun.

▶ The noted *biologist and author* departed from *her* prepared speech.

all ready / already

All ready is a two-word phrase meaning "completely prepared." *Already* is an **adverb** that means "before this time" or "previously." ("They were *all ready* to cancel the order; fortunately, we had *already* corrected the shipments.")

all right

All right means "all correct." ("The answers were *all right*.") In formal writing, it should not be used to mean "good" or "acceptable." It is always written as two words, with no hyphen; *alright* is nonstandard.

all together / altogether

All together means "all acting together" or "all in one place." ("The necessary instruments were *all together* on the tray.") *Altogether* means "entirely" or "completely." ("The trip was *altogether* unnecessary.")

allude / elude / refer

Allude means to make an indirect reference to something. ("The report simply *alluded* to the problem, rather than stating it explicitly.") *Elude* means to escape notice or detection. ("The leak *eluded* the inspectors.") *Refer* is used to indicate a direct reference to something. ("She *referred* to the merger during her presentation.")

allusion / illusion

An *allusion* is an indirect reference to something not specifically mentioned. ("The report made an *allusion* to metal fatigue in the support structures.") An *illusion* is a mistaken perception or a false image. ("County officials are under the *illusion* that the landfill will last indefinitely.")

allusions

An allusion is an indirect reference to something from past or current events, literature, or other familiar sources. The use of allusion promotes economical writing because it is a shorthand way of referring to a body of material in a few words or of helping to explain a new and unfamiliar process in terms of one that is familiar. In the following example, the writer sums up a description with an allusion to a well-known story. The allusion, with its implicit reference to "right standing up to might," concisely emphasizes the writer's point.

▶ As it currently exists, the review process involves the consumer's attorney sitting alone, usually without adequate technical assistance, faced by two or three government attorneys, two or three attorneys from CompuSystems, and large teams of experts who support the government and the corporation. The entire proceeding is reminiscent of David versus Goliath.

Be sure, of course, that your reader is familiar with the material to which you allude. Allusions should be used with restraint, especially in <u>international correspondence</u>. If overdone, allusions can lead to <u>affectation</u> or can be viewed merely as <u>clichés</u>. See also <u>technical writing style</u>.

almost / most

Do not use *most* as a colloquial substitute for *almost* in your writing.

▶ New shipments arrive ~~most~~ *almost* every day.

also

Also is an <u>adverb</u> that means "additionally." ("Two 5,000-gallon tanks are on-site, and several 2,500-gallon tanks are *also* available.") *Also* should not be used as a connective in the sense of "and."

▶ He brought the reports, the memos, ~~also~~ *and* the director's recommendations.

Avoid opening sentences with *also*. It is a weak transitional word that suggests an afterthought rather than planned writing.

▶ ~~Also~~, *In addition* he brought a cost analysis to support his proposal.

▶ ~~Also, he~~ *He also* brought a cost analysis to support his proposal.

ambiguity

A word or passage is ambiguous when it can be interpreted in two or more ways, yet it provides the reader with no certain basis for choosing among the alternatives.

▶ Mathematics is more valuable to an engineer than a computer. [Does that mean an engineer is more in need of mathematics than a computer is? Or does it mean that mathematics is more valuable to an engineer than a computer is?]

Ambiguity can take many forms, as in ambiguous **pronoun reference**.

AMBIGUOUS	Inadequate quality-control procedures have resulted in more equipment failures. This is our most serious problem at present. [Does *this* refer to *inadequate quality-control procedures* or to *equipment failures*?]
SPECIFIC	Inadequate quality-control procedures have resulted in more equipment failures. *These failures* are our most serious problem at present.
SPECIFIC	Inadequate quality-control procedures have resulted in more equipment failures. *Quality control* is our most serious problem at present.

Incomplete **comparison** and missing or misplaced **modifiers** (including **dangling modifiers**) cause ambiguity.

▶ Ms. Lee values rigid quality-control standards more than Mr. Rosenblum *does*. [Complete the comparison.]

▶ He lists his hobby as cooking. He is *also* especially fond of cocker spaniels. [Add the missing modifier.]

The placement of some modifiers enables them to be interpreted in either of two ways.

▶ She volunteered *immediately* to deliver the toxic substance.

By moving the word *immediately*, the meaning can be clarified.

▶ She *immediately* volunteered to deliver the toxic substance.

▶ She volunteered to deliver the toxic substance *immediately*.

Imprecise **word choice** (including faulty **idioms**) can cause ambiguity.

▶ The general manager has denied reports that the plant's recent fuel-allocation cut will be ~~restored~~ *rescinded*. [inappropriate word choice]

Various forms of **awkwardness** also can cause ambiguity.

amount / number

Amount is used with things that are thought of in bulk and that cannot be counted (mass **nouns**), as in "the *amount* of electricity." *Number* is used with things that can be counted as individual items (count nouns), as in "the *number* of employees."

ampersands

The ampersand (&) is a symbol sometimes used to represent the word *and*, especially in the names of organizations (Kirkwell & Associates). When you are writing the name of an organization in sentences, addresses, or references, spell out the word *and* unless the ampersand appears in the organization's official name on its letterhead stationery or Web site.

and/or

And/or means that either both circumstances are possible or only one of two circumstances is possible. This term is awkward and confusing because it makes the reader stop to puzzle over your distinction.

AWKWARD	Use A *and/or* B.
IMPROVED	Use A or B or both.

antonyms

An antonym is a word with a meaning opposite that of another word (*good/bad*, *wet/dry*, *fresh/stale*). Many pairs of words that look as if they are antonyms, such as *limit/delimit*, are not. When in doubt, consult thesauruses and print or online **dictionaries** for antonyms as well as **synonyms**.

apostrophes

An apostrophe (') is used to show possession or to indicate the omission of letters. Sometimes it is also used to avoid confusion with certain plurals of words, letters, and **abbreviations**.

Showing Possession

An apostrophe is used with an *s* to form the possessive case of some nouns (the *report's* title). For further advice on using apostrophes to show possession, see **possessive case**.

Indicating Omission

An apostrophe is used to mark the omission of letters or **numbers** in a **contraction** or a date (*can't, I'm, I'll*; the class of *'09*).

Forming Plurals

An apostrophe can be used in forming the plurals of letters, words, or lowercase abbreviations if confusion might result from using *s* alone.

▶ The search program does not find *a*'s and *i*'s.
▶ Do not replace all *of which*'s in the document.
▶ *I*'s need to be distinguished from the number 1.
▶ Check for any c.o.d.'s.

Generally, however, add only *s* in roman (or regular) type when referring to words as words or capital letters. See also **italics**.

▶ Five *and*s appear in the first sentence.
▶ The applicants received *A*s and *B*s in their courses.

Do not use an apostrophe for plurals of abbreviations with all capital letters (PDFs) or a final capital letter (ten Ph.D.s) or for plurals of numbers (7s, the late 1990s).

appendixes

An appendix, located at the end of a **formal report**, **proposal**, or other long document, supplements or clarifies the information in the body of the document. Appendixes (or *appendices*) can provide information that is too detailed or lengthy for the primary **audience** of the document. For example, an appendix could contain such material as **maps**, statistical analysis, **résumés** of key personnel involved in a proposed project, or other documents needed by secondary readers.

A document may have more than one appendix, with each providing only one type of information. When you include more than one appendix, arrange them in the order they are mentioned in the body of the document. Begin each appendix on a new page, and identify each with a letter, starting with the letter *A* (Appendix A: Sample Questionnaire). If you have only one appendix, title it simply "Appendix." List the titles and beginning page numbers of the appendixes in the **table of contents**.

application letters

When applying for a job, you usually need to submit both an application letter (also referred to as a *cover letter*) and a **résumé**. Employers may ask you to submit them by standard mail, online form, fax, or **e-mail**. See also **job search** and **letters**.

The application letter is essentially a sales letter in which you market your skills, abilities, and knowledge. Therefore, your application letter must be persuasive. The successful application letter accomplishes four tasks: (1) It catches the reader's attention favorably by describing how your skills will contribute to the organization, (2) it explains which particular job interests you and why, (3) it convinces the reader that you are qualified for the job by highlighting and interpreting the particularly impressive qualifications in your résumé, and (4) it requests an interview. See also **correspondence**, **interviewing for a job**, **persuasion**, and **salary negotiations**.

The sample application letters shown in Figures A–8 through A–10 follow the structure described in this entry. Each sample's **emphasis**, **tone**, and **style** are tailored to fit the applicant's experience and the particular **audience**. Note that the letter shown in Figure A–9 matches the résumé in Figure R–7 ("Joshua S. Goodman") and the letter sent as an e-mail in Figure A–10 matches the résumé in Figure R–8 ("Robert Mandillo").

Opening

In the opening paragraph, provide **context** by indicating how you heard about the position and name the specific job title or area. If you have been referred to a company by an employee, a career counselor, a professor, or someone else, be sure to say so. ("I have recently learned from Jodi Hammel, a graphic designer at Dyer/Khan, that you are recruiting....") Show enthusiasm by explaining why you are interested in the

36 application letters

From: Molly Sennett <sennett@execpr.com>
To: Bob Lupert <lupert@appliedsciences.com>
Sent: Monday, February 23, 2009 10:47 AM
Subject: Application for Summer Internship
Attachment: 📄 Sennett_Resume.pdf

Dear Mr. Lupert:

In the February 19 issue of the *Butler Gazette*, I learned that you have summer technical training internships available. This opportunity interests me because I have learned that Applied Sciences has one of the leading user-education centers in the country.

From your Web site, I have also learned that you value collaborative and leadership skills. My current project at Penn State University—a computer tutoring system that teaches LISP—has helped me to develop these skills. This tutoring system, which I helped develop as part of a student engineering team, is the first of its kind and is now being sold across the country to various corporations and universities. As part of the team, I solved specific problems and then helped test and revise our work to improve efficiency.

Pursuing degrees in Industrial Management and Computer Science has prepared me well to make valuable contributions to your goal of successful implementation of new software. Through my varied courses, which are described on my résumé, I have developed the ability to learn new skills quickly and independently and to interact effectively in a technical environment. I would look forward to applying these abilities to user education at Applied Sciences, Inc.

I would appreciate the chance to interview with you at your earliest convenience. If you would like additional information, please contact me at (435) 228-3490 any Tuesday or Thursday after 10 a.m. or e-mail me at <sennett@execpr.com>.

Thank you for your consideration.

Sincerely,

Molly Sennett

FIGURE A–8. Application Letter Sent as E-mail (College Student Applying for an Internship)

application letters

222 Morewood Ave.
Pittsburgh, PA 15212
April 16, 2009

Ms. Judith Castro, Director
Human Resources
Natural History Museum
1201 S. Figueroa Street
Los Angeles, CA 90015

Dear Ms. Castro:

I have recently learned from Jodi Hammel, a graphic designer at Dyer/Khan, that you are recruiting for a graphic designer in your Marketing Department. Your position interests me greatly because it offers me an opportunity both to fulfill my career goals and to promote the work of an internationally respected institution. Having participated in substantial volunteer activities at a local public museum, I am aware of the importance of your work.

I bring strong up-to-date academic and practical skills in multimedia tools and graphic arts production, as indicated in my enclosed résumé. Further, I have recent project management experience at Dyer/Khan, where I was responsible for the development of client brochures, newsletters, and posters. As project manager, I coordinated the project time lines, budgets, and production with clients, staff, and vendors.

My experience and contacts in the Los Angeles area media and entertainment community should help me make use of state-of-the-art design expertise. As you will see on my résumé, I have worked with the leading motion picture, television, and music companies—that experience should help me develop exciting marketing tools Museum visitors and patrons will find attractive. For example, I helped design an upgrade of the CGI logo for Paramount Pictures and was formally commended by the Director of Marketing.

Could we schedule a meeting at your convenience to discuss this position further? Call me any weekday morning at 412-555-1212 (cell) or e-mail me at <jgoodman@aol.com> if you have questions or need additional information. Thank you for your consideration.

Sincerely,

Joshua S. Goodman

Joshua S. Goodman

Enclosure: Résumé

FIGURE A–9. Application Letter (Recent Graduate Applying for a Graphic Design Job)

> Dear Ms. Smathers:
>
> During the recent NOMAD convention in Washington, Karen Jarrett, Director of Operations, informed me of a possible opening at Aerospace Technologies for a manager of new product development. My extensive background in engineering exhibit design and management makes me an ideal candidate for the position she described.
>
> I have been manager of the Exhibit Design Lab at Wright-Patterson Air Force Base for the past seven years. During that time, I received two Congressional Commendations for models of a space station laboratory and a docking/repair port. My experience in advanced exhibit design would give me a special advantage in helping develop AT's wind tunnel and aerospace models. Further, I have just learned this week that my exhibit design presented at NOMAD received a "Best of Show" Award.
>
> As described on the enclosed résumé, I not only have workplace management experience but also have recently received an MBA from the University of Dayton. As a student in the MBA program, I won the Luson Scholarship to complete my coursework as well as the Jonas Outstanding Student Award.
>
> I would be happy to discuss my qualifications in an interview at your convenience. Please telephone me at (937) 255-4137 or e-mail me at <mand@juno.com>. I look forward to talking with you.
>
> Sincerely,
>
> Robert Mandillo

FIGURE A–10. Application Letter (Applicant with Years of Experience)

job and demonstrate your initiative as well as your knowledge of the organization by relating your interest to some facet of the organization. ("This opportunity interests me because I have learned that Applied Sciences has one of the leading user-education centers in the country.")

Body

In the middle paragraphs, show through examples that you are qualified for the job. Limit each of these **paragraphs** to just one basic point that is clearly stated in the topic sentence. For example, your second paragraph might focus on work experience and your third paragraph on educational achievements. Do not just *tell* readers that you are qualified—*show* them by including examples and details. ("I helped design an upgrade of the CGI logo for Paramount Pictures and was formally commended by the Director of Marketing.") Highlight your achievements and refer to your enclosed résumé, but do not simply summarize your résumé. Indicate how your talents can make valuable contributions to the company.

Closing

In the final paragraph, request an interview. Let the reader know how to reach you by including your phone number or e-mail address. End with a statement of goodwill, even if only a "thank you."

Proofread your letter *carefully*. Research indicates that if employers notice even one spelling, grammatical, or mechanical error, they often eliminate candidates from consideration immediately. Such errors give employers the impression that you lack writing skills or that you are careless in the way you present yourself professionally. See also **proofreading**.

appositives

An appositive is a **noun** or noun **phrase** that follows and amplifies another noun or noun phrase. It has the same grammatical function as the noun it complements.

▶ Dennis Gabor, *the famous British scientist*, developed holography as an "exercise in serendipity."

▶ The famous British scientist *Dennis Gabor* developed holography as an "exercise in serendipity."

For detailed information on the use of **commas** with appositives, see **restrictive and nonrestrictive elements**.

If you are in doubt about the **case** of an appositive, check it by substituting the appositive for the noun it modifies. See also **pronouns**.

▶ My boss gave the two of us, Jim and ~~I~~ *me*, the day off.

[You would not say, "My boss gave *I* the day off."]

articles

Articles (*a*, *an*, *the*) function as **adjectives** because they modify the items they designate by either limiting them or making them more specific. Articles may be indefinite or definite.

articles

ESL TIPS for Using Articles

Whether to use a definite or an indefinite article is determined by what you can safely assume about your audience's knowledge. In each of these sentences, you can safely assume that the reader can clearly identify the noun. Therefore, use a definite article.

- *The* sun rises in the east.
  [The Earth has only one *sun*.]
- Did you know that yesterday was *the* coldest day of the year so far?
  [The modified noun refers to *yesterday*.]
- *The* man who left his briefcase in the conference room was in a hurry.
  [The relative phrase *who left his briefcase in the conference room* restricts and, therefore, identifies the meaning of *man*.]

In the following sentence, however, you cannot assume that the reader can clearly identify the noun.

- *A* package is on the way.
  [It is impossible to identify specifically what package is meant.]

A more important question for some nonnative speakers of English is when *not* to use articles. These generalizations will help. Do not use articles with the following:

Singular proper nouns

- Utah, Main Street, Harvard University, Mount Hood

Plural nonspecific count nouns (when making generalizations)

- Helicopters are the new choice of transportation for the rich and famous.

Singular mass nouns

- She loves chocolate.

Plural count nouns used as complements

- Those women are physicians.

See also **English as a second language**.

The indefinite articles, *a* and *an*, denote an unspecified item.

- *A* package was delivered yesterday. [*not* a specific package]

The choice between *a* and *an* depends on the sound rather than on the letter following the article, as described in the entry **a/an**.

The definite article, *the*, denotes a particular item.

▶ *The* package was delivered yesterday. [*one* specific package]

Do not omit all articles from your writing in an attempt to be concise. Including articles costs nothing; eliminating them makes reading more difficult. (See also **telegraphic style**.) However, do not overdo it. An article can be superfluous.

▶ Fill with *a* half *a* pint of fluid.
[Choose one article and eliminate the other.]

Do not capitalize articles in titles except when they are the first word ("*The Scientist* reviewed *Saving the Humboldt Penguin*").

as / because / since

As, *because*, and *since* are commonly used to mean "because." To express cause, *because* is the strongest and most specific connective; *because* is unequivocal in stating a causal relationship. (*Because* she did not have an engineering degree, she was not offered the job.)

Since is a weak substitute for *because* as a connective to express cause. However, *since* is an appropriate connective when the emphasis is on circumstance, condition, or time rather than on cause and effect. (*Since* it went public, the company has earned a profit every year.)

As is the least definite connective to indicate cause; its use for that purpose is best avoided. See also **subordination**.

Avoid colloquial, nonstandard, or wordy phrases sometimes used instead of *as*, *because*, or *since*. See also **as much as / more than**, **as such**, **as well as**, **conciseness**, and **due to / because of**.

PHRASE	REPLACE WITH
being as, being that	because, since
inasmuch as, insofar as	since, because
on account of	because
on the grounds of/that	because
due to the fact that	because, since

as much as / more than

The phrases *as much as* and *more than* are sometimes incorrectly combined, especially when intervening phrases delay the completion of the phrase.

- ► The engineers had as much, ~~if not more,~~ influence in planning the program ~~than~~ the accountants.
  - *as* (insert before "much")
  - *, if not more* (insert after "much")
  - *as* (replace "than")

as such

The phrase *as such* is seldom useful and should be omitted.

- ► Style sheets~~, as such,~~ are useful in designing Web pages.

as well as

Do not use *as well as* with *both*. The two expressions have similar meanings; use one or the other and adjust the verb as needed.

- ► Both General Motors ~~as well as~~ *and* Ford ~~is~~ *are* marketing hybrid vehicles.
- ► ~~Both~~ General Motors as well as Ford is marketing hybrid vehicles.

audience

Considering the needs of your audience is crucial to achieving your **purpose**. When you are writing to a specific reader, for example, you may find it useful to visualize a reader sitting across from you as you write. (See **correspondence**.) Likewise, when writing to an audience composed of relatively homogeneous **readers**, you might create an image of a composite reader and write for *that* reader. In such cases, using the **"you" viewpoint** and an appropriate **tone** will help you meet the needs of your readers as well as achieve an effective **technical writing style**. For meeting the needs of an audience composed of listeners, see **presentations**.

Analyzing Your Audience's Needs

The first step in analyzing your audience is to determine the readers' needs relative to your purpose and goals. Ask key questions during **preparation**.

- Who specifically is your reader? Do you have multiple readers? Who needs to see or use the document?

- What do your readers already know about your subject? What are your readers' attitudes about the subject? (Skeptical? Supportive? Anxious? Bored?)
- What particular information about your readers (experience, training, and work habits, for example) might help you write at the appropriate level of detail?
- What does the **context** suggest about meeting the readers' expectations for content or **layout and design**?
- Do you need to adapt your message for international readers? If so, see **global communication**, **global graphics**, and **international correspondence**.

In the workplace, your readers are usually less familiar with the subject than you are. You have to be careful, therefore, when writing on a topic that is unique to your area of specialization. Be sensitive to the needs of those whose training or experience lies in other areas; provide definitions of nonstandard terms and explanations of principles that you, as a specialist, take for granted. See also **defining terms**.

Writing for Diverse Audiences

For documents aimed at multiple audiences with different needs, consider segmenting the document for different groups of readers: an executive summary for top managers, an appendix with detailed data for technical specialists, and the body for those readers who need to make decisions based on a detailed discussion. See also **formal reports** and **proposals**.

When you have multiple audiences with various needs but cannot segment your document, first determine your primary or most important readers—such as those who will make decisions based on the document—and be sure to meet their needs. Then, meet the needs of secondary readers, such as those who need only some of the document's contents, as long as you do not sacrifice the needs of your primary readers. See also **persuasion** and "Five Steps to Successful Writing" (page xv).

augment / supplement

Augment means to increase or magnify in size, degree, or effect. ("Our retirees can *augment* their incomes through consulting.") *Supplement* means to add something to make up for a deficiency. ("The patient should *supplement* his diet with vitamins.")

average / median / mean

The *average* (or arithmetic *mean*) is determined by adding two or more quantities and dividing the sum by the number of items totaled. For example, if one report is 10 pages, another is 30 pages, and a third is 20 pages, their *average* length is 20 pages. It is incorrect to say that "each report averages 20 pages" because each report is a specific length.

▶ ~~Each report averages~~ *The three reports average* 20 pages.

The *median* is the middle number in a sequence of numbers. For example, the *median* of the series 1, 3, 4, 7, 8 is 4.

awhile / a while

The **adverb** *awhile* means "for a short time." The **preposition** *for* should not precede *awhile* because *for* is inherent in the meaning of *awhile*. The two-word noun **phrase** *a while* means "a period of time."

▶ Wait ~~for~~ awhile before testing the sample.

▶ Wait for ~~awhile~~ *a while* before testing the sample.

awkwardness

Any writing that strikes readers as awkward—that is, as forced or unnatural—impedes their understanding. The following checklist and the entries indicated will help you smooth out most awkward passages.

> **WRITER'S CHECKLIST** Eliminating Awkwardness
>
> ✔ Strive for **clarity** and **coherence** during **revision**.
> ✔ Check for **organization** to ensure your writing develops logically.
> ✔ Keep **sentence construction** as direct and simple as possible.
> ✔ Use **subordination** appropriately and avoid needless **repetition**.
> ✔ Correct any **logic errors** within your sentences.
> ✔ Revise for **conciseness** and avoid **expletives** where possible.
> ✔ Use the active **voice** unless you have a justifiable reason to use the passive voice.
> ✔ Eliminate jammed or misplaced **modifiers** and, for particularly awkward constructions, apply the tactics in **garbled sentences**.

bad / badly

Bad is the **adjective** form that follows such linking **verbs** as *feel* and *look*. ("We don't want to look *bad* at the meeting.") *Badly* is an **adverb**. ("The iron frame corroded *badly*.") To say "I feel *badly*" would mean, literally, that your sense of touch is impaired. See also **good / well**.

be sure to

The **phrase** *be sure and* is colloquial and unidiomatic when used for *be sure to*. See also **idioms**.

▶ When you sign the contract, be sure ~~and~~ *to* keep a copy.

beside / besides

Besides, meaning "in addition to" or "other than," should be carefully distinguished from *beside*, meaning "next to" or "apart from." ("*Besides* two of us from Radiology, three interns from Neurology stood *beside* the director during the ceremony.")

between / among

Between is normally used to relate two items or persons. ("The alloy offers a middle ground *between* durability and cost.") *Among* is used to relate more than two. ("The subcontracting was distributed *among* three firms.") *Amongst* is a variant, chiefly British spelling.

between you and me

The expression *between you and I* is incorrect. Because the **pronouns** are **objects** of the **preposition** *between*, the objective form of the personal pronoun (*me*) must be used. See also **case**.

▶ Between you and I̸, Joan should be promoted. [*me* inserted]

bi- / semi-

When used with periods of time, *bi-* means "two" or "every two," as in *bimonthly*, which means "once in two months." When used with periods of time, *semi-* means "half of" or "occurring twice within a period of time." *Semimonthly* means "twice a month." Both *bi-* and *semi-* normally are joined with the following element without a space or **hyphen**.

biannual / biennial

In conventional usage, *biannual* means "twice during the year," and *biennial* means "every other year." See also **bi- / semi-**.

biased language

Biased language refers to words and expressions that offend because they make inappropriate assumptions or stereotypes about gender, ethnicity, physical or mental disability, age, or sexual orientation. Biased language, which is often used unintentionally, can defeat your **purpose** by damaging your credibility.

◆ **ETHICS NOTE** The easiest way to avoid bias is simply not to mention differences among people unless the differences are relevant to the discussion. Keep current with accepted usage and, if you are unsure of the appropriateness of an expression or the tone of a passage, have several colleagues review the material and give you their assessments. ◆

Sexist Language

Sexist language can be an outgrowth of sexism, the arbitrary stereotyping of men and women — it can breed and reinforce inequality. To avoid sexism in your writing, treat men and women equally, and do not make assumptions about traditional or occupational roles. Accordingly, use nonsexist occupational descriptions in your writing.

INSTEAD OF	CONSIDER
chairman, chairwoman	chair, chairperson
foreman	supervisor, manager
man-hours	staff hours, worker hours
policeman, policewoman	police officer
salesman, saleswoman	salesperson

Use parallel terms to describe men and women.

INSTEAD OF	USE
ladies and men	ladies and gentlemen; women and men
man and wife	husband and wife
Ms. Jones and Bernard Weiss	Ms. Jones and Mr. Weiss; Mary Jones and Bernard Weiss

Sexism can creep into your writing by the unthinking use of male or female **pronouns** where a reference could apply equally to a man or a woman. One way to avoid such usage is to rewrite the sentence in the plural.

▶ ~~Every employee~~ *All employees* will have ~~his manager~~ *their managers* sign ~~his travel voucher.~~ *their travel vouchers.*

Other possible solutions are to use *his or her* instead of *his* alone or to omit the pronoun completely if it is not essential to the meaning of the sentence. See also **he / she** and **pronoun reference**.

Other Types of Biased Language

Identifying people by racial, ethnic, or religious categories is simply not relevant in most workplace writing. Telling readers that an accountant is Native American or an attorney is Jewish almost never conveys useful information. Linking a profession or characteristic to race or ethnicity reinforces stereotypes, implying that it is either rare or expected for a person of a particular background to have achieved a certain position.

Consider how you refer to people with disabilities. If you refer to "a disabled employee," you imply that the part (*disabled*) is as significant as the whole (*employee*). Use "an employee with a disability" instead. Similarly, the preferred usage is "a person who uses a wheelchair" rather than "a wheelchair-bound person," an expression that inappropriately equates the wheelchair with the person. Likewise, references to a person's age can be inappropriate, as in expressions like "middle-aged manager" or "young Web designer."

In most workplace writing, such issues are simply not relevant. Of course, in some contexts, race, ethnicity, religion, disability, or age should be identified. For example, if you are writing an Equal Employment Opportunity Commission report about your firm's hiring practices, the racial composition of the workforce is relevant. In such cases, you need to present the issues in ways that respect and do not demean the individuals or groups to which you refer. See also **ethics in writing**.

bibliographies

A bibliography is an alphabetical list of books, articles, Web sources, and other works that have been consulted in preparing a document or that are useful for reference purposes. A bibliography provides a convenient alphabetical listing of sources in a standardized form for readers interested in getting further information on the topic or in assessing the scope of the research.

Whereas a list of references or works cited refers to works actually cited in the text, a bibliography also includes works consulted for general background information. For information on using various citation styles, see **documenting sources**.

Entries in a bibliography are listed alphabetically by the author's last name. If an author is unknown, the entry is alphabetized by the first word in the title (other than *A*, *An*, or *The*). Entries also can be arranged by subject and then ordered alphabetically within those categories.

An annotated bibliography includes complete bibliographic information about a work (author, title, place of publication, publisher, and publication date) followed by a brief description or evaluation of what the work contains. The following is an annotation of a book on style and format:

> Sabin, William A. *The Gregg Reference Manual: A Manual of Style, Grammar, Usage, and Formatting*. 10th ed. New York: McGraw Hill/Irwin, 2005.
>
> > The tenth edition of this standard 688-page office guide "is intended for anyone who writes, edits, or prepares material for distribution or publication" (p. viii). The book begins with a section titled How to Look Things Up, which includes a reference to an online index at *www.gregg.com*. Part 1, Grammar, Usage, and Style, contains 11 sections on topics such as punctuation, abbreviations, spelling, and usage. Part 2, Techniques and Formats, contains seven sections on such topics as proofreading,

correspondence (letters, memos, and e-mail), and other elements of manuscript preparation. Part 3, References, includes four appendixes: several essays on style, rules for alphabetic filing, a glossary of grammatical terms, and an alphabetical list offering appropriate pronunciation for words used in business, legal, and professional contexts.

blogs

A blog* is a Web-based journal in which "bloggers" can post entries (displayed from the most recent to the initial posting) that express opinions, provide information, and respond to other bloggers on subjects of mutual interest. An individual, for example, might create a blog not only to document life experiences, including amusing stories or reactions to current events, but also to market career skills and even post a **résumé** for a **job search**. A group of cancer patients might create a blog to share their treatment and recovery processes with other Web readers, who can then post their own comments or advice.

The usefulness of a blog stems in large part from the resources that bloggers offer their readers to help them access additional information, such as by posting links to relevant Web sites and documents as well as to video, audio, and other interactive files.

Organizational Blogs

Organizations create blogs to help meet such goals as attracting and retaining clients or customers, promoting goodwill, obtaining valuable feedback on their products and services, and even developing a sense of community among their customers and employees. Unlike Web sites, which usually present a formal perspective of an organization, blogs provide an informal and interactive forum for organizations to communicate directly with target **audiences**. Organizations can create both external and internal blogs.

External blogs are publicly available on the Internet both for an organization's customers or clients and for executives, spokespeople, or employees to share their views. The informal environment of blogs helps to build credibility for an organization because customers know they can "chat" with an organization's representative and exchange current information instead of relying on information in published

---

*The word *blog* is a shortened form of *Web log*; estimates suggest that there are over 70 million blogs. One way to start your own blog is through Google Blogger at *www.blogger.com/start*.

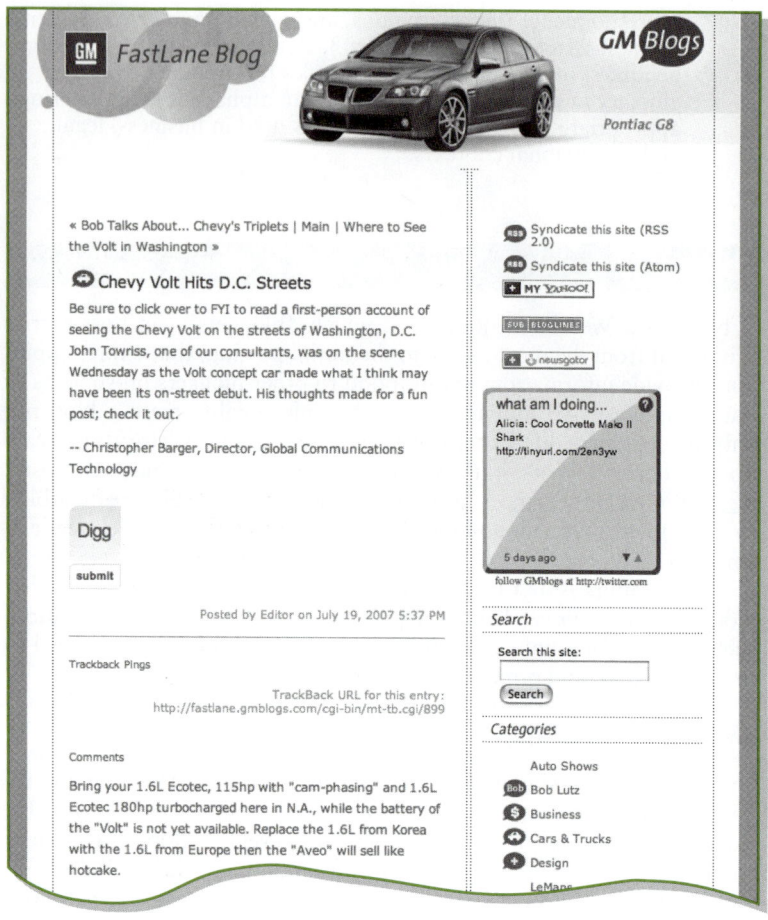

FIGURE B–1. Corporate Blog

documents and on Web sites. For example, General Motors (GM) has an external blog in which GM employees can post the latest information about their products, such as where customers can test drive a new concept car, as well as read and respond to the opinions from bloggers about those products. Figure B–1 shows a corporate blog that includes comments from the GM Director of Global Communications Technology and postings by bloggers.

Internal blogs are usually created for employees and can be accessed only through the organization's intranet. Such internal blogs may serve as **newsletters** and achieve the following organizational goals:

- Provide feedback opportunities through a forum to openly discuss issues or concerns
- Encourage employee participation in upcoming events or initiatives
- Share employee and community news, such as birth announcements and charitable events

Writing for Organizational Blogs

Write blog entries in an informal, personal **style** that uses **contractions**, second **person**, and active **voice**.

▶ "Check out the latest concept for our new dashboards—we've added more space. Is the console large enough to hold your Starbucks and iPod? Tell us what you think."

When writing blog entries, keep your sentences and paragraphs short. As in **writing for the Web**, use **conciseness** and brief entries to help readers scan the text to find information that is interesting or relevant to them. Use bulleted **lists**, **italics**, and other **layout and design** elements, such as boldface and white space, to organize the content. Headlines that are short and direct also catch attention and increase the blog's visual appeal and readability.

Provide access to additional information from your blog. For example, if you post a blog entry about your company's opinion of a report released from the Food and Drug Administration (FDA), include a Web link to the report on the FDA's Web site or post the report as a downloadable document. Clearly identify the sources of your ideas and attribute **quotations** accordingly.

The Wall Street Journal
▶ ~~A national newspaper~~ reported today that Smith Company's stock will likely rise.

Tim Smith
▶ ~~A guy~~ in Marketing said the company picnic will start at noon.

As a representative of your organization, sign your posting with your full name and your position title. Unlike outside bloggers who often use pseudonyms, identifying yourself helps maintain your credibility as well as your company's. Much of the appeal of a corporate blog is that customers can communicate directly with identifiable company representatives, not with anonymous people whose credentials within the company are unknown.

ETHICS NOTE Because organizations expect employees to assume full responsibility for the content they post on a company blog, you must maintain high ethical standards.

- Do not post information that is confidential, proprietary, or sensitive to your employer.
- Do not attack competitors or other individuals (such as employees, other bloggers, or shareholders).
- Do not post content that is profane, libelous, or harassing, or that violates the privacy of others. See also **biased language**.
- Obtain permission, as needed, for any material that is protected by **copyright**. See also **plagiarism**. ✦

Before publishing your entry to the blog, proofread carefully for grammar, typos, and formatting errors. Also obtain your supervisor's approval of the content before publishing to ensure that the content, **tone**, and Web links are compatible with company policies. See also **proofreading**.

both . . . and

Statements using the *both . . . and* construction should always be balanced grammatically and logically. See also **parallel structure**.

▶ To succeed in engineering, you must be able *both* to develop writing skills *and* ~~mastering~~ *to master* mathematics.

brackets

The primary use of brackets ([ ]) is to enclose a word or words inserted by the writer or editor into a quotation.

▶ The text stated, "Hypertext systems can be categorized as either modest [not modifiable] or robust [modifiable]."

Brackets are used to set off a parenthetical item within parentheses.

▶ We must credit Emanuel Foose (and his brother Emilio [1912–1982]) for founding the institute.

Brackets are also used to insert the Latin word *sic*, which is a scholarly **abbreviation** that indicates a writer has quoted material exactly as it appears in the original, even though it contains a misspelled or wrongly used word. See also **quotations**.

▶ The contract states, "Tinted windows will be installed to protect against son [*sic*] damage."

brainstorming

Brainstorming, a form of free association used to generate ideas about a topic, can be done individually or in groups. Brainstorming can stimulate creative thinking and reveal fresh perspectives and new connections. When preparing to write, jot down as many random ideas as you can think of about the topic. When a group brainstorms, designate a person to note ideas the group suggests. Do not stop to analyze ideas or hold back looking for only the "best" ideas; just note everything that comes to mind. After compiling a list of initial ideas, ask *what*, *when*, *who*, *where*, *how*, and *why* for each idea, then list additional details that those questions bring to mind. When you run out of ideas, analyze each one you recorded, discarding those that are redundant. Then group the items in the most logical order, based on the **purpose** of the document and the needs of the **audience** to create a tentative **outline** of the document. Although the outline will be sketchy and incomplete, it will show where further brainstorming or research is needed and provide a framework for any new details that additional research yields. (See **outlining**.)

Many writers find a technique called *clustering* (also called *mind mapping*), as shown in Figure B–2, helpful in recording and organizing

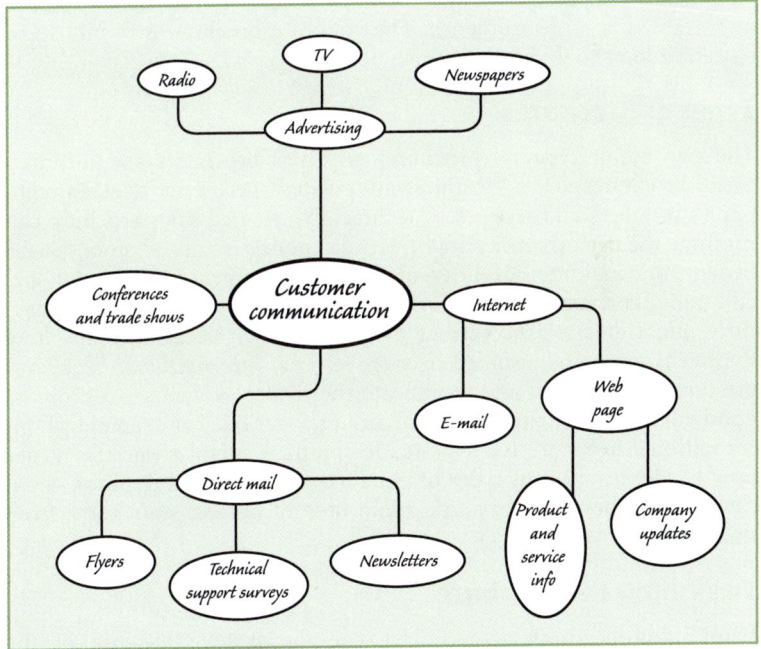

FIGURE B–2. Cluster Map from a Brainstorming Session

ideas created during a brainstorming session. To cluster, begin with a blank sheet of paper or a flip chart. Think of a key term that best characterizes your topic and put it in a circle at the center of the paper. Figure B–2 shows brainstorming about the best way to communicate with customers, so that topic could be "Customer communication." Then think of subtopics most closely related to the main topic. In Figure B–2, the main topic ("Customer communication") might lead to such subtopics as "Advertising," "Internet," and "Direct mail." Draw circles or boxes around the subtopics, connecting each to the center circle, like spokes to a wheel hub. Repeat the exercise for each subtopic. In Figure B–2, for example, one subtopic is "Internet," which could stimulate additional subtopics such as "E-mail" and "Web page." Continue the process until you exhaust all possible ideas. The resulting "map" will show clusters of terms grouped around the central concept from which you can create a topic outline.

brochures

Brochures are printed publications that promote the products and services offered by a business or that promote the image of a business or an organization by providing general or specific technical information important to a target **audience**. The goal of a brochure is to inform or to persuade or to do both. See also **persuasion**.

Types of Brochures

The two major types of brochures are sales brochures and informational brochures. *Sales brochures* are created specifically to sell a company's products and services. A technically oriented sales brochure can promote the performance data of various models of a product or list the benefits or capabilities of different types of equipment available for specific jobs. For example, the brochure of a microwave oven manufacturer might describe the various models and their accessories and how the ovens are to be installed over stove tops. *Informational brochures* are created to inform and to educate the reader as well as to promote goodwill and to raise the profile of an organization. For example, an informational brochure for a pesticide company might show the reader how to identify various types of pests like silverfish and termites, detail the damage they can do, and explain how to protect your home from them.

Designing the Brochure

Before you begin to write, determine the specific **purpose** of the brochure—to provide information about a service? to sell a product?

You must also identify your target audience—general reader? expert? potential or existing client? Understanding your purpose, audience, and **context** is essential to creating content and a design that will be both appropriate and persuasive to your target audience. To help you develop an effective design and to stimulate your thinking, gather sample brochures for products or services similar to yours.

Cover Panel. The main goal of the cover panel is to gain the audience's attention. It should clearly identify the organization or product being promoted and provide a carefully selected visual image or headline geared toward the interest of the audience. Keep the amount of text to a minimum—for example, a statement about the organization's mission and success or a brief promotional quotation from a satisfied customer.

Figure B–3 shows the cover panel of a brochure produced by the National Cancer Institute, which hopes to attract health-care professionals to an online course on conducting clinical trials in their practices. The cover photo features a health-care professional who appears to be counseling a patient; the text briefly defines the course, provides a Web address, and identifies the sponsoring organization.

First Inside Panel. The first inside panel of a brochure should again identify the organization and attract the reader with headlines and brief, readable content, such as that used in advertising. The first inside panel of the National Cancer Institute brochure uses boldface **headings** to highlight brief descriptions of the course and those participants who may benefit.

Subsequent Panels. Subsequent panels should describe the product or service with the reader's needs in mind, clearly stating the benefits and solutions to

FIGURE B–3. Brochure (Cover Panel)

problems that your product or service offers. Include relevant and accurate supporting facts and **visuals**. You might further establish credibility with a brief company or product history or with a list of current clients or testimonials. Use subheadings and bulleted **lists** to break up the text and highlight key points. In the final panel, be clear about the action you want the reader to take, such as calling to order the product or visiting a Web site for more information. In Figure B–4, the second inside panel describes the course objectives and tells readers how to register.

Design Style and Unity. Develop a style that unifies and complements the content of your brochure, and use it consistently throughout. For example, a brochure for a cruise line would feature many attractive photos of passengers and scenery. A brochure related to a serious med-

What is "Incorporating Cancer Clinical Trials Into Your Practice?"

It's a free Web-based tutorial for health care professionals that emphasizes the importance of participating in cancer clinical trials, either through patient referral or becoming a clinical trials investigator.

The course begins with a brief overview of cancer clinical trials and explains ways one can become involved in clinical trials. It continues with practical information and guidance for professionals interested in referring patients to, or conducting, clinical trials.

Participants will hear from experienced clinicians about some of the issues they have faced and how these issues were addressed.

Who should take this course?

The course is designed for health care professionals who are new to the clinical trials process, including oncologists, family physicians, internists, nurses, and other clinical staff. Specifically, it is for those clinicians that are referring patients to cancer clinical trials for the first time, and clinical research staff who are new to conducting clinical trials in their practices.

What are the course objectives?

At the end of the course, participants should be able to:

- Describe the role of clinical research in advancing cancer care
- Explain how clinical trials are conducted
- Describe the sponsorship of clinical trials
- Demonstrate how to locate clinical trials
- Understand how clinical trial sites are set up, staffed, and funded
- Make successful patient referrals to clinical trials
- Know how a physician may become a cancer clinical trials investigator
- Understand regulatory, record keeping, and reporting requirements in conducting clinical trials
- Identify key issues in communication among the patient, referring clinician, and clinical trials investigator

To register for this course, go to **http://cme.cancer.gov**, where you will be given further instructions.

Participants may receive continuing medical education (CME) credits or contact hours for course completion.

FIGURE B–4. Brochure (Inside Panels)

ical subject, as in Figures B–3 and B–4, might use images of professionals in workplace settings. See also **layout and design**.

> **WRITER'S CHECKLIST** Designing a Brochure
>
> - Collect other brochures to stimulate your thinking as you consider the goals and content for your brochure and each panel.
> - Create a rough sketch that maps out the content of each panel to help you select visuals, color schemes, and the number of panels.
> - Experiment with margins, spacing, and the arrangement and amount of text on each panel; make appropriate changes to content or length and allow adequate white space for readability.
> - Experiment with fonts and formatting, such as enlarging the first letter of the first word in a paragraph, but do not overuse unusual fonts or alternative styles, such as running type vertically.
> - Consider color choices: black and white, which is cheaper than color, can be effective; in some circumstances, however, color is a must.
> - Evaluate the impact of the design with your content. Have you adequately considered the needs of your audience?
> - Consider using a professional printer and high-quality paper, depending on your budget and the scale of your project.

> **WEB LINK** Sample Brochures
>
> For links to Web sites with sample brochures, see *bedfordstmartins.com/alredtech* and select *Links for Handbook Entries*.

bulleted lists (*see* lists)

buzzwords

Buzzwords are words or phrases that suddenly become popular and, because of an intense period of overuse, lose their freshness and preciseness. They may become popular through their association with technology, popular culture, or even sports.

▶ interface [as a <u>verb</u>]   deliverables   slam dunk
    impact [as a verb]   24/7   data dump
    cyberslackers   dot-com   action items

B

Obviously, the words in this example are appropriate in the right context.

> We must establish an *interface* between the computer and the satellite hardware.
> [*Interface* is appropriately used as a **noun**.]

When writers needlessly shift from the normal function of a word, however, they often create a buzzword that is imprecise.

> We must ~~interface~~ *cooperate* with the Radiology Department.
> [*Interface* is inappropriately used as a verb; *cooperate* is more precise.]

We include such words in our vocabulary because they *seem* to give force and vitality to our language. Actually, buzzwords often sound pretentious in technical writing. See also **word choice**.

WEB LINK | **Buzzwords**

Former newspaper editor John Walston offers BuzzWhack, a lighthearted site "dedicated to demystifying buzzwords." See *bedfordstmartins.com/alredtech* and select *Links for Handbook Entries*.

can / may

In writing, *can* refers to capability ("I *can* have the project finished today"). *May* refers to possibility ("I *may* be in Boston on Monday") or permission ("*May* I leave early?").

cannot

Cannot is one word.

▶ We ~~can not~~ *cannot* meet the deadline specified in the contract.

capitalization

DIRECTORY
Proper Nouns   59
Common Nouns   60
First Words   60
Specific Groups   60
Specific Places   60
Specific Institutions, Events, Concepts   61
Titles of Works   61
Professional and Personal Titles   62
Abbreviations and Letters   62
Miscellaneous Capitalizations   62

The use of capital, or uppercase, letters is determined by custom. Capital letters are used to call attention to certain words, such as proper **nouns** and the first word of a sentence. Use capital letters carefully because they can affect a word's meaning (march / March, china / China) and because a spell checker would fail to identify such an error.

Proper Nouns

Capitalize proper nouns that name specific persons, places, or things (Pat Wilde, Peru, Technical Writing 205, Microsoft). Proper nouns that form other **parts of speech** are often capitalized (Keynesian economics, Boolean function, Balkanize), but you should check a current **dictionary**.

Common Nouns

Common nouns name general classes or categories of people, places, things, concepts, or qualities rather than specific ones and are not capitalized (technical writing class, company, person, country).

First Words

The first letter of the first word in a sentence is always capitalized. ("Of the plans submitted, ours is best.") The first word after a **colon** is capitalized when the colon introduces two or more sentences (independent **clauses**) or if the colon precedes a statement requiring special emphasis.

▶ The meeting will address only one issue: What is the firm's role in environmental protection?

If a subordinate element follows the colon or if the thought is closely related, use a lowercase letter following the colon.

▶ We kept working for one reason: the approaching deadline.

The first word of a complete sentence in **quotation marks** is capitalized.

▶ Albert Einstein stated, "Imagination is more important than knowledge."

The first word in the salutation (Dear Mr. Smith:) and complimentary close (Sincerely yours,) are capitalized, as are the names of the recipients. See also **letters**.

Specific Groups

Capitalize the names of ethnic groups, religions, and nationalities (Native American, Christianity, Mongolian). Do not capitalize the names of social and economic groups (middle class, unemployed).

Specific Places

Capitalize the names of all political divisions (Ward Six, Chicago, Cook County, Illinois) and geographical divisions (Europe, Asia, North America, the Middle East). Do not capitalize geographic features unless they are part of a proper name.

▶ The mountains in some areas, such as the *Great Smoky Mountains*, make radio transmission difficult.

The words *north*, *south*, *east*, and *west* are capitalized when they refer to sections of the country. They are not capitalized when they refer to directions.

▶ I may relocate further *west*, but my family will remain in the South.

Capitalize the names of stars, constellations, and planets.

▶ North Star, Andromeda, Saturn

Do not capitalize *earth*, *sun*, and *moon* except when they are referred to formally as astronomical bodies.

▶ My workday was so long that I saw the *sun* rise over the lake and the *moon* appear as darkness settled over the *earth*.

▶ The various effects of the *Sun* on *Earth* and the *Moon* were discussed at the symposium.

Specific Institutions, Events, Concepts

Capitalize the names of institutions, organizations, and associations (U.S. Department of Health and Human Services). An organization usually capitalizes the names of its internal divisions and departments (Aeronautics Division, Human Resources Department). Types of organizations are not capitalized unless they are part of an official name (a technical communication association; Society for Technical Communication). Capitalize historical events (the Great Depression of the 1930s). Capitalize words that designate holidays, specific periods of time, months, or days of the week (Labor Day, the Renaissance, January, Monday). Do not capitalize seasons of the year (spring, summer, autumn, winter).

Capitalize the scientific names of classes, families, and orders but not the names of species or English derivatives of scientific names (Mammalia/mammal, Carnivora/carnivorous).

Titles of Works

Capitalize the initial letters of the first, last, and major words in the title of a book, an article, a play, or a film. Do not capitalize **articles** (*a*, *an*, *the*), coordinating **conjunctions** (*and*, *but*), or **prepositions** unless they begin or end the title (*The Lives of a Cell*). Capitalize prepositions within **titles** when they contain five or more letters (*Between*, *Within*, *Until*, *After*), unless you are following a style that recommends otherwise. The same rules apply to the subject lines of **e-mails** or **memos**.

Professional and Personal Titles

Titles preceding proper names are capitalized (Ms. Berger, Senator Lieberman). Appositives following proper names normally are not capitalized (Joseph Lieberman, *senator* from Connecticut). However, the word *President* is often capitalized when it refers to the chief executive of a national government. See **appositives**.

Job titles used with personal names are capitalized (Ho-shik Kim, *Laboratory Director*). Job titles used without personal names are not capitalized. ("The *laboratory director* will meet us tomorrow.") Use capital letters to designate family relationships only when they occur before a name (my uncle, Uncle Fred).

Abbreviations and Letters

Capitalize **abbreviations** if the words they stand for would be capitalized, such as B.S.Ch.E. (Bachelor of Science in Chemical Engineering). Capitalize letters that serve as names or indicate shapes (vitamin B, T-square, U-turn, I-beam).

Miscellaneous Capitalizations

The first word of a complete sentence enclosed in **dashes**, **brackets**, or **parentheses** is not capitalized when it appears as part of another sentence.

▶ We must improve our safety record this year (accidents last year were up 10 percent).

Certain units, such as parts and chapters of books and rooms in buildings, when specifically identified by number, are capitalized (Chapter 5, Ch. 5; Room 72, Rm. 72). Minor divisions within such units are not capitalized unless they begin a sentence (page 11, verse 14, seat 12).

case

DIRECTORY

Subjective Case 63	Appositives 65
Objective Case 64	Determining the Case of Pronouns 65
Possessive Case 64	

Grammatical case indicates the functional relationship of a **noun** or a **pronoun** to the other words in a sentence. Nouns change form only in the possessive case; pronouns may show change for the subjective, the objective, or the possessive case.

The case of a noun or pronoun is always determined by its function in a **phrase**, **clause**, or sentence. If it is the subject of a phrase, clause, or sentence, it is in the subjective case; if it is an **object** in a phrase, clause, or sentence, it is in the objective case; if it reflects possession or ownership and modifies a noun, it is in the possessive case. Figure C–1 is a table of pronouns in the subjective, objective, and possessive cases. See also **sentence construction**.

SINGULAR	SUBJECTIVE	OBJECTIVE	POSSESSIVE
First person	I	me	my, mine
Second person	you	you	your, yours
Third person	he, she, it	him, her, it	his, her, hers, its
PLURAL	**SUBJECTIVE**	**OBJECTIVE**	**POSSESSIVE**
First person	we	us	our, ours
Second person	you	you	your, yours
Third person	they	them	their, theirs

FIGURE C–1. Pronoun-Case Chart

The subjective case can indicate the person or thing acting ("*He* sued the vendor"), the person or thing acted upon ("*He* was sued by the vendor"), or the topic of description ("*He* is the vendor"). The objective case can indicate the thing acted on ("The vendor sued *him*") or the person or thing acting but in the objective position ("The vendor was sued by *him*"). (See also **voice**.) The possessive case indicates the person or thing owning or possessing something ("It was *his* company"). See also **modifiers**.

Subjective Case

A pronoun is in the subjective case (also called *nominative case*) when it represents the person or thing acting or is the receiver of the action even though it is in the subject position.

▶ *I* wrote a proposal.

▶ *I* was praised for my proposal writing.

A linking **verb** links a pronoun to its antecedent to show that they identify the same thing. Because they represent the same thing, the pronoun is in the subjective case even when it follows the verb, which makes it a subjective complement.

- *He* is the head of the Quality Control Group. [subject]
- The head of the Quality Control Group is *he*. [subjective complement]

The subjective case is used after the words *than* and *as* because of the understood (although unstated) portion of the clauses in which those words appear.

- George is as good a designer as *I* [am].
- Our subsidiary can do the job better than *we* [can].

Objective Case

A pronoun is in the objective case (also called the *accusative case*) when it indicates the person or thing receiving the action that is expressed by a verb in the active voice.

- They selected *me* to attend the conference.

Pronouns that follow action verbs (which excludes all forms of the verb *be*) must be in the objective case. Do not be confused by an additional name.

- The company promoted *me* in July.
- The company promoted John and *me* in July.

A pronoun is in the objective case when it is the object of a gerund or **preposition** or the subject of an infinitive.

- Between *you* and *me*, his facts are questionable. [objects of a preposition]
- Many of *us* attended the conference. [object of a preposition]
- Training *him* was the best thing I could have done. [object of a gerund]
- We asked *them* to return the deposit. [subject of an infinitive]

English does not differentiate between direct objects and indirect objects; both require the objective form of the pronoun. See also **complements**.

- The interviewer seemed to like *me*. [direct object]
- They wrote *me* a letter. [indirect object]

Possessive Case

A noun or pronoun is in the possessive case when it represents a person, place, or thing that possesses something. To make a singular noun possessive, add *'s* (the *manufacturer's* robotic inventory system). With plural nouns that end in *s*, show the possessive by placing an apostrophe after the *s* that forms the plural (a *managers'* meeting). For other guidelines, see **possessive case**.

Appositives

An **appositive** is a noun or noun phrase that follows and amplifies another noun or noun phrase. Because it has the same grammatical function as the noun it complements, an appositive should be in the same case as the noun with which it is in apposition.

▶ Two nurses, Jim Knight and *I*, were asked to review the patient's chart. [subjective case]

▶ The group leader selected two members to represent the department—Mohan Pathak and *me*. [objective case]

Determining the Case of Pronouns

One test to determine the proper case of a pronoun is to try it with some transitive verb such as *resembled* or *hit*. If the pronoun would logically precede the verb, use the subjective case; if it would logically follow the verb, use the objective case.

▶ *She* [*He, They*] resembled her father. [subjective case]

▶ Angela resembled *him* [*her, them*]. [objective case]

In the following example, try omitting the noun to determine the case of the pronoun.

SENTENCE	(*We / Us*) pilots fly our own airplanes.
INCORRECT	*Us* fly our own airplanes. [This incorrect usage is obviously wrong.]
CORRECT	*We* fly our own airplanes. [This correct usage sounds right.]

To determine the case of a pronoun that follows *as* or *than*, mentally add the words that are omitted but understood.

▶ The other technical writer is not paid as well as *she* [is paid]. [You would not write, "*Her* is paid."]

▶ His partner was better informed than *he* [was informed]. [You would not write, "*Him* was informed."]

If pronouns in compound constructions cause problems, try using them singly to determine the proper case.

SENTENCE	(*We/Us*) and the clients are going to lunch.
CORRECT	*We* are going to lunch.
	[You would not write, "*Us* are going to lunch."]

For advice on when to use *who* and *whom*, see **who / whom**.

cause-and-effect method of development

The cause-and-effect **method of development** is a common strategy to explain why something happened or why you think something will happen. The goal of the cause-and-effect method of development is to make as plausible as possible the relationship between a situation and either its cause or its effect. The conclusions you draw about the relationships should be based on evidence you have gathered. Like all methods, this one is often used in combination with others. If you were examining a problem with multiple causes, for example, you might combine cause-and-effect with **order-of-importance method of development** as you examine each cause and its effect.

Evaluating Evidence

Because not all evidence you gather will be of equal value, keep in mind the following guidelines:

- *Your facts and arguments should be relevant to your topic.* Be careful not to draw a conclusion that your evidence does not lead to or support. You may have researched some statistics, for example, showing that an increasing number of Americans are licensed to fly small airplanes. You cannot use that information as evidence for a decrease in new car sales in the United States — the evidence does not lead to that conclusion.
- *Your evidence should be adequate.* Incomplete evidence can lead to false conclusions.
  - ▶ Driver-training classes do not help prevent auto accidents. Two people I know who completed driver-training classes were involved in accidents.

  A thorough investigation of the usefulness of driver-training classes in keeping down the accident rate would require more than one or two examples. It would require a systematic comparison of the driving records for a representative sample of drivers who had completed driver training and those who had not.
- *Your evidence should be representative.* If you conduct a survey to obtain your evidence, do not solicit responses only from individu-

als or groups whose views are identical to yours; be sure you obtain responses from a diverse population.

- *Your evidence should be demonstrable.* Two events that occur close to each other in time or place may or may not be causally related. For example, that new traffic signs were placed at an intersection and the next day an accident occurred does not necessarily prove that the signs caused the accident. You must demonstrate the relationship between the two events with pertinent facts and arguments. See **logic errors**.

Linking Causes to Effects

To show a true relationship between a cause and an effect, you must demonstrate that the existence of the one *requires* the existence of the other. It is often difficult to establish beyond any doubt that one event was the cause of another event. More often, a result will have more than one cause. As you research a subject, your task is to determine which cause or causes are most plausible.

When several probable causes are equally valid, report your findings accordingly, as in the following excerpt from an article on the use of an energy-saving device called a furnace-vent damper. The damper is a metal plate that fits inside the flue or vent pipe of a natural-gas or fuel-oil furnace to allow poisonous gases to escape up the flue. Tests run on several dampers showed a number of probable causes for their malfunctioning.

> ▶ One damper was sold without proper installation instructions, and another was wired incorrectly. Two of the units had slow-opening dampers (15 seconds) that prevented the [furnace] burner from firing. And one damper jammed when exposed to a simulated fuel temperature of more than 700 degrees.
> —Don DeBat, "Save Energy but Save Your Life, Too," *Family Safety*

The investigator located more than one cause of damper malfunctions and reported on them. Without such a thorough account, recommendations to prevent malfunctions would be based on incomplete evidence.

center on

Use the phrase *center on* in writing, not *center around*. ("The experiments *center on* the new discovery.") Often the idea intended by *center on* is better expressed by other words.

> ▶ The hearings on computer security ~~centered on~~ *dealt with* access codes.

chronological method of development

The chronological **method of development** arranges the events under discussion in sequential order, emphasizing time as it begins with the first event and continues chronologically to the last. **Trip reports, laboratory reports**, work schedules, some **minutes of meetings,** and certain **trouble reports** are among the types of writing in which information is organized chronologically. Chronological order is typically used in **narration**.

The trouble report shown in Figure C–2 describes the development of a fire at an industrial facility. After providing important background information, the writer presents both the events and the steps taken in chronological order.

cite / sight / site

Cite means "acknowledge" or "quote an authority." ("The speaker *cited* several ecology experts.") *Sight* is the ability to see. ("He feared that he might lose his *sight*.") *Site* is a plot of land (a construction *site*) or the place where something is located (a Web *site*).

clarity

Clarity is essential to effective communication with your **readers**. You cannot achieve your **purpose** or a goal like **persuasion** without clarity. Many factors contribute to clarity, just as many other elements can defeat it.

A logical **method of development** and an outline will help you avoid presenting your reader with a jumble of isolated thoughts. A method of development and an outline that puts your thoughts into a logical, meaningful sequence brings **coherence** as well as **unity** to your writing. Clear **transition** contributes to clarity by providing the smooth flow that enables the reader to connect your thoughts with one another without conscious effort. See also **outlining**.

Proper **emphasis** and **subordination** are mandatory if you want to achieve clarity. If you do not use those two complementary techniques wisely, all your clauses and sentences will appear to be of equal importance. Your reader will only be able to guess which are most important, which are least important, and which fall somewhere in between. The **pace** at which you present your ideas is also important to clarity; if the

Memorandum

To: Charles Artmier, Chairman, Safety Committee

From: Willard Ricke, Plant Superintendent, Sequoia Plant *WR*

Date: November 17, 2009

Subject: Fire at our Sequoia Plant in Ardville

The following description provides an account of the conditions and development of the fire at our Sequoia Plant in Ardville. Because this description will be incorporated in the report that the Safety Committee will prepare, please let me know if any details should be added or clarified.

Three piles of scrap paper were located about 100 feet south of Building A and 150 feet west of Building B at our Sequoia Chemical Plant in Ardville. Building A, which consisted of one story and a partial attic, was used in part as an electrical shop and in part for equipment storage. Six pallets of Class I flammable liquids in 55-gallon drums were temporarily stored just outside Building A, along its west wall.

A spark created in the electrical shop in Building A started a fire. The night watchman first noticed the fire in the electrical shop at about 6:00 a.m. and immediately called the fire department and the plant superintendent. Before the fire department arrived, the fire spread to the drums of flammable liquid just outside the west wall and quickly engulfed Building A in flames. The fire then spread to the scrap paper outside, with a 40 mph wind blowing it in the direction of Building B.

During this time, the watchman began to hose down the piles of scrap paper between the buildings to try to keep the fire from reaching Building B. He also set up lawn sprinklers between the fire and Building B as part of his attempt to protect Building B from the fire. However, he saw smoke coming from Building B at 6:15 a.m., in spite of his best efforts to protect it.

The volunteer fire department, which was 20 miles away, reached the scene at 6:57 a.m., nearly an hour after the fire was discovered. It is estimated that the fire department's pumps were started after the fire had been burning in Building B for at least 20 minutes.

Building A was destroyed, along with its contents, and the fire burned into the hollow joisted roof of Building B, which sustained a $250,000 loss. . . .

FIGURE C–2. Chronological Method of Development

pace is not carefully adjusted to both the topic and the reader, your writing will appear cluttered and unclear.

Point of view establishes through whose eyes or from what vantage point the reader views the subject. A consistent point of view is essential to clarity; if you inappropriately switch from the first person to the third person in midsentence, you are certain to confuse your reader.

Precise **word choice** contributes to clarity and helps eliminate **ambiguity** and **awkwardness**. **Vague words**, **clichés**, poor use of **idiom**, and inappropriate **usage** detract from clarity. That **conciseness** is a requirement of clearly written communication should be evident to anyone who has ever attempted to decipher an insurance policy or a legal contract. For the sake of clarity, remove unnecessary words from your writing.

clauses

A clause is a group of words that contains a subject and a predicate and that functions as a sentence or as part of a sentence. (See **sentence construction**.) Every subject-predicate word group in a sentence is a clause, and every sentence must contain at least one independent clause; otherwise, it is a **sentence fragment**.

A clause that could stand alone as a simple sentence is an *independent clause*. ("*The scaffolding fell* when the rope broke.") A clause that could not stand alone if the rest of the sentence were deleted is a *dependent* (or *subordinate*) *clause*. ("I was at the St. Louis branch *when the decision was made*.")

Dependent clauses are useful in making the relationship between thoughts clearer and more succinct than if the ideas were presented in a series of simple sentences or compound sentences.

FRAGMENTED	The recycling facility is located between Millville and Darrtown. Both villages use it. [The two thoughts are of approximately equal importance.]
SUBORDINATED	The recycling facility, *which is located between Millville and Darrtown*, is used by both villages. [One thought is subordinated to the other.]

Subordinate clauses are especially effective for expressing thoughts that describe or explain another statement. Too much **subordination**, however, can be confusing and foster wordiness. See also **conciseness**.

▶ He selected instructors whose classes ~~had a slant that was~~ *were* specifically directed ~~toward students who intended to go into engineering~~ *at engineering students.*

A clause can be connected with the rest of its sentence by a coordinating **conjunction**, a subordinating conjunction, a relative **pronoun**, or a conjunctive **adverb**.

- It was 500 miles to the facility, *so* we made arrangements to fly. [coordinating conjunction]
- Mission control will need to be alert *because* at launch the space shuttle could be damaged by flying debris. [subordinating conjunction]
- Robert M. Fano was the scientist *who* developed the earliest multiple-access computer system at MIT. [relative pronoun]
- We arrived in the evening; *nevertheless*, we began the tour of the facility. [conjunctive adverb]

clichés

Clichés are expressions that have been used for so long that they are no longer fresh but come to mind easily because they are so familiar. Clichés are often wordy as well as vague and can be confusing, especially to speakers of **English as a second language**. A better, more direct word or phrase is given for each of the following clichés.

INSTEAD OF	USE
all over the map	scattered, unfocused
the game plan	strategy, schedule
last but not least	last, finally

Some writers use clichés in a misguided attempt to appear casual or spontaneous, just as other writers try to impress readers with **buzzwords**. Although clichés may come to mind easily while you are **writing a draft**, eliminate them during **revision**. See also **affectation**, **conciseness**, and **international correspondence**.

coherence

Writing is coherent when the relationships among ideas are clear to readers. The major components of coherent writing are a logical sequence of related ideas and clear transitions between these ideas. See also **clarity** and **organization**.

Presenting ideas in a logical sequence is the most important requirement in achieving coherence. The key to achieving a logical sequence is the use of a good outline. (See **outlining**.) An outline forces you to establish a beginning, a middle, and an end. That structure contributes greatly to coherence by enabling you to experiment with sequences and lay out the most direct route to your **purpose** without digressing.

Thoughtful **transition** is also essential; without it, your writing cannot achieve the smooth flow of sentence to sentence and **paragraph** to paragraph that results in coherence.

Check your draft carefully for coherence during **revision**. If possible, have someone else review your draft for how well it expresses the relationships between ideas. Such help is especially important in **collaborative writing** where team-member contributions need to be blended into a coherent document. See also **unity**.

collaborative writing

Collaborative writing occurs when two or more writers work together to produce a single document for which they share responsibility and decision-making authority, as is often the case with **proposals**. Collaborative writing teams are formed when (1) the size of a project or the time constraints imposed on it require collaboration, (2) the project involves multiple areas of expertise, or (3) the project requires the melding of divergent views into a single perspective that is acceptable to the whole team or to another group. Many types of collaborations are possible, from the collaboration of a primary writer with a variety of contributors and reviewers to a highly interactive collaboration in

> **DIGITAL TIP**
>
> ### Wikis for Collaborative Documents
>
> Wikis are Web sites in which the content and organization can be quickly and easily edited by authorized users. Wikis allow collaborators in different geographic locations to coauthor documents by providing them with basic editing functions and the ability to post comments or questions. A writer may post a document draft to a wiki for collaborators to edit the draft. The edits are tracked so that all users can see who revised what and when, thus reducing problems with tracking multiple versions. Wikis can be used also for sharing and distributing information, managing projects, and providing communication spaces for clients and customers. For more detailed information on wikis, go to *bedfordstmartins.com/alredtech* and select *Digital Tips*, "Wikis for Collaborative Documents."

which everyone on a team plays a relatively equal role in shaping the document.*

Tasks of the Collaborative Writing Team

The collaborating team strives to achieve a compatible working relationship by dividing the work in a way that uses each writer's expertise and experience to its advantage. The team should also designate a coordinator who will guide the team members' activities, organize the project, and ensure **coherence** and consistency within the document. The coordinator's duties can be determined by mutual agreement or, if the team often works together, assigned on a rotating basis.

Planning. The team conceptualizes the document to be produced as the members collectively identify the **audience**, **purpose**, **context**, and **scope** of the project. See also **meetings** and "Five Steps to Successful Writing" (page xv).

At this stage, the team establishes a project plan that may include guidelines for communication among team members, version control (naming, dating, and managing document drafts), review procedures, and writing **style** standards that team members are expected to follow. The plan includes a schedule with due dates for completing initial research tasks, outlines, drafts, reviews, revisions, and the final document. Milestone deadlines must be met, even if the drafts are not as polished as the individual writers would like: One missed deadline can delay the entire project.

Research and Writing. At this stage, the team completes initial research tasks, elicits reports from team members, creates a broad outline of the document (see **outlining**), and assigns writing tasks to individual team members, based on their expertise and the outline. Depending on the project, each team member further researches an assigned segment of the document, expands and develops the broad outline, and produces a draft from a detailed outline. See also **research** and **writing a draft**.

Reviewing. Keeping the audience's needs and the document's purpose in mind, each team member critically yet diplomatically reviews the other team members' drafts, from the overall **organization** to the **clarity** of each **paragraph**, and offers advice to help improve the writer's work.

---

*For a discussion of various forms of collaboration, see Paul Benjamin Lowry, Aaron Curtis, and Michelle René Lowry, "Building a Taxonomy and Nomenclature of Collaborative Writing to Improve Interdisciplinary Research and Practice," *Journal of Business Communication* 41.1 (January 2004): 66–99.

Team members can easily solicit feedback by sharing electronic files. Tracking and commenting features allow the reviewer to show the suggested changes without deleting the original text.

> **DIGITAL TIP**
>
> **Reviewing Collaborative Documents**
>
> Some Adobe Acrobat® applications and many word-processing packages have options for providing feedback on and tracking changes in the drafts of collaborators' documents. You can add text or voice annotations within the text, allowing collaborators to read or hear your comments and edits and to accept or reject each suggested change. For more on this topic, see *bedfordstmartins.com/alredtech* and select *Digital Tips*, "Reviewing Collaborative Documents."

Revising. In this stage, individual writers evaluate their colleagues' reviews and accept, reject, or build on their suggestions. Then, the team coordinator can consolidate all drafts into a final master copy and maintain and evaluate it for consistency and coherence. See also **revision**.

Conflict

As you collaborate, be ready to tolerate some disharmony, but temper it with mutual respect. Team members may not agree on every subject, and differing perspectives can easily lead to conflict, ranging from mild differences over minor points to major showdowns. However, creative differences resolved respectfully can energize the team and, in fact, strengthen a finished document by compelling writers to reexamine assumptions and issues in unanticipated ways. See also **listening**.

> **WRITER'S CHECKLIST** **Writing Collaboratively**
>
> ✔ Designate one person as the team coordinator.
> ✔ Identify the audience, purpose, context, and scope of the project.
> ✔ Create a project plan, including a schedule and standards.
> ✔ Create a working outline of the document.
> ✔ Assign segments or tasks to each team member.
> ✔ Research and write drafts of document segments.
> ✔ Follow the schedule: due dates for drafts, revisions, and final versions.
> ✔ Use the agreed-upon standards for style and format.
> ✔ Exchange segments for team member reviews.

> **Writer's Checklist: Writing Collaboratively (continued)**
> - Revise segments as needed.
> - Meet the established deadlines.

colons

The colon (:) is a mark of introduction that alerts readers to the close connection between the preceding statement and what follows.

Colons in Sentences

A colon links independent **clauses** to words, **phrases**, clauses, or lists that identify, rename, explain, emphasize, amplify, or illustrate the sentence that precedes the colon.

- Two topics will be discussed: *the new lab design and the revised safety procedures.* [phrases that identify]
- Only one thing will satisfy Mr. Sturgess: *our finished report.* [appositive (renaming) phrase for **emphasis**]
- Any organization is confronted with two separate, though related, information problems: *It must maintain an effective internal communication system and an effective external communication system.* [clause to amplify and explain]
- Heart patients should make key lifestyle changes: *stop smoking, exercise regularly, eat a low-fat diet, and reduce stress.* [list to identify and illustrate]

Colons with Salutations, Titles, Citations, and Numbers

A colon follows the salutation in business **letters**, even when the salutation refers to a person by first name.

- Dear Professor Jeffers: *or* Dear Georgia:

Colons separate titles from subtitles and separate references to sections of works in citations. See also **documenting sources**.

- *Writing That Works: Communicating Effectively on the Job*
- Genesis 10:16 [chapter 10, verse 16]

Colons separate numbers in time references and indicate numerical ratios.

▶ 9:30 a.m. [9 hours and 30 minutes]
▶ The cement is mixed with water and sand at 5:3:1.
[The colon is read as the word *to*.]

Punctuation and Capitalization with Colons

A colon always goes outside **quotation marks**.

▶ This was the real meaning of the manager's "suggestion": Cooperation within our department must improve.

As this example shows, the first word after a colon may be capitalized if the statement following the colon is a complete sentence and functions as a formal statement or question. If the element following the colon is subordinate, however, use a lowercase letter to begin that element. See also **capitalization**.

▶ We have only one way to stay within our present budget: to reduce expenditures for research and development.

Unnecessary Colons

Do not place a colon between a **verb** and its **objects**.

▶ Three fluids that clean pipettes are: water, alcohol, and acetone.

Likewise, do not use a colon between a **preposition** and its object.

▶ I may be transferred to: Tucson, Boston, or Miami.

Do not insert a colon after *including*, *such as*, or *for example* to introduce a simple list.

▶ Office computers should not be used for activities such as: personal e-mail, Web surfing, Internet shopping, and playing computer games.

One common exception is made when a verb or preposition is followed by a stacked **list**; however, it may be possible to introduce the list with a complete sentence instead.

▶ *The following corporations* :
 ~~Corporations that~~ manufacture computers ~~include:~~
 Apple   Compaq   Dell
 Gateway   IBM   Sony

comma splice

A comma splice is a grammatical error in which two independent <u>clauses</u> are joined by only a <u>comma</u>.

> INCORRECT   It was 500 miles to the facility, we arranged to fly.

A comma splice can be corrected in several ways.

1. Substitute a <u>semicolon</u>, a semicolon and a conjunctive <u>adverb</u>, or a comma and a coordinating <u>conjunction</u>.
   - It was 500 miles to the facility; we arranged to fly.
   - It was 500 miles to the facility; *therefore*, we arranged to fly.
   - It was 500 miles to the facility, *so* we arranged to fly.
2. Create two sentences.
   - It was 500 miles to the facility. We arranged to fly.
3. Subordinate one clause to the other. (See <u>subordination</u>.)
   - *Because it was 500 miles to the facility*, we arranged to fly.

See also <u>sentence construction</u> and <u>sentence faults</u>.

commas

DIRECTORY

Linking Independent Clauses 78
Enclosing Elements 78
Introducing Elements 79
Separating Items in a Series 80
Clarifying and Contrasting 81
Showing Omissions 81
Using with Numbers and Names 81
Using with Other Punctuation 82
Avoiding Unnecessary Commas 83

Like all <u>punctuation</u>, the comma (,) helps <u>readers</u> understand the writer's meaning and prevents <u>ambiguity</u>. Notice how the comma helps make the meaning clear in the second example.

> AMBIGUOUS   To be successful nurses with M.S. degrees must continue to learn.
> CLEAR   To be successful, nurses with M.S. degrees must continue to learn.
> [The comma makes clear where the main part of the sentence begins.]

Do not follow the old myth that you should insert a comma wherever you would pause if you were speaking. Although you would pause wherever you encounter a comma, you should not insert a comma wherever you might pause. Effective use of commas depends on an understanding of **sentence construction**.

Linking Independent Clauses

Use a comma before a coordinating **conjunction** (*and*, *but*, *or*, *nor*, and sometimes *so*, *yet*, and *for*) that links independent **clauses**.

▶ The new microwave disinfection system was delivered, *but* the installation will require an additional week.

However, if two independent clauses are short and closely related—and there is no danger of confusing the reader—the comma may be omitted. Both of the following examples are correct.

▶ The cable snapped and the power failed.

▶ The cable snapped, and the power failed.

Enclosing Elements

Commas are used to enclose nonessential information in nonrestrictive clauses, **phrases**, and parenthetical elements. See also **restrictive and nonrestrictive elements**.

▶ Our new factory, *which began operations last month*, should add 25 percent to total output. [nonrestrictive clause]

▶ The technician, *working quickly and efficiently*, finished early. [nonrestrictive phrase]

▶ We can, *of course*, expect their lawyer to call us. [parenthetical element]

Yes and *no* are set off by commas in such uses as the following:

▶ I agree with you, *yes*.

▶ *No*, I do not think we can finish by the deadline.

A **direct address** should be enclosed in commas.

▶ You will note, *Jeff*, that the surface of the brake shoe complies with the specification.

An **appositive** phrase (which re-identifies another expression in the sentence) is enclosed in commas.

▶ Our company, *Envirex Medical Systems*, won several awards last year.

Interrupting parenthetical and transitional words or phrases are usually set off with commas. See also **transition**.

▶ The report, *therefore*, needs to be revised.

Commas are omitted when the word or phrase does not interrupt the continuity of thought.

▶ I *therefore* suggest that we begin construction.

For other means of punctuating parenthetical elements, see **dashes** and **parentheses**.

Introducing Elements

Clauses and Phrases. Generally, place a comma after an introductory clause or phrase, especially if it is long, to identify where the introductory element ends and the main part of the sentence begins.

▶ *Because we have not yet contained the new strain of influenza*, we recommend vaccination for high-risk patients.

A long modifying phrase that precedes the main clause should always be followed by a comma.

▶ *During the first series of field-performance tests at our Colorado proving ground*, the new engine failed to meet our expectations.

When an introductory phrase is short and closely related to the main clause, the comma may be omitted.

▶ *In two seconds* a 5°C temperature rise occurs in the test tube.

A comma should always follow an absolute phrase, which modifies the whole sentence.

▶ *The tests completed*, we organized the data for the final report.

Words and Quotations. Certain types of introductory words are followed by a comma. One example is a transitional word or phrase (*however*, *in addition*) that connects the preceding clause or sentence with the thought that follows.

▶ *Furthermore*, steel can withstand a humidity of 99 percent, provided that there is no chloride or sulfur dioxide in the atmosphere.

▶ *For example*, this change will make us more competitive in the global marketplace.

When an **adverb** closely modifies the **verb** or the entire sentence, it should not be followed by a comma.

▶ *Perhaps* we can still solve the turnover problem. *Certainly* we should try.
[*Perhaps* and *certainly* closely modify each statement.]

A proper **noun** used in an introductory direct address is followed by a comma, as is an **interjection** (such as *oh*, *well*, *why*, *indeed*, *yes*, and *no*).

▶ *Nancy*, enclosed is the article you asked me to review.
[direct address]

▶ *Indeed*, I will ensure that your request is forwarded. [interjection]

Use a comma to separate a direct **quotation** from its introduction.

▶ Morton and Lucia White *said*, "People live in cities but dream of the countryside."

Do not use a comma when giving an indirect quotation.

▶ Morton and Lucia White *said that* people dream of the countryside, even though they live in cities.

Separating Items in a Series

Although the comma before the last item in a series is sometimes omitted, it is generally clearer to include it.

▶ Random House, Bantam, Doubleday, and Dell were individual publishing companies.
[Without the final comma, "Doubleday and Dell" might refer to one company or two.]

Phrases and clauses in coordinate series are also punctuated with commas.

▶ Plants absorb noxious gases, act as receptors of dirt particles, and cleanse the air of other impurities.

When phrases or clauses in a series contain commas, use **semicolons** rather than commas to separate the items.

▶ Our new products include amitriptyline, which has sold very well; and cholestyramine, which was just introduced.

When **adjectives** modifying the same noun can be reversed and make sense, or when they can be separated by *and* or *or*, they should be separated by commas.

▶ The aircraft featured a *modern*, *sleek*, *swept-wing* design.

When an adjective modifies a phrase, no comma is required.

▶ She was investigating the *damaged inventory-control system*. [The adjective *damaged* modifies the phrase *inventory-control system*.]

Never separate a final adjective from its noun.

▶ He is a conscientious, honest, reliable̷ worker.

Clarifying and Contrasting

Use a comma to separate two contrasting thoughts or ideas.

▶ The project was finished on time, but not within the budget.

Use a comma after an independent clause that is only loosely related to the dependent clause that follows it or that could be misread without the comma.

▶ I should be able to finish the plan by July, even though I lost time because of illness.

Showing Omissions

A comma sometimes replaces a verb in certain elliptical constructions.

▶ Some were punctual; *others, late*. [The comma replaces *were*.]

It is better, however, to avoid such constructions in technical writing.

Using with Numbers and Names

Commas are conventionally used to separate distinct items. Use commas between the elements of an address written on the same line (but not between the state and the ZIP Code).

▶ Kristen James, 4119 Mill Road, Dayton, Ohio 45401

A full date that is written in month-day-year format uses a comma preceding and following the year.

▶ November 30, 2020, is the payoff date.

Do not use commas for dates in the day-month-year format, which is used in many parts of the world and by the U.S. military. See also **international correspondence**.

▶ Note that 30 November 2020 is the payoff date.

No commas are used when showing only the month and year or month and day in a **date**.

▶ The target date of May 2012 is optimistic, so I would like to meet on March 4 to discuss our options.

Use commas to separate the elements of Arabic numbers.

▶ 1,528,200 feet

However, because many countries use the comma as the decimal marker, use spaces or periods rather than commas in international documents.

▶ 1.528.200 meters *or* 1 528 200 meters

A comma may be substituted for the colon in the salutation of a personal **letter** or **e-mail**. Do not, however, use a comma in the salutation of a business letter or e-mail, even if you use the person's first name.

▶ Dear Marie, [personal letter or e-mail]
▶ Dear Marie: [business letter or e-mail]

Use commas to separate the elements of geographical names.

▶ Toronto, Ontario, Canada

Use a comma to separate names that are reversed (Smith, Alvin) and commas with professional **abbreviations**.

▶ Jim Rogers Jr., M.D., chaired the conference.
[*Jr.* or *Sr.* does not require a comma.]

Using with Other Punctuation

Conjunctive adverbs (*however*, *nevertheless*, *consequently*, *for example*, *on the other hand*) that join independent clauses are preceded by a **semicolon** and followed by a comma. Such adverbs function both as **modifiers** and as connectives.

▶ The idea is good; *however*, our budget is not sufficient.

As shown earlier in this entry, use semicolons rather than commas to separate items in a series when the items themselves contain commas.

When a comma should follow a phrase or clause that ends with words in parentheses, the comma always appears outside the closing parenthesis.

▶ Although we left late (at 7:30 p.m.), we arrived in time for the keynote address.

▶ Commas always go inside **quotation marks**.

▶ The operator placed the discharge bypass switch at "*normal*," which triggered a second discharge.

Except with abbreviations, a comma should not be used with a dash, an **exclamation mark**, a **period**, or a **question mark**.

▶ "Have you finished the project?~~,~~" she asked.

Avoiding Unnecessary Commas

A number of common writing errors involve placing commas where they do not belong. As stated earlier, such errors often occur because writers assume that a pause in a sentence should be indicated by a comma.

Do not place a comma between a subject and verb or between a verb and its **object**.

▶ The conditions at the test site in the Arctic~~,~~ made accurate readings difficult.

▶ She has often said~~,~~ that one company's failure is another's opportunity.

Do not use a comma between the elements of a compound subject or compound predicate consisting of only two elements.

▶ The director of the design department~~,~~ and the supervisor of the quality-control section were opposed to the new schedules.

▶ The design director listed five major objections~~,~~ and asked that the new schedule be reconsidered.

Do not include a comma after a coordinating conjunction such as *and* or *but*.

▶ The chairperson formally adjourned the meeting, but~~,~~ the members of the committee continued to argue.

Do not place a comma before the first item or after the last item of a series.

▶ The new products we are considering include~~,~~ calculators, scanners, and cameras.

▶ It was a fast, simple, inexpensive~~,~~ process.

Do not use a comma to separate a prepositional phrase from the rest of the sentence unnecessarily.

▶ We discussed the final report~~,~~ on the new project.

compare / contrast

When you *compare* things, you point out similarities or both similarities and differences. ("He *compared* the two brands before making his choice.") When you *contrast* things, you point out only the differences. ("Their speaking styles *contrasted* sharply.") In either case, you compare or contrast only things that are part of a common category.

When *compare* is used to establish a general similarity, it is followed by *to*. ("He *compared* our receiving the grant *to* winning a marathon.") When *compare* is used to indicate a close examination of similarities or differences, it is followed by *with*. ("We *compared* the features of the new copier *with* those of the current one.")

Contrast is normally followed by *with*. ("The new policy *contrasts* sharply *with* the earlier one.") When the **noun** form of *contrast* is used, one speaks of the *contrast between* two things or of one thing being *in contrast to* the other.

▶ There is a sharp *contrast between* the old and new policies.

▶ The new policy is *in* sharp *contrast to* the earlier one.

comparison

When you are making a comparison, be sure that both or all of the elements being compared are clearly evident to your **reader**.

▶ The Nicom 3 software is better *than the Nicom 2 software*.

The things being compared must be of the same kind.

▶ A hard-side briefcase offers more protection than *a* fabric *briefcase*.

Be sure to point out the parallels or differences between the things being compared. Do not assume your reader will know what you mean.

▶ Washington is farther from Boston than *it is from* Philadelphia.

A double comparison in the same sentence requires that the first comparison be completed before the second one is stated.

▶ The discovery of electricity was one of the great ~~if not the greatest~~ scientific discoveries in history*, if not the greatest*.

Do not attempt to compare things that are not comparable.

▶ Agricultural experts advise that ~~storage space is reduced by 40 percent compared with~~ *requires 40 percent less storage space than loose hay requires* baled hay.

[*Storage space* is not comparable to *baled hay*.]

comparison method of development

As a **method of development**, comparison points out similarities and differences between the elements of your subject. The comparison method of development can help readers in an **audience** understand a difficult or an unfamiliar subject by relating it to a simpler or more familiar one.

You must first determine the basis for the **comparison**. For example, if you were comparing bids from contractors for a remodeling project at your company, you most likely would compare such factors as price, previous experience, personnel qualifications, availability at a time convenient for you, and completion date. Once you have determined the basis or bases for comparison, you can determine the most effective way to structure your comparison: whole by whole or part by part.

In the *whole-by-whole method*, all the relevant characteristics of one item are examined before all the relevant characteristics of the next item. The description of typical woodworking glues in Figure C–3 is organized according to the whole-by-whole method. It describes each type of glue and all its characteristics before moving to the next one. This description would be useful for those readers who wish to learn about all types of wood glues.

If your **purpose** is to help readers consider the various characteristics of all the glues, the information might be arranged according to the *part-by-part method of comparison*, in which the relevant features of the items are compared one by one (as shown in Figure C–4). The part-by-part method could accommodate further comparison—such as temperature ranges, special warnings, and common use, in this case.

Comparisons can also be made effectively with the use of **tables**, as shown in Figure C–5. The advantage of a table is that it provides a quick reference, allowing readers to see and compare all the information at once. The disadvantage is that a table cannot convey as much related detailed information as a narrative description.

> *White glue* is the most useful all-purpose adhesive for light construction, but it cannot be used on projects that will be exposed to moisture, high temperature, or great stress. Wood that is being joined with white glue must remain in a clamp until the glue dries, which takes about 30 minutes.
>
> *Aliphatic resin glue* has a stronger and more moisture-resistant bond than white glue. It must be used at temperatures above 50°F. The wood should be clamped for about 30 minutes. . . .
>
> *Plastic resin glue* is the strongest of the common wood adhesives. It is highly moisture resistant, though not completely waterproof. Sold in powdered form, this glue must be mixed with water and used at temperatures above 70°F. It is slow setting, and the joint should be clamped for four to six hours. . . .
>
> *Contact cement* is a very strong adhesive that bonds so quickly it must be used with great care. It is ideal for mounting sheets of plastic laminate on wood. It is also useful for attaching strips of veneer to the edges of plywood. Because this adhesive bonds immediately when two pieces are pressed together, clamping is not necessary, but the parts to be joined must be carefully aligned before being placed together. Most brands are flammable, and the fumes can be harmful if inhaled. To meet current safety standards, this type of glue must be used in a well-ventilated area, away from flames or heat.

FIGURE C–3. Whole-by-Whole Method of Comparison

> Woodworking adhesives are rated primarily according to their bonding strength, moisture resistance, and setting time.
>
> *Bonding strength* is categorized as very strong, moderately strong, or adequate for use with little stress. Contact cement and plastic resin glue bond very strongly, while aliphatic resin glue bonds moderately strongly. White glue provides a bond least resistant to stress.
>
> The *moisture resistance* of woodworking glues is rated as high, moderate, or low. Plastic resin glue and contact cement are highly moisture resistant, aliphatic resin glue is moderately moisture resistant, and white glue is least moisture resistant.
>
> *Setting time* for the glues varies from an immediate bond to a four-to-six-hour bond. Contact cement bonds immediately and requires no clamping. Because the bond is immediate, surfaces being joined must be carefully aligned before being placed together. White glue and aliphatic resin glue set in 30 minutes; both require clamping to secure the bond. Plastic resin, the strongest wood glue, sets in four to six hours and also requires clamping.

FIGURE C–4. Part-by-Part Method of Comparison

	White Glue	Aliphatic Resin Glue	Plastic Resin Glue	Contact Cement
Bonding Strength	Low	Moderate	High	High
Moisture Resistance	Low	Moderate	High	High
Setting Time	Thirty minutes	Thirty minutes	Four to six hours	Bonds on contact
Common Uses	Light construction	General purpose	General purpose	Laminate and veneer to wood

FIGURE C–5. Comparison Using a Table to Illustrate Key Differences

complaint letters

A complaint **letter** (or **e-mail**) describes a problem that the writer requests the recipient to solve. The **tone** of a complaint letter or e-mail is important; the most effective ones do not sound complaining. If your message reflects only your annoyance and anger, you may not be taken seriously. Assume that the recipient will be conscientious in correcting the problem. However, anticipate reader reactions or rebuttals. See **audience**.

▶ I reviewed my user manual's "safe operating guidelines" carefully before I installed the device.
  [This assures readers you followed instructions.]

Without such explanations, readers may be tempted to dismiss your complaint. Figure C–6 shows a complaint letter that details a billing problem. Although the circumstances and severity of the problem may vary, effective complaint letters generally follow this pattern:

1. Identify the problem or faulty item(s) and include relevant invoice numbers, part names, and dates. Include a copy of the receipt, bill, or contract, and keep the original for your records.
2. Explain logically, clearly, and specifically what went wrong, especially for a problem with a service. (Avoid guessing why you *think* some problem occurred.)
3. State what you expect the reader to do to solve the problem.

Begin by checking to see if the company's Web site provides instructions for submitting a complaint. Otherwise, for large organizations, you may address your complaint to Customer Service. In smaller organizations, you might write to a vice president in charge of sales or service, or directly to the owner. As a last resort, you may find that sending

> Subject: ST3 Diagnostic Scanners
>
> On July 10, I ordered nine ST3 Diagnostic Scanners (order # ST3-1179R). The scanners were ordered from your customer Web site.
>
> On August 3, I received seven HL monitors from your parts warehouse in Newark, New Jersey. I immediately returned those monitors with a note indicating that a mistake had been made. However, not only have I failed to receive the ST3 scanners that I ordered, but I have also been billed repeatedly for the seven monitors.
>
> I have enclosed a copy of my confirmation e-mail, the shipping form, and the most recent bill. If you cannot send me the scanners I ordered by November 2, please cancel my order.
>
> Sincerely,

FIGURE C–6. Complaint Letter

copies of a complaint letter to more than one person in the company will get faster results. See also **adjustment letters** and **refusal letters**.

complement / compliment

Complement means "anything that completes a whole" (see also **complements**). It is used as either a **noun** or a **verb**.

▶ A *complement* of four employees would bring our staff up to its normal level. [noun]

▶ The two programs *complement* one another perfectly. [verb]

Compliment means "praise." It too is used as either a noun or a verb.

▶ The manager's *compliment* boosted staff morale. [noun]

▶ The manager *complimented* the staff on its efficient job. [verb]

complements

A complement is a word, **phrase**, or **clause** used in the predicate of a sentence to complete the meaning of the sentence.

- Pilots fly *airplanes*. [word]
- To live is *to risk death*. [phrase]
- John knew *that he would be late*. [clause]

Four kinds of complements are generally recognized: direct **object**, indirect object, objective complement, and subjective complement. See also **sentence construction**.

A *direct object* is a **noun** or noun equivalent that receives the action of a transitive **verb**; it answers the question *What?* or *Whom?* after the verb.

- I designed *a Web site*. [noun phrase]
- I like *to work*. [verbal]
- I like *it*. [pronoun]
- I like *what I saw*. [noun clause]

An *indirect object* is a noun or noun equivalent that occurs with a direct object after certain kinds of transitive verbs such as *give*, *wish*, *cause*, and *tell*. It answers the question *To whom or what?* or *For whom or what?*

- We should buy the *office* a *scanner*.
  [*Scanner* is the direct object, and *office* is the indirect object.]

An *objective complement* completes the meaning of a sentence by revealing something about the object of its transitive verb. An objective complement may be either a noun or an **adjective**.

- They call him *a genius*. [noun phrase]
- We painted the building *white*. [adjective]

A *subjective complement*, which follows a linking verb rather than a transitive verb, describes the subject. A subjective complement may be either a noun or an adjective.

- His sister is *a consultant*. [noun phrase follows linking verb *is*]
- His brother is *ill*. [adjective follows linking verb *is*]

compose / constitute / comprise

Compose and *constitute* both mean "make up the whole." The parts *compose* or *constitute* the whole. ("The nine offices *compose* the division. Unethical activities *constitute* cause for dismissal.") *Comprise* means "include," "contain," or "consist of." The whole *comprises* the parts. ("The division *comprises* nine offices.")

compound words

A compound word is made from two or more words that function as a single concept. A compound may be hyphenated, written as one word, or written as separate words.

▶ high-energy nevertheless post office
  low-level online Web site

If you are not certain whether a compound word should use a **hyphen**, check a dictionary.

Be careful to distinguish between compound words (*greenhouse*) and words that simply appear together but do not constitute compound words (*green house*). For plurals of compound words, generally add *s* to the last letter (*bookcases* and *Web sites*). However, when the first word of the compound is more important to its meaning than the last, the first word takes the *s* (*editors in chief*). Possessives are formed by adding *'s* to the end of the compound word (the *editor in chief's* desk, the *pipeline's* diameter, the *post office's* hours). See also **possessive case**.

conciseness

Conciseness means that extraneous words, **phrases**, **clauses**, and sentences have been removed from writing without sacrificing **clarity** or appropriate detail. Conciseness is not a synonym for brevity; a long **report** may be concise, while its **abstract** may be brief and concise. Conciseness is always desirable, but brevity may or may not be desirable in a given passage, depending on the writer's **purpose**. Although concise sentences are not guaranteed to be effective, wordy sentences always sacrifice some of their readability and **coherence**.

Causes of Wordiness

Modifiers that repeat an idea implicit or present in the word being modified contribute to wordiness by being redundant. See also **reason is [because]**.

basic essentials *completely* finished
final outcome *present* status

Coordinated synonyms that merely repeat each other contribute to wordiness.

each and every   *basic and fundamental*
finally and for good *first and foremost*

Excess qualification also contributes to wordiness.

perfectly clear   *completely* accurate

Expletives, relative **pronouns**, and relative **adjectives**, although they have legitimate purposes, often result in wordiness.

WORDY   *There are* [expletive] many Web designers *who* [relative pronoun] are planning to attend the conference, *which* [relative pronoun] is scheduled for May 13–15.

CONCISE Many Web designers plan to attend the conference scheduled for May 13–15.

Circumlocution (a long, indirect way of expressing things) is a leading cause of wordiness. See also **gobbledygook**.

WORDY   The payment to which a subcontractor is entitled should be made promptly so that in the event of a subsequent contractual dispute we, as general contractors, may not be held in default of our contract by virtue of nonpayment.

CONCISE Pay subcontractors promptly. Then, if a contractual dispute occurs, we cannot be held in default of our contract because of nonpayment.

When conciseness is overdone, writing can become choppy and ambiguous. (See also **telegraphic style**.) Too much conciseness can produce a style that is not only too brief but also too blunt, especially in **correspondence**.

WRITER'S CHECKLIST Achieving Conciseness

Wordiness is understandable when you are **writing a draft**, but it should not survive **revision**.

- Use **subordination** to achieve conciseness.

    - The *five-page* financial report was carefully documented. ~~and it covered five pages.~~

Writer's Checklist: Achieving Conciseness (continued)

- Avoid **affectation** by using simple words and phrases.

  WORDY It is the policy of the company to provide Internet access to enable employees to conduct the online communication necessary to discharge their responsibilities; such should not be utilized for personal communications or nonbusiness activities.

  CONCISE Employee Internet access should be used only for appropriate company business.

- Eliminate redundancy.

  WORDY Postinstallation testing, which is offered to all our customers at no further cost to them whatsoever, is available with each Line Scan System One purchased from this company.

  CONCISE Free postinstallation testing is offered with each Line Scan System One.

- Change the passive **voice** to the active voice and the indicative **mood** to the imperative mood whenever possible.

  WORDY Bar codes normally are used when an order is intended to be displayed on a computer, and inventory numbers normally are used when an order is to be placed with the manufacturer.

  CONCISE Use bar codes to display the order on a computer, and use inventory numbers to place the order with the manufacturer.

- Eliminate or replace wordy introductory phrases or pretentious words and phrases (*in the case of*, *it may be said that*, *it appears that*, *needless to say*).

REPLACE	WITH
in order to, with a view to	to
due to the fact that, for the reason that, owing to the fact that, the reason for	because
by means of, by using, in connection with, through the use of	by, with
at this time, at this point in time, at present, at the present	now, currently

- Do not overuse **intensifiers**, such as *very*, *more*, *most*, *best*, *quite*, *great*, *really*, and *especially*. Instead provide specific and useful details.

Use the search-and-replace command to find and revise wordy expressions, including *to be* and unnecessary helping **verbs** such as *will*.

conclusions

The conclusion of a document ties the main ideas together and can clinch a final significant point. This final point may, for example, make a prediction or a judgment, summarize the key findings of a study, or recommend a course of action. Figure C–7 is a conclusion from a proposal to reduce health-care costs by increasing employee fitness through health-club subsidies. Notice that it makes a recommendation, summarizes key points, and points out several benefits of implementing the recommendation.

Conclusion and Recommendation

[Makes a recommendation] I recommend that ABO, Inc., participate in the corporate membership program at AeroFitness Clubs, Inc., by subsidizing employee memberships. Offering this benefit to employees will demonstrate ABO's commitment to the importance of a healthy workforce. Club membership allows employees at all five ABO warehouses to participate in the program. The more employees who participate, the greater the long-term savings in ABO's health-care costs.

[Summarizes key points] Enrolling employees in the corporate program at AeroFitness would allow them to receive a one-month, free trial membership. Those interested in continuing could then join the club and pay half of the one-time membership fee of $900 and receive a 30 percent discount on the $600 yearly fee. The other half of the membership fee ($450) would be paid for by ABO. If employees leave the company, they would have the option of purchasing ABO's share of the membership to continue at AeroFitness or selling their half of the membership to another ABO employee wishing to join AeroFitness.

[Points to benefits] Implementing this program will help ABO, Inc., reduce its health-care costs while building stronger employee relations by offering employees a desirable benefit. If this proposal is adopted, I have some additional thoughts about publicizing the program to encourage employee participation that I would be pleased to share.

FIGURE C–7. Conclusion

conclusions

The way you conclude depends on your **purpose**, the needs of your **audience**, and the **context**. For example, a lengthy sales **proposal** might conclude persuasively with a summary of the proposal's salient points and the company's relevant strengths. The following examples are typical concluding strategies.

RECOMMENDATION
Our findings suggest that you need to alter your marketing to adjust to the changing demographics for your products.

SUMMARY
As this report describes, we would attract more recent graduates with the following strategies:
1. Establish a Web site where students can register and submit online résumés.
2. Increase our advertising in local student newspapers and our attendance at college career fairs.
3. Expand our local co-op program.

JUDGMENT
Based on the scope and degree of the tornado's damage, the current construction code for roofing on light industrial facilities is inadequate.

IMPLICATION
Although our estimate calls for a substantially higher budget than in the three previous years, we believe that it is reasonable given our planned expansion.

PREDICTION
Although I have exceeded my original estimate for equipment, I have reduced my labor costs; therefore, I will easily stay within the original bid.

The concluding statement may merely present ideas for consideration, call for action, or deliberately provoke thought.

IDEAS FOR CONSIDERATION
The new prices become effective the first of the year. Price adjustments are routine for the company, but some of your customers will not consider them acceptable. Please bear in mind the needs of both your customers and the company as you implement these price adjustments.

CALL FOR ACTION
Please send us a check for $250 now if you wish to keep your account active. If you have not responded to our previous letters

because of some special hardship, I will be glad to work out a solution with you.

THOUGHT-PROVOKING STATEMENT
Can we continue to accept the losses incurred by inefficiency? Or should we take the necessary steps to control it now?

Be especially careful not to introduce a new topic when you conclude. A conclusion should always relate to and reinforce the ideas presented earlier in your writing. Moreover, the conclusions must be consistent with what the **introduction** promised the report would examine (its purpose) and how it would do so (its method).

For guidance about the location of the conclusion section in a report, see **formal reports**. For letter and other short closings, see **correspondence** and entries on specific types of documents throughout this book.

conjunctions

A conjunction connects words, **phrases**, or **clauses** and can also indicate the relationship between the elements it connects.

A *coordinating conjunction* joins two sentence elements that have identical functions. The coordinating conjunctions are *and*, *but*, *or*, *for*, *nor*, *yet*, and *so*.

- Nature *and* technology affect petroleum prices. [joins two **nouns**]
- To hear *and* to listen are two different things. [joins two phrases]
- I would like to include the survey, *but* that would make the report too long. [joins two clauses]

Coordinating conjunctions in the titles of books, articles, plays, and movies should not be capitalized unless they are the first or last word in the title.

- Our library contains *Consulting and Financial Independence* as well as *So You Want to Be a Consultant?*

Occasionally, a conjunction may begin a sentence; in fact, conjunctions can be strong transitional words and at times can provide **emphasis**. See also **transition**.

- I realize that the project is more difficult than expected and that you have encountered staffing problems. *But* we must meet our deadline.

Correlative conjunctions are used in pairs. The correlative conjunctions are *either . . . or, neither . . . nor, not only . . . but also, both . . . and,* and *whether . . . or.*

▶ The inspector will arrive *either* on Wednesday *or* on Thursday.

A *subordinating conjunction* connects sentence elements of different relative importance, normally independent and dependent clauses. Frequently used subordinating conjunctions are *so, although, after, because, if, where, than, since, as, unless, before, that, though,* and *when.*

▶ I left the lab *after* finishing the tests.

A *conjunctive adverb* has the force of a conjunction because it joins two independent clauses. The most common conjunctive **adverbs** are *however, moreover, therefore, further, then, consequently, besides, accordingly, also,* and *thus.*

▶ The engine performed well in the laboratory; *however,* it failed under road conditions.

connotation / denotation

The *denotations* of a word are its literal meanings, as defined in a dictionary. The *connotations* of a word are its meanings and associations beyond its literal definitions. For example, the denotations of *Hollywood* are "a district of Los Angeles" and "the U.S. movie industry as a whole"; its connotations for many are "glamour, opulence, and superficiality."

Often words have particular connotations for **audiences** within professional groups and organizations. Choose words with both the most accurate denotations and the most appropriate connotations for the **context**. See also **defining terms** and **word choice**.

consensus

Because *consensus* means "harmony of opinion" among most of those in a group, the phrases *consensus of opinion* and *general consensus* defeat **conciseness**. The word *consensus* can be used to refer only to a group, never to one or two people.

▶ The ~~general~~ consensus ~~of opinion~~ among researchers is that the drug is ineffective.

content management

Content management refers to the processes that organizations use to manage the creation, tracking, and production of content, such as text files, images, and other digital forms. A primary goal of content management is to make it easier for multiple writers to access, share, and use such content. Content management aims to reduce the duplication of effort, increase efficiency, and maintain consistent messages across organizational documents (Web sites, printed text, and other media).

Writing for Content Management

Writing in a content-management environment challenges writers to think of **audience**, **purpose**, and **context** broadly. The content you create for instructions or marketing copy, for example, may be reused on a Web site, in **brochures**, and in technical **manuals**. As a result, you must either write a product description with a **tone**, **style**, and level of technical detail appropriate for all audiences or write the content so that it can be easily altered for other contexts.

Writers who work with content management often develop *information models* that visually depict the guidelines, much like **flowcharts**, for organizing and managing the content. These guidelines may direct writers to code a passage or other content with "metadata" (terms, symbols, or number combinations) so that it can be shared across an organization. If you write for content management, then, you must understand your organization's system for coding and preparing content that can be reused in other documents. The technologies developed to achieve the goals of content management are referred to as *content-management systems*.

Content-Management Systems

Content-management systems are integrated software tools that help organizations control the processes they use to manage and publish content for organizational Web sites, widely used documents, interactive forms, and other media used across an organization. Types of content-management systems include the following:

- *Web Content-Management System*: Provides tools for creating, updating, and managing Web-site content. Some systems include functions that allow Web managers to monitor workflow and maintain the entire Web site.
- *Document-Management System*: Stores and tracks electronic documents or images of paper documents, allowing organizations to replace their hard-copy files with sophisticated indexing, browsing,

and searching capabilities. Document-management systems often include workflow tools that help writers track and manage changes to documents during review cycles, such as in collaborative writing projects.

- *Component Content-Management System*: Manages content for any media at the paragraph, sentence, and even word level.* For example, a safety warning might state, "Don't stick a fork in the toaster." With a component content-management system, a writer can edit this warning to say "Don't stick any silverware into the toaster" and the system could automatically update all instances of this warning in an organization's manuals, help files, brochures, Web pages, and other documents.
- *Enterprise Content-Management System*: Blends the functions of Web, document, and often component content-management systems to create an all-in-one system for an entire organization. Such systems aim to generate and organize content for various publications, media, and other important applications, such as e-mail, financial record keeping, and human resource activities.

WEB LINK	Content-Management Systems

For links to Web sites and books that provide information on content management and content-management systems, see *bedfordstmartins.com/alredtech* and select *Links for Handbook Entries*.

context

Context is the environment or circumstances in which writers produce documents and within which readers interpret their meanings. Everything is written in a context, as illustrated in many entries and examples throughout this book. This entry considers the significance of context for documents and suggests how you can be aware of it as you write. See also **audience**.

The context for any document, such as a **proposal** or **report**, is determined by interrelated events or circumstances both inside and outside an organization.† For example, when you write a proposal to fund

---

*Such content-management systems are sometimes called "granularized" systems because they manage such small pieces of data (or "granules").

†For a detailed illustration of specific documents that have been affected by particular contexts, see Linda Driskill, "Understanding the Writing Context in Organizations," in *Central Works in Technical Communication*, ed. Johndan Johnson-Eilola and Stuart Selber (New York: Oxford University Press, 2004), 55–69.

a project within your company, the economic condition of that company is part of the context that will determine how your proposal is received. If the company has recently laid off a dozen employees, its management may not be inclined to approve a proposal to expand its operations—regardless of how well the proposal is written.

When you correspond with someone, the events that prompted you to write shape the context of the message and will affect what you say and how you say it. If you write to a customer in response to a complaint, for example, the **tone** and approach of your message will be determined by the context—what you find when you investigate the issue. Is your company fully or partly at fault? Has the customer incorrectly used a product? contributed to a problem? (See also **adjustment letters**.) If you write a **manual** for auto-service technicians, other questions will reveal the context. What are the lighting and other physical conditions in garages? Will these physical conditions affect the **layout and design** of the manual? What potentially dangerous situations might the technicians encounter?

Assessing Context

Each time you write, the context needs to be clearly in your mind so that your document will achieve its **purpose**. The following questions are starting points to help you become aware of the context, how it will influence your approach and your readers' interpretation of what you have written, and how it will affect the decisions you need to make during the writing process. See also "Five Steps to Successful Writing" (page xv).

- What is your professional relationship with your readers, and how might that affect the tone, **style**, and **scope** of your writing? What is "the story" behind the immediate reason you are writing; that is, what series of events or perhaps previous documents led to your need to write?
- What is the preferred medium of your readers? See also **selecting the medium**.
- What specific factors (such as competition, finance, and regulation) are recognized within your organization or department as important?
- What is the corporate culture in which your readers work, and what are the key values that you might find in its mission statement?
- What are the professional relationships among the specific readers who will receive the document?
- What current events within or outside an organization or a department may influence how readers interpret your writing?

- What national cultural differences might affect your readers' expectations or interpretations of the document? See also **global communication**.

As these questions suggest, context is very specific each time you write and often involves, for example, the history of a specific organization or your past dealings with individual readers.

Signaling Context

Because context is so important, it is often helpful to remind your reader in some way of the context for your writing, as in the following opening for a cover message to a proposal.

> ▶ During our meeting on improving quality last week, you mentioned that we have previously required usability testing only for documents going to high-profile clients because of the costs involved. The idea occurred to me that we might try less extensive usability testing for many of our other clients. Because you asked for suggestions, I have proposed in the attached document a method of limited usability testing for a broad range of clients in order to improve overall quality.

Of course, as described in **introductions**, providing context for a reader may require only a brief background statement or short reminders.

> ▶ Several weeks ago, a financial adviser noticed a recurring problem in the software developed by Datacom Systems. Specifically, error messages repeatedly appeared when, in fact, no specific trouble . . .

> ▶ Jane, as I promised in my e-mail yesterday, I've attached the personnel budget estimates for the next fiscal year.

As the last example suggests, always provide some context when you send an **e-mail** with attachments.

continual / continuous

Continual implies "happening over and over" or "frequently repeated." ("Writing well requires *continual* practice.") *Continuous* implies "occurring without interruption" or "unbroken." ("The *continuous* roar of the machinery was deafening.")

contractions

A contraction is a shortened spelling of a word or phrase with an **apostrophe** substituting for the missing letter or letters (*cannot/can't*; *have not/haven't*; *will not/won't*; *it is/it's*). Contractions are often used in speech and informal writing; they are generally not appropriate in **reports**, **proposals**, and formal **correspondence**. See also **technical writing style**.

copyright

Copyright establishes legal protection for literary, dramatic, musical, artistic, and other intellectual works in printed or electronic form; it gives the copyright owner exclusive rights to reproduce, distribute, perform, or display a work. Copyright protects all original works from the moment of their creation, regardless of whether they are published or contain a notice of copyright (©).

▶ **ETHICS NOTE** If you plan to reproduce copyrighted material in your own publication or on your Web site, you must obtain permission from the copyright holder. To do otherwise is a violation of U.S. law. ◆

Permissions

To seek permission to reproduce copyrighted material, you must write to the copyright holder. In some cases, it is the author; in other cases, it is the editor or publisher of the work. For Web sites, read the site's "terms-of-use" information (if available) and e-mail your request to the appropriate party. State specifically which portion of the work you wish to reproduce and how you plan to use it. The copyright holder has the right to charge a fee and specify conditions and limits of use.

Exceptions

Some print and Web material—including text, **visuals**, and other digital forms—may be reproduced without permission. The rules governing copyright can be complex, so it is prudent to check carefully the copyright status of anything you plan to reproduce.

- *Educational material.* A small amount of material from a copyrighted source may be used for educational purposes (such as classroom handouts) without permission or payment as long as the use satisfies the "fair-use" criteria, as described at the U.S. Copyright Office Web site, *www.copyright.gov*.

- *Company boilerplate.* Employees often borrow material freely from in-house manuals, reports, and other company documents to save time and ensure consistency. Using such "boilerplate" or "repurposed" material is not a copyright violation because the company is considered the author of works prepared by its employees on the job. See also **purpose**.
- *Public domain material.* Works created by or for the U.S. government and not classified or otherwise protected are in the public domain—that is, they are not copyrighted. The same is true for older written works when their copyright has lapsed or never existed. Be aware that some works in the public domain may include "value-added" features, such as introductions, visuals, and indexes, that are copyrighted separately from the original work.
- *Copyleft Web material.* Some public access Web sites, such as *Wikipedia*, follow the "copyleft" principle and grant permission to freely copy, distribute, or modify material.*

▶ **ETHICS NOTE** Even when you use material that may be reproduced or published without permission, you must nonetheless give appropriate credit to the source from which the material is taken, as described in **plagiarism** and **documenting sources**. ◆

correspondence

DIRECTORY
Audience and Writing Style 103
Goodwill and the "You" Viewpoint 104
Writer's Checklist: Using Tone to Build Goodwill 105
Good-News and Bad-News Patterns 106
Openings and Closings 108
Clarity and Emphasis 109
Writer's Checklist: Correspondence and Accuracy 110

Correspondence in the workplace—whether through **e-mail**, **letters**, or **memos**—requires many of the same steps that are described in "Five Steps to Successful Writing" (page xv). As you prepare an e-mail, for

---

*"Copyleft" is a play on the word *copyright* and is the effort to free materials from many of the restrictions of copyright. See http://en.wikipedia.org/wiki/Copyleft.

example, you might study previous messages (research) and then list or arrange the points you wish to cover (organization) in an order that is logical for your **audience**. See also **selecting the medium**.

Corresponding with others in the workplace also requires that you focus on both establishing or maintaining a positive working relationship with your **readers** and conveying a professional image of yourself and your organization.

Audience and Writing Style

Effective correspondence uses an appropriate conversational **style**. To achieve that style, imagine your reader sitting across from you and write to the reader as if you were talking face to face. Take into account your reader's needs and feelings. Ask yourself, "How might I feel if I received this letter or e-mail?" and then tailor your message accordingly. Remember, an impersonal and unfriendly message to a customer or client can tarnish the image of you and your business, but a thoughtful and sincere one can enhance it.

Whether you use a formal or an informal writing style depends entirely on your reader and your **purpose**. You might use an informal (or a casual) style, for example, with a colleague you know well and a formal (or restrained) style with a client you do not know.

CASUAL	It worked! The new process is better than we had dreamed.
RESTRAINED	You will be pleased to know that the new process is more effective than we had expected.

You will probably find yourself using the restrained style more frequently than the casual style. Remember that an overdone attempt to sound casual or friendly can sound insincere. However, do not adopt so formal a style that your writing reads like a legal contract. **Affectation** not only will irritate and baffle readers but also can waste time and produce costly errors.

AFFECTED	Per our dialogue yesterday, we no longer possess an original copy of the brochure requested. Please be advised that a PDF copy is attached to this e-mail.
IMPROVED	We are out of original copies of the brochure we discussed yesterday, so I am attaching a PDF copy to this e-mail.

The improved version is not only clearer and less stuffy but also more concise. See also **technical writing style** and **conciseness**.

Goodwill and the "You" Viewpoint

Write concisely, but do not be so blunt that you risk losing the reader's goodwill. Responding to a vague written request with "Your request was unclear" or "I don't understand" could offend your reader. Instead, establish goodwill to encourage your reader to provide the information you need.

> ▶ I will be glad to help, but I need more information to find the report you requested. Specifically, can you give me the report's title, release date, or number?

Although this version is a bit longer, it is more tactful and will elicit a faster response. See also **telegraphic style**.

You can also build goodwill by emphasizing the reader's needs or benefits. Suppose you received a refund request from a customer who forgot to enclose the receipt with the request. In a response to that customer, you might write the following:

> **WEAK** We must receive the sales receipt before we can process a refund.
> [The writer's needs are emphasized: "*We* must."]

If you consider how you might keep the customer's goodwill, you could word the request this way:

> **IMPROVED** Please mail or fax the sales receipt so that we can process your refund.
> [This is polite, but the writer's needs are still emphasized: "so that *we* can process."]

You can put the reader's needs and interests foremost by writing from the reader's perspective. Often, doing so means using the words *you* and *your* rather than *we*, *our*, *I*, and *mine*—a technique called the **"you" viewpoint**. Consider the following revision:

> **EFFECTIVE** So that you can receive your refund promptly, please mail or fax the sales receipt.
> [The reader's needs are emphasized with *you* and *your*.]

This revision stresses the reader's benefit and interest. By emphasizing the reader's needs, the writer will be more likely to accomplish the purpose: to get the reader to act. See also **positive writing**.

If overdone, however, goodwill and the "you" viewpoint can produce writing that is fawning and insincere. Messages that are full of excessive praise and inflated language may be ignored—or even resented—by the reader.

EXCESSIVE PRAISE	You are just the kind of astute client that deserves the finest service that we can offer — and you deserve our best deal. Knowing how carefully you make decisions, I know you'll think about the advantages of using our consulting service.
REASONABLE	From our earlier correspondence, I understand your need for reliable service. We strive to give all our priority clients our full attention, and after you have reviewed our proposal I am confident you will appreciate our "five-star" consulting option.

WRITER'S CHECKLIST   Using Tone to Build Goodwill

Use the following guidelines to achieve a **tone** that builds goodwill with your recipients.

- Be respectful, not demanding.

DEMANDING	Submit your answer in one week.
RESPECTFUL	I would appreciate your answer within one week.

- Be modest, not arrogant.

ARROGANT	My attached report is thorough, and I'm sure that you won't be able to continue without it.
MODEST	The attached report contains details of the refinancing options that I hope you will find useful.

- Be polite, not sarcastic.

SARCASTIC	I just now received the shipment we ordered six months ago. I'm sending it back — we can't use it now. Thanks a lot!
POLITE	I am returning the shipment we ordered on March 12. Unfortunately, it arrived too late for us to be able to use it.

- Be positive and tactful, not negative and condescending.

NEGATIVE	Your complaint about our prices is way off target. Our prices are definitely not any higher than those of our competitors.
TACTFUL	Thank you for your suggestion concerning our prices. We believe, however, that our prices are competitive with or lower than those of our competitors.

Good-News and Bad-News Patterns

Although the relative directness of correspondence may vary, it is generally more effective to present good news directly and bad news indirectly, especially if the stakes are high.* This principle is based on the fact that readers form their impressions and attitudes very early and that you as a writer may want to subordinate the bad news to reasons that make the bad news understandable. Further, if you are writing **international correspondence**, understand that far more cultures are generally indirect in business messages than are direct.

Consider the thoughtlessness of the direct rejection in Figure C–8. Although the message is concise and uses the pronouns *you* and *your*, the writer does not consider how the recipient is likely to feel as she reads the rejection. Its pattern is (1) the bad news, (2) an explanation, and (3) the closing.

Dear Ms. Mauer:

Your application for the position of Cardiovascular Technician at Southtown Cardiac Associates has been rejected. We have found someone more qualified than you.

Sincerely,

FIGURE C–8. A Poor Bad-News Message

A better general pattern for bad news is (1) an opening that provides **context** (often called a "buffer"), (2) an explanation, (3) the bad news, and (4) a goodwill closing. (See also **refusal letters**.) The opening introduces the subject and establishes a professional tone. The body provides an explanation by reviewing the facts that make the bad news understandable. Although bad news is never pleasant, information that either puts the bad news in perspective or makes it seem reasonable promotes goodwill between the writer and the reader. The closing should reinforce a positive relationship through goodwill or helpful information. Consider, for example, the courteous rejection shown in Figure C–9. It carries the same disappointing news as does the letter in Figure C–8, but the writer is careful to thank the reader for her time

---

*Gerald J. Alred, "'We Regret to Inform You': Toward a New Theory of Negative Messages," in *Studies in Technical Communication*, ed. Brenda R. Sims (Denton: University of North Texas and NCTE, 1993), 17–36.

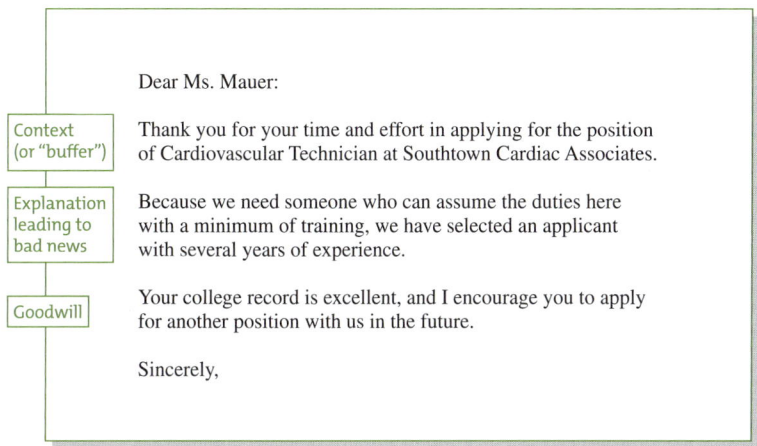

FIGURE C–9. A Courteous Bad-News Message

and effort, explain why she was not accepted for the job, and offer her encouragement.

This pattern can also be used in relatively short e-mail messages and memos. Consider the unintended secondary message the following notice conveys:

WEAK   It has been decided that the office will be open the day after Thanksgiving.

"It has been decided" not only sounds impersonal but also communicates an authoritarian, management-versus-employee tone. The passive voice also suggests that the decision-maker does not want to say "I have decided" and thus accept responsibility. One solution is to remove the first part of the sentence.

IMPROVED   The office will be open the day after Thanksgiving.

The best solution, however, would be to suggest both that there is a good reason for the decision and that employees are privy to (if not a part of) the decision-making process.

EFFECTIVE   Because we must meet the December 15 deadline for submitting the Bradley Foundation proposal, the office will be open the day after Thanksgiving.

By describing the context of the bad news first (the need to meet the deadline), the writer focuses on the reasoning behind the decision to work. Employees may not necessarily like the message, but they will at least understand that the decision is not arbitrary and is tied to an important deadline.

Presenting good news is, of course, easier. Present good news in your opening. By doing so, you increase the likelihood that the reader will pay careful attention to details, and you achieve goodwill from the start. The pattern for good-news messages should be (1) a good-news opening, (2) an explanation of facts, and (3) a goodwill closing. Figure C–10 is an example of an effective good-news message.

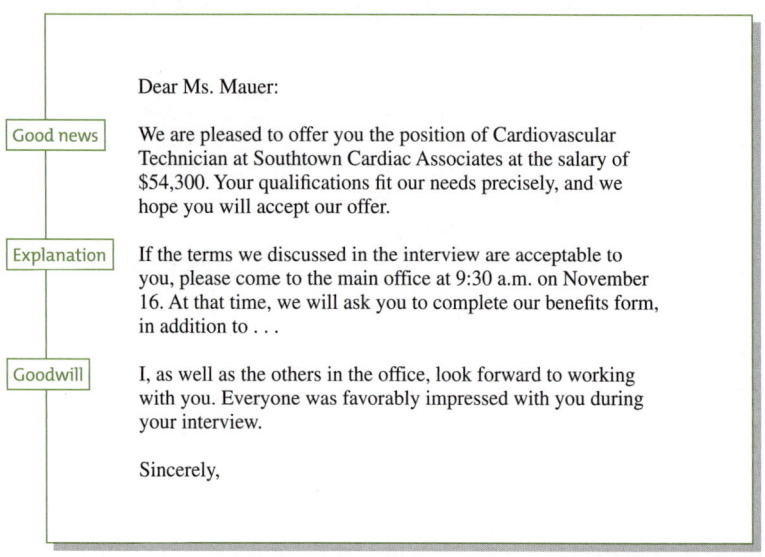

FIGURE C–10. A Good-News Message

Openings and Closings

Although methods of development vary, the opening of any correspondence should identify the subject and often the main point of the message.

E-MAIL Attached is the final installation report, which I hope you can review by Monday, December 7. You will notice that the report includes . . .

LETTER Enclosed with this letter is your refund check for the defective compressor, number AJ50172. I have also enclosed a catalog of our latest . . .

MEMO ACM Electronics has asked us to prepare a comprehensive set of brochures for its Milwaukee office by August 10, 2009. We have worked with similar firms in the past . . .

When your reader is not familiar with the subject or with the background of a problem, you may provide an introductory paragraph before stating the main point of the message. Doing so is especially important in correspondence that will serve as a record of crucial information. Generally, longer or complex subjects benefit most from more thorough **introductions**. However, even when you are writing a short message about a familiar subject, remind readers of the context. In the following example, words that provide context are shown in *italics*.

> ▶ *As Maria Lopez recommended*, I reviewed the office reorganization plan. I like most of the features; however, . . .

Do not state the main point first when (1) readers are likely to be highly skeptical or (2) key readers, such as managers or clients, may disagree with your position. In those cases, a more persuasive tactic is to state the problem or issue first, then present the specific points supporting your final recommendation, as discussed in the earlier section on bad-news messages. See also **persuasion**.

Your closing can accomplish many important tasks, such as building positive relationships with readers, encouraging colleagues and employees, and letting recipients know what you will do or what you expect of them.

> ▶ I will discuss the problem with the marketing consultant and let you know by Wednesday what we are able to change.

Although routine statements are sometimes unavoidable ("If you have further questions, please let me know"), try to make your closing work for you by providing specific prompts to which the reader can respond. See also **conclusions**.

> ▶ Thanks again for the report, and let me know if you want me to send you a copy of the test results.

Clarity and Emphasis

A clear message is one that is adequately developed and emphasizes your main points. The following example illustrates how adequate development is crucial to the **clarity** of your message.

VAGUE	Be more careful on the loading dock.
DEVELOPED	To prevent accidents on the loading dock, follow these procedures: 1. Check . . . 2. Load only . . . 3. Replace . . .

Although the first version is concise, it is not as clear and specific as the "developed" revision. Do not assume your readers will know what you mean: vague messages are easily misinterpreted.

Lists. Vertically stacked words, **phrases**, and other items with numbers or bullets can effectively highlight such information as steps in sequence, materials or parts needed, key or concluding points, and recommendations. As described in the entry **lists**, make sure you provide context. Be careful not to overuse lists—a message that consists almost entirely of lists is difficult to understand because it forces readers to connect separate and disjointed items. Further, lists lose their effectiveness when they are overused.

Headings. **Headings** are particularly useful because they call attention to main topics, divide material into manageable segments, and signal a shift in topic. Readers can scan the headings and read only the section or sections appropriate to their needs.

Subject Lines. Subject lines for e-mails, memos, and some letters announce the topic and focus of the correspondence. Because they also aid filing and later retrieval, they must be specific and accurate.

VAGUE	Subject: Tuition Reimbursement
VAGUE	Subject: Time-Management Seminar
SPECIFIC	Subject: Tuition Reimbursement for Time-Management Seminar

Capitalize all major words in a subject line except articles, prepositions, and conjunctions with fewer than five letters (unless they are the first or last words). Remember that the subject line should not substitute for an opening that provides context for the message.

WRITER'S CHECKLIST   Correspondence and Accuracy

- Begin by establishing your purpose, analyzing your reader's needs, determining your **scope**, and considering the context.
- Prepare an outline, even if it is only a list of points to be covered in the order you want to cover them (see **outlining**).
- Write the first draft (see **writing a draft**).
- Allow for a cooling-off period prior to **revision** or seek a colleague's advice, especially for correspondence that addresses a problem.
- Revise the draft, checking for key problems in clarity and **coherence**.
- Use the appropriate or standard format (see **e-mail**, **letters**, and **memos**).

Writer's Checklist: Correspondence and Accuracy (continued)

- ✔ Check for accuracy: Make sure that all facts, figures, and dates are correct.
- ✔ Use effective **proofreading** techniques to check for **punctuation**, **grammar**, **spelling**, and appropriate **usage**.
- ✔ Consider who should receive a copy of the message and in what order the names or e-mail addresses should be listed (alphabetize if rank does not apply).
- ✔ Remember that when you sign a letter or send a memo or an e-mail, you are accepting responsibility for it.

cover letters (or transmittals)

A cover **letter**, **memo**, or **e-mail** accompanies a document (such as a **proposal**), an electronic file, or other material. It identifies an item that is being sent, the person to whom it is being sent, the reason that it is being sent, and any content that should be highlighted for **readers** (see **purpose**). A cover letter provides a permanent record for both the writer and the reader. For cover letters to **résumés**, see **application letters**.

The example in Figure C–11 is concise, but it also includes details such as how the information for the **report** was gathered.

Dear Mr. Hammersmith:

Attached is the report estimating our energy needs for the year as requested by John Brenan, Vice President, on September 4.

The report is a result of several meetings with the manager of plant operations and her staff and an extensive survey of all our employees. The survey was delayed by the transfer of key staff in Building A. We believe, however, that the report will provide the information you need in order to furnish us with a cost estimate for the installation of your Mark II Energy Saving System.

We would like to thank Diana Biel of ESI for her assistance in preparing the survey. If you need any more information, please let me know.

Sincerely,

FIGURE C–11. Cover Message

credible / creditable

Something is *credible* if it is believable. ("The statistics in this report are *credible*.") Something is *creditable* if it is worthy of praise or credit. ("The lead engineer did a *creditable* job.")

criteria / criterion

Criterion is a singular **noun** meaning "an established standard for judging or testing." *Criteria* and *criterions* are both acceptable plural forms of *criterion*, but *criteria* is generally preferred.

critique

A *critique* is a written or an oral evaluation of something. Avoid using *critique* as a **verb** meaning "criticize."

▶ Please ⌃critique⌃ his job description.
     prepare a   *of*

dangling modifiers

Phrases that do not clearly and logically refer to the correct **noun** or **pronoun** are called *dangling modifiers*. Dangling modifiers usually appear at the beginning of a sentence as an introductory **phrase**.

DANGLING	*While eating lunch*, the computer malfunctioned. [*Who* was eating lunch?]
CORRECT	While *I* was eating lunch, the computer malfunctioned.

Dangling modifiers can appear at the end of the sentence as well.

DANGLING	The program gains efficiency *by eliminating the superfluous instructions*. [*Who* eliminates the superfluous instructions?]
CORRECT	The program gains efficiency *when you* eliminate the superfluous instructions.

To correct a dangling modifier, add the appropriate subject to either the dangling modifier or the main **clause**.

DANGLING	After finishing the research, the proposal was easy to write. [The appropriate subject is *I*, but it is not stated in either the dangling phrase or the main clause.]
CORRECT	After *I* finished the research, the proposal was easy to write. [The pronoun *I* is now the subject of an introductory clause.]
CORRECT	After finishing the research, *I* found the proposal easy to write. [The pronoun *I* is now the subject of the main clause.]

For a discussion of misplaced modifiers, see **modifiers**.

dashes

The dash (—) can perform all the punctuation duties of linking, separating, and enclosing. The dash, sometimes indicated by two consecutive **hyphens**, can also indicate the omission of letters. (Mr. A— admitted his error.)

Use the dash cautiously to indicate more **emphasis**, informality, or abruptness than the other punctuation marks would show. A dash can emphasize a sharp turn in thought.

▶ The project will end May 15—unless we receive additional funding.

A dash can indicate an emphatic pause.

▶ The project will begin—after we are under contract.

Sometimes, to emphasize contrast, a dash is used with *but*.

▶ We completed the survey quickly—*but* the results were not accurate.

A dash can be used before a final summarizing statement or before repetition that has the effect of an afterthought.

▶ It was hot near the heat-treating ovens—steaming hot.

Such a statement may also complete the meaning of the **clause** preceding the dash.

▶ We try to write as we speak—or so we believe.

Dashes set off parenthetical elements more sharply and emphatically than **commas**. Unlike dashes, **parentheses** tend to deemphasize what they enclose. Compare the following sentences:

▶ Only one person—the president—can authorize such activity.

▶ Only one person, the president, can authorize such activity.

▶ Only one person (the president) can authorize such activity.

Dashes can be used to set off parenthetical elements that contain commas.

▶ Three of the applicants—John Evans, Rosalita Fontiana, and Kyong-Shik Choi—seem well qualified for the job.

The first word after a dash is capitalized only if it is a proper **noun**.

data

In formal and scholarly writing, *data* is generally used as a plural, with *datum* as the singular form. In much informal writing, however, *data* is considered a collective singular **noun**. Base your decision on whether your **audience** should consider the data as a single collection or as a group of individual facts. Whatever you decide, be sure that your **pronouns** and **verbs** agree in number with the selected **usage**. See also **agreement** and **English, varieties of**.

▶ These *data are* persuasive. *They indicate* a need for additional research questions. [formal]

▶ The attached *data is* confidential. *It is* the result of a survey of employee records. [less formal]

dates

In the United States, full dates are generally written in the month-day-year format, with a comma preceding and following the year.

▶ November 30, 2020, is the payoff date.

Do not use **commas** in the day-month-year format, which is used in many parts of the world and by the U.S. military.

▶ Note that 30 November 2020 is the payoff date.

No commas are used when showing only the month-year or month-day in a date.

▶ The target date of May 2012 is optimistic, so I would like to meet on March 4 to discuss our options.

When writing days of the month without the year, use the cardinal number ("March 4") rather than the ordinal number ("March 4th"). Of course, in speech or **presentations**, use the ordinal number ("March fourth").

Avoid the strictly numerical form for dates (11/6/12) because the date is not always immediately clear, especially in **international correspondence**. In many countries, 11/6/12 means June 11, 2012, rather than November 6, 2012. Writing out the name of the month makes the entire date immediately clear to all readers.

Centuries often cause confusion with **numbers** because their spelled-out forms, which are not capitalized, do not correspond with their numeral designations. The twentieth century, for example, is the 1900s: 1900–1999.

When the century is written as a **noun**, do not use a **hyphen**.

▶ During the twentieth century, technology transformed business practices.

When the centuries are written as **adjectives**, however, use hyphens.

▶ Twenty-first-century technology relies on dependable power sources.

defective / deficient

If something is *defective*, it is faulty. ("The wiring was *defective*.") If something is *deficient*, it is lacking or is incomplete in an essential component. ("The patient's diet was *deficient* in calcium.")

defining terms

Defining key terms and concepts is often essential for **clarity**. Terms can be defined either formally or informally, depending on your **purpose**, your **audience**, and the **context**.

A *formal definition* is a form of classification. You define a term by placing it in a category and then identifying the features that distinguish it from other members of the same category.

TERM	CATEGORY	DISTINGUISHING FEATURES
An *annual* is	a plant	that completes its life cycle, from seed to natural death, in one growing season.

An *informal definition* explains a term by giving a more familiar word or phrase as a **synonym**.

▶ Plants have a *symbiotic*, or *mutually beneficial*, relationship with certain kinds of bacteria.

State definitions positively; focus on what the term *is* rather than on what it is not.

NEGATIVE	In a legal transaction, *real property* is not personal property.
POSITIVE	*Real property* is legal terminology for the right or interest a person has in land and the permanent structures on that land.

For a discussion of when negative definitions are appropriate, see **definition method of development**.

Avoid circular definitions, which merely restate the term to be defined and therefore fail to clarify it.

CIRCULAR   *Spontaneous combustion* is fire that begins spontaneously.

REVISED   *Spontaneous combustion* is the self-ignition of a flammable material through a chemical reaction.

In addition, avoid "is when" and "is where" definitions. Such definitions fail to include the category and are too indirect.

▶ A *biopsy* is ~~when~~ *a medical procedure in which* a tissue sample is removed for testing.

In technical writing, as illustrated in this example, definitions often contain the purpose ("testing") of what is defined (*biopsy*).

definite / definitive

Definite and *definitive* both apply to what is precisely defined, but *definitive* more often refers to what is complete and authoritative. ("Once we receive a *definite* proposal, our attorney can provide a *definitive* legal opinion.")

definition method of development

Definition is often essential to **clarity** and accuracy. Although **defining terms** may be sufficient, sometimes definitions need to be expanded through (1) extended definition, (2) definition by analogy, (3) definition by cause, (4) definition by components, (5) definition by exploration of origin, and (6) negative definition. See also **methods of development**.

Extended Definition

When you need more than a simple definition to explain an idea, use an extended definition, which explores a number of qualities of the item being defined. How an extended definition is developed depends on your **audience** and on the complexity of the subject. Readers familiar with a topic might be able to handle a long, fairly complex definition, whereas readers less familiar with a topic might require simpler language and more basic information.

The easiest way to give an extended definition is with specific examples. Examples give readers easy-to-picture details that help them see and thus understand the term being defined.

▶ Form, which is the shape of landscape features, can best be represented both by small-scale features, such as *trees* and *shrubs*, and by large-scale elements, such as *mountains* and *mountain ranges*.

Definition by Analogy

Another useful way to define a difficult concept, especially when you are writing for nonspecialists, is to use an **analogy**. An analogy can help the reader understand an unfamiliar term by showing its similarities with a more familiar term. In the following description of radio waves in terms of their length (long) and frequency (low), notice how the writer develops an analogy to show why a low frequency is advantageous.

▶ The low frequency makes it relatively easy to produce a sound wave having virtually all its power concentrated at one frequency. Think, for example, of a group of people lost in a forest. If they hear sounds of a search party in the distance, they all will begin to shout for help in different directions. Not a very efficient process, is it? But suppose all of the energy that went into the production of this noise could be concentrated into a single shout or whistle. Clearly the chances that the group will be found would be much greater.

Definition by Cause

Some terms are best defined by an explanation of their causes. In the following example from a professional journal, a nurse describes an apparatus used to monitor blood pressure in severely ill patients. Called an *indwelling catheter*, the device displays blood-pressure readings on an oscilloscope and on a numbered scale. Users of the device, the writer explains, must understand what a *dampened wave form* is.

▶ The dampened wave form, the smoothing out or flattening of the pressure wave form on the oscilloscope, is usually caused by an obstruction that prevents blood pressure from being freely transmitted to the monitor. The obstruction may be a small clot or bit of fibrin at the catheter tip. More likely, the catheter tip has become positioned against the artery wall and is preventing the blood from flowing freely.

Definition by Components

Sometimes a formal definition of a concept can be made simpler by breaking the concept into its component parts. In the following example, the formal definition of *fire* is given in the first paragraph, and the component parts are given in the second.

FORMAL DEFINITION	Fire is the visible heat energy released from the rapid oxidation of a fuel. A substance is "on fire" when the release of heat energy from the oxidation process reaches visible light levels.
COMPONENT PARTS	The classic fire triangle illustrates the elements necessary to create fire: *oxygen*, *heat*, and *burnable material* (*fuel*). Air provides sufficient oxygen for combustion; the intensity of the heat needed to start a fire depends on the characteristics of the burnable material. A burnable substance is one that will sustain combustion after an initial application of heat to start the combustion.

Definition by Exploration of Origin

Under certain circumstances, the meaning of a term can be clarified and made easier to remember by an exploration of its origin. Medical terms, because of their sometimes unfamiliar Greek and Latin roots, benefit especially from an explanation of this type. Tracing the derivation of a word also can be useful when you want to explain why a word has favorable or unfavorable associations, particularly if your goal is to influence your reader's attitude toward an idea or activity. See also **persuasion**.

▶ Efforts to influence legislation generally fall under the head of *lobbying*, a term that once referred to people who prowled the lobbies of houses of government, buttonholing lawmakers and trying to get them to take certain positions. Lobbying today is all of this, and much more, too. It is a respected—and necessary—activity. It tells the legislator which way the winds of public opinion are blowing, and it helps inform [legislators] of the implications of certain bills, debates, and resolutions [that they must face].
—Bill Vogt, *How to Build a Better Outdoors*

Negative Definition

In some cases, it is useful to point out what something is not to clarify what it is. A negative definition is effective only when the reader is familiar with the item with which the defined item is contrasted. If you say "*x* is not *y*," your readers must understand the meaning of *y* for the explanation to make sense. In a crane operator's manual, for instance, a negative definition is used to show that, for safety reasons, a hydraulic crane cannot be operated in the same manner as a lattice boom crane.

▶ A hydraulic crane is *not* like a lattice boom crane [a friction machine] in one very important way. In most cases, the safe lifting capacity of a lattice boom crane is based on the *weight needed to tip the machine*. Therefore, operators of friction machines sometimes

depend on signs that the machine might tip to warn them of impending danger. This practice is very dangerous with a hydraulic crane.
— *Operator's Manual* (Model W-180), Harnischfeger Corporation

description

The key to effective description is the accurate presentation of details, whether for simple or complex descriptions. In Figure D–1, notice that the simple description contained in the purchase order includes five specific details (in addition to the part number) structured logically.

PURCHASE ORDER

PART NO.	DESCRIPTION	QUANTITY
IW 8421	Infectious-waste bags, 12″ × 14″, heavy-gauge polyethylene, red double closures with self-sealing adhesive strips	5 boxes containing 200 bags per box

FIGURE D–1. Simple Description

Complex descriptions, of course, involve more details. In describing a mechanical device, for example, describe the whole device and its function before giving a detailed description of how each part works. The description should conclude with an explanation of how each part contributes to the functioning of the whole.

In descriptions intended for readers who are unfamiliar with the topic, details are crucial. For such an **audience**, show or demonstrate (as opposed to "tell") primarily through the use of images and details. Notice the use of color, shapes, and images in the following description of a company's headquarters. The writer assumes that the reader knows such terms as *colonial design* and *haiku fountain*.

▶ Their corporate headquarters, which reminded me of a rural college campus, are located north of the city in a 90-acre lush green wooded area. The complex consists of five three-story buildings of colonial design. The buildings are spaced about 50 feet apart and are built in a U shape surrounding a reflection pool that frames a striking haiku fountain.

You can also use analogy, as described in **figures of speech**, to explain unfamiliar concepts in terms of familiar ones, such as "U shape" in the previous example or "armlike block" in Figure D–2.

description

The *die block assembly* shown in Figure 1 consists of two machined block sections, eight code punch pins, and a feed punch pin. The larger section, called the die block, is fashioned of a hard noncorrosive beryllium-copper alloy. It houses the eight code punch pins and the smaller feed punch pin in nine finely machined guide holes.

> Number of features and relative size differences noted

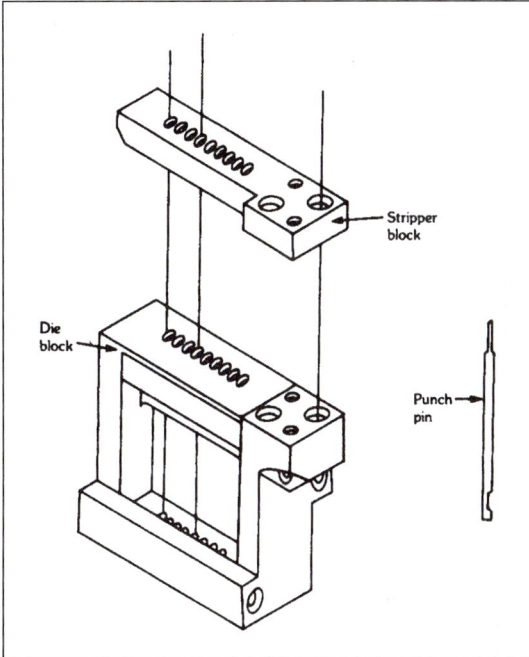

Figure 1. Die Block Assembly

> Figure of speech

The guide holes in the upper part of the die block are made smaller to conform to the thinner tips of the feed punch pins. Extending over the top of the die block and secured to it at one end is a smaller, armlike block called the *stripper block*. The stripper block is made from hardened tool steel, and it also has been drilled through with nine finely machined guide holes. It is carefully fitted to the die block at the factory so that its holes will be precisely above those in the die block and so that the space left between the blocks will measure .015″ (± .003″). The residue from the punching operation, called chad, is pushed out through the top of the stripper block and guided out of the assembly by means of a plastic *residue chad collector* and *chad collector extender*.

> Definition

FIGURE D–2. Complex Description

Visuals can be powerful aids in descriptive writing, especially when they show details too intricate to explain completely in words. The example in Figure D–2 uses an illustration to help describe a mechanical assembly for an assembler or repairperson. Note that the description concentrates on the number of pieces—their sizes, shapes, and dimensions—and on their relationship to one another to perform their function. It also specifies the materials of which the hardware is made. Because the description is illustrated (with identifying labels) and is intended for technicians who have been trained on the equipment, it does not require the use of bridging devices to explain the unfamiliar in relation to the familiar. Even so, an important term is defined ("chad"), and crucial alignment dimensions are specified (± .003"). The illustration is labeled appropriately and integrated with text of the description.

design (*see* layout and design)

despite / in spite of

Although there is no literal difference between *despite* and *in spite of*, *despite* suggests an effort to avoid blame.

▶ *Despite* our best efforts, the plan failed.
  [We are not to blame for the failure.]

▶ *In spite of* our best efforts, the plan failed.
  [We did everything possible, but failure overcame us.]

Despite and *in spite of* (both meaning "notwithstanding") should not be blended into *despite of*.

▶ ~~Despite of~~ *In spite of* our best efforts, the plan failed.

diagnosis / prognosis

Because they sound somewhat alike, these words are often confused with each other. *Diagnosis* means "an analysis of the nature of something" or "the conclusions reached by such analysis." ("The meteorological *diagnosis* was that the pollution resulted from natural contaminants.") *Prognosis* means "a forecast or prediction." ("The meteorologist's *prognosis* is that contaminant levels will increase over the next ten years.")

dictionaries

Dictionaries give more than just information about the meanings of words. As illustrated in Figure D–3, they often provide words' etymologies (origins and history), forms, pronunciations, **spellings**, uses as **idioms**, and functions as **parts of speech**. For certain words, a dictionary lists **synonyms** and may also provide illustrations, if appropriate, such as tables, maps, photographs, and drawings.

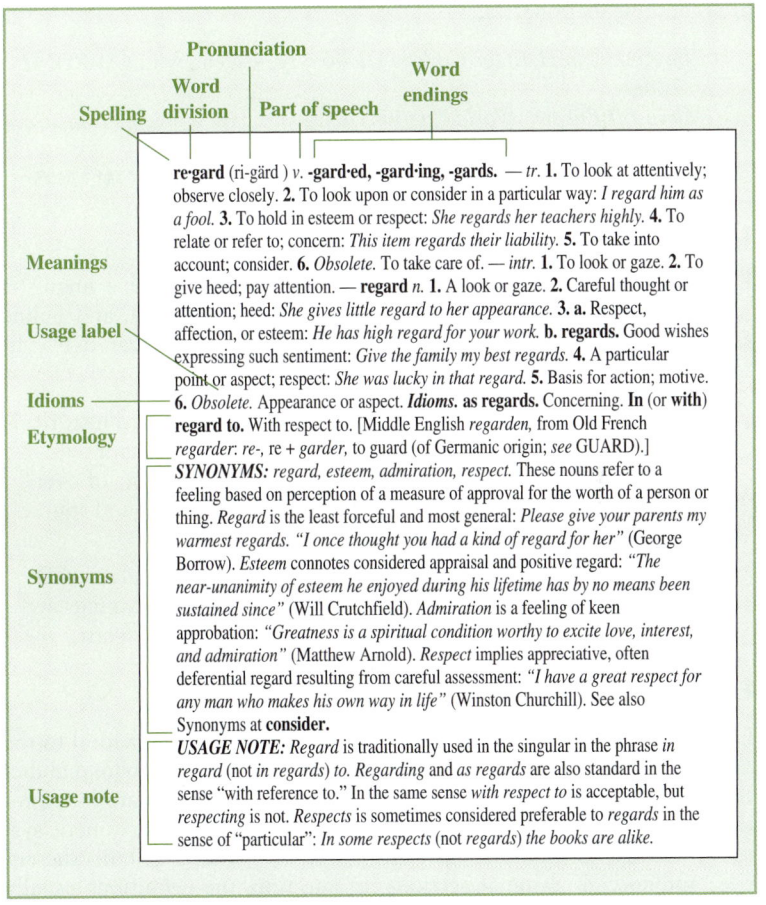

FIGURE D–3. Dictionary Entry

> **WEB LINK** Online Dictionaries
>
> Dictionaries available on the Web often include a human-voice pronunciation of words and thesaurus links. For a list of current online dictionaries, see *bedfordstmartins.com/alredtech* and select *Links for Handbook Entries*.

Abridged Dictionaries

Abridged or desk dictionaries contain words commonly used in schools and offices. There is no single "best" dictionary, but you should choose the most recent edition with upward of 200,000 entries. The following are reputable dictionaries.

> *The American Heritage College Dictionary*, 4th ed., with CD-ROM, 2006
> *Microsoft Encarta World English Dictionary*, free online edition at encarta.msn.com
> *Random House Webster's College Dictionary with CD-ROM*, 2005

Unabridged Dictionaries

Unabridged dictionaries provide complete and authoritative linguistic information. Because the size and cost of the printed, CD, and online subscription versions are impractical for many individual users, libraries make available these important reference sources.

- *The Oxford English Dictionary*, 2nd ed., is the standard historical dictionary of the English language. Its 20 volumes contain over 500,000 words and give the chronological developments of over 240,000 words, providing numerous examples of uses and sources. It is also available on CD-ROM and online.
- *Webster's Third New International Dictionary*, Unabridged, CD-ROM edition, contains over 450,000 entries. Word meanings are listed in historical order, with the current meaning given last.

ESL Dictionaries

English-as-a-second-language (ESL) dictionaries are more helpful to the nonnative speaker than are regular English dictionaries or bilingual dictionaries. The pronunciation symbols in ESL dictionaries are based on the international phonetic alphabet rather than on English phonetic systems, and useful grammatical information is included in both the entries and special grammar sections. In addition, the definitions usually are easier to understand than those in regular English dictionaries; for example, a regular English dictionary defines *opaque* as "impervious to the passage of light," while an ESL dictionary defines the word as "not allowing light to pass through." The definitions in ESL dictionaries also are usually more thorough than those in bilingual dictionaries. For example, a bilingual dictionary might indicate that *obstacle* and *blockade* are synonymous—but not indicate that only *obstacle* can be used for abstract meanings. (Lack of money can be an *obstacle* [not a *blockade*] to a college education.)

The following dictionaries and references provide helpful information for nonnative speakers of English.

> *Longman Advanced American Dictionary* with CD-ROM
> *Longman Dictionary of American English* with CD-ROM
> *Longman American Idioms Dictionary* with CD-ROM, Karen Stern
> *Oxford American Wordpower Dictionary for Learners of English*, Ruth Urbom

Subject Dictionaries

For the meanings of words too specialized for a general dictionary, a subject dictionary is useful. Subject dictionaries define terms used in a particular field, such as business, geography, architecture, or consumer affairs. Definitions in subject dictionaries are generally more detailed and comprehensive than those found in general dictionaries, but they are written in language that can be understood by nonspecialists. One well-known example is *Stedman's Medical Dictionary* (Lippincott Williams & Wilkins).

differ from / differ with

Differ from suggests that two things are not alike. ("Our earlier proposal *differs from* the current one.") *Differ with* indicates disagreement between persons. ("The architect *differed with* the contractor on the proposed site.")

different from / different than

In formal writing, the preposition *from* is used with *different*. ("The product I received is *different from* the one I ordered.") *Different than* is used when it is followed by a clause. ("The actual cost was *different than* we estimated in our proposal.")

direct address

Direct address refers to a sentence or phrase in which the person being spoken or written to is explicitly named. It is often used in **presentations** and in **e-mail** messages. Notice that the person's name in a direct address is set off by **commas**.

- *John*, call me as soon as you arrive at the airport.
- Call me, *John*, as soon as you arrive at the airport.

discreet / discrete

Discreet means "having or showing prudent or careful behavior." ("Because the matter was personal, he asked Bob to be *discreet*.") *Discrete* means something is "separate, distinct, or individual." ("The court ordered tests on three *discrete* samples.")

disinterested / uninterested

Disinterested means "impartial, objective, unbiased."

- Like good judges, researchers should be passionately interested in the problems they tackle but completely *disinterested* when they seek to solve those problems.

Uninterested means simply "not interested."

- Despite Asha's enthusiasm, her manager remained *uninterested* in the project.

division-and-classification method of development

An effective **method of development** for a complex subject is either to divide it into manageable parts and then discuss each part separately (division) or to classify (or group) individual parts into appropriate categories and discuss each category separately (classification). See also **instructions** and **process explanation**.

Division

You might use division to describe a physical object, such as the parts of a copy machine; to examine an organization, such as a company; or to explain the components of a system, such as the Internet. The emphasis in division as a method of development is on breaking down a complex whole into a number of like units—it is easier to consider smaller units and to examine the relationship of each to the other than to attempt to discuss the whole. The basis for division depends, of course, on your subject and your **purpose**.

If you were a financial planner describing the types of mutual funds available to your investors, you could divide the variety available into three broad categories: money-market funds, bond funds, and stock funds. Such division would be accurate, but it would be only a first-level grouping of a complex whole. The three broad categories could, in turn, be subdivided into additional groups based on investment strategy, as follows:

Money-market funds
- Taxable money market
- Tax-exempt money market

Bond funds
- Taxable bonds
- Tax-exempt bonds
- Balanced (mix of stocks and bonds)

Stock funds
- Balanced
- Equity income
- Domestic growth
- Growth and income
- International growth
- Small capitalization
- Aggressive growth
- Specialized

Specialized stock funds could be further subdivided as follows:

Specialized funds
- Communications
- Energy
- Environmental services
- Financial services
- Gold
- Health services
- Technology
- Utilities
- Worldwide capital goods

Classification

The process of classification is the grouping of a number of units (such as people, objects, or ideas) into related categories, as in Figure D–4. Consider the following list:

| triangular file | steel tape ruler | needle-nose pliers |
| vise | pipe wrench | keyhole saw |

mallet	tin snips	C-clamps
rasp	hacksaw	plane
glass cutter	ball-peen hammer	steel square
spring clamp	claw hammer	utility knife
crescent wrench	folding extension ruler	slip-joint pliers
crosscut saw	tack hammer	utility scissors

To group the items in the list, you would first determine what they have in common. The most obvious characteristic they share is that they all belong in a carpenter's tool chest (Figure D–4). With that observation as a starting point, you can begin to group the tools into related cate-

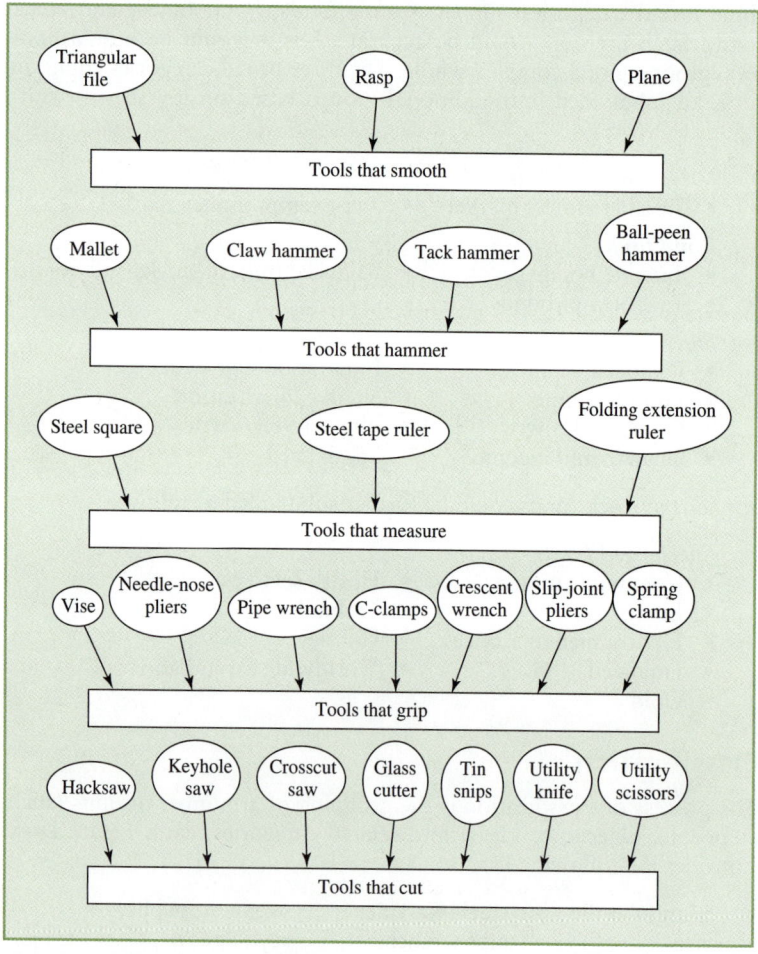

FIGURE D–4. Classification (Tools Placed into Categories)

gories. Pipe wrenches belong with slip-joint pliers because both tools grip objects. The rasp and the plane belong with the triangular file because all three tools smooth rough surfaces. By applying this kind of thinking to all the items in the list, you can group (classify) the tools according to function.

To classify a subject, you must first sort the individual items into the largest number of comparable groups. For explaining the functions of carpentry tools, the classifications (or groups) in Figure D–4 (smoothing, hammering, measuring, gripping, and cutting) are excellent. For recommending which tools a new homeowner should buy first, however, those classifications are not helpful—each group contains tools that a new homeowner might want to purchase right away. To give homeowners advice on purchasing tools, you probably would classify the types of repairs they most likely will have to do (plumbing, painting, etc.). That classification could serve as a guide to tool purchase.

Once you have established the basis for the classification, apply it consistently, putting each item in only one category. For example, it might seem logical to classify needle-nose pliers as both a tool that cuts and a tool that grips because most needle-nose pliers have a small section for cutting wires. However, the primary function of needle-nose pliers is to grip. So listing them only under "tools that grip" would be consistent with the basis used for listing the other tools.

documenting sources

DIRECTORY
APA Documentation   132
   APA In-Text Citations   132
   APA Documentation Models   132
   APA Sample Pages   137
IEEE Documentation   139
   IEEE In-Text Citations   139
   IEEE Documentation Models   139
   IEEE Sample Pages   143
MLA Documentation   145
   MLA In-Text Citations   145
   MLA Documentation Models   145
   MLA Sample Pages   151

Documenting sources achieves three important purposes:

- It allows readers to locate and consult the sources used and to find further information on the subject.

- It enables writers to support their assertions and arguments in such documents as **proposals**, **reports**, and **trade journal articles**.
- It helps writers to give proper credit to others and thus avoid **plagiarism** by identifying the sources of facts, ideas, **visuals**, **quotations**, and paraphrases. See also **paraphrasing**.

This entry shows citation models and sample pages for three principal documentation systems: APA, IEEE, and MLA. The following examples compare these three styles for citing a book by one author: *Project Management in New Product Development* by Bruce T. Barkley, which was published in 2007 by McGraw-Hill in New York, New York.

- The American Psychological Association (APA) system of citation is often used in the social sciences. It is referred to as an author-date method of documentation because parenthetical in-text citations and a reference list (at the end of the paper) in APA style emphasize the author(s) and date of publication so that the currency of the research is clear.

APA IN-TEXT CITATION

(Author's Last Name, Year)

(Barkley, 2007)

APA REFERENCES ENTRY

Author's Last Name, Initials. (Year). *Title in italics*. City, State (abbreviated) or Country of Publication: Publisher.

Barkley, B. T. (2007). *Project management in new product development*. New York, NY: McGraw-Hill.

- The system set forth in the *IEEE (Institute of Electrical and Electronics Engineers, Inc.) Standards Style Manual* is often used for the production of technical documents and standards in areas ranging from computer engineering, biomedical technology, and telecommunications to electric power, aerospace, and consumer electronics. The IEEE system is referred to as a number-style method of documentation because bibliographical reference numbers and a numbered bibliography identify the sources of information.

IEEE IN-TEXT CITATION

[Bibliographical reference number corresponding to a bibliography entry]

[B1]

IEEE BIBLIOGRAPHY ENTRY

[Bibliographical reference number] Author's Last Name, First and Middle Initial (or Full First Name), *Title in Italics*. Place of Publication: Publisher, Date of Publication, Pages.

[B1] Barkley, B. T., *Project Management in New Product Development.* New York: McGraw-Hill, 2007, pp. 112–125.

- The Modern Language Association (MLA) system is used in literature and the humanities. MLA style uses parenthetical in-text citations and a list of works cited, and it places greater importance on the pages on which cited information can be found than on the publication date.

MLA IN-TEXT CITATION

(Author's Last Name Page Number)

(Barkley 162)

MLA WORKS-CITED ENTRY

Author's Last Name, First Name. *Title Italicized.* Place of Publication: Publisher, Date of Publication. Medium of publication.

Barkley, Bruce T. *Project Management in New Product Development.* New York: McGraw, 2007. Print.

These systems are described in full detail in the following style manuals:

American Psychological Association. *Publication Manual of the American Psychological Association.* 6th ed. Washington: APA, 2010. See also *www.apastyle.org*.

IEEE Standards Style Manual. 2007 ed. New York: IEEE, 2007. See also *http://standards.ieee.org/guides/style*.

MLA Handbook for Writers of Research Papers. 7th ed. New York: MLA, 2009. See also *www.mlahandbook.org*.

For additional bibliographic advice and documentation models for types of sources not included in this entry, consult these style manuals or those listed in "Web Link: Other Style Manuals and Documentation Systems" on page 153. See also **bibliographies** and **research**.

APA Documentation

APA In-Text Citations. Within the text of a paper, APA parenthetical documentation gives a brief citation—in parentheses—of the author, year of publication, and a relevant page number if it helps locate a passage in a lengthy document.

- Technology has the potential to produce a transformational impact on human life that will enable the human brain to reach beyond its current limitations (Kurzweil, 2006).

- According to Kurzweil (2006), we will witness a "pace of technological change that will be so rapid, its impact so deep, that human life will be irreversibly transformed" (p. 7).

When APA parenthetical citations are needed midsentence, place them after the closing quotation marks and continue with the rest of the sentence.

- In short, "the Singularity" (Kurzweil, 2006, p. 9) is a blending of human biology and technology that will help us develop beyond our human limitations.

If the APA parenthetical citation follows a block quotation, place it after the final punctuation mark.

- . . . a close collaboration with the nursing staff and the hospital bed safety committee is essential. (Jackson, 2008)

When a work has two authors, cite both names joined by an ampersand: (Hinduja & Nguyen, 2008). For the first citation of a work with three to six authors, include all names. For subsequent citations and for works with more than six authors, include only the last name of the first author followed by *et al.* (not italicized and with a period after *al.*). When two or more works by different authors are cited in the same parentheses, list the citations alphabetically and use semicolons to separate them: (Hinduja & Nguyen, 2008; Townsend, 2007).

APA Documentation Models. In reference lists, APA requires that the first word of book and article titles be capitalized and all subsequent words be lowercased. Exceptions include the first word after a colon or dash and proper nouns.

PRINTED BOOKS

Single Author

Taleb, N. N. (2007). *The black swan: The impact of the highly improbable.* New York, NY: Random House.

Multiple Authors

Jones, E., Haenfler, R., & Johnson, B. (2007). *Better world handbook: Small changes that make a big difference.* Gabriola Island, British Columbia, Canada: New Society.

Multiple Books by Same Author

List the works in chronological order.

Gawande, A. (2002). *Complications: A surgeon's notes on an imperfect science.* New York, NY: Picador.

Gawande, A. (2007). *Better: A surgeon's notes on performance.* New York, NY: Picador.

Corporate Author

Microsoft Corporation. (2007). *Windows Vista: The new experience.* Redmond, WA: Microsoft Press.

Edition Other Than First

Kouzes, J. M., & Posner, B. Z. (2007). *The leadership challenge* (4th ed.). New York, NY: Wiley.

Multivolume Work

Knuth, D. E. (1998). *The Art of Computer Programming* (Vols. 1–3). Reading, MA: Addison-Wesley.

Work in an Edited Collection

Sen, A. (2007). Education and standards of living. In R. Curren (Ed.), *Philosophy of education: An anthology* (pp. 95–101). Malden, MA: Blackwell.

Encyclopedia or Dictionary Entry

Gibbard, B. G. (2007). Particle detector. In *World Book encyclopedia* (Vol. 15, pp. 202–203). Chicago, IL: World Book.

ARTICLES IN PRINTED PERIODICALS

Magazine Article

Austen, B. (2009, August). End of the road: After Detroit, the wreck of an American dream. *Harper's Magazine, 317*(8), 26–36.

Journal Article

Valentine, S., & Fleischman, G. (2008). Ethics programs, perceived corporate social responsibility and job satisfaction. *Journal of Business Ethics, 77,* 159–172.

Article with an Unknown Author

Plant genome organization. (2008). *Science 320*, 498–501.

Newspaper Article

Chazan, G. (2007, November 29). Can wind power find footing in the deep? *Wall Street Journal,* p. B1.

ELECTRONIC SOURCES

Entire Web Site

The APA recommends that, at minimum, a reference to a Web site should provide an author (whenever possible), the date of publication or update (use "(n.d.)" if no date is available), the title, and an address (URL) that links directly to the document or section. If the content could change or be deleted, such as on a Web site, include the retrieval date also. (The retrieval date is not necessary for content with a fixed publication date, such as a journal article.) On the rare occasion that you need to cite multiple pages of a Web site (or the entire site), provide a URL that links to the site's homepage. No periods follow URLs.

Society for Technical Communication. (2008). Retrieved from http://www.stc.org

Online Book

Use this form for books made available online or for e-books.

Sowell, T. (2007). *Basic economics: A common sense guide to the economy* (3rd ed.). Retrieved from http://books.google.com/

Short Work from a Web Site, with an Author

DuVander, A. (2006, June 29). Cookies make the Web go 'round. Retrieved from http://www.webmonkey.com/webmonkey/06/26/index3a.html

Short Work from a Web Site, with a Corporate or an Organizational Author

General Motors. (2007). Company profile. Retrieved from http://www.gm.com/corporate/about/company.jsp

Short Work from a Web Site, with an Unknown Author

Timeline: Alaska pipeline chronology. (2006, April 4). Retrieved from http://www.pbs.org/wgbh/amex/pipeline/timeline/index.html

Article or Other Work from a Database

Gaston, N., & Kishi, T. (2007). Part-time workers doing full-time work in Japan. *Journal of the Japanese and International Economies, 21*, 434–454. doi:10.1016/j.jjie.2006.04.001

Article in an Online Periodical, Not Available in Print

Gruener, W. (2008, April 1). Intel fires up new Atom processors. *TG Daily*. Retrieved from http://www.tgdaily.com/content/view/36735/135/

Article Posted on a Wiki

Datastream guides. (2007, June 5). *Biz Wiki*. Retrieved May 6, 2008, from http://www.library.ohiou.edu/subjects/bizwiki/index.php/Datastream_Guides

E-mail Message

E-mail messages are not cited in an APA reference list. They can be cited in the text as follows: "According to J. D. Kahl (personal communication, October 2, 2007), Web pages need to reflect. . . ."

Online Posting (Lists, Forums, Discussion Boards)

Harris, S. (2007, December 5). Camera manual comments [Online forum comment]. Retrieved from http://www.techwr-l.com/techwhirl/

Blog Entry

Gwozdz, G. L. (2005, December 5). Deductibility of 529 plans [Web log post]. Retrieved from http://glgcpa.blogspot.com/

Publication on CD-ROM

Tapscott, D., & Williams, A. D. (2007). *Wikinomics: How mass collaboration changes everything* [CD-ROM]. Old Saybrook, CT: Tantor.

MULTIMEDIA SOURCES (ELECTRONIC AND PRINT)

Film, Video, or Podcast

Tracy, B. (Host), & Jeffreys, M. (2007). *Doubling your productivity* [DVD]. Waterford, MI: Seminars on DVD.

Iowa Public Television. (2007, December 12). National debt impact on national security [Video file]. Retrieved from http://www.youtube.com/watch?v=hW1UCvqWY2E

Radio or Television Program

Vigeland, T. (Host). (2007, December 5). Plug-in hybrids need more juice. *Marketplace* [Radio broadcast]. Los Angeles, CA: American Public Media. Retrieved from http://marketplace.publicradio.org/display/web/2007/12/05/plug_in_cars/

Kurzweil, R. (2006, November 5). In Depth: Ray Kurzweil. *In Depth* [Television broadcast]. Washington, DC: C-Span.

OTHER SOURCES

Visual from Secondary Source

The APA recommends citing the source in which the visual (table or figure) appeared, without mentioning the visual specifically in the reference list. The in-text citation should include a source line (Source: author's last name, year, page number) placed directly below the visual. See also **visuals** and Ethics Note on page 153.

Published Interview

Pitney, J. (2008, January). Q & A [Interview]. *Automobile, 22*(1), 66.

Personal Communications

Personal communications such as lectures, letters, interviews, and e-mail messages are generally not cited in an APA reference list. They can be cited in the text as follows: "According to J. D. Kahl (personal communication, October 2, 2007), Web pages need to reflect. . . ."

Brochure or Pamphlet

Library of Congress, U.S. Copyright Office. (2007). *Copyright basics* [Brochure]. Washington, DC: U.S. Government Printing Office.

Government Document

U.S. Department of Labor, Bureau of Labor Statistics. (2007). National census of fatal occupational injuries in 2006 (Report No. USDL 07-1202). Washington, DC: U.S. Department of Labor.

Report

Ditch, W. (2007). *XML-based office document standards*. Bristol, England: Higher Education Funding Council for England.

Unpublished Data

Garcia, A. (2008). [Ohio information technology resources, by county]. Unpublished raw data.

APA Sample Pages

> **Shortened title and page number.**

ETHICS CASES   14

> **One-inch margins. Text double-spaced.**

This report examines the nature and disposition of the 3,458 ethics cases handled companywide by CGF's ethics officers and managers during 2007. The purpose of such reports is to provide the Ethics and Business Conduct Committee with the information necessary for assessing the effectiveness of the first year of CGF's Ethics Program (Davis, Marks, & Tegge, 2004). According to Matthias Jonas (2004), recommendations are given for consideration "in planning for the second year of the Ethics Program" (p. 152).

The Office of Ethics and Business Conduct was created to administer the Ethics Program. The director of the Office of Ethics and Business Conduct, along with seven ethics officers throughout CGF, was given the responsibility for the following objectives, as described by Rossouw (2000):

> **Long quote indented five to seven spaces, double-spaced, without quotation marks.**

> Communicate the values, standards, and goals of CGF's Program to employees. Provide companywide channels for employee education and guidance in resolving ethics concerns. Implement companywide programs in ethics awareness and recognition. Employee accessibility to ethics information and guidance is the immediate goal of the Office of Business Conduct in its first year. (p. 1543)

The purpose of the Ethics Program, established by the Committee, is to "promote ethical business conduct through open communication and compliance with company ethics standards" (Jonas, 2006, p. 89). To accomplish this purpose, any ethics policy must ensure confidentiality and anonymity for employees who raise genuine ethics concerns. The procedure developed at CGF guarantees that employees can

> **In-text citation gives name, date, and page number.**

FIGURE D–5. APA Sample Page from Report

ETHICS CASES 21

References

Davis, W. C., Marks, R., & Tegge, D. (2004). *Working in the system: Five new management principles*. New York, NY: St. Martin's Press.

Jonas, M. (2004). The Internet and ethical communication: Toward a new paradigm. *Journal of Ethics and Communication, 32*, 147–177.

Jonas, M. (2006). Ethics in organizational communication: A review of the literature. *Journal of Ethics and Communication, 29*, 79–99.

National Science Foundation. (2007). *Conflicts of interest and standards of ethical conduct*. Arlington, VA: Author.

Rossouw, G. J. (2000). Business ethics in South Africa. *Journal of Business Ethics, 16*, 1539–1547.

Schipper, F. (2007). Transparency and integrity: Contrary concepts? In K. Homann, P. Koslowski, and C. Luetge (Eds.), *Globalisation and business ethics* (pp. 101–118). Burlington, VT: Ashgate.

Smith, T. (Reporter), & Lehrer, J. (Host). (2006). *NewsHour business ethics anthology* [DVD]. Encino, CA: Business Training Media.

FIGURE D–6. APA Sample List of References

IEEE Documentation

IEEE In-Text Citations. IEEE bracketed bibliographical reference numbers within the text of a paper provide sequential numbers that correspond to the full citation in the bibliography: [A1], [A2], etc. The letter within the brackets specifies the annex, or appendix, where the bibliography is located (Annex A, Annex B, etc.). In IEEE style, the bibliography is always placed in an annex. Note that "Annex A" is usually reserved for standards (or *normative references*), whereas "Annex B" is reserved for typical in-text (or *informative*) citations.

▶ As Peterson writes, preparing a videotape of measurement methods is cost effective and can expedite training [B1].

▶ The results of these studies have led even the most conservative managers to adopt technologies that will "catapult the industry forward" [B2].

The first bibliographic reference cited in the document should be marked with a footnote that reads as follows.

▶ [1]The numbers in brackets correspond to those of the bibliography in Annex B.

IEEE Documentation Models

BOOKS

Single Author

[B1] Taleb, N. N., *The Black Swan: The Impact of the Highly Improbable*. New York: Random House, 2007, pp. 165–189.

Multiple Authors

[B2] Jones, E., Haenfler, R., and Johnson, B., *Better World Handbook: Small Changes That Make a Big Difference*. Gabriola Island, BC: New Society, 2007, pp. 129–142.

Corporate Author

[B3] Microsoft Corporation, *Windows Vista: The New Experience*. Redmond, WA: Microsoft Press, 2007, pp. 29–44.

Edition Other Than First

[B4] Kouzes, J. M., and Posner, B. Z., *The Leadership Challenge*, 4th ed. New York: Wiley, 2007, pp. 221–247.

Multivolume Work

[B5] Knuth, D. E., *The Art of Computer Programming*, vol. 2. Reading, MA: Addison-Wesley, 1998, pp. 35–51.

Work in an Edited Collection

[B6] Sen, A. "Education and standards of living," in *Philosophy of Education: An Anthology*, R. Curren, Ed. Malden, MA: Blackwell, 2007, pp. 95–101.

Encyclopedia or Dictionary Entry

[B7] *World Book Encyclopedia*, 2007 ed., s.v. "particle detector."

ARTICLES IN PERIODICALS (*See also* Electronic Sources)

Magazine Article

[B8] McGirt, E., "Facebook opens up," *Fast Company*, pp. 54–89, Nov. 2007.

Journal Article

[B9] Valentine, S., and Fleischman, G., "Ethics programs, perceived corporate social responsibility and job satisfaction," *Journal of Business Ethics*, vol. 77, pp. 159–172, 2008.

Article with an Unknown Author

[B10] "Plant genome organization," *Science*, pp. 498–501, Apr. 2008.

Newspaper Article

[B11] Chazan, G., "Can wind power find footing in the deep?" *Wall Street Journal*, sec. B, 29 Nov. 2007.

INTERNET SOURCES

The *IEEE Standards Style Manual*, 2007 ed., includes general information when citing information found on the Internet: If a document listed in a bibliography is accessed from the Internet, the *Style Manual* recommends that the document title, date, version, or other pertinent information should be listed, followed by a footnote that gives the Internet location. The URL should be the most stable location whenever possible "to avoid inadvertent or intentional changes that would affect the site name," so you should use the index to the page rather than the page itself.

Because the editors of the *Style Manual* suggest that readers consult the *Chicago Manual of Style* for more information on how to list various electronic and multimedia sources, the following models are based on the guidelines set forth in the *Chicago Manual of Style*, 15th ed.

Entire Web Site

[B12] Society for Technical Communication. *Society for Technical Communication*. 2008. [Footnote should follow with www.stc.org.]

Short Work from a Web Site, with an Author
[B13] DuVander, A., "Cookies make the web go 'round." *Webmonkey*. 29 June 2006. [Footnote should follow with www.webmonkey.com.]

Short Work from a Web Site, with a Corporate Author
[B14] General Motors. "Company Profile." *General Motors*. 2007. [Footnote should follow with www.gm.com.]

Short Work from a Web Site, with an Unknown Author
[B15] "Timeline: Alaska pipeline chronology." *The Alaska Pipeline*. 4 Apr. 2006. [Footnote should follow with www.pbs.org.]

Article or Other Work from a Database
[B16] Gaston, N., and Kishi, T., "Part-time workers doing full-time work in Japan." *Journal of the Japanese and International Economies*. 21 (2007): 435–454. [Footnote should follow with search.ebscohost.com.]

Article in an Online Periodical
[B17] Gruener, W., "Intel fires up new Atom processor." *TG Daily* 29 Apr. 2008. [Footnote should follow with www.tgdaily.com.]

MULTIMEDIA SOURCES (Print and Electronic)
Map or Chart
[B18] *Middle East*. Map. Chicago: Rand, 2007.

Film or Video
[B19] "Doubling Your Productivity." *Seminars on DVD*. Hosted by Brian Tracy, directed by Michael Jeffreys. Waterford, MI, 2007.

Television Interview
[B20] Khazaee, M., Interview by Charlie Rose. *The Charlie Rose Show*. Public Broadcasting System, 6 Dec. 2007.

OTHER SOURCES
E-mail
The *IEEE Standards Style Manual* does not provide advice for citing personal communications such as conversations, letters, and e-mail; the *Chicago Manual of Style*, 15th edition, suggests that personal communications are usually cited in the text rather than listed in a bibliography.

Published Interview
[B21] Pitney, J., "Q & A," Interview. *Automobile*, p. 66, Jan. 2008.

Personal Interview

[B22] Andersen, E., Interview by author. Tape recording. San Jose, CA, 29 Nov. 2007.

Personal Letter

Personal communications such as letters, e-mail, and messages from discussion groups and electronic bulletin boards are not cited in an IEEE bibliography.

Brochure or Pamphlet

[B23] Library of Congress, U.S. Copyright Office, *Copyright Basics*. Washington, DC: GPO, 2007, pp. 1–3.

Government Document

[B24] U.S. Department of Labor, Bureau of Labor Statistics, *National census of fatal occupational injuries in 2006*. Washington, DC: U.S. Department of Labor, 2007.

Report

[B25] Ditch, W., *XML-Based Office Document Standards*. Bristol, England: Higher Education Funding Council for England, 2007.

IEEE Sample Pages

Standards in Business, June 2008

This report examines the nature and disposition of the 3,458 ethics cases handled companywide by CGF's ethics officers and managers during 2007. The purpose of such reports is to provide the Ethics and Business Conduct Committee with the information necessary for assessing the effectiveness of the first year of CGF's Ethics Program. According to Matthias Jonas, recommendations are given for consideration "in planning for the second year of the Ethics Program" [B1].[1]

The Office of Ethics and Business Conduct was created to administer the Ethics Program. The director of the Office of Ethics and Business Conduct was given the responsibility for the following objectives, as described by Rossouw:

> Communicate the values, standards, and goals of CGF's Program to employees. Provide companywide channels for employee education and guidance in resolving ethics concerns. Implement companywide programs in ethics awareness and recognition. Employee accessibility to ethics information and guidance is the immediate goal of the Office of Business Conduct in its first year [B2].

The purpose of the Ethics Program, according to Jonas, is to "promote ethical business conduct through open communication and compliance with company ethics standards" [B3].

Major ethics cases were defined as those situations potentially involving serious violations of company policies or illegal conduct. Examples of major ethics cases included cover-up of defective workmanship or use of defective parts in products; discrimination in hiring and promotion; involvement in monetary or other kickbacks; sexual harassment; disclosure of proprietary or company information; theft; and use of corporate Internet resources for inappropriate purposes, such as conducting personal business, gambling, or access to pornography.

The effectiveness of CGF's Ethics Program during the first year of implementation is most evidenced by (1) the active participation of employees in the program and the 3,458 contacts employees made regarding ethics concerns through the various channels available to them, and (2) the action taken in the cases reported by employees,

---

[1] The numbers in brackets correspond to those in the bibliography in Annex B.

14

FIGURE D–7. IEEE Sample Page from Report

Standards in Business, June 2008

Annex B
(informative)
Bibliography

[B1] Jonas, M., "The internet and ethical communication: Toward a new paradigm," *Journal of Ethics and Communication*, vol. 32, pp. 147–177, Fall 2004.

[B2] Rossouw, G. J., "Business ethics in South Africa," *Journal of Business Ethics*, vol. 16, pp. 1539–1547, 1997.

[B3] Jonas, M., "Ethics in organizational communication: A review of the literature," *Journal of Ethics and Communication*, vol. 29, pp. 77–99, Summer 2006.

[B4] Davis, W. C., Marks, R., and Tegge, D., *Working in the System: Five New Management Principles*. New York: St. Martin's, 2001, pp. 18–42.

[B5] Schipper, F., "Transparency and integrity: Contrary concepts?" in *Globalisation and Business Ethics*, Karl Homann, Peter Koslowski, and Christoph Luetge, Eds. Burlington, VT: Ashgate, 2007, pp. 101–118.

[B6] Sariolgholam, M., Interview by author. Tape recording. Berkeley, CA, 29 Jan. 2008.

[B7] United States. National Science Foundation, *Conflicts of Interest and Standards of Ethical Conduct*. Arlington, VA: National Science Foundation, 2007, pp. 1–10.

[B8] International Business Ethics Institute. *International Business Ethics Institute.**

[B9] *NewsHour Business Ethics Anthology*. DVD. Hosted by Jim Lehrer. Business Training Media, Encino, CA, 2006.

---

*Available at www.business-ethics.org.

FIGURE D–8. IEEE Sample Bibliography

MLA Documentation

MLA In-Text Citations. The MLA parenthetical citation within the text of a paper gives a brief citation—in parentheses—of the author and relevant page numbers, separated only by a space. When citing Web sites where no author or page reference is available, provide a short title of the work in parentheses.

- We will witness a "pace of technological change that will be so rapid, its impact so deep, that human life will be irreversibly transformed" (Kurzweil 7).

- Kurzweil predicts that technology has the potential to produce a transformational impact on human life that will enable the human brain to reach beyond its current limitations (7).
  [If the author is cited in the text, include only the page number(s) in parentheses.]

- In 1810, Peter Durand invented the can, which was later used to provide soldiers and explorers canned rations and ultimately "saved legions from sure starvation" ("Forgotten Inventors").

If the parenthetical citation refers to a long, indented quotation, place it outside the punctuation of the last sentence.

- . . . a close collaboration with the nursing staff and the hospital bed safety committee is essential. (Jackson 208)

If no author is named or if you are using more than one work by the same author, give a shortened version of the title in the parenthetical citation, unless you name the title in the text (a "signal phrase"). A citation for Quint Studer's book *Results That Last: Hardwiring Behaviors That Will Take Your Company to the Top* would appear as (Studer, *Results* 93).

MLA Documentation Models

PRINTED BOOKS

Single Author

Taleb, Nassim Nicholas. *The Black Swan: The Impact of the Highly Improbable.* New York: Random, 2007. Print.

Multiple Authors

Jones, Ellis, Ross Haenfler, and Brett Johnson. *Better World Handbook: Small Changes That Make a Big Difference.* Gabriola Island, BC: New Society, 2007. Print.

Multiple Books by Same Author

List the works in alphabetical order by title.

Gawande, Atul. *Better: A Surgeon's Notes on Performance*. New York: Picador, 2007. Print.

---. *Complications: A Surgeon's Notes on an Imperfect Science*. New York: Picador, 2002. Print.

Corporate Author

Microsoft Corporation. *Windows Vista: The New Experience*. Redmond: Microsoft Press, 2007. Print.

Edition Other Than First

Kouzes, James M., and Barry Z. Posner. *The Leadership Challenge*. 4th ed. New York: Wiley, 2007. Print.

Multivolume Work

Knuth, Donald E. *The Art of Computer Programming*. 3 vols. Reading, MA: Addison-Wesley, 1998. Print.

Work in an Edited Collection

Sen, Amartya. "Education and Standards of Living." *Philosophy of Education: An Anthology*. Ed. Randall Curren. Malden: Blackwell, 2007. 95-101. Print.

Encyclopedia or Dictionary Entry

Gibbard, Bruce G. "Particle Detector." *World Book Encyclopedia*. 2007 ed. Print.

ARTICLES IN PRINTED PERIODICALS (*See also* Electronic Sources)
Magazine Article

McGirt, Ellen. "Facebook Opens Up." *Fast Company* Nov. 2007: 54-89. Print.

Journal Article

Valentine, Sean, and Gary Fleischman. "Ethics Programs, Perceived Corporate Social Responsibility, and Job Satisfaction." *Journal of Business Ethics* 77 (2008): 159-72. Print.

Article with an Unknown Author

"Plant Genome Organization." *Science* Apr. 2008: 498-501. Print.

Newspaper Article

Chazan, Guy. "Can Wind Power Find Footing in the Deep?" *Wall Street Journal* 29 Nov. 2007: B1+. Print.

ELECTRONIC SOURCES

Entire Web Site

The MLA recommends that a reference to a Web site should provide the author's name (whenever possible), the title of the site (italicized), the name of the sponsoring organization, a publication date (posted or updated), the medium of publication ("Web"), and a retrieval date.

Society for Technical Communication. 2008. Soc. for Technical Communication. Web. 18 Mar. 2008.

Online Book

Sowell, Thomas. *Basic Economics: A Common Sense Guide to the Economy*. 3rd ed. New York: Basic, 2008. *Google Book Search*. Web. 12 Jan. 2008.

Short Work from a Web Site, with an Author

DuVander, Adam. "Cookies Make the Web Go 'Round." *Webmonkey*. 29 June 2006. Web. 20 Dec. 2007.

Short Work from a Web Site, with a Corporate or an Organizational Author

General Motors. "Company Profile." *General Motors*. General Motors, 2009. Web. 14 July 2009.

Short Work from a Web Site, with an Unknown Author

"Timeline: Alaska Pipeline Chronology." *The Alaska Pipeline*. PBS, 4 Apr. 2006. Web. 22 Mar. 2008.

Article or Other Work from a Database

Gaston, Noel, and Tomoka Kishi. "Part-time Workers Doing Full-time Work in Japan." *Journal of the Japanese and International Economies* 21 (2007): 435-54. *Business Source Premier*. Web. 25 Nov. 2007.

Article in an Online Periodical

Gruener, Wolfgang. "Intel Fires Up New Atom Processors." *TG Daily*. DD&M Inc., 1 Apr. 2008. Web. 29 Apr. 2008.

Article Posted on a Wiki

"Datastream Guides." Article posted to wiki. *The Biz Wiki*. Ohio University Libraries, 5 June 2007. Web. 6 May 2008.

E-mail Message

Kalil, Ari. "Technical Support Survey." Message to the author. 12 Jan. 2008. E-mail.

Online Posting (Lists, Forums, Discussion Boards)

Harris, Sandy. "Camera Manual Comments." *TECHWR-L*. 5 Dec. 2007. Web. 8 Feb. 2008.

Blog Entry

Ojala, Marydee. "EPA Comes to SLA." *Infotoday Blog*. 7 June 2007. Web. 12 Sept. 2007.

Publication on CD-ROM

Tapscott, Don, and Anthony D. Williams. "Wikinomics." *Wikinomics: How Mass Collaboration Changes Everything*. Old Saybrook: Tantor, 2007. CD-ROM.

MULTIMEDIA SOURCES (Print and Electronic)

Film, Video, or Podcast

Doubling Your Productivity. Perf. Brian Tracy. Dir. Michael Jeffreys. Waterford: Seminars on DVD, 2007. DVD.

"National Debt Impact on National Security." *Des Moines Register Republican Presidential Debate*. Iowa Public Television. 12 Dec. 2007. *Iptv.org*. Web. 13 Dec. 2007.

Radio or Television Program

"Plug-in Hybrids Need More Juice." *Marketplace*. Host Tess Vigeland. American Public Media, 5 Dec. 2007. Web. 11 Feb. 2008.

"In Depth: Ray Kurzweil." *In Depth*. Host Pedro Echevarria. C-Span. 5 Nov. 2006. Television.

Television Interview

Khazaee, Mohammad. Interview by Charlie Rose. *The Charlie Rose Show*. PBS. WNET, New York, 6 Dec. 2007. Television.

documenting sources

OTHER SOURCES

Visual from a Secondary Source

Tables and figures (graphs, charts, maps, photographs, and drawings) are classified as visuals in MLA style. Cite the artist or author (if available), title of visual, type of visual, *the original publication date (if available)*, and the institution where it is located (if applicable). Follow with identifying information about the source in which the visual appeared. See also **visuals**.

Source: Jo Mackiewicz, "Compliments and Criticisms in Book Reviews about Business Communication." *Journal of Business and Technical Communication* 21.2 (2007): 205. Print.

Mackiewicz, Jo. "Criticism Mitigation Strategies." Chart. "Compliments and Criticisms in Book Reviews about Business Communication." *Journal of Business and Technical Communication* 21.2 (2007): 205. [Citation from print source.]

"Global Warming Effects." Map. *National Geographic*. Apr. 2007. 22 Jan. 2008. [Citation from Web.]

See also Ethics Note on page 153.

Interview

Pitney, Jack. "Q & A." Interview. *Automobile* Jan. 2008: 66. Print.

Personal Interview

Andersen, Elizabeth. Personal interview. 29 Nov. 2007.

Personal Letter

Pascatore, Monica. Letter to the author. 10 Apr. 2008. TS.

Brochure or Pamphlet

Library of Congress. US Copyright Office. *Copyright Basics*. Washington: GPO, 2007. Print.

Government Document

United States. Dept. of Labor. Bureau of Labor Statistics. *National Census of Fatal Occupational Injuries in 2006*. Rept. No. USDL 07-1202. Washington: US Dept. of Labor, 2007. Print.

Report

Ditch, Walter. *XML-Based Office Document Standards*. Bristol, Eng.: Higher Educ. Funding Council for England, 2007. Print.

Lecture or Speech

Nicholas, Ilene M. Demarest Hall, Hobart and William Smith Colls., Geneva, NY. 3 May 2008. Lecture.

Yaden, Ryan. "Evolving Architecture around the Globe." Boston Soc. of Architects Lecture Series. Boston Public Lib. 28 May 2008. Lecture.

MLA Sample Pages

> Litzinger 14
>
> *[Author's last name and page number.]*
>
> This report examines the nature and disposition of the 3,458 ethics cases handled companywide by CGF's ethics officers and managers during 2007. The purpose of such reports is to provide the Ethics and Business Conduct Committee with the information necessary for assessing the effectiveness of the first year of CGF's Ethics Program (Davis, Marks, and Tegge 142). According to Matthias Jonas, recommendations are given for consideration "in planning for the second year of the Ethics Program" ("Internet" 152).
>
> *[One-inch margins. Text double-spaced.]*
>
> The Office of Ethics and Business Conduct was created to administer the Ethics Program. The director of the Office of Ethics and Business Conduct, along with seven ethics officers throughout CGF, was given the responsibility for the following objectives, as described by Rossouw:
>
>> Communicate the values, standards, and goals of CGF's Program to employees. Provide companywide channels for employee education and guidance in resolving ethics concerns. Implement companywide programs in ethics awareness and recognition. Employee accessibility to ethics information and guidance is the immediate goal of the Office of Business Conduct in its first year. (1543)
>
> *[Long quote indented one inch (or 10 spaces), double-spaced, without quotation marks.]*
>
> The purpose of the Ethics Program, according to Jonas, is to "promote ethical business conduct through open communication and compliance with company ethics standards" ("Ethics" 89). To accomplish this purpose, any ethics policy must ensure confidentiality for anyone
>
> *[In-text citations give author name and page number. Title used when multiple works by same author cited.]*

FIGURE D–9. MLA Sample Page from Report

Litzinger 21

Works Cited

Davis, W. C., Roland Marks, and Diane Tegge. *Working in the System: Five New Management Principles*. New York: St. Martin's, 2004. Print.

International Business Ethics Institute. 2007. Inter. Business Ethics Inst., 14 Jan. 2008. Web. 19 June 2008.

Jonas, Matthias. "Ethics in Organizational Communication: A Review of the Literature." *Journal of Ethics and Communication* 29 (2006): 79-99. Print.

---. "The Internet and Ethical Communication: Toward a New Paradigm." *Journal of Ethics and Communication* 32 (2004): 147-77. Print.

Rossouw, George J. "Business Ethics in South Africa." *Journal of Business Ethics* 16 (2000): 1539-47. Print.

Sariolgholam, Mahmood. Personal interview. 29 Jan. 2008.

Schipper, Fritz. "Transparency and Integrity: Contrary Concepts?" *Globalisation and Business Ethics*. Ed. Karl Homann, Peter Koslowski, and Christoph Luetge. Burlington: Ashgate, 2007. 101-18. Print.

Smith, Terence. "Legislating Ethics." *NewsHour Business Ethics Anthology*. Host Jim Lehrer. Encino: Business Training Media, 2006. DVD.

United States. Natl. Science Foundation. *Conflicts of Interest and Standards of Ethical Conduct*. NSF Manual No. 15. Arlington: Natl. Science Foundation, 2007. Print.

FIGURE D–10. MLA Sample List of Works Cited

▶ **ETHICS NOTE** Visuals under copyright require permission from copyright owners before they can be reproduced. Contact the publisher to find out what needs permission, how best to obtain it, and what information copyright owners require in source lines and reference entries. Source lines almost always require mention of permission from the copyright owner to reproduce. ✦

> **WEB LINK** | **Other Style Manuals and Documentation Systems**
>
> Many professional societies, publishing companies, and other organizations publish manuals that prescribe bibliographic reference formats for their publications or for publications in their fields. For example, the American Medical Association (AMA) publishes the *American Medical Association Manual of Style*. For links to style manuals and other documentation resources, see *bedfordstmartins.com/alredtech* and select *Links for Handbook Entries*.

double negatives

A double negative is the use of an additional negative word to reinforce an expression that is already negative. In writing and speech, avoid such constructions.

    UNCLEAR   We don't have none.
                     [This sentence literally means that we have some.]
    CLEAR   We have none.

Barely, *hardly*, and *scarcely* cause problems because writers sometimes do not recognize that those words are already negative.

▶ The corporate policy ~~doesn't~~ hardly covers the problem.

Not unfriendly, *not without*, and similar constructions are not double negatives because in such constructions two negatives are meant to suggest the gray area between negative and positive meanings. Be careful how you use such constructions; they can be confusing to the reader and should be used only if they serve a purpose.

▶ He is *not unfriendly*.
   [He is neither hostile nor friendly.]

▶ It is *not without* regret that I offer my resignation.
   [I have mixed feelings rather than only regret.]

The correlative **conjunctions** *neither* and *nor* may appear together in a clause without creating a double negative, so long as the writer does not attempt to use the word *not* in the same clause.

▶ It was ~~not,~~ *neither,* as a matter of fact, ~~neither~~ his duty nor his desire to dismiss the employee.

▶ It was not, as a matter of fact, ~~neither~~ *either* his duty ~~nor~~ *or* his desire to dismiss the employee.

▶ She ~~did not~~ neither ~~care~~ *cared* about nor ~~notice~~ *noticed* the error.

Negative forms are full of traps that often entice writers into **logic errors**, as illustrated in the following example:

ILLOGICAL   The book reveals *nothing* that has *not* already been published in some form, but some of it is, I believe, relatively unknown.

In this sentence, "some of it" logically can refer only to "*nothing* that has *not* already been published." The sentence can be corrected by stating the idea in more **positive writing**.

LOGICAL   Everything in the book has been published in some form, but some of it is, I believe, relatively unknown.

drawings

A drawing can show an object's appearance and illustrate the steps in procedures or **instructions**. It can emphasize the significant parts or functions of a device or product, omit what is not significant, and focus on details or relationships that a **photograph** cannot reveal. Think about your need for drawings during your **preparation** and **research**. Include them in your outline, indicating approximately where each should appear throughout the outline with "drawing of . . ." enclosed in brackets. For advice on integrating drawings into your text, see **outlining** and **visuals**.

The types of drawings discussed in this entry are conventional line drawings, exploded-view drawings, cutaway drawings, schematic diagrams, and clip-art images.

A conventional line drawing is appropriate if your **audience** needs an overview of a series of steps or an understanding of an object's appearance or construction, as in Figure D–11.

An exploded-view drawing, like that in Figure D–12, can be useful when you need to show the proper sequence in which parts fit together or to show the details of individual parts. Figure D–12 shows owners of a Xerox machine how to safely unpack the machine and its key parts.

A cutaway drawing, like the one in Figure D–13, can be useful when you need to show the internal parts of a device or structure and illustrate their relationship to the whole.

Schematic diagrams (or schematics) are drawings that portray the operation of electrical and mechanical systems, using lines and symbols rather than physical likenesses. Schematics are useful in instructions or descriptions—for example, when you need to emphasize the relationships among parts without giving their precise proportions.

The schematic diagram in Figure D–14 depicts the process of trapping particulates from coal-fired power plants. Note that the particulate filter is depicted both symbolically and as an enlarged drawing to show relevant structural details. Schematic symbols generally represent equipment, gauges, and electrical or electronic components.

If you need only general-interest images to illustrate **newsletters** and **brochures** or to create **presentation** slides, use noncopyrighted clip-art drawings from your word-processing program or from thousands of noncopyrighted Web and library sources. Figure D–15 shows examples of clip-art images.

▶ **ETHICS NOTE** Do not use drawings from copyrighted sources without permission and proper documentation, especially from the Web—the same copyright laws that apply to printed material also apply to Web-based graphics. See also **copyright**, **documenting sources**, and **plagiarism**. ✦

PULL

the pin: Some extinguishers require releasing a lock latch, pressing a puncture lever, or taking another first step.

FIGURE D–11. Conventional Line Drawing Illustrating Instructions

Installation

As you unpack the WorkCentre, familiarize yourself with its contents. After the WorkCentre is installed, and the Ready Indicator is lit, the WorkCentre is ready to make copies.

IMPORTANT: Save the carton and packing materials. They should be used to repack the WorkCentre if it has to be shipped for servicing or in case you move.

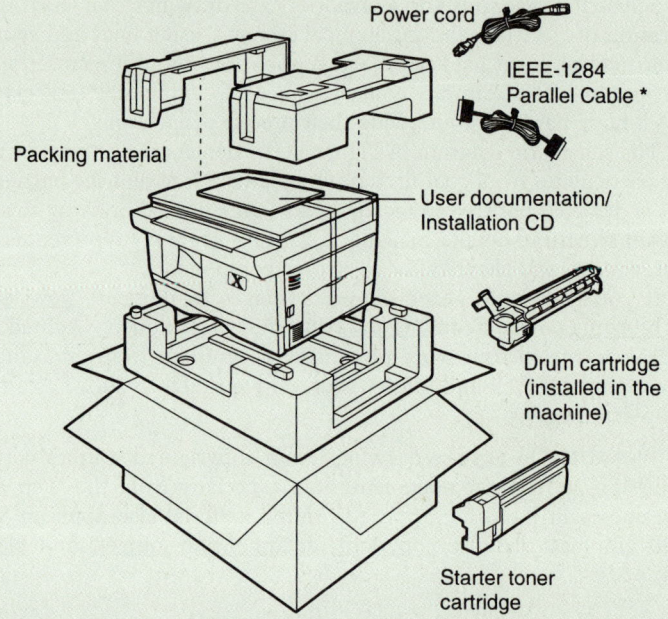

* **Note:** To ensure reliability of the WorkCentre, use the IEEE-1284 compliant parallel cable that is supplied with the machine. Only cables labeled "IEEE-1284" can be used with your WorkCentre.

Figure D–12. Exploded-View Drawing

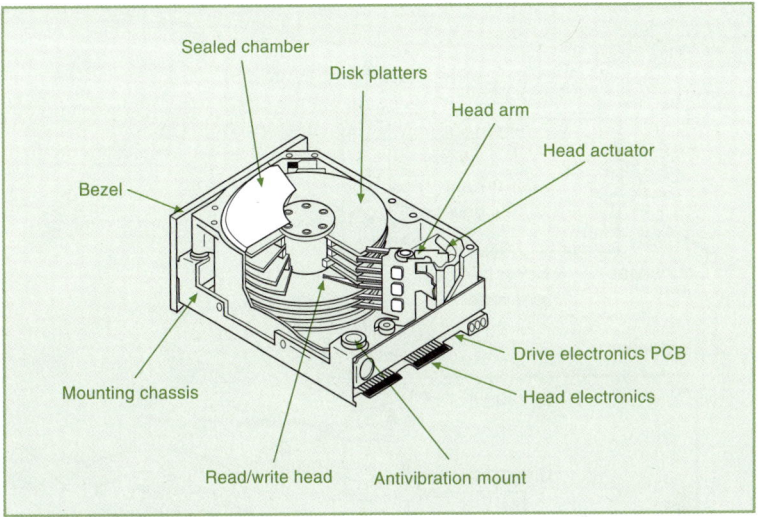

FIGURE D–13. Cutaway Drawing

WRITER'S CHECKLIST Creating and Using Drawings

- Select the type of drawing based on your **purpose** and on what your readers might need or expect.
- Seek the help of graphics specialists for drawings that require a high degree of accuracy and precision.
- Show equipment and other objects from the point of view of the person who will use them.
- When illustrating a subsystem, show its relationship to the larger system of which it is a part.
- Draw the parts of an object in proportion to one another and identify any parts that are enlarged or reduced.
- When a sequence of drawings is used to illustrate a process, arrange them from left to right or from top to bottom on the page.
- Label parts in the drawing so that the text references to them are clear and consistent.
- Depending on the complexity of what is shown, label the parts themselves, as in Figure D–12, or use a key, as in Figure G–10 on page 239.

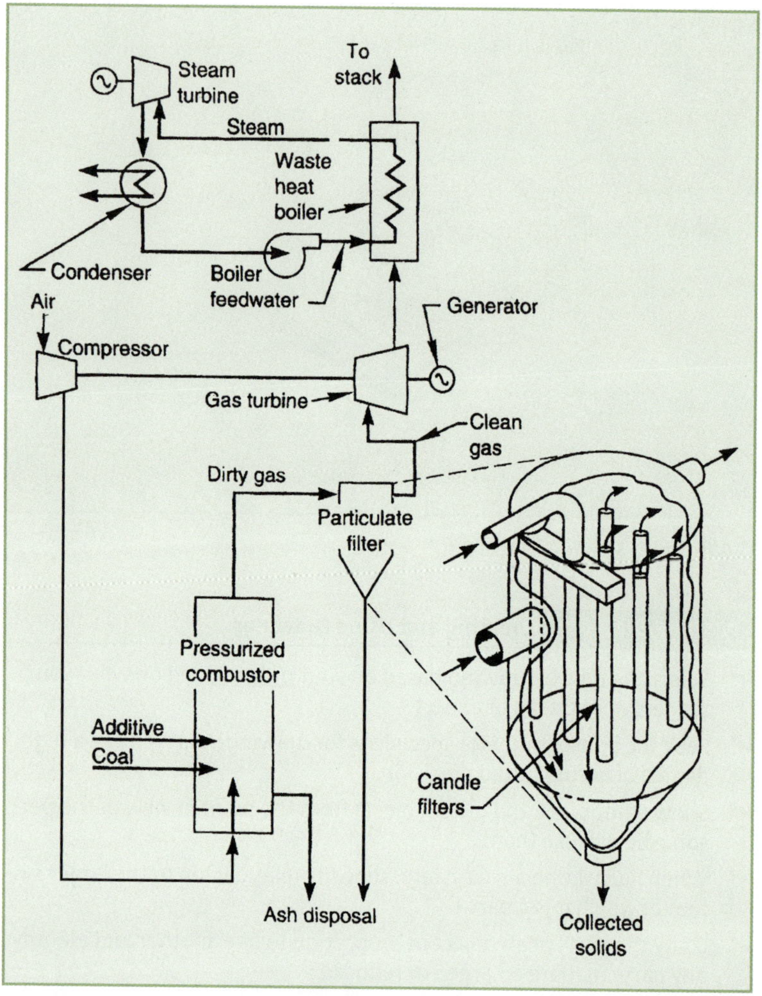

FIGURE D–14. Schematic Diagram

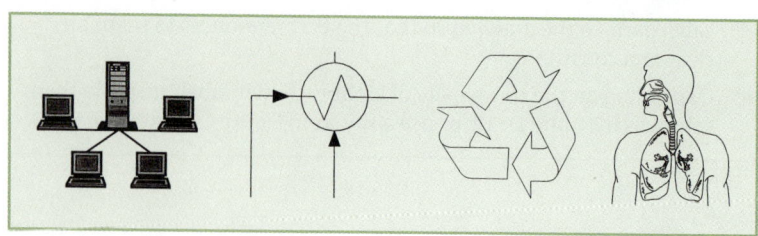

FIGURE D–15. Clip-Art Images

due to / because of

Due to (meaning "caused by") is acceptable following a linking **verb**.

▶ His absence was *due to* a work-related injury.

Due to is not acceptable, however, when it is used with a nonlinking verb to replace *because of*.

▶ He left work ~~due to~~ *because of* illness.

E

each

When *each* is used as a subject, it takes a singular **verb** or **pronoun**. ("*Each* of the reports *is* to be submitted ten weeks after *it* is assigned.") When *each* refers to a plural subject, it takes a plural verb or pronoun. ("The reports *each have* company logos on *their* title pages.") See also **agreement** and **conciseness**.

economic / economical

Economic refers to the production, development, and management of material wealth. ("Technology infrastructure has an *economic* impact on communities.") *Economical* simply means "not wasteful or extravagant." ("Employees should be as *economical* as possible in their equipment purchases.")

editing (*see* revision *and* proofreading)

e.g. / i.e.

The abbreviation *e.g.* stands for the Latin *exempli gratia*, meaning "for example"; *i.e.* stands for the Latin *id est*, meaning "that is." Because the English expressions (*for example* and *that is*) are clear to all **readers**, avoid the Latin *e.g.* and *i.e.* **abbreviations** except to save space in notes and **visuals**. If you must use *i.e.* or *e.g.*, do not italicize either and punctuate them as follows. If *i.e.* or *e.g.* connects two independent clauses, a **semicolon** should precede it and a **comma** should follow it.

▶ The conference reflected international viewpoints; *i.e.*, Germans, Italians, Japanese, Chinese, and Americans gave presentations.

If *i.e.* or *e.g.* connects a **noun** and an **appositive**, a comma should precede it and follow it.

▶ The conference included speakers from five countries, *i.e.*, Germany, Italy, Japan, China, and the United States.

ellipses

An ellipsis is the omission of words from quoted material; it is indicated by three spaced **periods** called *ellipsis points* (. . .). When you use ellipsis points, omit original punctuation marks, unless they are necessary for **clarity** or the omitted material comes at the end of a quoted sentence.

ORIGINAL TEXT	"Promotional material sometimes carries a fee, particularly in high-volume distribution to schools, although prices for these publications are much lower than the development costs when all factors are considered."
WITH OMISSION AND ELLIPSIS POINTS	"Promotional material sometimes carries a fee . . . although prices for these publications are much lower than the development costs. . . ."

Notice in the preceding example that the final period is retained and what remains of the quotation is grammatically complete. When the omitted part of the quotation is preceded by a period, retain the period and add the three ellipsis points after it, as in the following example.

ORIGINAL TEXT	"Of the 172 major ethics cases reported, 57 percent were found to involve unsubstantiated concerns. Misinformation was the cause of unfounded concerns of misconduct in 72 cases. Forty-four cases, or 26 percent of the total cases reported, involved incidents partly substantiated by ethics officers as serious misconduct."
WITH OMISSION AND ELLIPSIS POINTS	"Of the 172 major ethics cases reported, 57 percent were found to involve unsubstantiated concerns. . . . Forty-four cases, or 26 percent of the total cases reported, involved incidents partly substantiated by ethics officers as serious misconduct."

Do not use ellipsis points when the beginning of a quoted sentence is omitted. Notice in the following example that the comma is dropped to prevent a grammatical error. See also **quotations**.

▶ The ethics report states that "26 percent of the total cases reported involved incidents partly substantiated by ethics officers as serious misconduct."

e-mail

DIRECTORY
Review and Confidentiality 162
Writer's Checklist: Observing Workplace Netiquette 163
Design Considerations 163
Salutations, Closings, and Signature Blocks 164
Writer's Checklist: Managing Your E-mail and Reducing Overload 166

E-mail (or *email*) functions in the workplace as a medium to exchange information and share electronic files with colleagues, clients, and customers. Although e-mail may take the form of informal notes, you should follow the writing strategy and style described in **correspondence** because e-mail messages often function as business **letters** to those outside organizations and **memos** to those within organizations. Of course, memos and letters can also be attached to e-mails. This entry reviews topics that are specifically related to the medium of e-mail. See also **selecting the medium**.

Review and Confidentiality

E-mail is such a quick and easy way to communicate that you need to avoid the temptation of sending a first draft without revision. As with all correspondence, your message should include all crucial details and be free of grammatical or factual errors, ambiguities, or unintended implications. See **proofreading**.

◆ **ETHICS NOTE** Keep in mind that e-mail can be intercepted by someone other than the intended recipient and that e-mail messages are never truly deleted. Most companies back up and save all their e-mail messages and are legally entitled to monitor e-mail use. Companies can be compelled, depending on circumstances, to provide e-mail and **instant messaging** logs in a court of law. Consider the content of all your messages in the light of these possibilities, and carefully review your text before you click "Send." ◆

Be especially careful when sending messages to superiors in your organization or to people outside the organization. Spending extra time reviewing your e-mail before you click "Send" can save you the embarrassment of misunderstandings caused by a careless message. One helpful strategy is to write the draft and revise your e-mail before filling in the "To" line with the address of your recipient. Be careful as well to observe the rules of netiquette (Inter*net* + *etiquette*) in the *Writer's Checklist* that follows.

| WRITER'S CHECKLIST | Observing Workplace Netiquette |

- ✓ Review your organization's policy regarding the appropriate use of e-mail.
- ✓ Maintain a high level of professionalism in your use of e-mail.
    - Do not forward jokes or *spam*, discuss office gossip, or use **biased language**.
    - Do not send *flames* (e-mails that contain abusive, obscene, or derogatory language) to attack someone.
    - Do not use an e-mail account with a clever or hobby-related address (yogalover@gja.com); e-mail addresses based on your last name are appropriate (jones23@gja.com).
- ✓ Provide a subject line that describes the topic and focus of your message (as described on page 110 in **correspondence**) to help recipients manage their e-mail.
- ✓ Adapt forwarded messages: revise the subject line to reflect the current content and cut irrelevant previous text, based on your **purpose** and **context**.
- ✓ Use the "cc:" (copy) and "bcc:" (blind copy) address lines thoughtfully and consider your organization's practice or protocol.
- ✓ Include a cover (or *transmittal*) message for all e-mail messages with attachments ("Attached is a copy of the report for your review...."). See **cover letters**.
- ✓ Send a "courtesy response" informing someone when you need a few days or longer to reply to a request.
- ✓ Do not write in ALL UPPERCASE LETTERS (called "shouting") or in all lowercase letters.
- ✓ Avoid e-mail abbreviations (BTW for *by the way*, for example) used in personal e-mail, chat rooms, and **instant messaging**.
- ✓ Do not use emoticons (keyboard characters used to create sideways faces conveying emotions) for business and professional e-mail.

Design Considerations

Some e-mail systems allow you to use typographical features, such as various fonts and bullets. These options increase your e-mail file size and may display unpredictably in other e-mail systems. Unless you are sure your recipient's software will display your formatted message correctly, set your e-mail software to send messages in "plain text" and use alternative highlighting devices. For example, capital letters or asterisks,

> **DIGITAL TIP**
>
> **Sending an E-mail Attachment**
>
> Large files, such as graphics, slow the transmission speed of e-mail messages, and the recipient's software or Internet provider may not be able to accept large attachments. Consider using a compression software utility, such as WinZip at *http://winzip.com*, which can reduce the file size by 80 percent or more. *Caution:* Viruses can be embedded in e-mail attachments, so regularly update your virus-scanning software and do not open attachments from anyone you do not know. See *bedfordstmartins.com/alredtech* and select *Digital Tips*, "Sending an E-mail Attachment."

used sparingly, can substitute for boldface, italics, and underlines as **emphasis**.

- ▶ Dr. Wilhoit's suggestions benefit doctors AND patients.
- ▶ Although the proposal is sound in *theory*, it will never work in *practice*.

Intermittent underlining can replace solid underlining or italics when referring to published works in an e-mail message.

- ▶ My report follows the format given in _Handbook of Technical Writing_.

If you find that you need many such substitutions, consider preparing a document that you attach to an e-mail. Doing so will allow you to use formatted elements, such as bulleted **lists** and **tables**, that do not transmit well in e-mail messages. Keep in mind the following additional design considerations when sending e-mail:

- Break the text into short paragraphs to avoid dense blocks of text.
- Consider providing an overview at the top in a brief paragraph for messages that run longer than a screen of text.
- Place your response to someone else's message at the beginning (or top) of the e-mail window so that recipients can see your response immediately.
- When replying to a message, quote only relevant parts. If your system does not distinguish the quoted text, note it with a greater-than symbol (>).

Salutations, Closings, and Signature Blocks

An e-mail can function as a letter, memo, or personal note; therefore, you must adapt your salutation and complimentary closing to your **au-**

dience. Unless your employer requires certain forms, use the following guidelines:

- When e-mail functions as a traditional letter, use the standard letter salutation (*Dear Ms. Tucker:* or *Dear Docuform Customer:*) and closing (*Sincerely,* or *Best wishes,*).
- When you send e-mail to individuals or small groups inside an organization, you may wish to adopt a more personal greeting (*Dear Andy,* or *Dear Project Colleagues,*).
- When e-mail functions as a personal note to a friend or close colleague, you can use an informal greeting or only a first name (*Hi Mike,* or *Hello Jenny,* or *Bill,*) and a closing (*Take care,* or *Best,*).

Be aware that, in some cultures, business correspondents do not use first names as freely as do American correspondents. See **international correspondence**.

Because e-mail does not provide letterhead with standard addresses and contact information, many companies and individual writers include signature blocks (also called *signatures*) at the bottom of their messages. Signature blocks, which writers can preprogram to appear on every e-mail they send, supply information that company letterhead usually provides as well as appropriate links to Web sites. If your organization requires a certain format, adhere to that standard. Otherwise, use the pattern shown in Figure E–1. For signature blocks, consider as well the following guidelines:

- Keep line length to 60 characters or fewer to avoid unpredictable line wraps.
- Test your signature block in plain-text e-mail systems to verify your format.
- Use highlighting cues, such as **hyphens**, equal signs, and white space, to separate the signature from the message.
- Avoid using quotations, aphorisms, proverbs, or other sayings from popular culture, religion, or poetry in professional signatures.

```
================================
Daniel J. Vasquez, Publications Manager    ←    Name and Title
Medical Information Systems                ←    Department or Division
TechCom Corporation                        ←    Company Name
P.O. Box 5413    Salinas, CA 93962         ←    Mailing Address
Office Phone 888-229-4511 (x 341)          ←    Phone Number
General Office Fax 888-229-1132            ←    Fax Number
www.tcc.com                                ←    Web Address (URL)
================================
```

FIGURE E–1. **E-mail Signature Block**

> **DIGITAL TIP**
>
> **Leaving an Away-from-Desk Message**
>
> Many e-mail systems allow you to create an away-from-desk automatic response. Your message should inform senders when you are expected back and, if necessary, whom they can contact in your absence. To learn how to set up an automatic response, see *bedfordstmartins.com/alredtech* and select *Digital Tips*, "Leaving an Away-from-Desk Message."

WRITER'S CHECKLIST Managing Your E-mail and Reducing Overload

Given the high volume of e-mail in business, you need to manage your e-mail strategically.*

- Avoid becoming involved in an e-mail exchange if a phone call or meeting would be more efficient.
- Consider whether an e-mail message could prompt an unnecessary response from the recipient and make clear to the recipient whether you expect a response.
- Send a copy ("cc:") of an e-mail only when the person copied, in fact, needs or wants the information.
- Review all messages on a subject before responding to avoid dealing with issues that are no longer relevant.
- Set priorities for reading e-mail by skimming sender names and subject lines as well as where you appear in a "cc:" (copy) and "bcc:" (blind copy) address line.
- Check e-mail addresses before sending an e-mail and keep your addresses current.
- Check your in-box regularly and try to clear it by the end of each day.
- Learn the advanced features of your system that can, for example, file messages as they arrive.
- Create e-mail folders using key topics and personal names to file messages.
- Copy yourself or save sent copies of important e-mail messages in your folders.
- Use the search command to find particular subjects and personal names.

*For understanding the causes of e-mail overload, see Gail Fann Thomas and Cynthia L. King, "Reconceptualizing E-mail Overload," *Journal of Business and Technical Communication* 20, no. 3 (July 2006): 252–87.

Writer's Checklist: Managing Your E-mail and Reducing Overload (continued)

- Print copies of messages or attachments that you need for meetings, files, or similar purposes.

emphasis

Emphasis is the principle of stressing the most important ideas in your writing. You can achieve emphasis as described below with position, climactic order, sentence length, sentence type, active **voice**, **repetition**, **intensifiers**, direct statements, long **dashes**, and mechanical devices.

Achieving Emphasis

Position. Place the idea in a conspicuous position. The first and last words of a sentence, **paragraph**, or document stand out in readers' minds.

▶ Moon craters are important to understanding the earth's history because they reflect geological history.

This sentence emphasizes *moon craters* simply because the term appears at the beginning of the sentence and *geological history* because it is at the end of the sentence. See also **subordination**.

Climactic Order. List the ideas or facts within a sentence in sequence from least to most important, as in the following example. See also **lists**.

▶ Over subsequent weeks the Human Resources Department worked diligently, management showed tact and patience, and the employees demonstrated remarkable support for the policy changes.

Sentence Length. Vary sentence length strategically. A very short sentence that follows a very long sentence or a series of long sentences stands out in the reader's mind, as in the short sentence ("We must raise our standards") that ends the following paragraph. See **sentence construction**.

▶ We have already reviewed the problem the quality-control department has experienced during the past year. We could continue to examine the causes of our problems and point an accusing finger at all the culprits beyond our control, but in the end it all leads to one simple conclusion. We must raise our standards.

Sentence Type. Vary sentences by using a compound sentence, a complex sentence, or a simple sentence. See **sentence variety**.

▶ The report submitted by the committee was carefully illustrated, and it covered five pages of single-spaced copy.
[This compound sentence carries no special emphasis; it contains two coordinate independent clauses.]

▶ The committee's report, which was carefully illustrated, covered five pages of single-spaced copy.
[This complex sentence emphasizes the size of the report.]

▶ The carefully illustrated report submitted by the committee covered five pages of single-spaced copy.
[This simple sentence emphasizes that the report was carefully illustrated.]

Active Voice. Use the active voice to emphasize the performer of an action: Make the performer the subject of the **verb**.

▶ Our department designed the new system.
[This sentence emphasizes *our department*, which is the performer and the subject of the verb, *designed*.]

Repetition. Repeat key terms, as in the use of the word *remains* and the phrase *come and go* in the following sentence.

▶ Similarly, atoms *come and go* in a molecule, but the molecule *remains*; molecules *come and go* in a cell, but the cell *remains*; cells *come and go* in a body, but the body *remains*; persons *come and go* in an organization, but the organization *remains*.
—Kenneth Building, *Beyond Economics*

Intensifiers. Although you can use intensifiers (*most*, *much*, *very*) for emphasis, this technique is so easily abused that it should be used with caution.

▶ The final proposal is *much* more persuasive than the first one.
[The intensifier *much* emphasizes the contrast.]

Direct Statements. Use direct statements, such as "most important," "foremost," or someone's name in a **direct address**.

▶ Most important, keep in mind that everything you do affects the company's bottom line.

▶ John, I believe we should rethink our plans.

Long Dashes. Use a dash to call attention to a particular word or statement.

▶ The job will be done—after we are under contract.

Mechanical Devices. Use *italics*, **bold type**, underlining, and CAPITAL LETTERS—but use them sparingly because overuse can create visual clutter. See also **capitalization**, **italics**, and **layout and design**.

English as a second language

DIRECTORY

Count and Mass Nouns 169	Present Perfect Verb Tense 172
Articles and Modifiers 170	Present Progressive Verb Tense 172
Gerunds and Infinitives 171	ESL Entries 172
Adjective Clauses 171	

Learning to write well in a second language takes a great deal of effort and practice. The most effective way to improve your command of written English is to read widely beyond the reports and professional articles your job requires, such as magazines, newspapers, articles, novels, biographies, and any other writing that interests you. In addition, listen carefully to native speakers on television, on radio, and in person. Do not hesitate to consult a native speaker of English, especially for important writing tasks, such as **e-mails**, **memos**, and **reports**. Focus on those particular areas of English that give you trouble. See also **global communication**.

Count and Mass Nouns

Count nouns refer to things that can be counted (tables, pencils, projects, reports). *Mass nouns* (also called *noncount nouns*) identify things that cannot be counted (electricity, air, loyalty, information). This distinction can be confusing with words like *electricity* and *water*. Although we can count kilowatt-hours of electricity and bottles of water, counting becomes inappropriate when we use the words *electricity* and *water* in a general sense, as in "*Water* is an essential resource." Following is a list of typical mass **nouns**.

anger	education	money	technology
biology	equipment	news	transportation
business	furniture	oil	water
clothing	health	precision	weather
coffee	honesty	research	work

The distinction between whether something can or cannot be counted determines the form of the noun to use (singular or plural), the kind of **article** that precedes it (*a*, *an*, *the*, or no article), and the kind of limiting **adjective** it requires (such as *fewer* or *less* and *much* or *many*). (See also **fewer / less**.) Notice that count and mass nouns are always common nouns, not proper nouns, such as the names of people.

Articles and Modifiers

The general rule is that every count noun must be preceded by an article (*a*, *an*, *the*), a demonstrative adjective (*this*, *that*, *these*, *those*), a possessive adjective (*my*, *your*, *her*, *his*, *its*, *their*), or some expression of quantity (such as *one*, *two*, *several*, *many*, *a few*, *a lot of*, *some*, *no*). The article, adjective, or expression of quantity appears either directly in front of the noun or in front of the whole noun phrase.

- Beth read *a* report last week. [article]
- *Those* reports Beth read were long. [demonstrative adjective]
- *Their* report was long. [possessive adjective]
- *Some* reports Beth read were long. [indefinite adjective]

The articles *a* and *an* are used with count nouns that refer to one item of the whole class of like items.

- Matthew has *a* pen.
  [Matthew could have any pen.]

The article *the* is used with nouns that refer to a specific item that both the reader and the writer can identify.

- Matthew has *the* pen.
  [Matthew has a specific pen that is known to both the reader and the writer.]

When making generalizations with count nouns, writers can either use *a* or *an* with a singular count noun or use no article with a plural count noun. Consider the following generalization using an article.

- *An* egg is a good source of protein.
  [any egg, all eggs, eggs in general]

However, the following generalization uses a plural count noun with no article.

- *Eggs* are good sources of protein.
  [any egg, all eggs, eggs in general]

When you are making a generalization with a mass noun, do not use an article in front of the mass noun.

▶ *Sugar* is bad for your teeth.

Gerunds and Infinitives

Nonnative writers of English are often puzzled by which form of a **verbal** (a **verb** used as another part of speech) to use when it functions as the direct object of a verb—or a **complement**. No structural rule exists for distinguishing between the use of an infinitive and a gerund as the object of a verb. Any specific verb may take an infinitive as its object, others may take a gerund, and yet others take either an infinitive or a gerund. At times, even the base form of the verb is used.

▶ He enjoys *working*. [gerund as a complement]

▶ She promised *to fulfill* her part of the contract.
[infinitive as a complement]

▶ The president had the manager *assign* her staff to another project.
[basic verb form as a complement]

To make such distinctions accurately, rely on what you hear native speakers use or what you read. You might also consult a reference book for ESL students.

Adjective Clauses

Because of the variety of ways adjective clauses are constructed in different languages, they can be particularly troublesome for nonnative writers of English. The following guidelines will help you form adjective clauses correctly.

Place an adjective clause directly after the noun it modifies.

▶ The tall woman ˰ is a vice president of the company ˰ who is standing across the room. *[with "who is standing across the room" inserted after "woman" and struck through after "company"]*

The adjective clause *who is standing across the room* modifies *woman*, not *company*, and thus comes directly after *woman*.

Avoid using a relative pronoun with another pronoun in an adjective clause.

▶ The man who ~~he~~ sits at that desk is my boss.

Present Perfect Verb Tense

In general, use the present perfect **tense** to refer to events completed in the past that have some implication for the present.

> PRESENT PERFECT   She *has revised* that report three times.
> [She might revise it again.]

When a specific time is mentioned, however, use the simple past.

> SIMPLE PAST   I *wrote* the letter yesterday morning.
> [The action, *wrote*, does not affect the present.]

Use the present perfect with a *since* or *for* phrase to describe actions that began in the past and continue in the present.

▶ This company *has been* in business *for* seventeen years.

▶ This company *has been* in business *since* 1990.

Present Progressive Verb Tense

The present progressive tense is especially difficult for those whose native language does not use this tense. The present progressive tense is used to describe some action or condition that is ongoing (or in progress) in the present and may continue into the future.

> PRESENT PROGRESSIVE   I *am searching* for an error in the document.
> [The search is occurring now and may continue.]

In contrast, the simple present tense more often relates to habitual actions.

> SIMPLE PRESENT   I *search* for errors in my documents.
> [I regularly search for errors, but I am not necessarily searching now.]

See ESL Tips for Using the Progressive Form in the entry **tense**.

ESL Entries

Most of the entries in this handbook may interest writers of English as a second language; however, the specific entries listed in the Contents by Topic under ESL Trouble Spots address issues that often cause problems.

WEB LINK   English as a Second Language

For Web sites and electronic grammar exercises intended for speakers of English as a second language, see *bedfordstmartins.com/alredtech* and select *Links for Handbook Entries* and *Exercise Central*.

English, varieties of

Written English includes two broad categories: standard and nonstandard. Standard English is used in business, industry, government, education, and all professions. It has rigorous and precise criteria for capitalization, punctuation, spelling, and usage. Nonstandard English does not conform to such criteria; it is often regional in origin, or it reflects the special usages of a particular ethnic or social group. As a result, although nonstandard English may be vigorous and colorful, its usefulness as a means of communication is limited to certain contexts and to people already familiar and comfortable with it in those contexts. It rarely appears in printed material except for special effect. Nonstandard English is characterized by inexact or inconsistent **capitalization**, **punctuation**, **spelling**, diction, and **usage** choices.

Colloquial English

Colloquial English is spoken English or writing that uses words and expressions common to casual conversation. ("We need to get him up to speed.") Colloquial English is appropriate to some kinds of writing (personal letters, notes, some **e-mail**) but not to most workplace writing.

Dialectal English

Dialectal English is a social or regional variety of the language that is comprehensible to people of that social group or region but may be incomprehensible to outsiders. Dialect, which is usually nonstandard English, involves distinct **word choice**, grammatical forms, and pronunciations. For example, residents of southern Louisiana who descended from French colonists speak a dialect often referred to as Cajun.

Localisms

A localism is a regional wording or phrasing. For example, a large sandwich on a long split roll is variously known throughout the United States as a *hero*, *hoagie*, *grinder*, *poor boy*, *submarine*, and *torpedo*. Such words normally should be avoided in workplace writing because not all readers will be familiar with the local meanings.

Slang

Slang is an informal vocabulary composed of **figures of speech** and colorful words used in humorous or extravagant ways. There is no objective test for slang, and many standard words are given slang applications. For instance, slang may be a familiar word used in a new way (*chill* meaning "relax") or it may be a new word (*wonk* meaning "someone who works or studies excessively").

Most slang is short-lived and has meaning only for a narrow **audience**. Sometimes, however, slang becomes standard because the word fills a legitimate need. *Skyscraper* and *date* (as in "go on a date"), for example, were once considered slang expressions. Nevertheless, although slang may be valid in informal and personal writing or fiction, it generally should be avoided in workplace writing. See also **jargon** and **technical writing style**.

environmental impact statements

Environmental impact statements (EISs) are documents that describe the anticipated environmental effects of large-scale construction projects, such as bridges, dams, roads, wind farms, power plants, and waste-disposal sites, before work on them begins. EISs report on both the types (such as ecological, economic, historic, and cultural) and magnitudes of the expected impacts. In addition to highlighting these probable effects, EISs are required to describe alternatives to the proposed project, such as whether to locate a structure or site elsewhere or to take no action. See also **proposals**.

Background and Scope

EISs are required by the National Environmental Policy Act (NEPA), which mandates that federal agencies evaluate the probable environmental effects of projects under their control. NEPA established the Council on Environmental Quality, a board that set guidelines for EISs to assess the following:

- Direct and indirect environmental impact of the proposed project
- Interference with other activities (such as access or traffic)
- Energy and resource requirements
- Conservation and reparation potential
- Preservation of urban, historic, and cultural quality
- Ways to minimize damages

In addressing these topics, EISs are often multivolume documents that may cover such diverse topics as economics, geology, health physics, avian and wetland resources, protected species, noise and transportation impacts, air pollution, cultural and recreational resources, and many more, depending on the size and location of the proposed project.* Many U.S. cities and states also require EISs for planned projects under their jurisdiction that have probable environmental effects.

Audiences

Draft EISs must be available for public comment for a specified period (usually 60 days) before the final EIS is prepared and issued. During the comment period, many individuals and groups evaluate the draft and submit comments. These **audiences** include federal, state, and local officials; local populations near the proposed site of the project; legal experts; industry representatives; labor unions; Native American tribes (when affected); local and national environmental groups; and others. Each of these groups will include economic, legal, engineering, and biology experts, as well as members of the general public. The final EIS responds to comments made on the draft.

Writing Guidelines

EISs are aimed at multiple readers because of the scope of material they address. They are organized into sections by the areas evaluated (such as air, water, soil, socioeconomics, and transportation), with each broad area further subdivided. Socioeconomics, for example, encompasses demography, the local economic basis, local government finances, land use, community services, recreation, and more. Experts and other interested readers will focus on the areas of concern to them. To reach general readers, however, EISs also include an **executive summary** that provides an overview of the scope of the project, areas evaluated, areas of controversy, and issues to be resolved.

Given the diverse fields and backgrounds of EIS readers, writing the document in plain language is essential and is required by federal regulations. If you write a portion of an EIS as a subject expert, you cannot avoid the specialized terms and concepts in your field. However, strive to communicate in language that is clear and accurate. When you finish your draft, ask yourself if a friend or relative unfamiliar with your material could read and understand it. If you don't think so, revise it with that reader in mind. When you use essential technical terminology, either define it in the text or include it in a **glossary** of technical terms. And use consistent terminology, abbreviations, and units for measurements. If you use both metric and English units of measurement, ensure that they appear together consistently throughout: 5 kilometers (3 miles).

---

*For example, an Army Corps of Engineers' Draft Environmental Impact Statement for a proposed wind farm off the coast of Massachusetts ran to 4,000 pages in four volumes. (U.S. Army Corps of Engineers Press Release No. MA-2004-105, Nov. 8, 2004.)

▶ **ETHICS NOTE** Members of the public have a vital interest in proposed activities that may affect their health and safety. They also have the right through the EIS process to comment on these activities before they are under way. The public cannot participate in this process, however, unless they can understand the environmental impact statements designed to inform them of potential effects. To promote public understanding, many states have created plain-language laws. These laws require that legal and other documents be written in clear, understandable language. The federal government also requires that all its new regulations be written in plain language. Keep these aims in mind as you draft material for EISs. ✦

Collaborative Guidelines

Given the size, scope, and areas of expertise necessary to develop EISs, collaboration among technical and legal specialists is essential. To bring order to a project of such magnitude, it's especially important that someone on the writing team review all drafts to ensure that the correct version is being circulated to the team for comment and to impose overall editorial uniformity and consistency. Otherwise, the wrong draft of one or more sections may be inadvertently released to the public, or data in one section may not be expressed consistently with data in another section. See **collaborative writing**.

> **WEB LINK** Environmental Impact Statements
>
> All EISs issued by federal agencies since 2004 are listed at *www.epa.gov/compliance/nepa/eisdata.html*. For a listing of all EISs issued by federal agencies since 1969, go to *www.library.northwestern.edu/transportation/searcheis.html*. Specific EISs at this site are available through interlibrary loan. EISs issued by states are listed at *ceq.eh.doe.gov/nepa/epa/filings/state/usmap.htm*. EISs for counties and municipalities are often available at their Web sites under environmental affairs, environmental protection, or a similar organization. For links to Web sites that contain samples of EISs, see *bedfordstmartins.com/alredtech* and select *Links for Handbook Entries*.

equal / unique / perfect

Logically, *equal* (meaning "having the same quantity or value as another"), *unique* (meaning "one of a kind"), and *perfect* (meaning "a state of highest excellence") are words with absolute meanings and therefore should not be compared. However, colloquial usage of *more* and *most* as **modifiers** of *equal*, *unique*, and *perfect* is so common that an absolute prohibition on such use is impossible.

▶ Our system is more unique [*or* more perfect] than theirs.

Some writers try to overcome the problem by using *more nearly* (*more nearly equal, more nearly unique, more nearly perfect*). When clarity and preciseness are critical, the use of comparative degrees with *equal*, *unique*, and *perfect* can be misleading. It is best to avoid using comparative degrees with absolute terms. See also **comparison**.

MISLEADING	Ours is a *more equal* percentage split than theirs.
PRECISE	Our percentage split is 51–49; theirs is 54–46.

etc.

Etc. is an abbreviation for the Latin *et cetera*, meaning "and others" or "and so on." Therefore, do not use the redundant phrase *and etc.* Likewise, do not use *etc.* at the end of a series introduced by the phrases *such as* and *for example*—those phrases already indicate unnamed items of the same category. Use *etc.* with a logical progression (1, 2, 3, etc.) and when at least two items are named. Do not italicize *etc.*

▶ The sorting machine processes coins (~~for example~~ pennies, nickels, ~~and~~ etc.), and then packages them for redistribution.

Otherwise, avoid *etc.* because the reader may not be able to infer what other items a list might include.

VAGUE	He will bring note pads, paper clips, etc., to the trade show.
CLEAR	He will bring note pads, paper clips, and other office supplies to the trade show.

ethics in writing

Ethics refers to the choices we make that affect others for good or ill. Ethical issues are inherent in writing and speaking because what we write and say can influence others. Further, how we express ideas affects our audience's perceptions of us and our organization's ethical stance. See also **audience**.

▶ **ETHICS NOTE** Obviously, no book can describe how to act ethically in every situation, but this entry describes some typical ethical lapses to watch for during **revision**.* In other entries throughout this book, ethical issues are highlighted using the symbols surrounding this paragraph. ✦

Avoid language that attempts to evade responsibility. Some writers use the passive **voice** because they hope to avoid responsibility or obscure an issue: "It has been decided" [*Who* has decided?] or "Several mistakes were made" [*Who* made them?].

Avoid deceptive language. Do not use words with more than one meaning to circumvent the truth. Consider the company document that stated, "A nominal charge will be assessed for using our facilities." When clients objected that the charge was actually very high, the writer pointed out that the word *nominal* means "the named amount" as well as "very small." In that situation, clients had a strong case in charging that the company was attempting to be deceptive. Various **abstract words**, technical and legal **jargon**, and **euphemisms** are unethical when they are used to mislead readers or to hide a serious or dangerous situation, even though technical or legal experts could interpret them as accurate. See also **word choice**.

Do not deemphasize or suppress important information. Not including information that a reader would want to have, such as potential safety hazards or hidden costs for which a customer might be responsible, is unethical and possibly illegal. Likewise, do not hide information in dense paragraphs with small type and little white space, as is common in credit-card contracts. Use **layout and design** features such as type size, bullets, **lists**, and footnotes to highlight information that is important to readers.

Do not mislead with partial or self-serving information. For example, avoid the temptation to highlight a feature or service that readers would find attractive but that is available only with some product models or at extra cost. (See also **logic errors** and **positive writing**.) Readers could justifiably object that you have given them a false impression to sell them a product or service, especially if you also deemphasize the extra cost or other special conditions.

In general, treat others—individuals, companies, groups—with fairness and with respect. Avoid language that is biased, racist, or sexist or that perpetuates stereotypes. See also **biased language**.

In technical writing, guard against reporting false, fabricated, or plagiarized research and test results. As an author, a technical reviewer, or an editor, your ethical obligation is to correct or point out any misrepresentations of fact before publication, whether the publication is a

---

*Adapted from Brenda R. Sims, "Linking Ethics and Language in the Technical Communication Classroom," *Technical Communication Quarterly* 2.3 (Summer 1993): 285–99.

trade journal article, **test report**, or product handbook. (See also **laboratory reports**.) The stakes of such ethical oversights are high because of the potential risk to the health and safety of others. Those at risk can include unwary consumers and workers injured because of faulty products or unprotected exposure to toxic materials.

Finally, be aware that both **plagiarism** and violations of **copyright** not only are unethical but also can have serious professional and legal consequences for you in the classroom and on the job.

WRITER'S CHECKLIST Writing Ethically

Ask yourself the following questions:

- *Am I willing to take responsibility, publicly and privately, for what I have written?* Make sure you can stand behind what you have written.
- *Is the document or message honest and truthful?* Scrutinize findings and conclusions carefully. Make sure that the data support them.
- *Am I acting in my employer's, my client's, the public's, or my own best long-term interest?* Have someone outside your company review and comment on what you have written.
- *Does the document or message violate anyone's rights?* If information is confidential and you have serious concerns, consider a review by the company's legal staff or an attorney.
- *Am I ethically consistent in my writing?* Apply consistently the principles outlined here and those you have assimilated throughout your life to meet this standard.
- *What if everybody acted or communicated in this way?* If you were the intended reader, consider whether the message would be acceptable and respectful.

If the answers to these questions do not come easily, consider asking a trusted colleague to review and comment on what you have written.

WEB LINK Professional Codes of Ethics

Professional groups often create ethical codes of conduct for their members. One example is the National Society of Professional Engineers (NSPE), which represents individual engineering professionals and licensed engineers. For links to the NSPE's code as well as codes of ethics for other fields, see *bedfordstmartins.com/alredtech* and select *Links for Handbook Entries*.

euphemisms

A euphemism is an inoffensive substitute for a word or phrase that could be distasteful, offensive, or too blunt: *passed away* for *died*; *previously owned* or *preowned* for *used*; *lay off* or *restructure* for *fire* or *terminate* employees. Used judiciously, euphemisms can help you avoid embarrassing or offending someone.

✦ ETHICS NOTE Euphemisms can also hide the facts of a situation (*incident* or *event* for *accident*) or be a form of **affectation** if used carelessly. Avoid them especially in **international correspondence** and other forms of **global communication** where their meanings could be not only confusing but also misleading. See also **ethics in writing**. ✦

everybody / everyone

Both *everybody* and *everyone* are usually considered singular and take singular **verbs** and **pronouns**.

▶ *Everyone* here *leaves* at 4:30 p.m.

▶ *Everybody* at the meeting presented *his or her* individual assessment.

However, the meaning can be obviously plural.

▶ *Everyone* thought the report should be revised, and I really couldn't blame *them*.

Although normally written as one word, write it as two words if you wish to emphasize each individual in a group. ("*Every one* of the team members contributed to this discovery.") See also **agreement**.

exclamation marks

The exclamation mark (!) indicates strong feeling, urgency, elation, or surprise ("Hurry!" "Great!" "Wow!"). (See also **interjections**.) However, it cannot make an argument more convincing, lend force to a weak statement, or call attention to an intended irony.

An exclamation mark can be used after a whole sentence or an element of a sentence.

▶ This meeting—please note it well!—concerns our budget deficit.

When used with **quotation marks**, the exclamation mark goes outside, unless what is quoted is an exclamation.

▶ The paramedic shouted, "Don't touch the victim!" The bystander then, according to a witness, "jumped like a kangaroo"!

In **instructions** and **manuals**, the exclamation mark is often used in text and **visuals** for cautions and warnings ("Danger!" ⚠). See also **emphasis**.

executive summaries

An executive summary consolidates the principal points of a **report** or **proposal**. Executive summaries differ from **abstracts** in that **readers** scan abstracts to decide whether to read the work in full. However, an executive summary may be the only section of a longer work read by many readers, so it must accurately and concisely represent the original document. It should restate the document's **purpose**, **scope**, methods, findings, **conclusions**, and recommendations, as well as summarize how results were obtained or the reasons for the recommendations. Executive summaries tend to be about 10 percent of the length of the documents they summarize and generally follow the same sequence.

Write the executive summary so that it can be read independently of the report or proposal. Executive summaries may occasionally include a figure, table, or footnote if that information is essential to the summary. However, do not refer by number to figures, tables, or references contained elsewhere in the document. See Figure F–5 (pages 205–6), which is an executive summary of a report on the disposition of ethics cases in an aircraft corporation.

WRITER'S CHECKLIST Writing Executive Summaries

- Write the executive summary after you have completed the original document.
- Avoid or define terminology that may not be familiar to your intended **audience**.
- Spell out all uncommon symbols and **abbreviations**.
- Make the summary concise, but do not omit transitional words and phrases (*however, moreover, therefore, for example, next*).
- Include only information discussed in the original document.
- Place the executive summary at the very beginning of the body of the report, as described in **formal reports**.

expletives

An expletive is a word that fills the position of another word, phrase, or clause. *It* and *there* are common expletives.

▶ *It* is certain that this fuel pump will fail at high pressures.

In the example, the expletive *it* occupies the position of subject in place of the real subject, *that this fuel pump will fail at high pressures*. Expletives are sometimes necessary to avoid **awkwardness**, but they are commonly overused, and most sentences can be better stated without them.

▶ ~~There were many~~ *Many* files lost *were* when we converted to the new server.

In addition to its grammatical use, the word *expletive* means an exclamation or oath, especially one that is obscene.

explicit / implicit

An *explicit* statement is one expressed directly, with precision and clarity.

▶ He gave us *explicit* directions to the Wausau facility.

An *implicit* meaning is one that is not directly expressed.

▶ Although the CEO did not mention the lawsuit directly, the company's commitment to ethical practices was *implicit* in her speech.

exposition

Exposition, or *expository writing*, informs **readers** by presenting facts and ideas in direct and concise language; it usually relies less on colorful or figurative language than writing meant to be expressive or persuasive. Expository writing attempts to explain to readers what the subject is, how it works, and how it relates to something else. Exposition is aimed at the readers' understanding rather than at their imagination or emotions; it is a sharing of the writer's knowledge. Exposition aims to provide accurate, complete information and to analyze it for the readers. See also **audience**.

Because it is the most effective form of discourse for explaining difficult subjects, exposition is widely used in **reports**, **memos**, and other types of technical writing. To use exposition effectively, you must have a thorough knowledge of your subject. As with all writing, how much of that knowledge you convey depends on the reader's needs and your **purpose**.

fact

Expressions containing the word *fact* ("due to the *fact* that," "except for the *fact* that," "as a matter of *fact*," or "because of the *fact* that") are often wordy substitutes for more accurate terms.

▶ *Because*
~~Due to the fact that~~ the technical staff has a high turnover rate, our training program has suffered.

Do not use the word *fact* to refer to matters of judgment or opinion.

▶ *In my opinion,*
~~It is a fact that~~ our research has improved because we now have a capable technical staff.

The word *fact* is, of course, valid when facts are what is meant.

▶ Our tests uncovered numerous facts to support your conclusion.

See also **conciseness** and **logic errors**.

FAQs (Frequently Asked Questions)

An FAQ is a **list** of questions, paired with their answers, that readers will likely ask about products, services, or other information presented on a Web site or in customer-oriented documents. By presenting commonly sought information in one place, an FAQ saves readers from searching through an entire document or Web site to find what they need.

A good FAQ can help create a positive impression with customers or clients because you are acknowledging that their time is valuable. An FAQ also helps a company spend less time answering phone calls and **e-mail** questions by anticipating customer needs and providing important information in a simple, a logical, and an organized format. However, an FAQ is not a substitute for solving problems with a product or service. If customers are experiencing numerous problems because of a product design or programming flaw, for example, you need to work with your company's product developers to correct the problem rather than try to solve every potential problem within an FAQ.

Questions to Include

The first step in writing an FAQ is to determine the questions your customers may ask about your organization, products, or services. Put yourself in your customers' place and keep in mind that even if you view a question as too simple, some customers will find value in the answer to that question.

A good place to start is by asking customer service representatives what questions customers typically ask and the appropriate answers to those questions. If customers frequently ask about company stock information and request annual reports, for example, your FAQ list could include the question "How do I obtain a copy of your latest annual report?" This question can be followed with a brief answer that includes the name, phone number, and e-mail address of the person who distributes the annual reports. For a Web site, you can also provide a link to a copy of the current annual report that can be downloaded. See also **writing for the Web**.

Organization

After developing the list of questions and answers, organize the list so that readers can find the information they need quickly and easily. Study other FAQs that your readers might review for products or services similar to yours. Use these FAQs to find answers to questions: Can you find answers quickly, or do you need to scroll through many pages to find them? Consider how the questions are organized: Are they separated into logical categories or listed in random order? Is it easy to differentiate the question from the answer? Do the answers provide too little or too much information?*

List your questions in decreasing **order of importance** so that readers can obtain the most important information first. See **order-of-importance method of development**. If you have a number of questions that are related to a specific topic, such as investor relations, product returns, or completing forms, group them into categories and identify each category with a heading, such as "Investor Relations," "Shipping," and "Forms." You may also want to create a **table of contents** at the top of the FAQ page so that readers can find quickly the topics relevant to their interests.

---

*For examples of FAQs, search with QueryCAT at *www.querycat.com*, which offers the "largest database of frequently asked questions."

Placement

The location of your FAQ should enable readers to find answers quickly. For Web sites, an FAQ page is usually linked from the homepage either in a directory or with a text link for easy access. In small printed documents, such as **brochures**, FAQs are usually highlighted and placed after the standard information in the body of the document.

WRITER'S CHECKLIST — Developing an FAQ

- *Focus on your reader.* Write your questions and answers from a **"you" viewpoint** and with a positive, conversational **tone**.
- *Separate long FAQ lists.* Group related questions under topic **headings**. For long online FAQs, consider listing only questions with links to separate pages, each containing an individual question and answer.
- *Distinguish questions from answers.* Use boldface for questions and use white space to separate questions from answers. Use sparingly multiple colors, **italics**, or other formatting styles that can make the list difficult to read.
- *Keep questions and answers concise.* Keep questions brief and answers to no more than one paragraph. If a question has a long answer, add a link to a separate Web page or refer to an appropriate page number in a printed document.
- *Keep the list updated.* Review and update FAQs at least monthly—or more frequently if your content changes often. Make it possible for your customers to submit questions they would like to see added to the FAQ list.

---

fax (*see* selecting the medium)

feasibility reports

When organizations consider a new project—developing a new product or service, expanding a customer base, purchasing equipment, or moving operations—they first try to determine the project's chances for success. A feasibility report presents evidence about the practicality of a proposed project based on specific criteria. It answers such questions as the following: Is new construction or development necessary? Is sufficient staff available? What are the costs? Is funding available? What are the legal ramifications? Based on the findings of this analysis, the report offers logical conclusions and recommends whether the project should be carried out. When feasibility reports stress specific steps that should be taken as a result of a study of a problem or an issue, they are often referred to as *recommendation reports*. In the condensed feasibility report shown in Figure F–1, a consultant conducts a study to determine how to upgrade a company's computer system and Internet capability.

Introduction
The purpose of this report is to determine which of two proposed options would best enable Darnell Business Forms Corporation to upgrade its file servers and its Internet capacity to meet its increasing data and communication needs. . . .

Background. In October 2008, the Information Development Group put the MACRON System into operation. Since then, the volume of processing transactions has increased fivefold (from 1,000 to 5,000 updates per day). This increase has severely impaired system response time; in fact, average response time has increased from 10 seconds to 120 seconds. Further, our new Web-based client-services system has increased exponentially the demand for processing speed and access capacity.

Scope. We have investigated two alternative solutions to provide increased processing capacity: (1) purchase of an additional Aurora processor to supplement the one in operation, and (2) purchase of an Icardo 60 with expandable peripherals to replace the Aurora processor currently in operation. The two alternatives are evaluated here, according to both cost and expanded capacity for future operations.

Additional Aurora Processor
Purchasing a second Aurora processor would require increased annual maintenance costs, salary for a second computer specialist, increased energy costs, and a one-time construction cost for necessary remodeling and installing Internet connections.

Annual maintenance costs	$35,000
Annual costs for computer specialist	75,000
Annual increased energy costs	7,500
Total annual operating costs	$117,500
Construction cost (one-time)	50,000
Total first-year costs	$167,500

The installation and operation of another Aurora processor are expected to produce savings in system reliability and readiness.

FIGURE F–1. Feasibility Report

System Reliability. An additional Aurora would reduce current downtime periods from four to two per week. Downtime recovery averages 30 minutes and affects 40 users. Assuming that 50 percent of users require system access at a given time, we determined that the following reliability savings would result:

2 downtimes × 0.5 hours × 40 users × 50% × $50/hour overtime × 52 weeks = $52,000 annual savings.

[The feasibility report would also discuss the second option—purchase of the Icardo 60 and its long-term savings.]

Conclusion

A comparison of costs for both systems indicates that the Icardo 60 would cost $2,200 more in first-year costs.

	Aurora	Icardo 60
Net additional operating costs	$56,300	$84,000
One-time construction costs	50,000	24,500
First-year total	$106,300	$108,500

Installation of an additional Aurora processor would permit the present information-processing systems to operate relatively smoothly and efficiently. It would not, however, provide the expanded processing capacity that the Icardo 60 processor would for implementing new subsystems required to increase processing speed and Internet access.

Recommendation

The Icardo 60 processor should be purchased because of the long-term savings and because its additional capacity and flexibility will allow for greater expansion in the future.

FIGURE F–1. Feasibility Report (*continued*)

Before beginning to write a feasibility report, analyze the needs of the **audience** as well as the **context** and **purpose** of the study. Then write a purpose statement, such as "The purpose of this study is to determine the feasibility of expanding our Pacific Rim operations," to guide you or a collaborative team. See also **brainstorming** and **collaborative writing**.

Report Sections

Every feasibility report should contain an **introduction**, a body, a **conclusion**, and a recommendation. See also **proposals** and **formal reports**.

Introduction. The introduction states the purpose of the report, describes the circumstances that led to the report, and includes any pertinent background information. It may also discuss the **scope** of the report, any procedures or methods used in the analysis of alternatives, and any limitations of the study.

Body. The body of the report presents a detailed review of the alternatives for achieving the goals of the project. Examine each option according to specific criteria, such as cost and financing, availability of staff, and other relevant requirements, identifying the subsections with **headings** to guide readers.

Conclusion. The conclusion interprets the available options and leads to one option as the best or most feasible.

Recommendation. The recommendation section clearly presents the writer's (or team's) opinion on which alternative best meets the criteria as summarized in the conclusion.

WEB LINK | Feasibility Reports

For links to full feasibility reports, see *bedfordstmartins.com/alredtech* and select *Links for Handbook Entries*.

---

female / male

The terms *female* and *male* are usually restricted to scientific, legal, and medical contexts (a *female* suspect, a *male* patient). Keep in mind that the terms sound cold and impersonal. *Girl, woman,* and *lady* or *boy, man,* and *gentleman* may be acceptable substitutes in other **contexts**, but be sensitive to their connotations involving age, dignity, and social position. See also **biased language** and **connotation / denotation**.

few / a few

In certain contexts, *few* carries more negative overtones than does the phrase *a few*.

NEGATIVE The report contains *few* helpful ideas.
POSITIVE The report contains *a few* helpful ideas.

fewer / less

Fewer refers to items that can be counted (count **nouns**). ("*Fewer* employees retired than we expected.") *Less* refers to mass quantities or amounts (mass nouns). ("Because we had *less* rain this year, the crop yield decreased.") See also **English as a second language**.

figuratively / literally

Literally means "actually" and is often confused with *figuratively*, which means "metaphorically." To say that someone "*literally* turned green with envy" would mean that the person actually changed color.

- In the winner's circle the jockey was, *figuratively* speaking, ten feet tall.
- When he said, "Let's bury our competitors," he did not mean it *literally*.

Avoid the use of *literally* to reinforce the importance of something.

- She was ~~literally~~ the best of the applicants.

See also **intensifiers**.

figures of speech

A figure of speech is an imaginative expression that often compares two things that are basically not alike but have at least one thing in common. For example, if a device is cone-shaped and has an opening at the narrow end, you might say that it looks like a volcano.

Figures of speech can clarify the unfamiliar by relating a new concept to one with which readers are familiar. In that respect, they help establish understanding between the specialist and the nonspecialist. (See **audience**.) Figures of speech can help translate the abstract into the concrete; in the process of doing so, they can also make writing more colorful and graphic. (See also **abstract / concrete words**.) A figure of speech must make sense, however, to achieve the desired effect.

> ILLOGICAL Without the fuel of tax incentives, our economic engine would operate less efficiently.
> [It would not operate at all without fuel.]

Figures of speech also must be consistent to be effective.

> ▶ We must get our research program *back on course*, and we are counting on you to ~~carry the ball~~. *steer the effort.*

A figure of speech should not overshadow the point the writer is trying to make. In addition, it is better to use no figure of speech at all than to use a trite one. A surprise that comes "like a bolt out of the blue" seems stale and not much of a surprise at all. See also **clichés**.

Types of Figures of Speech

Analogies are comparisons that show the ways in which two objects or concepts are similar, often used to make one of them easier to understand. The following example explains a computer search technique by comparing it to the use of keywords in a dictionary.

> ▶ The search technique used in *indexed sequential processing* is similar to a search technique you might use to find the page on which a particular word is located in a dictionary. You might scan the keywords located at the top of each dictionary page that identify the first and last words on each page until you find the keywords that encompass the word you seek. Indexed sequential processing works the same way with computer files.

Hyperboles are gross exaggerations used to achieve an effect or **emphasis**.

> ▶ We were dead after working all night on the report.

Litotes are understatements, for emphasis or effect, achieved by denying the opposite of the point you are making.

> ▶ Over 1,600 pages is no small size for a book.

Metaphors are figures of speech that point out similarities between two things by treating them as though they were the same thing.

- The astronaut's *umbilical cord* carries life-sustaining oxygen for space-walking.

Metonyms are figures of speech that use one aspect of a thing to represent it, such as *the blue* for the sky and *wheels* for a car.

- The economist predicted a decrease in *hard-hat* jobs.

Personification is a figure of speech that attributes human characteristics to nonhuman things or abstract ideas. We might refer, for example, to the *birth* of a planet or apply emotions to machines.

- She said that she was frustrated with the *stubborn* security system.

Similes are direct comparisons of two essentially unlike things, linking them with the word *like* or *as*.

- Reconstructing the plane's fuselage following the accident was *like piecing together a jigsaw puzzle*.

Avoid figures of speech in **global communication** and **international correspondence** because people in other cultures may translate figures of speech literally and be confused by their meanings.

fine

When used in expressions such as "I feel *fine*" or "a *fine* day," *fine* is colloquial and, like the word *nice*, is often too vague for technical writing. Use the word *fine* to mean "refined," "delicate," or "pure."

- A *fine* film of oil covered the surface of the water.
- *Fine* crystal is made in Austria.
- The Court made a *fine* distinction between the two statutes.

first / firstly

First and *firstly* are both adverbs. Avoid *firstly* in favor of *first*, which sounds less stiff than *firstly*. The same is true of other ordinal **numbers**, such as *second*, *third*, and so on.

flammable / inflammable / nonflammable

Both *flammable* and *inflammable* mean "capable of being set on fire." Because the *in-* **prefix** usually causes the base word to take its opposite meaning (*incapable*, *incompetent*), use *flammable* instead of *inflammable* to avoid possible misunderstanding. ("The cargo of gasoline is *flammable*.") *Nonflammable* is the opposite, meaning "not capable of being set on fire." ("The asbestos suit was *nonflammable*.")

flowcharts

A flowchart is a diagram using symbols, words, or pictures to show the stages of a process in sequence from beginning to end. A flowchart provides an overview of a process and allows the **reader** to identify its essential steps quickly and easily. Flowcharts can take several forms. The steps might be represented by labeled blocks, as shown in Figure F–2; pictorial symbols, as shown in Figure F–3; or ISO (International Organization for Standardization) symbols, as shown in Figure F–4.

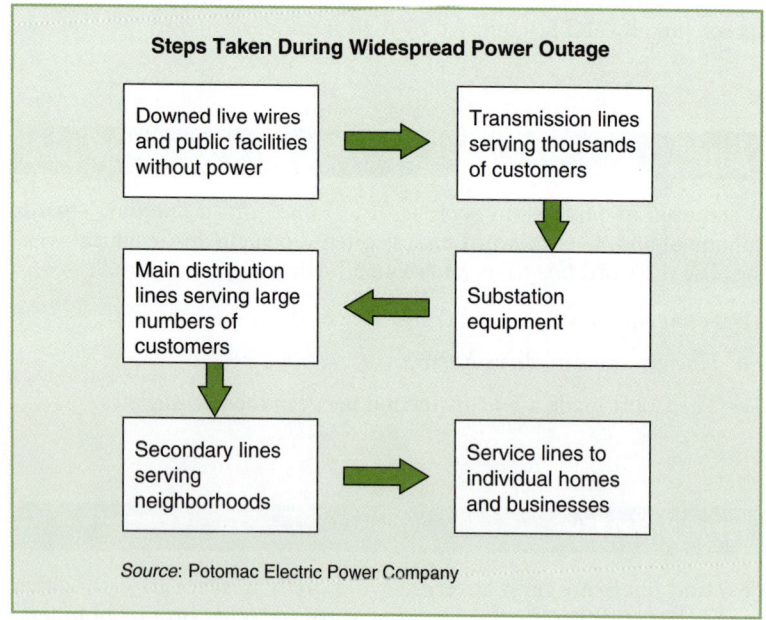

FIGURE F–2. Flowchart Using Labeled Blocks (Depicting Electric Utility Power Restoration Process)

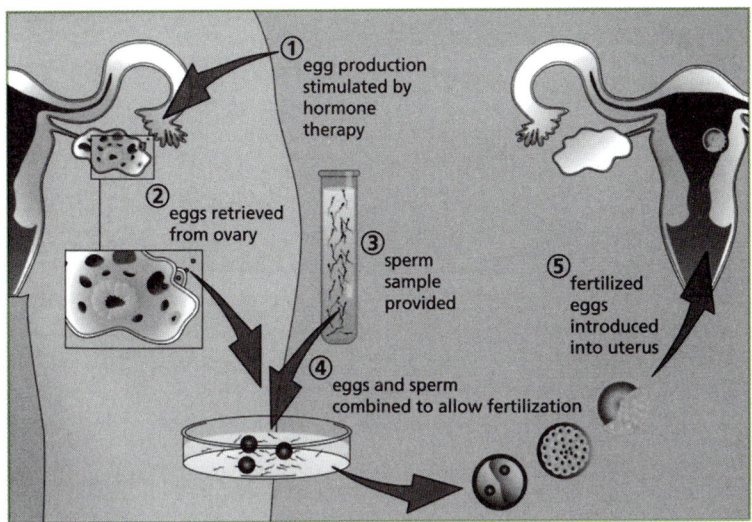

FIGURE F–3. Flowchart Using Pictorial Symbols. *Source*: FDA Consumer, Nov.—Dec. 2004. U.S. Food and Drug Administration.

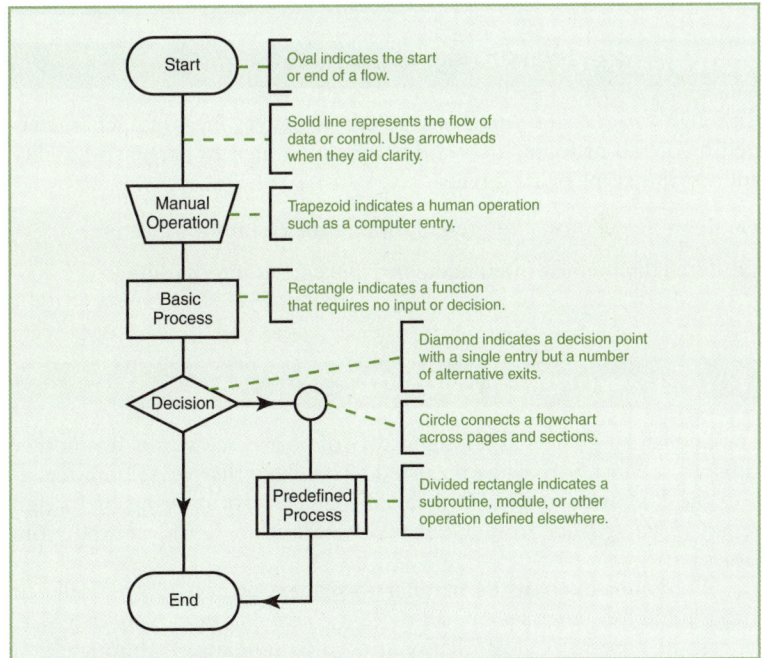

FIGURE F–4. Common ISO Flowchart Symbols (with Annotations)

WRITER'S CHECKLIST: Creating Flowcharts

- Label each step in the process or identify each step with labeled blocks, pictorial representations, or standardized symbols.
- Follow the standard flow directions: left to right and top to bottom. Indicate any nonstandard flow directions with arrows.
- Include a key (or callouts) if the flowchart contains symbols your **audience** may not understand.
- Use standardized symbols for flowcharts that document computer programs and other information-processing procedures, as detailed in *Information Processing—Documentation Symbols and Conventions for Data, Program and System Flowcharts, Program Network Charts, and System Resources Charts*, ISO 5807-1985 (E).

For advice on integrating flowcharts into your text, see **visuals**. See also **global graphics**.

footnotes (*see* documenting sources)

forceful / forcible

Although *forceful* and *forcible* are both **adjectives** meaning "characterized by or full of force," *forceful* is usually limited to persuasive ability and *forcible* to physical force.

▶ John made a *forceful* presentation at the committee meeting.
▶ Firefighters must often make *forcible* entries into buildings.

foreign words in English

The English language has a long history of borrowing words from other languages. Most borrowing occurred so long ago that we seldom recognize the borrowed terms (also called *loan words*) as being of foreign origin (*kindergarten* from German, *animal* from Latin, *church* from Greek).

Words not likely to be familiar to **readers** or not fully assimilated into the English language are set in **italics** (*sine qua non*, *coup de grâce*, *in res*, *in camera*). Most dictionaries offer guidance, although you should also be guided by the **context**. Even when foreign words have been fully assimilated, they often retain their diacritical marks (cliché, résumé, vis-à-vis). As foreign words are absorbed into English, their plural forms give way to English plurals (*agenda* becomes *agendas* and *formulae* becomes *formulas*).

Generally, foreign expressions should be used only if they serve a real need. (See also **e.g. / i.e.** and **etc.**) The overuse of foreign words in an attempt to impress your reader or achieve elegance is **affectation**. Effective communication can be accomplished only if your readers understand what you write. So choose foreign expressions only when they make an idea clearer or there is no English substitute (*Schadenfreude* for "pleasure taken from someone else's misfortune").

foreword / forward

Although the pronunciation is the same, the spellings and meanings of these two words are quite different. The word *foreword* is a **noun** meaning "introductory statement at the beginning of a book or other work."

▶ The director wrote a *foreword* for the proposal.

The word *forward* is an **adjective** or **adverb** meaning "at or toward the front."

▶ Sliding the throttle to the *forward* position [adjective] will cause the boat to move *forward*. [adverb]

formal reports

Formal reports are usually written accounts of major projects that require substantial **research**, and they often involve more than one writer. See also **collaborative writing**.

Most formal reports are divided into three primary parts—front matter, body, and back matter—each of which contains a number of elements. The number and arrangement of the elements may vary depending on the subject, the length of the report, and the kinds of material covered. Further, many organizations have a preferred style for formal reports and furnish guidelines for report writers to follow. If you are not required to follow a specific style, use the **format** recommended in this entry. The following list includes most of the elements a formal report might contain, in the order they typically appear. (The items shown with page numbers appear in the sample formal report on pages

201–17.) Often, a **cover letter** or **memo** precedes the front matter and identifies the report by title, the person or persons to whom it is being sent, the reason it was written, and any content that the **audience** considers important, as shown on page 201.

FRONT MATTER
Title Page (202)
Abstract (203)
Table of Contents (204)
List of Figures
List of Tables
Foreword
Preface
List of Abbreviations and Symbols

BODY
Executive summary (205)
Introduction (207)
Text (including headings) (210)
Conclusions (215)
Recommendations (216)
Explanatory Notes
References (or Works Cited) (217)

BACK MATTER
Appendixes
Bibliography
Glossary
Index

Front Matter

The front matter serves several functions: It gives readers a general idea of the writer's **purpose**, it gives an overview of the type of information in the report, and it lists where specific information is covered in the report. Not all formal reports include every element of front matter described here. A title page and table of contents are usually mandatory, but the **scope** of the report and its **context** as well as the intended audience determine whether the other elements are included.

Title Page. Although the formats of title pages may vary, they often include the following items:

- *The full title of the report.* The title describes the topic, scope, and purpose of the report, as described in **titles**.

- *The name of the writer(s), principal investigator(s), or compiler(s).* Sometimes contributors identify themselves by their job title in the organization or by their tasks in contributing to the report (Olivia Jones, Principal Investigator).
- *The date or dates of the report.* For one-time reports, the date shown is the date the report is distributed. For reports issued periodically (monthly, quarterly, or yearly), the subtitle shows the period that the report covers and the distribution date is shown elsewhere on the title page, as shown in Figure F–5 on page 202.
- *The name of the organization for which the writer(s) works.*
- *The name of the organization to which the report is being submitted.* This information is included if the report is written for a customer or client.

The title page should not be numbered, as in the example on page 202, but it is considered page i. The back of the title page, which is left blank and unnumbered, is considered page ii, and the abstract falls on page iii. The body of the report begins with Arabic number 1, and a new chapter or large section typically begins on a new right-hand (odd-numbered) page. Reports with printing on only one side of each sheet can be numbered consecutively regardless of where new sections begin. Center page numbers at the bottom of each page throughout the report.

Abstract.   An **abstract**, which normally follows the title page, highlights the major points of the report, as shown on page 203, enabling readers to decide whether to read the report.

Table of Contents.   A **table of contents** lists all the major sections or **headings** of the report in their order of appearance, as shown on page 204, along with their page numbers.

List of Figures.   All **visuals** contained in the report—**drawings**, **photographs**, **maps**, charts, and **graphs**—are labeled as figures. When a report contains more than five figures, list them, along with their page numbers, in a separate section, beginning on a new page immediately following the table of contents. Number figures consecutively with Arabic numbers.

List of Tables.   When a report contains more than five **tables**, list them, along with their titles and page numbers, in a separate section immediately following the list of figures (if there is one). Number tables consecutively with Arabic numbers.

Foreword. A foreword is an optional introductory statement about a formal report or publication that is written by someone other than the author(s). The foreword author is usually an authority in the field or an executive of the organization sponsoring the report. That author's name and affiliation appear at the end of the foreword, along with the date it was written. The foreword generally provides background information about the publication's significance and places it in the context of other works in the field. The foreword precedes the preface when a work has both.

Preface. The preface, another type of optional introductory statement, is written by the author(s) of the formal report. It may announce the work's purpose, scope, and context (including any special circumstances leading to the work). A preface may also specify the audience for a work, contain acknowledgments of those who helped in its preparation, and cite permission obtained for the use of copyrighted works. See also **copyright**.

List of Abbreviations and Symbols. When the report uses numerous **abbreviations** and symbols that readers may not be able to interpret, the front matter may include a section that lists symbols and abbreviations with their meanings.

Body

The body is the section of the report that provides context for the report, describes in detail the methods and procedures used to generate the report, demonstrates how results were obtained, describes the results, draws conclusions, and, if appropriate, makes recommendations.

Executive Summary. The body of the report begins with the **executive summary**, which provides a more complete overview of the report than an abstract does. See an example on pages 205–6 and review the entry cross-referenced above.

Introduction. The **introduction** gives readers any general information, such as the report's purpose, scope, and context necessary to understand the detailed information in the rest of the report (see pages 207–9).

Text. The text of the body presents, as appropriate, the details of how the topic was investigated, how a problem was solved, what alternatives were explored, and how the best choice among them was selected. This information is enhanced by the use of visuals, tables, and references that both clarify the text and persuade the reader. See also **persuasion**.

Conclusions. The **conclusions** section pulls together the results of the research and interprets the findings of the report, as shown on pages 215–16.

Recommendations. Recommendations, which are sometimes combined with the conclusions, state what course of action should be taken based on the earlier arguments and conclusions of the study, as are shown on pages 215–16.

Explanatory Notes. Occasionally, reports contain notes that amplify terms or points for some readers that might be a distraction for others. If such notes are not included as footnotes on the page where the term or point appears, they may appear in a "Notes" section at the end of the report.

References (or Works Cited). A list of references or works cited appears in a separate section if the report refers to or quotes directly from printed or online research sources. If your employer has a preferred reference style, follow it; otherwise, use one of the guidelines provided in the entry **documenting sources**. For a relatively short report, place a references or works-cited section at the end of the body of the report, as shown on page 217. For a report with a number of sections or chapters, place references or works cited at the end of each major section or chapter. In either case, title the reference or works-cited section as such and begin it on a new page. If a particular reference appears in more than one section or chapter, repeat it in full in each appropriate reference section. See also **plagiarism** and **quotations**.

Back Matter

The back matter of a formal report contains supplementary material, such as where to find additional information about the topic (bibliography), and expands on certain subjects (appendixes). Other back-matter elements define special terms (glossary) and provide information on how to easily locate information in the report (index). For very large formal reports, back-matter sections may be individually numbered (Appendix A, Appendix B).

Appendixes. An **appendix** clarifies or supplements the report with information that is too detailed or lengthy for the primary audience but is relevant to secondary audiences.

Bibliography. A **bibliography** lists alphabetically all of the sources that were consulted to prepare the report — not just those cited — and suggests additional resources that readers might want to consult.

Glossary. A **glossary** is an alphabetical list of specialized terms used in the report and their definitions.

Index. An index is an alphabetical list of all the major topics and subtopics discussed in the report. It cites the page numbers where discussion of each topic can be found and allows readers to find information on topics quickly and easily. The index is always the final section of a report. See also **indexing**.

> **DIGITAL TIP**
>
> **Creating Styles and Templates**
>
> In most word-processing programs, you can create styles—sets of formatting characteristics for text elements such as headings, paragraphs, and lists—that allow you to automate much of the formatting of your report. Once you specify your styles, save them as a template and use it each time you create a formal report. For step-by-step instructions, see *bedfordstmartins.com/alredtech* and select *Digital Tips*, "Creating Styles and Templates."

Sample Formal Report

Figure F–5 shows the typical sections of a formal report. Keep in mind that the number and arrangement of the elements vary, depending on the context, especially the requirements of an organization or a client.

CGF Aircraft Corporation Memo

To: Members of the Ethics and Business Conduct Committee

From: Susan Litzinger, Director of Ethics and Business Conduct *SL*

Date: March 4, 2009

Subject: Reported Ethics Cases, 2008

Enclosed is "Reported Ethics Cases: Annual Report, 2008." This report, required by CGF Policy CGF-EP-01, contains a review of the ethics cases handled by CGF ethics officers and managers during 2008, the first year of our Ethics Program. *[Identifies topic]*

The ethics cases reported are analyzed according to two categories: (1) major ethics cases, or those potentially involving serious violations of company policy or illegal conduct, and (2) minor ethics cases, or those that do not involve serious policy violations or illegal conduct. The report also examines the mode of contact in all of the reported cases and the disposition of the substantiated major ethics cases. *[Briefly summarizes content]*

It is my hope that this report will provide the Committee with the information needed to assess the effectiveness of the first year of CGF's Ethics Program and to plan for the coming year. Please let me know if you have any questions about this report or if you need any further information. I may be reached at (555) 211-2121 and by e-mail at sl@cgf.com. *[Offers contact information]*

Enc.

FIGURE F–5. Formal Report (Cover Memo). Reprinted and adapted by permission of Susan Litzinger, a student at Pennsylvania State University, Altoona.

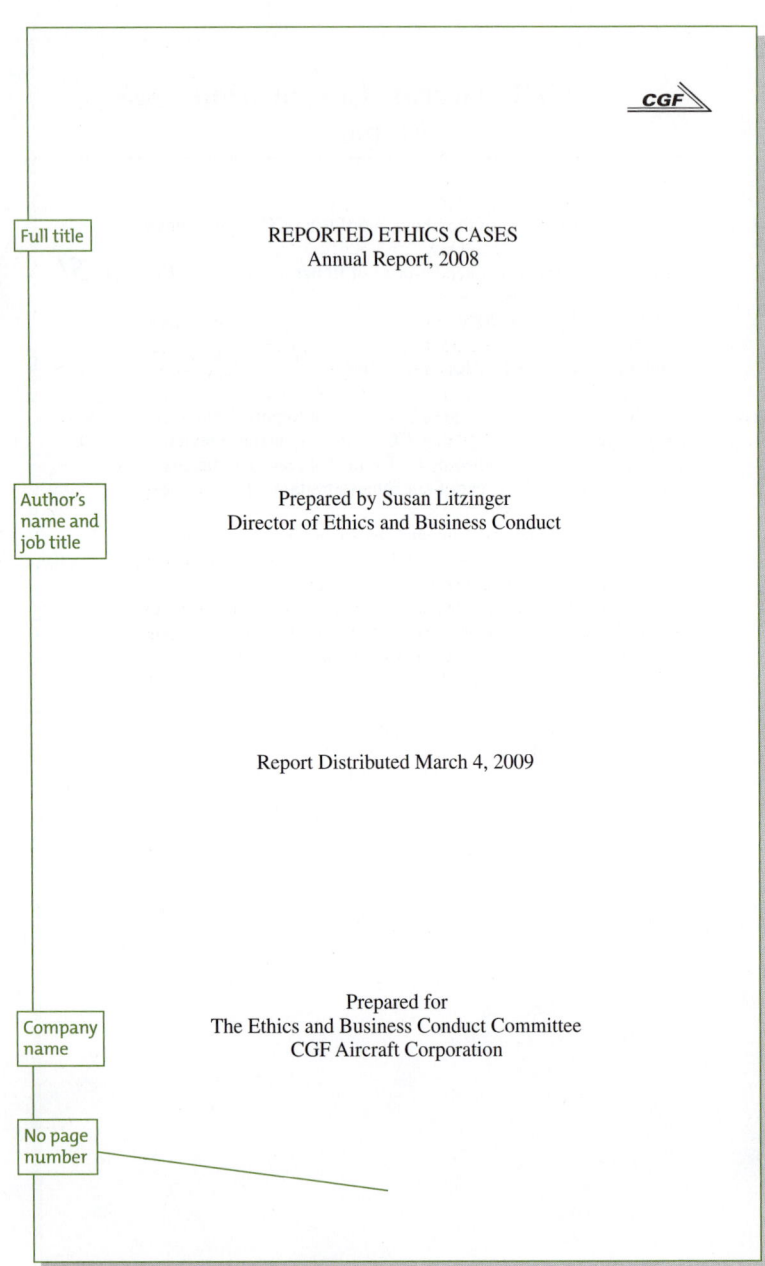

FIGURE F–5. Formal Report (*continued*) (Title Page)

Reported Ethics Cases — 2008

ABSTRACT

This report examines the nature and disposition of 3,458 ethics cases handled companywide by CGF Aircraft Corporation's ethics officers and managers during 2008. The purpose of this annual report is to provide the Ethics and Business Conduct Committee with the information necessary for assessing the effectiveness of the Ethics Program's first year of operation. Records maintained by ethics officers and managers of all contacts were compiled and categorized into two main types: (1) major ethics cases, or cases involving serious violations of company policies or illegal conduct, and (2) minor ethics cases, or cases not involving serious policy violations or illegal conduct. This report provides examples of the types of cases handled in each category and analyzes the disposition of 30 substantiated major ethics cases. Recommendations for planning for the second year of the Ethics Program are (1) continuing the channels of communication now available in the Ethics Program, (2) increasing financial and technical support for the Ethics Hotline, (3) disseminating the annual ethics report in some form to employees to ensure employee awareness of the company's commitment to uphold its Ethics Policies and Procedures, and (4) implementing some measure of recognition for ethical behavior to promote and reward ethical conduct.

FIGURE F–5. Formal Report (*continued*) (Abstract)

Reported Ethics Cases — 2008

TABLE OF CONTENTS

ABSTRACT iii

EXECUTIVE SUMMARY 1

INTRODUCTION 3

    Ethics and Business Conduct Policies and Procedures 3
    Confidentiality Issues 4
    Documentation of Ethics Cases 4
    Major/Minor Category Definition and Examples 5

ANALYSIS OF REPORTED ETHICS CASES 6

    Reported Ethics Cases, by Major/Minor Category 6
    Major Ethics Cases 6
    Minor Ethics Cases 8
    Mode of Contact 9

CONCLUSIONS AND RECOMMENDATIONS 11

WORKS CITED 13

iv

FIGURE F–5. Formal Report (*continued*) (Table of Contents)

Reported Ethics Cases—2008

EXECUTIVE SUMMARY

This report examines the nature and disposition of the 3,458 ethics cases handled by the CGF Aircraft Corporation's ethics officers and managers during 2008. The purpose of this report is to provide CGF's Ethics and Business Conduct Committee with the information necessary for assessing the effectiveness of the first year of the company's Ethics Program.

[States purpose]

Effective January 1, 2008, the Ethics and Business Conduct Committee (the Committee) implemented a policy and procedures for the administration of CGF's new Ethics Program. The purpose of the Ethics Program, established by the Committee, is to "promote ethical business conduct through open communication and compliance with company ethics standards." The Office of Ethics and Business Conduct was created to administer the Ethics Program. The director of the Office of Ethics and Business Conduct, along with seven ethics officers throughout the corporation, was given the responsibility for the following objectives:

[Provides background information]

- Communicate the values and standards for CGF's Ethics Program to employees.

- Inform employees about company policies regarding ethical business conduct.

- Establish companywide channels for employees to obtain information and guidance in resolving ethics concerns.

- Implement companywide ethics-awareness and education programs.

Employee accessibility to ethics information and guidance was available through managers, ethics officers, and an ethics hotline.

Major ethics cases were defined as those situations potentially involving serious violations of company policies or illegal conduct. Examples of major ethics cases included cover-up of defective workmanship or use of defective parts in products; discrimination in hiring and promotion; involvement in monetary or other kickbacks; sexual harassment; disclosure of proprietary or company information; theft; and use of corporate Internet resources for inappropriate purposes, such as conducting personal business, gambling, or access to pornography.

[Describes scope]

1

FIGURE F–5. Formal Report (*continued*) (Executive Summary)

Reported Ethics Cases—2008

Minor ethics cases were defined as including all reported concerns not classified as major ethics cases. Minor ethics cases were classified as informational queries from employees, situations involving coworkers, and situations involving management.

The effectiveness of CGF's Ethics Program during the first year of implementation is most evidenced by (1) the active participation of employees in the program and the 3,458 contacts employees made regarding ethics concerns through the various channels available to them, and (2) the action taken in the cases reported by employees, particularly the disposition of the 30 substantiated major ethics cases. Disseminating information about the disposition of ethics cases, particularly information about the severe disciplinary actions taken in major ethics violations, sends a message to employees that unethical or illegal conduct will not be tolerated.

Based on these conclusions, recommendations for planning the second year of the Ethics Program are (1) continuing the channels of communication now available in the Ethics Program, (2) increasing financial and technical support for the Ethics Hotline, the most highly used mode of contact in the ethics cases reported in 2008, (3) disseminating this report in some form to employees to ensure their awareness of CGF's commitment to uphold its Ethics Policies and Procedures, and (4) implementing some measure of recognition for ethical behavior, such as an "Ethics Employee of the Month" award to promote and reward ethical conduct.

2

FIGURE F–5. Formal Report (*continued*) (Executive Summary)

Reported Ethics Cases—2008

INTRODUCTION

This report examines the nature and disposition of the 3,458 ethics cases handled companywide by CGF's ethics officers and managers during 2008. The purpose of this report is to provide the Ethics and Business Conduct Committee with the information necessary for assessing the effectiveness of the first year of CGF's Ethics Program. Recommendations are given for the Committee's consideration in planning for the second year of the Ethics Program.

Ethics and Business Conduct Policies and Procedures

Effective January 1, 2008, the Ethics and Business Conduct Committee (the Committee) implemented Policy CGF-EP-01 and Procedure CGF-EP-02 for the administration of CGF's new Ethics Program. The purpose of the Ethics Program, established by the Committee, is to "promote ethical business conduct through open communication and compliance with company ethics standards" (CGF, "Ethics and Conduct").

The Office of Ethics and Business Conduct was created to administer the Ethics Program. The director of the Office of Ethics and Business Conduct, along with seven ethics officers throughout CGF, was given the responsibility for the following objectives:

- Communicate the values, standards, and goals of CGF's Ethics Program to employees.
- Inform employees about company ethics policies.
- Provide companywide channels for employee education and guidance in resolving ethics concerns.
- Implement companywide programs in ethics awareness, education, and recognition.
- Ensure confidentiality in all ethics matters.

Employee accessibility to ethics information and guidance became the immediate and key goal of the Office of Ethics and Business Conduct in its first year of operation. The following channels for contact were set in motion during 2008:

FIGURE F–5. Formal Report *(continued)* (Introduction)

Reported Ethics Cases — 2008

- Managers throughout CGF received intensive ethics training; in all ethics situations, employees were encouraged to go to their managers as the first point of contact.
- Ethics officers were available directly to employees through face-to-face or telephone contact, to managers, to callers using the Ethics Hotline, and by e-mail.
- The Ethics Hotline was available to all employees, 24 hours a day, seven days a week, to anonymously report ethics concerns.

Confidentiality Issues

CGF's Ethics Policy ensures confidentiality and anonymity for employees who raise genuine ethics concerns. Procedure CGF-EP-02 guarantees appropriate discipline, up to and including dismissal, for retaliation or retribution against any employee who properly reports any genuine ethics concern.

Documentation of Ethics Cases

The following requirements were established by the director of the Office of Ethics and Business Conduct as uniform guidelines for the documentation by managers and ethics officers of all reported ethics cases:

- Name, position, and department of individual initiating contact, if available
- Date and time of contact
- Name, position, and department of contact person
- Category of ethics case
- Mode of contact
- Resolution

Managers and ethics officers entered the required information in each reported ethics case into an ACCESS database file, enabling efficient retrieval and analysis of the data.

4

FIGURE F–5. Formal Report (*continued*) (Introduction)

Reported Ethics Cases—2008

Major/Minor Category Definition and Examples

Major ethics cases were defined as those situations potentially involving serious violations of company policies or illegal conduct. Procedure CGF-EP-02 requires notification of the Internal Audit and the Law departments in serious ethics cases. The staffs of the Internal Audit and the Law departments assume primary responsibility for managing major ethics cases and for working with the employees, ethics officers, and managers involved in each case.

Examples of situations categorized as major ethics cases:

- Cover-up of defective workmanship or use of defective parts in products
- Discrimination in hiring and promotion
- Involvement in monetary or other kickbacks from customers for preferred orders
- Sexual harassment
- Disclosure of proprietary customer or company information
- Theft
- Use of corporate Internet resources for inappropriate purposes, such as conducting private business, gambling, or gaining access to pornography

Minor ethics cases were defined as including all reported concerns not classified as major ethics cases. Minor ethics cases were classified as follows:

- Informational queries from employees
- Situations involving coworkers
- Situations involving management

> Organized by decreasing order of importance

FIGURE F–5. Formal Report (*continued*) (Introduction)

Reported Ethics Cases—2008

ANALYSIS OF REPORTED ETHICS CASES

Reported Ethics Cases, by Major/Minor Category

CGF ethics officers and managers companywide handled a total of 3,458 ethics situations during 2008. Of these cases, only 172, or 5 percent, involved reported concerns of a serious enough nature to be classified as major ethics cases (see Figure 1). Major ethics cases were defined as those situations potentially involving serious violations of company policy or illegal conduct.

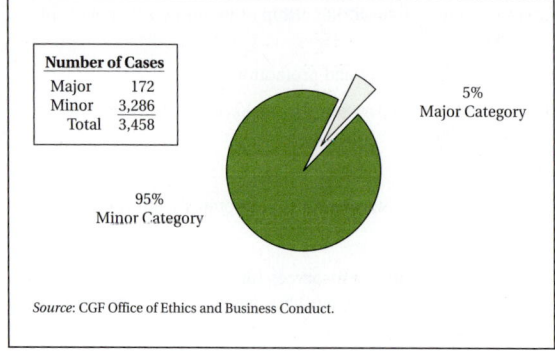

Figure 1.   Reported ethics cases, by major/minor category in 2008.

Major Ethics Cases

Of the 172 major ethics cases reported during 2008, 57 percent, upon investigation, were found to involve unsubstantiated concerns. Incomplete information or misinformation most frequently was discovered to be the cause of the unfounded concerns of misconduct in 98 cases. Forty-four cases, or 26 percent of the total cases reported, involved incidents partly substantiated by ethics

6

Reported Ethics Cases — 2008

officers as serious misconduct; however, these cases were discovered to also involve inaccurate information or unfounded issues of misconduct.

Only 17 percent of the total number of major ethics cases, or 30 cases, were substantiated as major ethics situations involving serious ethical misconduct or illegal conduct (CGF, "2008 Ethics Hotline Results") (see Figure 2).

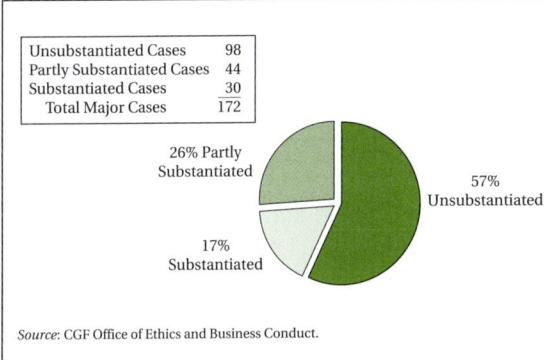

Figure 2.   Major ethics cases in 2008.

Of the 30 substantiated major ethics cases, seven remain under investigation at this time, and two cases are currently in litigation. Disposition of the remainder of the 30 substantiated reported ethics cases included severe disciplinary action in five cases: the dismissal of two employees and the demotion of three employees. Seven employees were given written warnings, and nine employees received verbal warnings (see Figure 3).

Reported Ethics Cases—2008

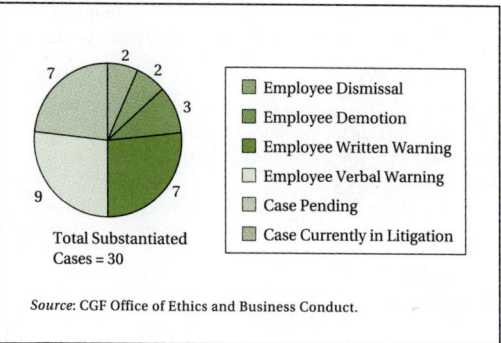

Figure 3. Disposition of substantiated major ethics cases in 2008.

Minor Ethics Cases

Minor ethics cases included those that did not involve serious violations of company policy or illegal conduct. During 2008, ethics officers and company managers handled 3,286 such cases. Minor ethics cases were further classified as follows:

- Informational queries from employees
- Situations involving coworkers
- Situations involving management

As might be expected during the initial year of the Ethics Program implementation, the majority of contacts made by employees were informational, involving questions about the new policies and procedures. These informational contacts comprised 65 percent of all contacts of a minor nature and numbered 2,148. Employees made 989 contacts regarding ethics concerns involving coworkers and 149 contacts regarding ethics concerns involving management (see Figure 4).

> Reports findings in detail

FIGURE F–5. Formal Report (*continued*) (Body)

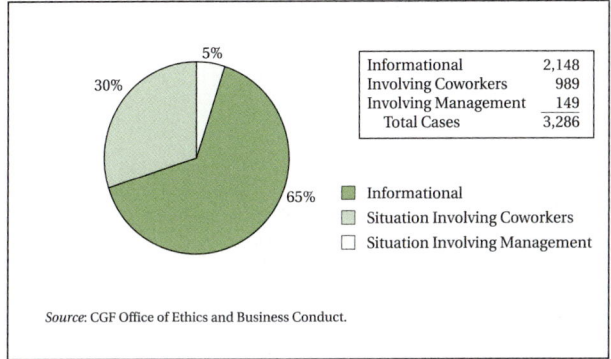

Figure 4. Minor ethics cases in 2008.

Mode of Contact

The effectiveness of the Ethics Program rested on the dissemination of information to employees and the provision of accessible channels through which employees could gain information, report concerns, and obtain guidance. Employees were encouraged to first go to their managers with any ethical concerns, because those managers would have the most direct knowledge of the immediate circumstances and individuals involved.

Other channels were put into operation, however, for any instance in which an employee did not feel able to go to his or her manager. The ethics officers companywide were available to employees through telephone conversations, face-to-face meetings, and e-mail contact. Ethics officers also served as contact points for managers in need of support and assistance in handling the ethics concerns reported to them by their subordinates.

The Ethics Hotline became operational in mid-January 2008 and offered employees assurance of anonymity and confidentiality. The Ethics Hotline was accessible to all employees on a 24-hour, 7-day basis. Ethics officers companywide took responsibility on a rotational basis for handling calls reported through the hotline.

> Assesses findings

Reported Ethics Cases — 2008

In summary, ethics information and guidance were available to all employees during 2008 through the following channels:

- Employee to manager
- Employee telephone, face-to-face, and e-mail contact with ethics officer
- Manager to ethics officer
- Employee Hotline

The mode of contact in the 3,458 reported ethics cases was as follows (see Figure 5):

- In 19 percent of the reported cases, or 657, employees went to managers with concerns.
- In 9 percent of the reported cases, or 311, employees contacted an ethics officer.
- In 5 percent of the reported cases, or 173, managers sought assistance from ethics officers.
- In 67 percent of the reported cases, or 2,317, contacts were made through the Ethics Hotline.

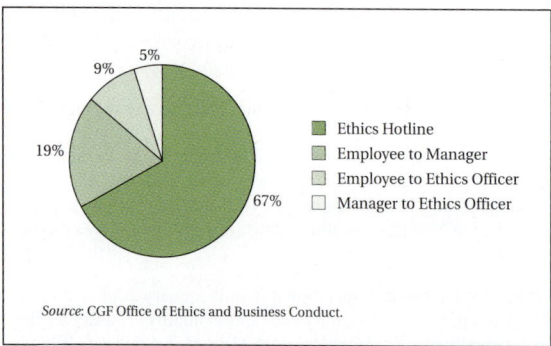

Source: CGF Office of Ethics and Business Conduct.

Figure 5.   Mode of contact in reported ethics cases in 2008.

Reported Ethics Cases — 2008

CONCLUSIONS AND RECOMMENDATIONS

The effectiveness of CGF's Ethics Program during the first year of implementation is most evidenced by (1) the active participation of employees in the program and the 3,458 contacts employees made regarding ethics concerns through the various channels available to them, and (2) the action taken in the cases reported by employees, particularly the disposition of the 30 substantiated major ethics cases. [Pulls together findings]

One of the 12 steps to building a successful Ethics Program identified by Frank Navran in *Workforce* magazine is an ethics communication strategy. Navran explains that such a strategy is crucial in ensuring [Uses sources for support]

> that employees have the information they need in a timely and usable fashion and that the organization is encouraging employee communication regarding the values, standards and the conduct of the organization and its members. (Navran 119)

The 3,458 contacts by employees during 2008 attest to the accessibility and effectiveness of the communication channels that exist in CGF's Ethics Program.

An equally important step in building a successful ethics program is listed by Navran as "Measurements and Rewards," which he explains as follows:

> In most organizations, employees know what's important by virtue of what the organization measures and rewards. If ethical conduct is assessed and rewarded, and if unethical conduct is identified and dissuaded, employees will believe that the organization's principals mean it when they say the values and code of ethics are important. (Navran 121)

[Long quotation in MLA style]

Disseminating information about the disposition of ethics cases, particularly information about the severe disciplinary actions taken in major ethics violations, sends a message to employees that unethical or illegal conduct will not be tolerated. Making public such actions taken in cases of ethical misconduct provides "a golden opportunity to make other employees aware that the behavior is unacceptable and why" (Ferrell, Fraedrich, and Ferrell 129). [Interprets findings]

FIGURE F–5. Formal Report (*continued*) (Conclusions and Recommendations)

Reported Ethics Cases — 2008

With these two points in mind, I offer the following recommendations for consideration for plans for the Ethics Program's second year:

- Continuation of the channels of communication now available in the Ethics Program
- Increased financial and technical support for the Ethics Hotline, the most highly used mode of contact in the reported ethics cases in 2008
- Dissemination of this report in some form to employees to ensure employees' awareness of CGF's commitment to uphold its Ethics Policy and Procedures
- Implementation of some measure of recognition for ethical behavior, such as an "Ethics Employee of the Month," to promote and reward ethical conduct

To ensure that employees see the value of their continued participation in the Ethics Program, feedback is essential. The information in this annual review, in some form, should be provided to employees. Knowing that the concerns they reported were taken seriously and resulted in appropriate action by Ethics Program administrators would reinforce employee involvement in the program. While the negative consequences of ethical misconduct contained in this report send a powerful message, a means of communicating the *positive* rewards of ethical conduct at CGF should be implemented. Various options for recognition of employees exemplifying ethical conduct should be considered and approved.

Continuation of the Ethics Program's successful 2008 operations, with the implementation of the above recommendations, should ensure the continued pursuit of the Ethics Program's purpose: "to promote a positive work environment that encourages open communication regarding ethics and compliance issues and concerns."

Annotations:
- Recommends specific steps
- Links recommendations to company goal

FIGURE F–5. Formal Report (*continued*) (Conclusions and Recommendations)

Reported Ethics Cases — 2008

> Section begins on a new page

Works Cited

CGF. "Ethics and Conduct at CGF Aircraft Corporation." 1 Jan. 2008. 11 Feb. 2009 <http://www.cgfac.com/aboutus/ethics.html>.

---. "2008 Ethics Hotline Investigation Results." 15 Jan. 2009. 11 Feb. 2009 <http://www.cgfac.com/html/ethics.html>.

> This report uses MLA style

Ferrell, O. C., John Fraedrich, and Linda Ferrell. <u>Business Ethics: Ethical Decision Making and Cases</u>. Boston: Houghton Mifflin, 2006.

Kelley, Tina. "Corporate Prophets, Charting a Course to Ethical Profits." <u>New York Times</u> 8 Feb. 1998: BU12.

Navran, Frank. "12 Steps to Building a Best-Practices Ethics Program." <u>Workforce</u> Sept. 1997: 117–22.

FIGURE F–5. Formal Report (*continued*) (Works Cited)

format

Format refers to both the organization of information in a document and the physical arrangement of information on the page.

In one sense, format refers to the fact that some types of job-related writing (such as **formal reports**, **proposals**, and various types of **correspondence**) are characterized by conventions that govern the **scope** and placement of information. For example, in formal reports, the **table of contents** precedes the preface but follows the title page and the **abstract**. Likewise, although variations exist, parts of **letters**—such as inside address, salutation, and complimentary closing—are arranged in standard patterns. See also **e-mail** and **memos**.

Format also refers to the general physical appearance of a finished document, whether printed or electronic. See also **layout and design** and **Web design**.

former / latter

Former and *latter* should be used to refer to only two items in a sentence or paragraph.

▶ The president and his aide emerged from the conference, the *former* looking nervous and the *latter* looking glum.

Because these terms make the reader look to previous material to identify the reference, however, they complicate reading and are best avoided.

forms

Forms allow you to gather information from respondents in a standardized print or online design that makes it easy for you to tabulate the responses and evaluate the information. See Figures F–6 and F–7 for two examples of forms.

◆ **ETHICS NOTE** Information gathered on forms can be sensitive or personal, so make sure to present questions in a way that is not invasive or illegal. Unless otherwise indicated on the form, the person filling out the form should have the expectation of confidentiality. If you are concerned about issues of confidentiality or legality, check your organization's policy or in a classroom seek your instructor's advice. ◆

104-M S A
Section 125 Flexible Spending Account (FSA) Claim Form

> [Form title]

Employee Name: _____

Social Security Number: _____-____-_____

> [Writing lines]

Name of Employer: _____

Employee Signature: _____

Complete section below for medical, dental, or vision reimbursement

> [Instructions]

CLAIM TYPE I: MEDICAL CARE ACCOUNT

Amount of Expense Incurred: $_____

Dates of Services: From: _____ To: _____

Complete section below for reimbursement of care for your dependent provided by a child-care facility, an adult dependent-care center, or a caretaker

> [Instructions]

CLAIM TYPE II: DEPENDENT-CARE ACCOUNT

Amount of Expense Incurred: $_____

Name of Dependent-Care Provider: _____

Provider Social Security or Federal ID Number: _____-____-_____

Mail or fax form with documentation to:
Specialized Benefit Services, Inc.
P. O. Box 498
Framingham, MA 01702
Fax: (508) 877-1182
For additional claim forms: www.sbsclaims.com

> [Mailing and contact information]

FIGURE F–6. Form (for a Medical Claim)

Figure F–7. Online Form with Typical Components

Preparing a Form

An effective form makes it easy for one person to supply information and for another person to retrieve, record, and interpret that information. Ideally, a form should be self-explanatory to someone seeing it for the first time. When preparing a form, determine the kind of information you are seeking and arrange the questions in a logical order. To ensure the usability of the form, test it with those from your target **audience** or others before publishing the final version. See also **questionnaires** and **usability testing**.

Choosing Online or Paper. Many organizations provide their forms not only on paper and as downloadable documents but also as interactive online forms. Forms especially well suited for online use include job applications, conference or seminar registrations, and order forms. Online forms standardize respondents' interfaces and link to databases that tabulate and interpret data. An online form can be programmed to ensure that all necessary fields are completed correctly before the form can be successfully submitted. Using online forms can eliminate the problems associated with distributing and collecting forms. However, using online forms can be difficult for people with limited computer literacy or access, so consider your audience carefully before opting to collect your responses online. See also **Web design**.

Writing Instructions. Place instructions at the beginning of the form or at the beginning of each section of the form and use **headings** or other design elements, such as the boldface type and shading in Figure F–6, to attract the readers' attention. When necessary, place instructions for distributing the copies of multiple-copy forms at the bottom of each page of the form. Instructions for submitting printed forms should be clearly indicated, as in Figure F–6. For printed forms, include space for a signature and a date. At the end of online forms, include a "submit" button that is programmed to record the data and to open a new page or send a confirming e-mail informing respondents that their responses have been successfully submitted.

Choosing Response Types. Forms should ask questions in ways that are best suited to the types of data you hope to collect. The two main types are open-ended questions and closed-ended questions.

- *Open-ended questions* allow respondents to answer questions in their own words. Such questions are most appropriate if you wish to elicit responses you may not have anticipated (as in a complaint form) or if there are too many possible answers to use a multiple-choice format. However, the responses to open-ended questions can be difficult to tabulate and analyze.
- *Closed-ended questions* provide a list of options from which the respondent can select, limiting the range of possible responses. When you want to make sure you receive a standardized, easy-to-tabulate response, use any of the types of closed-ended questions that follow:
  - *Multiple choice*: Choose one (or sometimes more) from a preset list of options.
  - *Ranked choice*: Rank items according to preference, such as selecting vacation days or choosing job assignments.
  - *Forced choice*: Choose between two preset options, such as yes/no or male/female.

Wording Questions. Questions are normally presented as short phrases or labels. Keep them brief, specific, and to the point; avoid unnecessary repetition by combining related information under an explanatory heading.

WORDY	What make of car (or vehicle) do you drive? _____
	What year was it manufactured? _____
	What model is it? _____
	What is the body style? _____
CONCISE	Vehicle Information
	Make _____ Year _____
	Model _____ Body Style _____

If a requested date is other than the date on which the form is being filled out, the label should read, for example, "Effective date" or "Date issued," rather than simply "Date." As in all writing, put yourself in your reader's place and imagine what sort of requests would be clear.

Sequencing Entries. The main portion of the form includes the entries that are required to obtain the necessary data. Arrange entries in an order that will be the most logical to the person filling out the form.

- Sequence entries to fit the subject matter. For instance, a form requesting reimbursement for travel expenses would logically begin with the first day of the week (or month) and end with the last day of the appropriate period.
- If the response to one item is based on the response to another item, be sure the items appear in the correct order.
- Group requests for related information together whenever possible.
- For ease of reading, arrange entries from left to right and from top to bottom.

Designing a Form

At the top of the form, clearly indicate preliminary information, such as the name of your organization, the title of your form, and any reference number. You can design computer-generated forms specific to your needs with form-design software, word-processing software, or markup languages (such as HTML or XML). However you prepare the final version of a form, pay particular attention to design details, especially to the placement of entry lines, labels, and the amount of space allowed for responses.

Entry Lines and Fields. A print form can be designed so that the person filling it out provides information on a writing line, in a writing block, or in square boxes. A *writing line* is simply a rule with a caption, such as the line for "Employee Name" shown in Figure F–6. A *writing block* is essentially the same as a writing line, except that each entry is enclosed in a ruled block, as shown in Figure F–8, making it unlikely for the respondent to associate a caption with the wrong line.

NAME		TELEPHONE
STREET ADDRESS		
CITY	STATE	ZIP CODE

FIGURE F–8. Writing Block for a Form

When it is possible to anticipate all likely responses, you can make the form easy to fill out by writing the question on the form, supplying a labeled box for each anticipated answer, and asking the respondent to check the appropriate boxes. Such a design also makes it easy to tabulate the data. Be sure your questions are both simple and specific.

▶ Would your department order another X2L Copier? Yes ☐ No ☐

For online forms, these functions are accomplished with form fields such as text boxes, option (or "radio") buttons, drop-down menus, lists, and checkboxes, which can be aligned using table cells or grouping. Each form field should have a label prompting users to type information or select from a list of options. Labels for text boxes, drop-down menus, and lists should be positioned to the left; labels for radio buttons and checkboxes should be positioned to the right. Be sure as well to indicate required form fields. See Figure F–7 for an example of spacing and alignment in an online form.*

Spacing. Provide enough space to enable the person filling out the form to enter the requested information. Insufficient writing or typing space makes it difficult for people to respond, resulting in responses that are hard to read, abbreviated, or incomplete. Reading responses that are too tightly spaced or that snake around the side of the form can cause errors when tabulating or interpreting data.

*For up-to-date information on designing online forms, see *www.stcsig.org/usability/topics/forms.html*.

fortuitous / fortunate

When an event is *fortuitous*, it happens by chance or accident and without plan. Such an event may be lucky, unlucky, or neutral.

▶ My encounter with the client in Denver was entirely *fortuitous*; I had no idea he was there.

When an event is *fortunate*, it happens by good fortune or happens favorably.

▶ Our chance meeting had a *fortunate* outcome.

fragments (*see* sentence fragments)

functional shift

Many words shift easily from one **part of speech** to another, depending on how they are used. When they do, the process is called a *functional shift*, or a shift in function.

▶ It takes ten minutes to *walk* from the laboratory to the emergency room. However, the long *walk* reduces efficiency.
[*Walk* shifts from **verb** to **noun**.]

▶ I talk to the Chicago office on the *phone* every day. He was concerned about the office *phone* expenses. He will *phone* the home office from London.
[*Phone* shifts from noun to **adjective** to verb.]

▶ *After* we discuss the project, we will begin work. *After* lengthy discussions, we began work. The partners worked well together forever *after*.
[*After* shifts from **conjunction** to **preposition** to **adverb**.]

Jargon is often the result of functional shifts. In hospitals, for example, an *attending physician* is often referred to simply as the "attending" (a shift from an adjective to a noun). Likewise, in nuclear plant construction, a *reactor containment building* is called a "containment" (a shift from an adjective to a noun). Do not shift the function of a word indiscriminately merely to shorten a phrase or an expression. See also **affectation**, **audience**, and **conciseness**.

garbled sentences

A garbled sentence is one that is so tangled with structural and grammatical problems that it cannot be repaired. Garbled sentences often result from an attempt to squeeze too many ideas into one sentence.

▶ My job objectives are accomplished by my having a diversified background which enables me to operate effectively and efficiently, consisting of a degree in computer science, along with twelve years of experience, including three years in Staff Engineering-Packaging sets a foundation for a strong background in areas of analyzing problems and assessing economical and reasonable solutions.

Do not try to patch such a sentence; rather, analyze the ideas it contains, list them in a logical sequence, and then construct one or more entirely new sentences. An analysis of the preceding example yields the following five ideas:

- My job requires that I analyze problems to find economical and workable solutions.
- My diversified background helps me accomplish my job.
- I have a computer-science degree.
- I have twelve years of job experience.
- Three of these years have been in Staff Engineering-Packaging.

Using those five ideas—together with **parallel structure**, **sentence variety**, **subordination**, and **transition**—the writer might have described the job as follows:

▶ My job requires that I analyze problems to find economical and workable solutions. Both my education and experience help me achieve this goal. Specifically, I have a computer-science degree and twelve years of job experience, three of which have been in the Staff Engineering-Packaging Department.

See also **clarity**, **mixed constructions**, and **sentence construction**.

gender

In English grammar, *gender* refers to the classification of **nouns** and **pronouns** as masculine, feminine, and neuter. The gender of most words can be identified only by the choice of the appropriate pronoun (*he*, *she*, *it*). Only these pronouns and a select few nouns (*buck/doe*) or noun forms (*heir/heiress*) reflect gender. Many such nouns have been replaced by single terms for both sexes. See also **biased language** and **he / she**.

Gender is important to writers because they must be sure that nouns and pronouns within a grammatical construction agree in gender. A pronoun, for example, must agree with its noun antecedent in gender. We refer to a woman as *she* or *her*, not as *it*; to a man as *he* or *him*, not as *it*; to a building as *it*, not as *he* or *she*. See also **agreement**.

ESL TIP for Assigning Gender

The English language has an almost complete lack of gender distinctions. That can be confusing for a nonnative speaker of English whose native language may assign gender. In the few cases in which English does make a gender distinction, there is a close connection between the assigning of gender and the sex of the subject. The few instances in which gender distinctions are made in English are summarized as follows:

Subject pronouns	he/she
Object pronouns	him/her
Possessive adjectives	his/her(s)
Some nouns	king/queen, boy/girl, bull/cow, etc.

When a noun, such as *doctor*, can refer to a person of either sex, you need to know the sex of the person to which the noun refers to determine the gender-appropriate pronoun.

▶ The doctor gave *her* patients advice.
   [Doctor is female.]

▶ The doctor gave *his* patients advice.
   [Doctor is male.]

When the sex of the noun antecedent is unknown, be sure to follow the guidelines for nonsexist writing in the entry **biased language**. (*Note*: Some English speakers refer to vehicles and countries as *she*, but contemporary usage favors *it*.)

general and specific methods of development

General and specific **methods of development** organize information either from general points to specific details or from specific details to a general conclusion. As with all methods of development, rarely does a writer rely on only one method throughout a document. Most documents blend methods and use combinations of methods.

General to Specific

The general-to-specific development (see Figure G–1) is especially useful for teaching **readers** about something with which they are not familiar because you can begin with generally known information and lead to new and increasingly specific details. This method can also be used to support a general statement with facts or examples that validate the statement. For example, if you begin your writing with the general statement "Companies that diversify are more successful than those that do not," you could follow that statement with examples and statistics

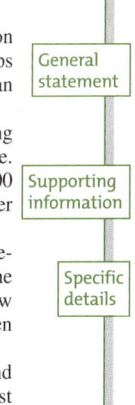

Subject: Expanding Our Supplier Base for Computer Chips

On the basis of information presented at the supply meeting on April 14, we recommend that the company initiate relationships with computer-chip manufacturers. Several events make such an action necessary. [General statement]

Our current supplier, Datacom, is experiencing growing pains and is having difficulty shipping the product on time. Specifically, we can expect a reduction of between 800 and 1,000 units per month for the remainder of this fiscal year. The number of units should stabilize at 15,000 units per month thereafter. [Supporting information]

Domestic demand for our computers continues to grow. Demand during the current fiscal year is up 500,000 units over the last fiscal year. Our sales projections for the next five years show that demand should peak next year at about 830,000 units given the consumer demand, which will increase exponentially. [Specific details]

Finally, our expansion into the Czech Republic and Kazakhstan markets will require additional shipments of at least 175,000 units per quarter for the remainder of this fiscal year. Sales Department projections put global computer sales at double that rate, or 350,000 units per fiscal year, for the next five years.

FIGURE G–1. General-to-Specific Method of Development

that prove to the reader that companies that diversify are, in fact, more successful than companies that do not.

A memo or short report organized entirely in a general-to-specific sequence discusses only one point. All other information in the document supports the general statement, as in Figure G–1 from a memo about locating additional computer-chip suppliers.

Specific to General

Specific-to-general development is especially useful when you wish to persuade skeptical readers of a general principle with an accumulation of specific details and evidence that reach a logical **conclusion**. It carefully builds its case, often with examples and analogies in addition to facts or statistics, and does not actually make its point until the end. (See also **order-of-importance method of development**.) Figure G–2 is an example of the specific-to-general method of development.

> **Specific details**
> Recently, a government agency studied the use of passenger-side air bags in 4,500 accidents involving nearly 7,200 front-seat passengers of the vehicles involved. Nearly all the accidents occurred on routes that had a speed limit of at least 40 mph. Only 20 percent of the adult front-seat passengers were riding in vehicles equipped with passenger-side air bags. Those riding in vehicles not equipped with passenger-side air bags were more than twice as likely to be killed as passengers riding in vehicles that were so equipped.
>
> **General conclusion**
> A conservative estimate is that 40 percent of the adult front-seat passenger-vehicle deaths could be prevented if all vehicles came equipped with passenger-side air bags. Children, however, should always ride in the backseat because other studies have indicated that a child can be killed by the deployment of an air bag. If you are an adult front-seat passenger in an accident, your chances of survival are far greater if the vehicle in which you are riding is equipped with a passenger-side air bag.

FIGURE G–2. Specific-to-General Method of Development

global communication

The prevalence of global communication technology and international markets means that the ability to communicate with **audiences** from varied cultural backgrounds is essential. The audiences for such communications include clients and customers as well as business partners and colleagues.

Many entries in this book, such as **meetings** and **résumés**, are based on U.S. cultural patterns. The treatment of such topics might be very different in other cultures where leadership styles, persuasive strategies, and even legal constraints differ. As illustrated in **international correspondence**, organizational patterns, forms of courtesy, and ideas about efficiency can vary significantly from culture to culture. What might be seen as direct and efficient in the United States could be considered blunt and even impolite in other cultures. The reasons behind these differing ways of viewing communication are complex. Researchers have found various ways to measure cultural differences through such concepts as the importance of saving face, perceptions of time, and individual versus group orientation.

Anthropologist Edward T. Hall, a pioneer in cross-cultural research, developed the concept of "contexting" to assess the predominant communication style of a culture.* By contexting, Hall means how much or how little an individual assumes another person understands about a subject under discussion. In a very low-context communication, the participants assume they share little knowledge and must communicate in great detail. Low-context cultures tend to assume little prior knowledge on the part of those with whom they communicate; thus, thorough documentation is important—written agreements (contracts) are expected, and rules are explicitly defined.

In a high-context communication, the participants assume they understand the **context** and thus depend on shared history (context) to communicate with each other. Thus, communications such as written contracts are not so important, while communication that relies on personal relationships and shared history is paramount. Of course, no culture is entirely high or low context; rather, these concepts can help you communicate more effectively to those in a particular culture.†

> **WRITER'S CHECKLIST** **Communicating Globally**

✔ Acknowledge diversity within your organization. Discussing the differing cultures within your company or region will reinforce the idea that people can interpret verbal and nonverbal communications differently.

---

*Edward Twitchell Hall and Mildred Reed Hall, *Understanding Cultural Differences: Germans, French and Americans* (Yarmouth, Maine: Intercultural Press, 1990).
†For an example of how contexting functions in a particular culture, see Gerald J. Alred, "Teaching in Germany and the Rhetoric of Culture," *Journal of Business and Technical Communication* 11.3 (July 1997): 353–78.

Writer's Checklist: Communicating Globally (continued)

- Invite global and intercultural communication experts to speak at your workplace. Companies in your area may have employees who could be resources for cultural discussions.
- Understand that the key to effective communication with global audiences is recognizing that cultural differences, despite the challenges they may present, offer opportunities for growth for both you and your organization.
- Consult with someone from your intended audience's culture. Many phrases, gestures, and visual elements are so subtle that only someone who is very familiar with the culture can explain the effect they may have on others from that culture. See also **global graphics**.

WEB LINK | **Intercultural Resources**

Intercultural Press publishes "books and training materials that help professionals, businesspeople, travelers, and scholars understand the meaning and diversity of culture." (See *www.interculturalpress.com*.) For other resources for global communication, see *bedfordstmartins.com/alredtech* and select *Links for Handbook Entries*.

---

global graphics

In a global business and technological environment, **graphs** and other **visuals** require the same careful attention given to other aspects of **global communication**. The complex cultural connotations of visuals challenge writers to think beyond their own experience when they are aiming for audiences outside their own culture.

Symbols, images, and even colors are not free from cultural associations—they depend on **context**, and context is culturally determined. For instance, in North America, a red cross is commonly used as a symbol for first aid or a hospital. In Muslim countries, however, a cross (red or otherwise) represents Christianity, whereas a crescent (usually green) signifies first aid or a hospital. A manual for use in Honduras could indicate "caution" by using a picture of a person touching a finger below the eye. In France, however, using that gesture would mean "You can't fool me."

Figure G–3 shows two different graphics depicting weight lifters. The drawing at the left may be appropriate for U.S. audiences and oth-

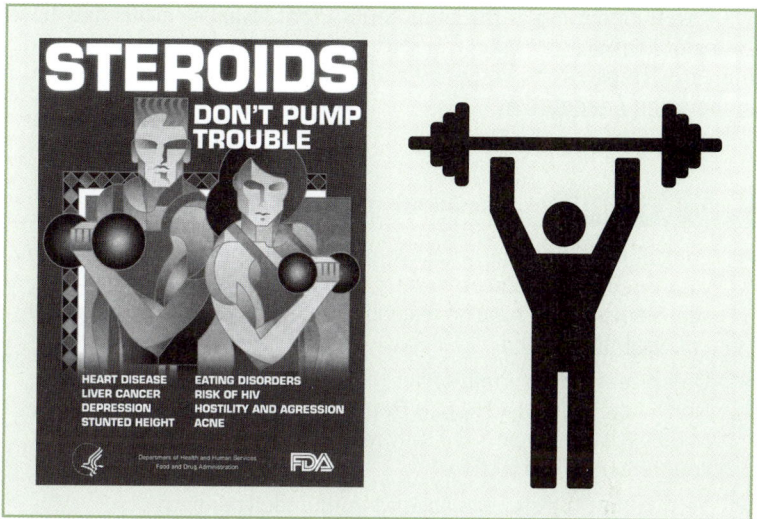

FIGURE G–3. Graphics for U.S. (left) and Global (right) Audiences

ers. However, that image would be highly inappropriate in many cultures where the image of a partially clothed man and woman in close proximity would be contrary to deeply held cultural beliefs and even laws about the public depiction of men and women. The drawing at the right in Figure G–3, however, depicts a weight lifter by using a neutral icon that avoids the connotations associated with more realistic images of people.

These examples suggest why the International Organization for Standardization (ISO) established agreed-upon symbols, such as those shown in Figure G–4, designed for public signs, guidebooks, and manuals.

FIGURE G–4. International Organization for Standardization Symbols

Careful attention to the connotations that visual elements may have for a global **audience** makes translations easier, prevents embarrassment, and earns respect for a company and its products and services. See also **connotation / denotation**.

WRITER'S CHECKLIST   Communicating with Global Graphics

- Consult with someone or test your use of graphics with people from your intended audience's country who understand the effect that visual elements will have on readers or listeners. See also **presentations** and **usability testing**.
- Organize visual information for the intended audience. For example, North Americans read visuals from left to right in clockwise rotation. Middle Eastern readers typically read visuals from right to left in counterclockwise rotation.
- Be sure that the graphics have no unintended religious implications.
- Carefully consider how you depict people in visuals. Nudity in advertising, for example, may be acceptable in some cultures, but in others showing even isolated bare body parts can alienate audiences.
- Use outlines or neutral abstractions to represent human beings. For example, use stick figures and avoid representing men and women.
- Examine how you display body positions in signs and visuals. Body positioning can carry unintended cultural meanings very different from your own. For example, some Middle Eastern cultures regard the display of the soles of one's shoes to be disrespectful and offensive.
- Choose neutral colors (or those you know are appropriate) for your graphics; generally, black-and-white and gray-and-white illustrations work well. Colors can be problematic. For example, in North America, Europe, and Japan, red indicates danger. In China, however, red symbolizes good fortune and joy.
- Check your use of punctuation marks, which are as language specific as symbols. For example, in North America, the question mark generally represents the need for information or the help function in a computer manual or program. In many countries, that symbol has no meaning at all.
- Create simple visuals and use consistent labels for all visual items. In most cultures, simple shapes with fewer elements are easier to read.

glossaries

A glossary is an alphabetical list of definitions of specialized terms used in a **formal report**, a manual, or other long document. You may want to include a glossary if some readers in your **audience** are not familiar with specialized or technical terms you use.

Keep glossary entries concise and be sure they are written in language that all your readers can understand.

▶ *Amplitude modulation*: Varying the amplitude of a carrier current with an audio-frequency signal.

Arrange the terms alphabetically, with each entry beginning on a new line. The definitions then follow the terms, dictionary style. In a formal report, the glossary begins on a new page and appears after the appendix(es) and bibliography.

Including a glossary does not relieve you of the responsibility of **defining terms** that your reader will not know when those terms are first mentioned in the text.

gobbledygook

Gobbledygook is writing that suffers from an overdose of traits guaranteed to make it stuffy, pretentious, and wordy. Such traits include the overuse of big and mostly **abstract words**, **affectation** (especially long variants), **buzzwords**, **clichés**, **euphemisms**, inappropriate **jargon**, stacked **modifiers**, and **vague words**. Gobbledygook is writing that attempts to sound official (officialese), legal (legalese), or scientific; it tries to make a "natural elevation of the geosphere's outer crust" out of a molehill. Consider the following statement from an auto-repair release form.

LEGALESE	I hereby authorize the above repair work to be done along with the necessary material and hereby grant you and/or your employees permission to operate the car or truck herein described on streets, highways, or elsewhere for the purpose of testing and/or inspection. An express mechanic's lien is hereby acknowledged on above car or truck to secure the amount of repairs thereto.
DIRECT	You have my permission to do the repair work listed on this work order and to use the necessary material. You may drive my vehicle to test its performance. I understand that you will keep my vehicle until I have paid for all repairs.

See also **clarity**, **conciseness**, and **word choice**.

good / well

Good is an **adjective**, and *well* is an **adverb**.

ADJECTIVE Janet presented a *good* plan.
ADVERB She presented the plan *well*.

Well also can be used as an adjective to describe health (a *well* child, *wellness* programs). See also **bad / badly**.

grammar

Grammar is the systematic description of the way words work together to form a coherent language. In that sense, it is an explanation of the structure of a language. However, grammar is popularly taken to mean the set of rules that governs how a language ought to be spoken and written. In that sense, it refers to the **usage** conventions of a language.

Those two meanings of grammar—how the language functions and how it ought to function—are easily confused. To clarify the distinction, consider the expression *ain't*. Unless used intentionally to add colloquial flavor, *ain't* is unacceptable because its use is considered nonstandard. Yet taken strictly as a **part of speech**, the term functions perfectly well as a verb. Whether it appears in a declarative sentence ("I *ain't* going") or an interrogative sentence ("*Ain't* I going?"), it conforms to the normal pattern for all verbs in the English language. Although readers may not approve of its use, they cannot argue that it is ungrammatical in such sentences.

To achieve **clarity**, you need to know both grammar (as a description of the way words work together) and the conventions of usage. Knowing the conventions of usage helps you select the appropriate over the inappropriate word or expression. (See also **word choice**.) A knowledge of grammar helps you diagnose and correct problems arising from how words and phrases function in relation to one another. For example, knowing that certain words and phrases function to modify other words and phrases gives you a basis for correcting those **modifiers** that are not doing their job. Understanding **dangling modifiers** helps you avoid or correct a construction that obscures the intended meaning. In short, an understanding of grammar and its special terminology is valuable chiefly

WEB LINK | Getting Help with Grammar

For helpful Web sites and grammar exercises, see *bedfordstmartins.com/alredtech* and select *Links for Handbook Entries* and *Exercise Central*.

because it enables you to recognize and correct problems so that you can communicate clearly and precisely. For a complete list of grammar entries, see the Contents by Topic on the inside front cover.

graphs

A graph presents numerical or quantitative data in visual form and offers several advantages over presenting data within the text or in **tables**. Trends, movements, distributions, comparisons, and cycles are more readily apparent in graphs than they are in tables. However, although graphs present data in a more comprehensible form than tables do, they are less precise. For that reason, some **audiences** may need graphs to be accompanied by tables that give exact data. The types of graphs described in this entry include line graphs, bar graphs, pie graphs, and picture graphs. For advice on integrating graphs within text, see **visuals**; for information about using presentation graphics, see **presentations**.

Line Graphs

A line graph shows the relationship between two variables or sets of numbers by plotting points in relation to two axes drawn at right angles (see Figure G–5). The vertical axis usually represents amounts, and the horizontal axis usually represents increments of time. Line

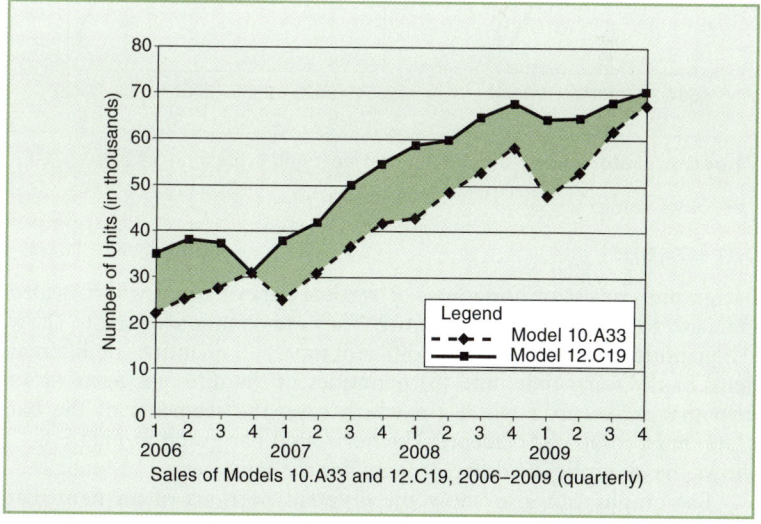

FIGURE G–5. Double-Line Graph (with Shading)

graphs that portray more than one set of variables (double-line graphs) allow for comparisons between two sets of data for the same period of time. You can emphasize the difference between the two lines by shading the space between them, as shown in Figure G–5.

ETHICS NOTE Be especially careful to proportion the vertical and horizontal scales so that they give a precise presentation of the data that is free of visual distortion. To do otherwise is not only inaccurate but potentially unethical. (See **ethics in writing**.) In Figure G–6, the graph on the left gives the appearance of a dramatic decrease in lab-test failures because the scale is unevenly compressed, with some of the years selectively omitted. The graph on the right represents the trend more accurately because the years are evenly distributed without omissions. ✦

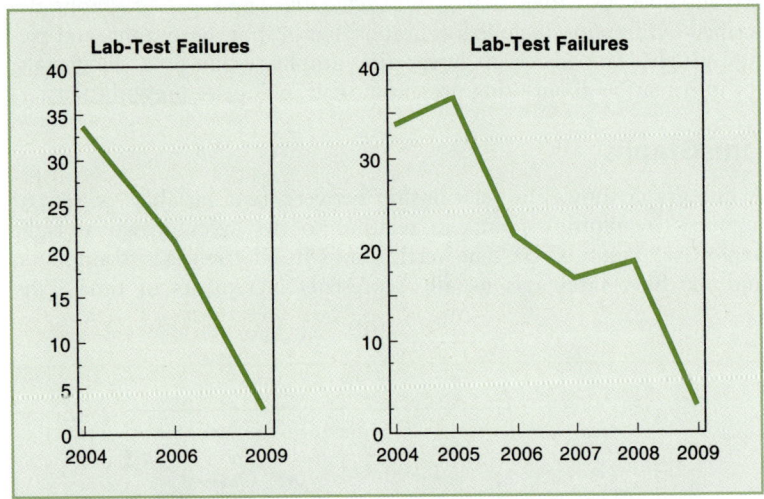

Figure G–6. Distorted (left) and Distortion-Free (right) Expressions of Data

Bar Graphs

Bar graphs consist of horizontal or vertical bars of equal width, scaled in length to represent some quantity. They are commonly used to show (1) quantities of the same item at different times, (2) quantities of different items at the same time, and (3) quantities of the different parts of an item that make up a whole (in which case, the segments of the bar graph must total 100 percent). The horizontal bar graph in Figure G–7 shows the quantities of different items for the same period of time.

Bar graphs can also show the different portions of an item that make up the whole, as shown in Figure G–8. Such a bar graph is divided

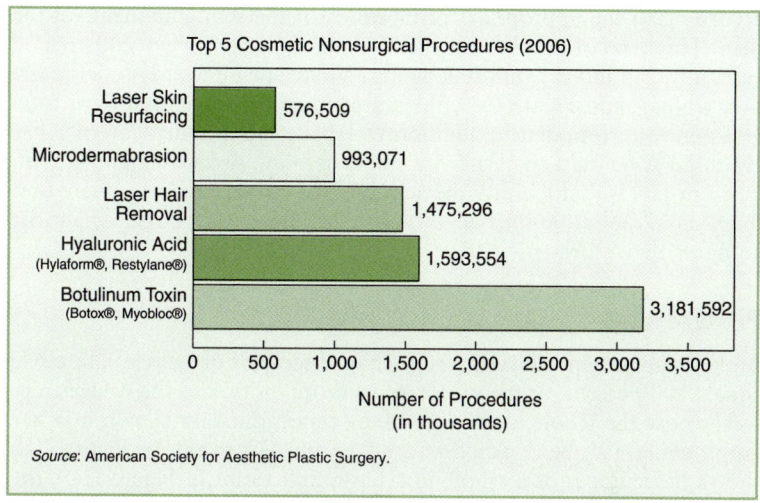

Figure G–7. Bar Graph (Quantities of Different Items During a Fixed Period)

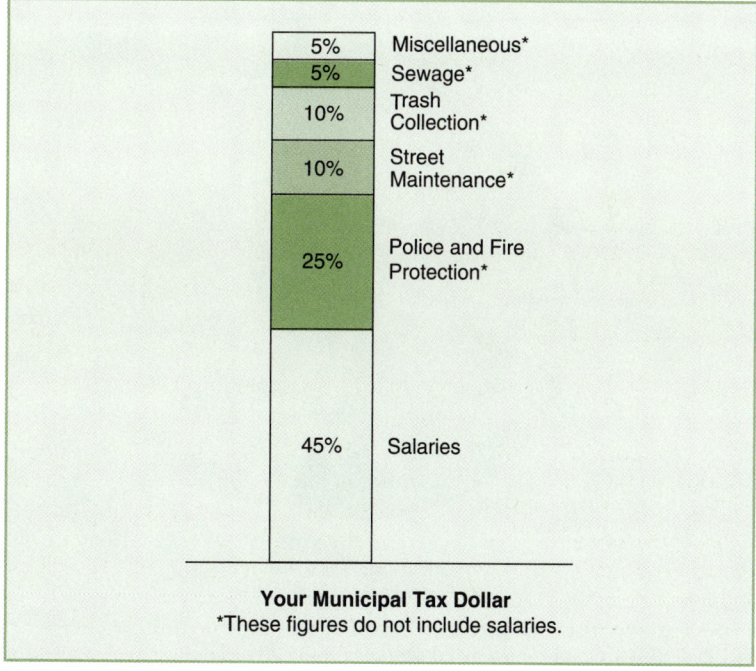

FIGURE G–8. Bar (Column) Graph (Showing the Parts That Make Up the Whole)

according to the appropriate proportions of the subcomponents of the item. This type of graph, also called a *column graph* when constructed vertically, can indicate multiple items. Where such items represent parts of a whole, as in Figure G–8, the segments in the bar graph must total 100 percent. Note that in addition to labels, each subdivision of a bar graph must be marked clearly by color, shading, or crosshatching, with a key or labels that identify the subdivisions represented. Be aware that three-dimensional graphs can make sections seem larger than the amounts they represent.

G Pie Graphs

A pie graph presents data as wedge-shaped sections of a circle. The circle equals 100 percent, or the whole, of some quantity, and the wedges represent how the whole is divided. Many times, the data shown in a bar graph could also be depicted in a pie graph. For example, Figure G–8 shows percentages of a whole in a bar-graph form. In Figure G–9, the same data are converted into a pie graph, dividing "Your Municipal Tax Dollar" into wedge-shaped sections that represent percentages. Pie graphs provide a quicker way of presenting information that can be

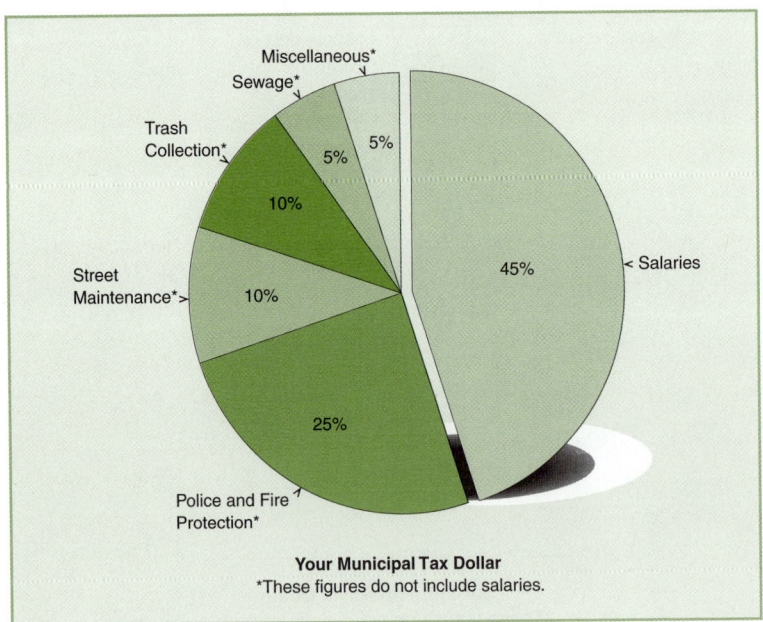

FIGURE G–9. Pie Graph (Showing Percentages of the Whole)

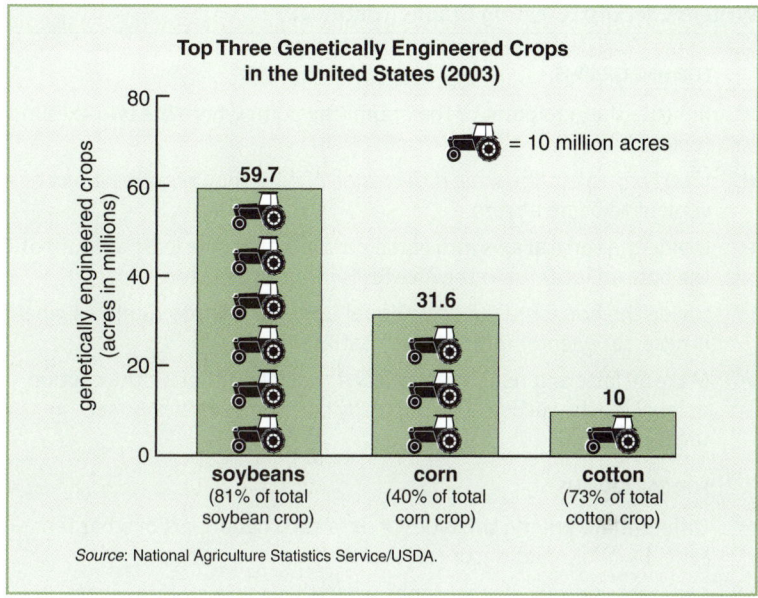

FIGURE G–10. Picture Graph

shown in a table; in fact, a table with a more-detailed breakdown of the same information often accompanies a pie graph.

Picture Graphs

Picture graphs are modified bar graphs that use pictorial symbols of the item portrayed. Each symbol corresponds to a specified quantity of the item, as shown in Figure G–10. Note that for precision and clarity, the picture graph includes the total quantity above the symbols.

WRITER'S CHECKLIST Creating Graphs

FOR ALL GRAPHS

- Use, as needed, a key or legend that lists and defines symbols (see Figure G–5).
- Include a source line under the graph at the lower left when the data come from another source.
- Place explanatory footnotes directly below the figure caption or label (see Figures G–8 and G–9).

Writer's Checklist: Creating Graphs (continued)

FOR LINE GRAPHS

- Indicate the zero point of the graph (the point where the two axes intersect).
- Insert a break in the scale if the range of data shown makes it inconvenient to begin at zero.
- Divide the vertical axis into equal portions, from the least amount at the bottom (or zero) to the greatest amount at the top.
- Divide the horizontal axis into equal units from left to right. If a label is necessary, center it directly beneath the scale.
- Make all lettering read horizontally if possible, although the caption or label for the vertical axis is usually positioned vertically (see Figure G–5).

FOR BAR GRAPHS

- Differentiate among the types of data each bar or part of a bar represents by color, shading, or crosshatching.
- Avoid three-dimensional graphs when they make sections seem larger than the amounts they represent.

FOR PIE GRAPHS

- Make sure that the complete circle is equivalent to 100 percent.
- Sequence the wedges clockwise from largest to smallest, beginning at the 12 o'clock position, whenever possible.
- Limit the number of items in the pie graph to avoid clutter and to ensure that the slices are thick enough to be clear.
- Give each wedge a distinctive color, pattern, shade, or texture.
- Label each wedge with its percentage value and keep all call-outs (labels that identify the wedges) horizontal.
- Detach a slice, as shown in Figure G–9, if you wish to draw attention to a particular segment of the pie graph.

FOR PICTURE GRAPHS

- Use picture graphs to add interest to **presentations** and documents (such as **newsletters**) that are aimed at wide audiences.
- Choose symbols that are easily recognizable. See also **global graphics**.
- Let each symbol represent the same number of units.
- Indicate larger quantities by using more symbols, instead of larger symbols, because relative sizes are difficult to judge accurately.

he / she

The use of either *he* or *she* to refer to both sexes excludes half of the population. (See also **biased language**.) To avoid this problem, you could use the phrases *he or she* and *his or her*. ("Whoever is appointed will find *his or her* task difficult.") However, *he or she* and *his or her* are clumsy when used repeatedly, as are *he/she* and similar constructions. One solution is to reword the sentence to use a plural **pronoun**; if you do, change the **nouns** or other pronouns to match the plural form.

▶ ~~The administrator~~ *Administrators* cannot do ~~his or her job~~ *their jobs* until ~~he or she understands~~ *they understand* the organization's culture.

In other cases, you may be able to avoid using a pronoun altogether.

▶ Everyone must submit ~~his or her~~ *an* expense report by Monday.

Of course, a pronoun cannot always be omitted without changing the meaning of a sentence.

Another solution is to omit troublesome pronouns by using the imperative **mood**.

▶ ~~Everyone must submit his or her~~ *Submit all* expense report*s* by Monday.

headers and footers

A *header* in a **formal report** or other document appears at the top of each page, and a *footer* appears at the bottom of each page. Although the information included in headers and footers varies greatly from one organization to the next, the header and footer shown in Figure H–1 are fairly typical.

Headers or footers should include at least the page number but may also include the document title, the topic (or subtopic) of a section, the date of the document, the names of the author or recipients, and other identifying information to help **readers** keep track of where they

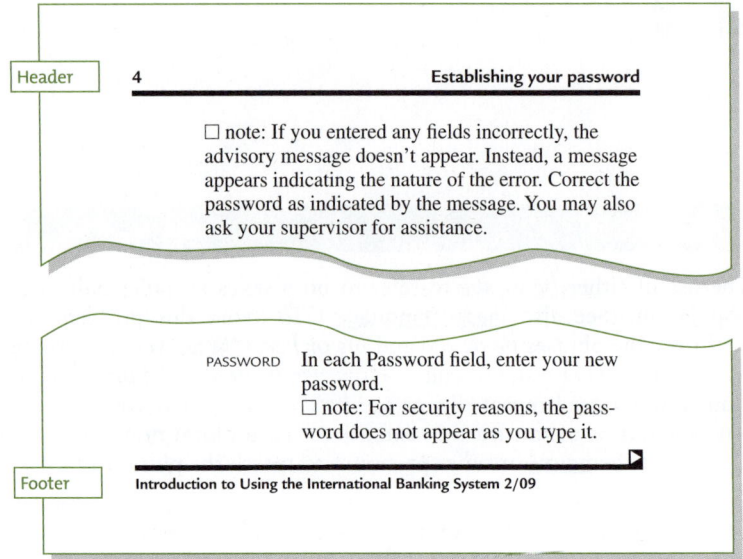

FIGURE H–1. Header and Footer

are in the document. Keep your headers and footers concise because too much information in them can create visual clutter. For examples of headers used in correspondence, see **letters** and **memos**.

headings

Headings (also called *heads*) are titles or subtitles that highlight the main topics and signal topic changes within the body of a document. Headings help **readers** find information and divide the material into comprehensible segments. A **formal report** or **proposal** may need several levels of headings (as shown in Figure H–2) to indicate major divisions, subdivisions, and even smaller units. If possible, avoid using more than four levels of headings. See also **layout and design**.

Headings typically represent the major topics of a document. In a short document, you can use the major divisions of your outline as headings; in a longer document, you may need to use both major and minor divisions.

General Heading Style

No one format for headings is correct. Often an organization settles on a standard format, which everyone in that organization follows. Some-

> **ENGINEERING AND MANUFACTURING
> PLANT LOCATION REPORT** ⟵ First-level head
>
> The committee initially considered 30 possible locations for the proposed new engineering and manufacturing plant. Of these, 20 were eliminated almost immediately for one reason or another (unfavorable zoning regulations, inadequate transportation infrastructure, etc.). Of the remaining ten locations, the committee selected for intensive study the three that seemed most promising: Chicago, Minneapolis, and Salt Lake City. We have now visited these three cities, and our observations and recommendations follow.
>
> **CHICAGO** ⟵ Second-level head
> Of the three cities, Chicago presently seems to the committee to offer the greatest advantages, although we wish to examine these more carefully before making a final recommendation.
>
> **Selected Location** ⟵ Third-level head
> Though not at the geographic center of the United States, Chicago is the demographic center to more than three-quarters of the U.S. population. It is within easy reach of our corporate headquarters in New York. And it is close to several of our most important suppliers of components and raw materials—those, for example, in Columbus, Detroit, and St. Louis. Several factors were considered essential to the location, although some may not have had as great an impact on the selection. . . .
>
> ***Air Transportation.***   Chicago has two major airports (O'Hare and Midway) and is contemplating building a third. Both domestic and international air-cargo service are available. . . . ⟵ Fourth-level heads
>
> ***Sea Transportation.***   Except during the winter months when the Great Lakes are frozen, Chicago is an international seaport. . . .
>
> ***Rail Transportation.***   Chicago is served by the following major railroads. . . .

FIGURE H–2. Headings Used in a Document

times a client for whom a report or proposal is being prepared requires a particular format. In the absence of specific guidelines, follow the system illustrated in Figure H–2.

Decimal Numbering System

Some documents, such as **specifications**, benefit from the decimal numbering system for ease of cross-referencing sections. The decimal numbering system uses a combination of numbers and decimal points to differentiate among levels of headings. The following example shows the correspondence between different levels of headings and the decimal numbers used:

1. FIRST-LEVEL HEADING
   1.1 Second-level heading
   1.2 Second-level heading
      1.2.1 Third-level heading
      1.2.2 Third-level heading
         1.2.2.1 Fourth-level heading
         1.2.2.2 Fourth-level heading
   1.3 Second-level heading
2. FIRST-LEVEL HEADING

Although decimal headings are indented in an outline or a **table of contents**, they are flush with the left margins when they function as headings in the body of a report. Every heading starts on a new line, with an extra line of space above and below the heading.

WRITER'S CHECKLIST   **Using Headings**

- ✔ Use headings to signal a new topic. Use a lower-level heading to indicate a new subtopic within the larger topic.
- ✔ Make headings concise but specific enough to be informative, as in Figure H–2.
- ✔ Avoid too many or too few headings or levels of headings; too many clutter a document, and too few fail to provide recognizable structure.
- ✔ Ensure that headings at the same level are of relatively equal importance and have **parallel structure**.
- ✔ Subdivide sections only as needed; when you do, try to subdivide them into at least two lower-level headings.
- ✔ Do not leave a heading as the final line of a page. If two lines of text cannot fit below a heading, start the section at the top of the next page.
- ✔ Do not allow a heading to substitute for discussion; the text should read as if the heading were not there.

hyphens

DIRECTORY
Hyphens with Compound Words   245
Hyphens with Modifiers   245
Hyphens with Prefixes and Suffixes   246
Hyphens and Clarity   246
Other Uses of the Hyphen   246
Writer's Checklist: Using Hyphens to Divide Words   246

The hyphen (-) is used primarily for linking and separating words and parts of words. The hyphen often improves the **clarity** of writing. The hyphen is sometimes confused with the **dash**, which may be indicated with two consecutive hyphens.

Hyphens with Compound Words

Some **compound words** are formed with hyphens (able-bodied, over-the-counter). Hyphens are also used with multiword **numbers** from twenty-one through ninety-nine and fractions when they are written out (three-quarters). Most current **dictionaries** indicate whether compound words are hyphenated, written as one word, or written as separate words.

Hyphens with Modifiers

Two- and three-word **modifiers** that express a single thought are hyphenated when they precede a **noun**.

▶ It was a *well-written* report.

However, a modifying phrase is not hyphenated when it follows the noun it modifies.

▶ The report was *well written*.

If each of the words can modify the noun without the aid of the other modifying word or words, do not use a hyphen (a *new laser* printer). If the first word is an **adverb** ending in *-ly*, do not use a hyphen (a *privately held* company). A hyphen is always used as part of a letter or number modifier (A-frame house, 17-inch screen).

In a series of unit modifiers that all have the same term following the hyphen, the term following the hyphen need not be repeated throughout the series; for greater smoothness and brevity, use the term only at the end of the series.

▶ The third-, fourth-, and fifth-floor laboratories were inspected.

Hyphens with Prefixes and Suffixes

A hyphen is used with a **prefix** when the root word is a proper noun (pre-Columbian, anti-American, post-Newtonian). A hyphen may be used when the prefix ends and the root word begins with the same vowel (re-enter, anti-inflammatory). A hyphen is used when *ex-* means "former" (ex-president, ex-spouse). A hyphen may be used to emphasize a prefix. ("He is *anti-change*.") The **suffix** *-elect* is hyphenated (president-elect).

Hyphens and Clarity

The presence or absence of a hyphen can alter the meaning of a sentence.

AMBIGUOUS We need a biological waste management system.

That sentence could mean one of two things: (1) We need a system to manage "biological waste," or (2) We need a "biological" system to manage waste.

CLEAR We need a *biological-waste* management system. [1]
CLEAR We need a biological *waste-management* system. [2]

To avoid confusion, some words and modifiers should always be hyphenated. *Re-cover* does not mean the same thing as *recover*, for example; the same is true of *re-sign* and *resign* and *un-ionized* and *unionized*.

Other Uses of the Hyphen

Hyphens are used between letters showing how a word is spelled.

▶ In his e-mail, he misspelled *believed* as b-e-l-e-i-v-e-d.

A hyphen can stand for *to* or *through* between letters and numbers (pages 44-46, the Detroit-Toledo Expressway, A-L and M-Z).

Hyphens are commonly used in telephone numbers (800-555-1212), Web addresses (computer-parts.com), file names (report-09.doc), and similar number/symbol combinations. See also **dates**.

> **WRITER'S CHECKLIST** Using Hyphens to Divide Words
>
> ✔ Do not divide one-syllable words.
> ✔ Divide words at syllable breaks, which you can determine with a dictionary.
> ✔ Do not divide a word if only one letter would remain at the end of a line or if fewer than three letters would start a new line.

Writer's Checklist: Using Hyphens to Divide Words (continued)

- ✔ Do not divide a word at the end of a page; carry it over to the next page.
- ✔ If a word already has a hyphen in its spelling, divide the word at the existing hyphen.
- ✔ Do not use a hyphen to break a URL or an e-mail address at the end of a line because it may confuse readers who could assume that the hyphen is part of the address.

idioms

An idiom is a group of words that has a special meaning apart from its literal meaning. Someone who "runs for office" in the United States, for example, need not be an athlete. The same candidate would "stand for office" in the United Kingdom. Because such expressions are specific to a culture, nonnative speakers must memorize them.

Idioms are often constructed with **prepositions** that follow **adjectives** (*similar to*), **nouns** (*need for*), and **verbs** (*approve of*). Some idioms can change meaning slightly with the preposition used, as in *agree to* ("consent") and *agree with* ("in accord"). The following are typical idioms that give nonnative speakers trouble.

call off [cancel]	hand in [submit]
call on [visit a client]	hand out [distribute]
cross out [draw a line through]	keep on [continue]
do over [repeat a task]	leave out [omit]
drop in on [visit unexpectedly]	look up [research a subject]
figure out [solve a problem]	put off [postpone]
find out [discover information]	run into [meet by chance]
get through with [finish]	run out of [deplete supply]
give up [quit]	watch out for [be careful]

Idioms often provide helpful shortcuts. In fact, they can make writing more natural and vigorous. Avoid them, however, if your writing is to be translated into another language or read in other English-speaking countries. Because no language system can fully explain such usages, a reader must check **dictionaries** or usage guides to interpret the meaning of idioms. See also **English as a second language** and **international correspondence**.

WEB LINK | **Prepositional Idioms**

For links to helpful lists of common pairings of prepositions with nouns, verbs, and adjectives, see *bedfordstmartins.com/alredtech* and select *Links for Handbook Entries*.

illegal / illicit

If something is *illegal*, it is prohibited by law. If something is *illicit*, it is prohibited by either law or custom. *Illicit* behavior may or may not be *illegal*, but it does violate social convention or moral codes and therefore usually has a clandestine or immoral **connotation**. ("The employee's *illicit* sexual behavior caused a scandal, but the company's attorney concluded that no *illegal* acts were committed.")

illustrations (*see* visuals)

imply / infer

If you *imply* something, you hint at or suggest it. ("Her e-mail *implied* that the project would be delayed.") If you *infer* something, you reach a conclusion based on evidence or interpretation. ("The manager *inferred* from the e-mail that the project would be delayed.")

in / into

In means "inside of"; *into* implies movement from the outside to the inside. ("The equipment was *in* the test chamber, so she reached *into* the chamber to adjust it.")

in order to

Most often, *in order to* is a meaningless filler phrase that is dropped into a sentence without thought. See also **conciseness**.

▶ ~~In order to~~ *To* start the engine, open the choke and throttle and then press the starter.

However, the phrase *in order to* is sometimes essential to the meaning of a sentence.

▶ If the vertical scale of a graph line would not normally show the zero point, use a horizontal break in the graph *in order to* include the zero point.

In order to also helps control the **pace** of a sentence, even when it is not essential to the meaning of the sentence.

▶ The committee must know the estimated costs *in order to* evaluate the feasibility of the project.

in terms of

When used to indicate a shift from one kind of language or terminology to another, the phrase *in terms of* can be useful.

▶ *In terms of* gross sales, the year has been relatively successful; however, *in terms of* net income, it has been discouraging.

When simply dropped into a sentence because it easily comes to mind, *in terms of* is meaningless **affectation**. See also **conciseness**.

▶ She was thinking ~~in terms~~ of subcontracting much of the work.

indexing

An index is an alphabetical list of all the major topics and sometimes subtopics in a written work. It cites the pages where each topic can be found and allows **readers** to find information on particular topics quickly and easily, as shown in Figure I–1. The index always comes at the very end of the work. Many Web sites also provide linked subject indexes to the content of the sites.

The key to compiling a useful index is selectivity. Instead of listing every possible reference to a topic, select references to passages where the topic is discussed fully or where a significant point is made about it. For index entries like those in Figure I–1, choose key terms that best

```
monitoring programs, 27–44  ─────────── Main entry
    aquatic, 42
    ecological, 40
    meteorological, 37  ─────────── Subentries
        operational, 39
        preoperational, 37  ─────── Sub-subentries
    radiological, 30
    terrestrial, 41, 43–44
    thermal, 27  ──────────────────── Subentries
```

FIGURE I–1. Index Entry (with Main Entry, Subentries, and Sub-subentries)

represent a topic. Key terms are those words or phrases that a reader would most likely look for in an index. For example, the key terms in a reference to the development of legislation about **environmental impact statements** would probably be *legislation* and *environmental impact statement*, not *development*. In selecting terms for index entries, use chapter or section titles only if they include such key terms. For index entries on **tables** and **visuals**, use the words from their titles that will function as key terms a reader might seek. Create alphabetical Web-site indexes from links to topics in subsites throughout the larger site.

Compiling an Index

Do not attempt to compile an index until the final manuscript is completed because terminology and page numbers will not be accurate before then. The best way to manually compile a list of topics is to read through your written work from the beginning; each time a key term appears in a significant context, note the term and its page number. An index entry can consist solely of a main entry and its page number.

▶ aquatic monitoring programs, 42

An index entry can also include a main entry, subentries, and even sub-subentries, as shown in Figure I–1. A subentry indicates the pages where a specific subcategory or subdivision of the main topic can be found. When you have compiled a list of key terms for the entire work, sort the main entries alphabetically, then sort all subentries and sub-subentries alphabetically beneath their main entries. To help indexers with this process, the American Society of Indexers lists software available for indexing at their Web site, *www.asindexing.org*.

Wording Index Entries

The first word of an index entry should be the principal word because the reader will look for topics alphabetically by their main words. Selecting the right word to list first is easier for some topics than for others. For instance, *tips on repairing electrical wire* would not be a suitable index entry because a reader looking for information on electrical wire would not look under the word *tips*. Ordinarily, an entry with two keywords, such as *electrical wire*, should be indexed under each word (*electrical wire* and *wire, electrical*). A main index entry should be written as a **noun** or a noun **phrase** rather than as an **adjective** alone or a **verb**.

▶ electrical wire, 20–22
    grounding, 21
    insulation, 20
    repairing, 22
    size, 21

> **DIGITAL TIP**
>
> **Creating an Index**
>
> Word-processing software can provide a quick and an efficient way to create an alphabetical subject index of your document. First, review the document to identify entries: the words, phrases, figure captions, or symbols that you wish to index. Then, following your software's instructions, highlight and code these entries so that they can be arranged in the index. Your draft index will still need careful review, but using the software to create the first draft can save time. For further instructions, see *bedfordstmartins.com/alredtech* and select *Digital Tips*, "Creating an Index."

Cross-Referencing

Cross-references in an index help readers find other related topics in the text. A reader looking up *technical writing*, for example, might find cross-references to *report* or *manual*. Cross-references do not include page numbers; they merely direct readers to another main index entry where they can find page numbers. The two kinds of cross-references are *see* references and *see also* references.

See references are most commonly used with topics that can be identified by several different terms. Listing the topic page numbers by only one of the terms, the indexer then lists the other terms throughout the index as *see* references.

> ▶ economic costs. *See* benefit-cost analyses

See references also direct readers to index entries where a topic is listed as a subentry.

> ▶ L-shaped fittings. *See* elbows, L-shaped fittings

See also references indicate other entries that include additional information on a topic.

> ▶ ecological programs, 40–49. *See also* monitoring programs

WRITER'S CHECKLIST **Indexing**

- ✔ Use lowercase for the first words and all subsequent words of main entries, subentries, and sub-subentries unless they are proper nouns or would otherwise be capitalized. See **capitalization**.
- ✔ Use italics for the cross-reference terms *see* and *see also*.

Writer's Checklist: Indexing (continued)

- Place each subentry in the index on a separate line, indented from its main entry. Indent sub-subentries from the preceding subentry. Indentations allow readers to scan a column quickly for pertinent subentries or sub-subentries.
- Separate entries from page numbers with commas.
- Format the index economically with double columns, as is done in the index to this book.

indiscreet / indiscrete

Indiscreet means "lacking in prudence or sound judgment." ("His public discussion of the proposed merger was *indiscreet*.") *Indiscrete* means "not divided or divisible into parts." ("The separate departments, once combined, become *indiscrete*.") See also **discreet / discrete**.

individual

Individual is most appropriate when used as an **adjective** to distinguish a single person from a group. ("The *individual* employee's obligations are detailed in the policy manual.") Using *individual* as a **noun** is an **affectation**. Use *people* or another appropriate term.

▶ Several ~~individuals on~~ ^members of^ the committee did not vote.

inquiries and responses

The purpose of writing inquiry **letters** or **e-mail** messages is to obtain responses to requests or specific questions, as in Figure I–2, which shows a college student who is requesting information from an official at a power company. Inquiries may either benefit the reader (as in requests for information about a product that a company has advertised) or benefit the writer (as in the student's inquiry in Figure I–2). Inquiries that primarily benefit the writer require the use of **persuasion** and special consideration of the needs of your **audience**. See also **correspondence**.

> Dear Ms. Metcalf:
>
> I am an architecture student at the University of Dayton, and I am working with a team of students to design an energy-efficient house for a class project. I am writing to request information on heating systems based on the specifications of our design. We would appreciate any information you could provide.
>
> The house we are designing contains 2,000 square feet of living space (17,600 cubic feet) and meets all the requirements in your brochure "Insulating for Efficiency." We need the following information, based on the southern Ohio climate:
>
> - The proper-size heat pump for such a home.
> - The wattage of the supplemental electrical heating units required.
> - The estimated power consumption and rates for those units for one year.
>
> We will be happy to send you our preliminary design report. If you have questions or suggestions, contact me at kparsons@fly.ud.edu or call 513-229-4598.
>
> Thank you for your help.

FIGURE I–2. Inquiry

Writing Inquiries

Inquiries need to be specific, clear, and concise in order to receive a prompt, helpful reply.

- Phrase your request so that the reader will immediately know the type of information you are seeking, why you need it, and how you will use it.
- If possible, present questions in a numbered or bulleted **list** to make it easy for your reader to respond to them.
- Keep the number of questions to a minimum to improve your chances of receiving a prompt response.
- Offer some inducement for the reader to respond, such as promising to share the results of what you are doing. See also **"you" viewpoint**.
- Promise to keep responses confidential, when appropriate.

In the closing, thank the reader for taking the time to respond. In addition, make it convenient for the recipient to respond by providing your contact information, such as a phone number or an e-mail address, as shown in Figure I–2.

Responding to Inquiries

When you receive an inquiry, determine whether you have both the information and the authority to respond. If you are the right person in your organization to respond, answer as promptly as you can, and be sure to answer every inquiry or question asked, as shown in Figure I–3. How long and how detailed your response should be depends on the nature of the question and the information the writer provides.

If you have received an inquiry that you feel you cannot answer, find out who can and forward the inquiry to that person. Notify the writer that you have forwarded the inquiry. The person who replies to a forwarded inquiry should state in the first paragraph of the response who has forwarded the original inquiry, as shown in Figure I–3.

Dear Ms. Parsons:

Jane Metcalf forwarded to me your inquiry of March 11 about the house that your architecture team is designing. I can estimate the heating requirements of a typical home of 17,600 cubic feet as follows:

- For such a home, we would generally recommend a heat pump capable of delivering 40,000 BTUs, such as our model AL-42 (17 kilowatts).
- With the AL-42's efficiency, you don't need supplemental heating units.
- Depending on usage, the AL-42 unit averages between 1,000 and 1,500 kilowatt-hours from December through March. To determine the current rate for such usage, check with Dayton Power and Light Company.

I can give you an answer that would apply specifically to your house based on its particular design (such as number of stories, windows, and entrances). If you send me more details, I will be happy to provide more precise figures for your interesting project.

Sincerely,

FIGURE I–3. Response to an Inquiry

inside / inside of

In the phrase *inside of*, the word *of* is redundant and should be omitted.

▶ The switch is just inside ~~of~~ the door.

Using *inside of* to mean "in less time than" is colloquial and should be avoided in writing.

▶ They were finished ~~inside of~~ *in less than* an hour.

insoluble / insolvable

The words *insoluble* and *insolvable* are sometimes used interchangeably to mean "incapable of being solved." *Insoluble* also means "incapable of being dissolved."

▶ Until yesterday, the production problem seemed *insolvable*.

▶ *Insoluble* fiber passes through the intestines largely intact.

instant messaging

Instant messaging (IM) is a text-based communications medium that fills a niche between telephone calls and **e-mail** messages. It allows both real-time communications, like a phone call, and the transfer of text or other files, like an e-mail message. It is especially useful to cell-phone users who are working at sites without access to e-mail. See also **selecting the medium**.

To set up routine IM exchanges, add the user names of the people you exchange messages with to your list of contacts (or buddy list). Choose a screen name that your colleagues will recognize. If you use IM routinely as part of your job, create an "away" message that signals when you are not available for IM interactions.

When writing instant messages, keep your messages simple and to the point, covering only one subject in each message to prevent confusion and inappropriate responses. Because screen space is often limited and speed is essential, many who send instant messages use abbreviations and shortened spellings ("u" for "you" and "L8R" for "later"). Be sure that your **reader** will understand such abbreviations; when in doubt, avoid them.*

*Many online sites, such as *www.netlingo.com*, define IM and other such abbreviations.

In Figure I–4, the manager of a software development company in Maine ("Diane") is exchanging instant messages with a business partner in the Netherlands ("Andre"). Notice that the correspondents use an informal style that includes personal and professional abbreviations with which both are familiar ("NP" for "no problem"; and "QSG" for Quick Start Guide). These messages demonstrate how IM can not only

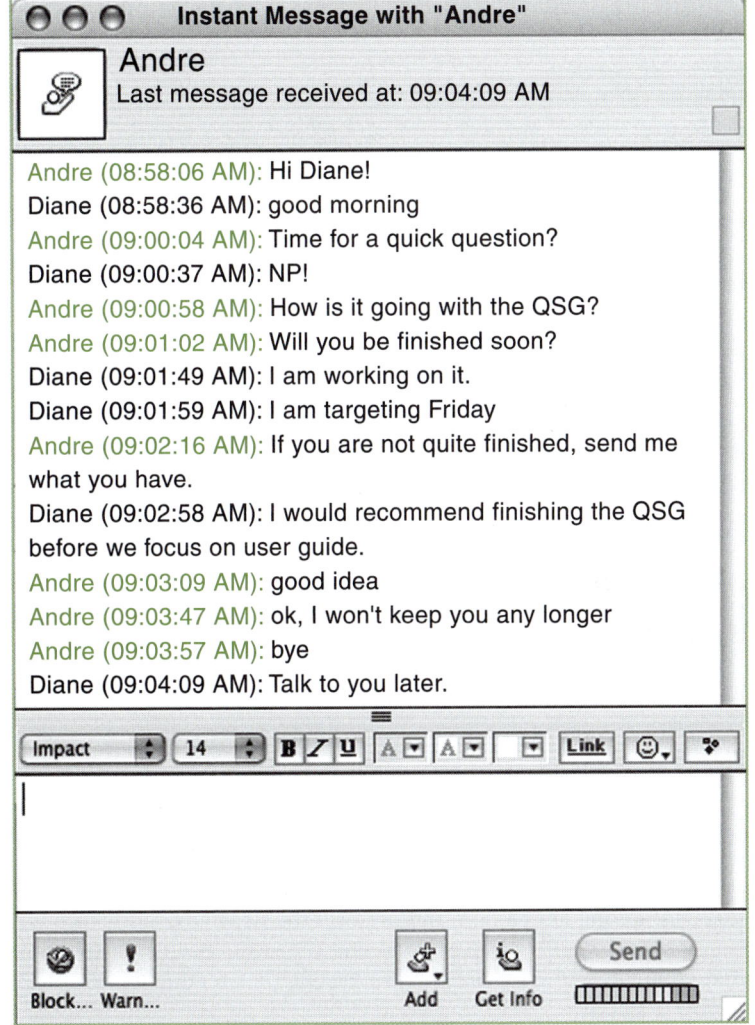

FIGURE I–4. Instant-Message Exchange

help exchange information quickly but also build rapport among distant colleagues and team members. The exchange also demonstrates why IM is not generally appropriate for many complex messages or for more formal circumstances, such as when establishing new professional relationships.

ETHICS NOTE Be sure to follow your employer's IM policies, such as any limitations on sending personal messages during work hours or requirements concerning confidentiality. If no specific policy exists, assume that personal use of IM is not appropriate in your workplace. ✦

> **WRITER'S CHECKLIST** **Instant Messaging Privacy and Security**
>
> ✓ Organize your contact, or buddy, lists to separate workplace contacts from family and friends so that you do not inadvertently send personal messages to professional associates and vice versa. Consider restricting buddy lists on professional IM accounts to business associates.
>
> ✓ Learn the options, capabilities, and security limitations of your IM system and set the preferences that best suit your use of the system.
>
> ✓ Be especially alert to the possibilities of virus infections and security risks with messages, attachments, access to buddy lists, and other privacy issues.
>
> ✓ Save significant IM exchanges (or logs) for your future reference.
>
> ✓ Be aware that instant messages can be saved by your recipients and may be archived by your employer. (See the Ethics Note on page 162.)
>
> ✓ Do not use professional IM for gossip or inappropriate exchanges.

instructions

Instructions that are clear and easy to follow prevent miscommunication and help **readers** complete tasks effectively and safely. To write effective instructions, you must thoroughly understand the process, system, or device you are describing. Often, you must observe someone as he or she completes a task and perform the steps yourself before you begin to write. Finally, keep in mind that the most effective instructions often combine written elements and **visuals** that reinforce each other. See also **manuals**.

Writing Instructions

Consider your audience's level of knowledge. If all your readers have good backgrounds in the topic, you can use fairly specialized terms. If that is not the case, use plain language or include a **glossary** for specialized terms that you cannot avoid. See also **audience**.

Clear and easy-to-follow instructions are written as commands in the imperative **mood**, active **voice**, and (whenever possible) present **tense**.

▶ *Raise the*
~~The~~ access lid ~~will be raised by the operator.~~

Although **conciseness** is important in instructions, **clarity** is essential. You can make sentences shorter by leaving out some **articles** (*a*, *an*, *the*), some **pronouns** (*you*, *this*, *these*), and some **verbs**, but such sentences may result in **telegraphic style** and be harder to understand. For example, the first version of the following instruction for placing a document in a scanner tray is confusing.

CONFUSING	Place document in tray with printed side facing opposite.
CLEAR	Place the document in the document tray with the printed side facing away from you.

One good way to make instructions easy to follow is to divide them into short, simple steps in their proper sequence. Steps can be organized with words (*first*, *next*, *finally*) that indicate time or sequence.

▶ *First*, determine the problem the customer is having with the computer. *Next*, observe the system in operation. *At that time*, question the operator until you are sure that the problem has been explained completely. *Then* analyze the problem and make any necessary adjustments.

You can also use numbers, as in the following:

▶ 1. Connect each black cable wire to a brass terminal.
2. Attach one 4-inch green jumper wire to the back.
3. Connect the jumper wire to the bare cable wire.

Consider using the numbered- or bulleted-list feature of your word-processing software to create sequenced steps. See **lists**.

Plan ahead for your reader. If the instructions in step 2 will affect a process in step 9, say so in step 2. Sometimes your instructions have to make clear that two operations must be performed simultaneously. Either state that fact in an **introduction** to the specific instructions or include both operations in one step.

CONFUSING	1. Hold down the CONTROL key.
	2. Press the RETURN key before releasing the CONTROL key.
CLEAR	1. While holding down the CONTROL key, press the RETURN key.

If your instructions involve many steps, break them into stages, each with a separate heading so that each stage begins again with step 1. Using **headings** as dividers is especially important if your reader is likely to be performing the operation as he or she reads the instructions.

Illustrating Instructions

Illustrations should be developed together with the text, especially for complex instructions that benefit from visuals that foster **clarity** and conciseness. Such visuals as **drawings**, **flowcharts**, **maps**, and **photographs** enable your reader to identify relationships more easily than long explanations. Some instructions, such as those for products sold internationally, use only visuals. The entry **global graphics** describes visuals that avoid culture-specific connotations and are often used in such instructions.

Consider the **layout and design** of your instructions to most effectively integrate visuals. Highlight important visuals as well as text by making them stand out from the surrounding text. Consider using boxes and boldface or distinctive headings. Experiment with font style, size, and color to determine which devices are most effective.

The instructions in Figure I–5 guide the reader through the steps of streaking a saucer-sized disk of material (called *agar*) used to grow bacteria colonies. The purpose is to thin out the original specimen (the *inoculum*) so that the bacteria will grow in small, isolated colonies. This section could be part of other larger instructional documents for which streaking is only one step among others.

> **WEB LINK** | **Writing Instructions**
>
> For links to additional advice on writing instructions and models, see *bedfordstmartins.com/alredtech* and select *Links for Handbook Entries*.

Warning Readers

Alert your readers to any potentially hazardous materials (or actions) before they reach the step for which the material is needed. Caution readers handling hazardous materials about any requirements for special clothing, tools, equipment, and other safety measures. Highlight warnings, cautions, and precautions to make them stand out visually

> **STREAKING AN AGAR PLATE**
>
> *Distribute the inoculum over the surface of the agar in the following manner:*
> 1. Beginning at one edge of the saucer, thin the inoculum by streaking back and forth over the same area several times, sweeping across the agar surface until approximately one-quarter of the surface has been covered. *Sterilize the loop in an open flame.*
> 2. Streak at right angles to the originally inoculated area, carrying the inoculum out from the streaked areas onto the sterile surface with only the first stroke of the wire. Cover half of the remaining sterile agar surface. *Sterilize the loop.*
> 3. Repeat as described in Step 2, covering the remaining sterile agar surface.
>
>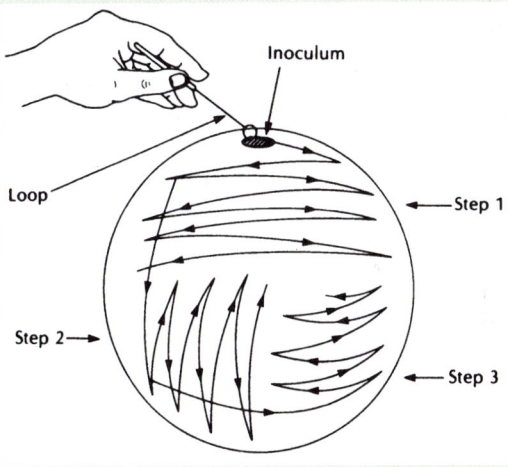

FIGURE I–5. Illustrated Instructions

from the surrounding text. Present warning notices in a box, in all uppercase letters, in large and distinctive fonts, or in color. Experiment with font style, size, and color to determine which devices are most effective.

Figure I–6 shows a warning from an instruction manual for the use of a gas grill. Notice that a drawing supports the text of the warning.

WARNING

- The assembler/owner is responsible for the assembly, installation, and maintenance of the grill.
- Use the grill outdoors only.
- Do not let children operate or play near your grill.
- Keep the grill area clear and free from materials that burn, gasoline, bottled gas in any form, and other flammable vapors and liquids.
- Do not block holes in bottom and back of grill.
- Visually check burner flames on a regular basis.
- Use the grill in a well-ventilated space. Never use in an enclosed space, carport, garage, porch, patio, or building made of combustible construction, or under overhead construction.
- Do not install your grill in or on recreational vehicles and/or boats.
- Keep grills a distance of 36", or 3 ft. (approximately 1 m), from buildings to reduce the possibility of fire or heat damage to materials.

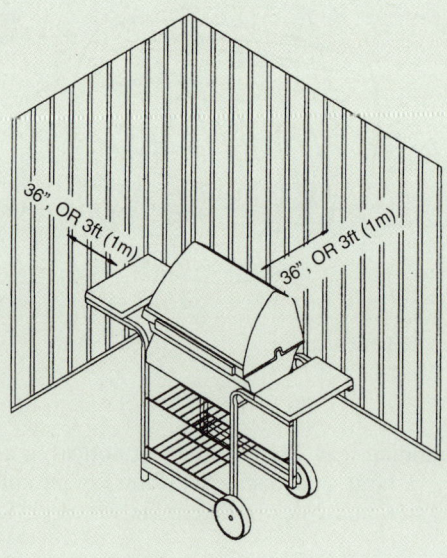

FIGURE I–6. Warning in a Set of Instructions

Various standards organizations and agencies publish guidelines on the use of terminology, colors, and symbols in instructions and labeling. Two widely influential organizations that publish such guidelines are the American National Standards Institute (*www.ansi.org*) and the International Organization for Standardization (*www.iso.org*).

Testing Instructions

To test the accuracy and clarity of your instructions, ask someone who is not familiar with the task to follow your directions. A first-time user can spot missing steps or point out passages that should be worded more clearly. As you observe your tester, note any steps that seem especially confusing and revise accordingly. See **usability testing**.

> **WRITER'S CHECKLIST** Writing Instructions
>
> ✔ Use the imperative mood and the active voice.
> ✔ Use short sentences and simple present tense as much as possible.
> ✔ Avoid technical terminology and **jargon** that your readers might not know, including undefined **abbreviations**.
> ✔ Do not use elegant variation (two different words for the same thing). See also **affectation**.
> ✔ Eliminate any **ambiguity**.
> ✔ Use effective visuals and highlighting devices.
> ✔ Include appropriate warnings and cautions.
> ✔ Verify that measurements, distances, times, and relationships are precise and accurate.
> ✔ Test your instructions by having someone else follow them while you observe.

insure / ensure / assure

Insure, *ensure*, and *assure* all mean "make secure or certain." *Assure* refers to people, and it alone has the connotation of setting a person's mind at rest. ("I *assure* you that the equipment will be available.") *Ensure* and *insure* mean "make secure from harm." Only *insure* is widely used in the sense of guaranteeing the value of life or property.

▶ We need all the data to *ensure* the success of the project.
▶ We should *insure* the contents of the warehouse.

intensifiers

Intensifiers are **adverbs** that **emphasize** degree, such as *very*, *quite*, *rather*, *such*, and *too*. Although intensifiers serve a legitimate and necessary function, unnecessary intensifiers can weaken your writing. Eliminate those that do not make an obvious contribution or replace them with specific details.

> The team learned the ~~very~~ good news that it had been awarded a ~~rather substantial monetary~~ $10,000 prize for its design.

Some words (such as *perfect*, *impossible*, and *final*) do not logically permit intensification because, by definition, they do not allow degrees of comparison. Although **usage** often ignores that logical restriction, to ignore it is to defy the basic meanings of such words. See also **adjectives**, **conciseness**, and **equal / unique / perfect**.

interface

An *interface* is a surface that provides a common boundary between two bodies or areas. The bodies or areas may be physical ("the *interface* of a piston and a cylinder") or conceptual ("the *interface* of mathematics and economics"). Do not use *interface* as a substitute for the verb *cooperate*, *interact*, or even *work*. See **affectation** and **buzzwords**.

interjections

An interjection is a word or phrase standing alone or inserted into a sentence to exclaim or to command attention. Grammatically, it has no connection to the sentence. An interjection can be strong (*Hey*! *Ouch*! *Wow*!) or mild (*oh*, *well*, *indeed*). A strong interjection is followed by an **exclamation mark**.

> *Wow*! Profits more than doubled last quarter.

A weak interjection is followed by a **comma**.

> *Well*, we need to rethink the proposal.

An interjection inserted into a sentence usually requires a comma before it and after it.

▶ We must, *indeed*, rethink the proposal.

Because they get their main expressive force from sound, interjections are more common in speech than in writing. Use them sparingly.

international correspondence

Business **correspondence** varies among national cultures. Organizational patterns, persuasive strategies, forms of courtesy, formality, and ideas about efficiency differ from country to country. For example, in the United States, direct, concise correspondence may demonstrate courtesy by not wasting another person's time. In many other countries, such directness and brevity may seem rude to **readers**, suggesting that the writer wishes to spend as little time as possible corresponding with the reader. (See **audience** and **tone**.) Likewise, where a U.S. writer might consider one brief **letter** or **e-mail** sufficient to communicate a request, a writer in another country may expect an exchange of three or four longer letters to pave the way for action.

Cultural Differences in Correspondence

When you read correspondence from businesspeople in other cultures or countries, be alert to differences in such features as customary expressions, openings, and closings. Japanese business writers, for example, traditionally use indirect openings that reflect on the season, compliment the reader's success, and offer hopes for the reader's continued prosperity. Consider deeper issues as well, such as how writers from other cultures express bad news. Japanese writers traditionally express negative messages, such as **refusal letters**, indirectly to avoid embarrassing the recipient. Such cultural differences are often based on perceptions of time, face-saving, and traditions. The features and communication styles of specific national cultures are complex; the entry **global communication** provides information and resources for cross-cultural study. See also **global graphics**.

Cross-Cultural Examples

Figures I–7 and I–8 on pages 266 and 267 are a draft and a final version of a letter written by an American businessman to a Japanese businessman. The opening and closing of the draft in Figure I–7 do not include enough of the politeness strategies that are important in Japanese culture, and the informal salutation inappropriately uses the recipient's first name (*Dear Ichiro:*). This draft also contains **idioms** (*looking forward, company family*), **jargon** (*transport will be holding . . .*), **contractions**

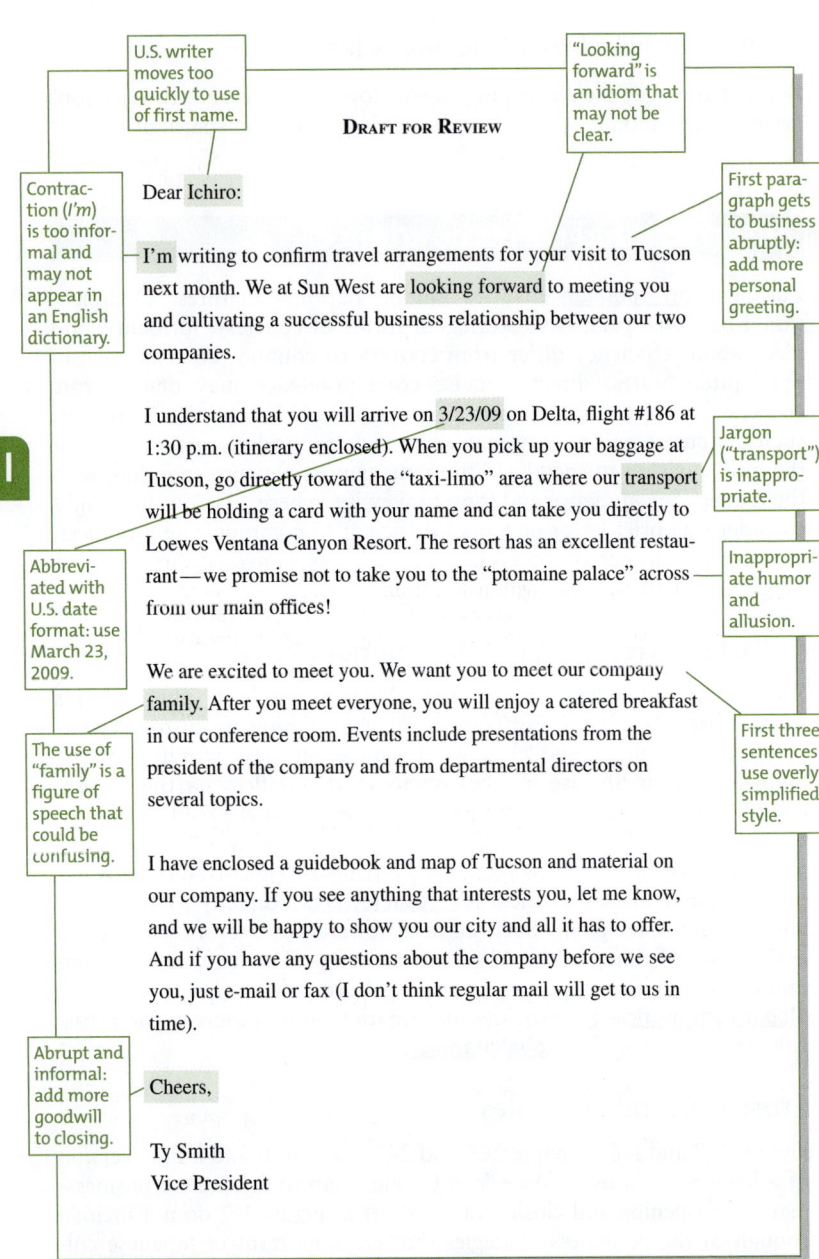

FIGURE I–7. Inappropriate International Correspondence (Draft Marked for Revision)

Sun West Corporation, Inc.

2565 North Armadillo
Tucson, AZ 85719
Phone: (602) 555-6677
Fax: (602) 555-6678   sunwest.com

March 4, 2009

Ichiro Katsumi
Investment Director
Toshiba Investment Company
1-29-10 Ichiban-cho
Tokyo 105, Japan

Dear Mr. Katsumi:

I hope that you and your family are well and prospering in the new year. We at Sun West Corporation are very pleased that you will be coming to visit us in Tucson this month. It will be a pleasure to meet you, and we are very gratified and honored that you are interested in investing in our company.

So that we can ensure that your stay will be pleasurable, we have taken care of all of your travel arrangements. You will

- Depart Narita–New Tokyo International Airport on Delta Airlines flight #75 at 1700 on March 23, 2009.
- Arrive at Los Angeles International Airport at 1050 local time and depart for Tucson on Delta flight #186 at 1205.
- Arrive at Tucson International Airport at 1330 local time on March 23.
- Depart Tucson International Airport on Delta flight #123 at 1845 on March 30.
- Arrive in Salt Lake City, Utah, at 1040 and depart on Delta flight #34 at 1115.
- Arrive in Portland, Oregon, at 1210 local time and depart on Delta flight #254 at 1305.
- Arrive in Tokyo at 1505 local time on March 31.

If you need additional information about your travel plans or information on Sun West Corporation, please call, fax, or e-mail me directly at tsmith@sunwest.com. That way, we will receive your message in time to make the appropriate changes or additions.

FIGURE I–8. Appropriate International Correspondence

Mr. Ichiro Katsumi   2   March 4, 2009

After you arrive in Tucson, a chauffeur from Skyline Limousines will be waiting for you at Gate 12. He or she will be carrying a card with your name, will help you collect your luggage from the baggage claim area, and will then drive you to the Loewes Ventana Canyon Resort. This resort is one of the most prestigious in Tucson, with spectacular desert views, high-quality amenities, and one of the best golf courses in the city. The next day, the chauffeur will be back at the Ventana at 0900 to drive you to Sun West Corporation.

We at Sun West Corporation are very excited to meet you and introduce you to all the staff members of our hardworking and growing company. After you meet everyone, you will enjoy a catered breakfast in our conference room. At that time, you will receive a schedule of events planned for the remainder of your trip. Events include presentations from the president of the company and from departmental directors on

- The history of Sun West Corporation
- The uniqueness of our products and current success in the marketplace
- Demographic information and the benefits of being located in Tucson
- The potential for considerable profits for both our companies with your company's investment

We encourage you to read through the enclosed guidebook and map of Tucson. In addition to events planned at Sun West Corporation, you will find many natural wonders and historical sites to see in Tucson and in Arizona in general. If you see any particular event or place that you would like to visit, please let us know. We will be happy to show you our city and all it has to offer.

Again, we are very honored that you will be visiting us, and we look forward to a successful business relationship between our two companies.

Sincerely,

Ty Smith

Ty Smith
Vice President

Enclosures (2)

FIGURE I–8. Appropriate International Correspondence (*continued*)

(*I'm, don't*), informal language (*just e-mail or fax, Cheers*), and humor and allusion (*"ptomaine palace" across from our main offices*).

Compare that letter to the one in Figure I–8, which is written in language that is courteous, literal, and specific. This revised letter begins with concern about the recipient's family and prosperity because that opening honors traditional Japanese patterns in business correspondence. The letter is free of slang, idioms, and jargon. The sentences are shorter than in the draft, bulleted lists break up the paragraphs, contractions are avoided, months are spelled out, and 24-hour clock time is used.

When writing for international readers, rethink the ingrained habits that define how you express yourself, learn as much as you can about the cultural expectations of others, and focus on politeness strategies that demonstrate your respect for readers. Doing so will help you achieve **clarity** and mutual understanding with international readers.

> **WEB LINK** Google's International Directory
>
> Google's International Business and Trade Directory provides an excellent starting point for searching the Web for information related to customs, communication, and international standards. See *bedfordstmartins.com/alredtech* and select *Links for Handbook Entries*.

> **WRITER'S CHECKLIST** Writing International Correspondence
>
> - Observe the guidelines for courtesy, such as those in the *Writer's Checklist: Using Tone to Build Goodwill* in **correspondence** on page 105.
> - Write clear and complete sentences: Unusual word order or rambling sentences will frustrate and confuse readers. See **garbled sentences**.
> - Avoid an overly simplified style that may offend or any **affectation** that may confuse the reader. See also **English as a second language**.
> - Avoid humor, irony, and sarcasm; they are easily misunderstood outside their cultural **context**.
> - Do not use idioms, jargon, slang expressions, unusual **figures of speech**, or **allusions** to events or attitudes particular to American life.
> - Consider whether necessary technical terminology can be found in abbreviated English-language dictionaries; if it cannot, carefully define such terminology.
> - Do not use contractions or **abbreviations** that may not be clear to international readers.

Writer's Checklist: Writing International Correspondence (continued)

- Avoid inappropriate informality, such as using first names too quickly.
- Write out **dates**, whether in the month-day-year style (June 11, 2006, not 6/11/06) used in the United States or the day-month-year style (11 June 2006, not 11/6/06) used in many other parts of the world.
- Specify time zones or refer to international standards, such as Greenwich Mean Time (GMT) or Coordinated Universal Time (UTC).
- Use international measurement standards, such as the metric system (18°C, 14 cm, 45 kg, and so on) where possible.
- Ask someone from your intended audience's culture or with appropriate expertise to review your draft before you complete your final **proofreading**.

interviewing for information

Interviewing others who have knowledge of your subject is often an essential method of **research** in technical writing. The process of interviewing for information includes determining the proper person to interview, preparing for the interview, conducting the interview, and expanding your notes soon after the interview.

Determining the Proper Person to Interview

Many times, your subject or **purpose** logically points to the proper person to interview for information. For example, if you were writing about using the Web to market a software-development business, you would want to interview someone with extensive experience in Web marketing as well as someone who has built a successful business developing software. The following sources can help you determine the appropriate person to interview: (1) workplace colleagues or faculty in appropriate academic departments, (2) local chapters of professional societies, (3) information from "contact" or "about" links at company or organization Web sites, and (4) targeted Internet searches, such as using relevant domains (.edu, .gov, and .org).

Preparing for the Interview

Before the interview, learn as much as possible about the person you are going to interview and the organization for which he or she works. When you contact the prospective interviewee, explain who you are, why you would like an interview, the subject and purpose of the interview, and generally how much time it will take. You should also ask permission if you plan to record the interview and let your interviewee know that you will allow him or her to review your draft.

After you have made the appointment, prepare a list of questions to ask your interviewee. Avoid vague, general questions. A question such as "Do you think the Web would be useful for you?" is too general to elicit useful information. It is more helpful to ask specific but open-ended questions, such as the following: "Many physicians in your specialty are using the Web to answer routine patient questions. How might providing such information on your Web site affect your relationship with your patients?"

Conducting the Interview

Arrive promptly for the interview and be prepared to guide the discussion. During the interview, take only memory-jogging notes that will help you recall the conversation later; do not ask your interviewee to slow down so that you can take detailed notes. As the interview is reaching a close, take a few minutes to skim your notes and ask the interviewee to clarify anything that is ambiguous.

WRITER'S CHECKLIST   Interviewing Successfully

- Be pleasant but purposeful. You are there to get information, so don't be timid about asking leading questions on the subject.
- Use the list of questions you have prepared, starting with the less complex and difficult aspects of the topic to get the conversation started, and then going on to the more challenging aspects.
- Let your interviewee do most of the talking. Remember that the interviewee is the expert. See also **listening**.
- Be objective. Don't offer your opinions on the subject. You are there to get information, not to debate.
- Ask additional questions as they arise.
- Don't get sidetracked. If the interviewee strays too far from the subject, ask a specific question to direct the conversation back on track.
- If you use an audiocassette or a digital voice recorder, do not let it lure you into relaxing so that you neglect to ask crucial questions.
- After thanking the interviewee, ask permission to contact him or her again to clarify a point or two as you complete your interview notes.

Expanding Your Notes Soon After the Interview

Immediately after leaving the interview, use your memory-jogging notes to help you mentally review the interview and expand those notes. Do not postpone this step. No matter how good your memory is, you will forget some important points if you do not complete this step at once. See also **note-taking**.

Interviewing by Phone or E-mail

When an interviewee is not available for a face-to-face meeting, an interview by phone may be useful. Most of the principles for conducting face-to-face interviews apply to phone interviews; be aware, however, that phone calls do not offer the important nonverbal cues of face-to-face meetings. Further, taking notes can be challenging while speaking on a phone, although a phone headset or a high-quality speakerphone can alleviate that problem.

As an alternative to a face-to-face or phone interview, consider an e-mail interview. Such an "interview," however, lacks the spontaneity and the immediacy of an in-person or a phone conversation. If e-mail is the only option, the interviewing principles in this entry can help you elicit useful responses. Before you send any questions, make sure that your contact is willing to participate and respond to follow-up requests for clarification. As a courtesy, give the respondent a general idea of the number of questions you plan to ask and the level of detail you expect. When you send the questions, ask for a reasonable deadline from the interviewee ("Would you be able to send your response by . . . ?"). See also **e-mail**.

interviewing for a job

Job interviews can take place in person, by phone, or by teleconference. They may last 30 minutes, an hour, or several hours. Sometimes, an initial job interview is followed by a series of additional interviews that can last a half or full day. Often, just one or two people conduct a job interview, but at times a group of four or more might do so. Because it is impossible to know exactly what to expect, it is important that you be well prepared. See also **application letters**, **job search**, and **résumés**.

Before the Interview

The interview is not a one-way communication. It presents you with an opportunity to ask questions of your potential employer. Before the interview, learn everything you can about the organization by answering for yourself such questions as the following:

- What kind of organization (profit, nonprofit, government) is it?
- How diversified are its activities or branches?
- Is it a locally owned business?
- Does it provide a product or service? If so, what kind?
- How large is the business? How large are its assets?
- Is the owner self-employed? Is the company a subsidiary of a larger operation? Is it expanding?
- How long has the company been in business?
- Where will I fit in?

You can obtain information from current employees, the Internet, the company's publications, and the business section of back issues of local newspapers. The company's Web site may help you learn about the company's size, sales volume, product line, credit rating, branch locations, subsidiary companies, new products and services, building programs, and similar information. You may also conduct **research** using a company's annual reports and other publications, such as *Moody's Industrials*, *Dun and Bradstreet*, *Standard and Poor's*, and *Thomas' Register*, as well as other business reference sources a librarian might suggest. Ask your interviewer about what you cannot find through your own research. Doing so demonstrates your interest and allows you to learn more about your potential employer.

Try to anticipate the questions your interviewer might ask, and prepare your answers in advance. Be sure you understand a question before answering it, and avoid responding too quickly with a rehearsed answer—be prepared to answer in a natural and relaxed manner. Interviewers typically ask the following questions:

- What are your short-term and long-term occupational goals?
- Where do you see yourself five years from now?
- What are your major strengths and weaknesses?
- Do you work better with others or alone?
- How do you spend your free time?
- What accomplishment are you particularly proud of? Describe it.
- Why are you leaving your current job?
- Why do you want to work for this organization?
- Why should I hire you?
- What salary and benefits do you expect?

Many employers use behavioral interviews. Rather than traditional, straightforward questions, the behavioral interview focuses on asking the candidate to provide examples or respond to hypothetical situations. Interviewers who use behavior-based questions are looking for specific examples from your experience. Prepare for the behavioral interview by recollecting challenging situations or problems that you successfully resolved. Examples of behavior-based questions include the following:

- Tell me about a time when you experienced conflict while on a team.
- If I were your boss and you disagreed with a decision I made, what would you do?
- How have you used your leadership skills to bring about change?
- Tell me about a time when you failed and what you learned from the experience.

Arrive for your interview on time or even 10 or 15 minutes early—you may be asked to fill out an application or other paperwork before you meet your interviewer. Always bring extra copies of your résumé, samples of your work (if applicable), and a list of references and contact information. If you are asked to complete an application form, read it carefully before you write and proofread it when you are finished. The form provides a written record for company files and indicates to the company how well you follow directions and complete a task.

During the Interview

The interview actually begins before you are seated: What you wear and how you act make a first impression. In general, dress simply and conservatively, avoid extremes in fragrance and cosmetics, and be well groomed.

Behavior. First, thank the interviewer for his or her time, express your pleasure at meeting him or her, and remain standing until you are offered a seat. Then sit up straight (good posture suggests self-assurance), look directly at the interviewer, and try to appear relaxed and confident. During the interview, you may find yourself feeling a little nervous. Use that nervous energy to your advantage by channeling it into the alertness that you will need to listen and respond effectively. Do not attempt to take extensive notes or use a laptop computer during the interview. You can jot down a few facts and figures on a small pad, but keep your focus on the interviewer. See also **listening**.

Responses. When you answer questions, do not ramble or stray from the subject. Say only what you must to answer each question properly and then stop, but avoid giving just yes or no answers—they usually do not allow the interviewer to learn enough about you. Some interviewers allow a silence to fall just to see how you will react. The burden of conducting the interview is the interviewer's, not yours—and he or she may interpret your rush to fill a void in the conversation as a sign of insecurity. If such a silence makes you uncomfortable, be ready to ask an intelligent question about the company.

If the interviewer overlooks important points, bring them up. However, let the interviewer mention salary first. Doing so yourself may indicate that you are more interested in the money than the work. However, make sure you are aware of prevailing salaries and benefits in your field. See **salary negotiations**.

Interviewers look for a degree of self-confidence and an applicant's understanding of the field, as well as genuine interest in the field, the company, and the job. Ask questions to communicate your interest in the job and company. Interviewers respond favorably to applicants who can communicate and present themselves well.

Conclusion. At the conclusion of the interview, thank the interviewer for his or her time. Indicate that you are interested in the job (if true) and try to get an idea of the company's hiring time frame. Reaffirm friendly contact with a firm handshake.

After the Interview

After you leave the interview, jot down the pertinent information you obtained, as it may be helpful in comparing job offers. As soon as possible following a job interview, send the interviewer a note of thanks in a brief **letter** or **e-mail**. Such notes often include the following:

- Your thanks for the interview and to individuals or groups that gave you special help or attention during the interview
- The name of the specific job for which you interviewed
- Your impression that the job is attractive, if true
- Your confidence that you can perform the job well
- An offer to provide further information or answer further questions

Figure I–9 shows a typical example of follow-up correspondence.

If you are offered a job you want, accept the offer verbally and write a brief letter of acceptance as soon as possible—certainly within a week. If you do not want the job, write a refusal letter or e-mail, as described in **acceptance / refusal letters**.

> Dear Mr. Itsuru:
>
> Thank you for the informative and pleasant interview we had yesterday. Please extend my thanks to Mr. Ragins of Human Resources as well.
>
> I came away from our meeting most favorably impressed with Calcutex Industries. I find the position of software designer to be an attractive one and feel confident that my qualifications would enable me to perform the duties to everyone's advantage.
>
> If I can answer any further questions, please let me know.
>
> Sincerely,
>
> Wilson Hathaway

FIGURE I–9. Follow-up Correspondence

introductions

DIRECTORY

Routine Openings 277
Opening Strategies 277
  Objective 277
  Problem Statement 277
  Scope 277
  Background 278
  Summary 278
  Interesting Detail 278
  Definition 279
  Anecdote 279
  Quotation 279
  Forecast 279
  Persuasive Hook 280
Full-Scale Introductions 280
Manuals and Specifications 280

Every document must have either an opening or an introduction. In general, **correspondence** and routine **reports** need only an opening; more complex reports and other longer documents need an introduction. An opening usually simply focuses the reader's attention on your topic and then proceeds to the body of your document. A formal introduction, however, sets the stage by providing necessary information to understand the discussion that follows in the body. Introductions are required for such documents as **formal reports** and major **proposals**. For a discussion of comparable sections for Web sites, see **writing for the Web** and **Web design**. See also **conclusions**.

Routine Openings

When your **audience** is familiar with your topic or if what you are writing is brief or routine, then a simple opening will provide adequate **context**, as shown in the following examples.

LETTER
Dear Mr. Ignatowski:
You will be happy to know that we have corrected the error in your bank balance. The new balance shows . . .

MEMO
To date, 18 of the 20 specimens your department submitted for analysis have been examined. Our preliminary analysis indicates . . .

E-MAIL
Jane, as I promised in my e-mail yesterday, I've attached the engineering division budget estimates for fiscal year 2008.

Opening Strategies

Opening strategies are aimed at focusing the readers' attention and motivating them to read the entire document.

Objective. In reporting on a project, you might open with a statement of the project's objective so that the readers have a basis for judging the results.

▶ The primary goal of this project was to develop new techniques to solve the problem of waste disposal. Our first step was to investigate . . .

Problem Statement. One way to give readers the perspective of your report is to present a brief account of the problem that led to the study or project being reported.

▶ Several weeks ago a manager noticed a recurring problem in the software developed by Datacom Systems. Specifically, error messages repeatedly appeared when, in fact, no specific trouble. . . . After an extensive investigation, we found that Datacom Systems . . .

For proposals or formal reports, of course, problem statements may be more elaborate and a part of the full-scale introduction, which is discussed later in this entry.

Scope. You may want to present the **scope** of your document in your opening. By providing the parameters of your material, the limitations of the subject, or the amount of detail to be presented, you enable your readers to determine whether they want or need to read your document.

▶ This pamphlet provides a review of the requirements for obtaining a private pilot's license. It is not intended as a textbook to prepare you for the examination itself; rather, it outlines the steps you need to take and the costs involved.

Background. The background or history of a subject may be interesting and lend perspective and insight to a subject. Consider the following example from a newsletter describing the process of oil drilling:

▶ From the bamboo poles the Chinese used when the pyramids were young to today's giant rigs drilling in hundreds of feet of water, there has been considerable progress in the search for oil. But whether in ancient China or a modern city, underwater or on a mountaintop, the objective of drilling has always been the same — to manufacture a hole in the ground, inch by inch.

Summary. You can provide a summary opening by describing in abbreviated form the results, conclusions, or recommendations of your article or report. Be concise: Do not begin a summary by writing "This report summarizes...."

CHANGE	This report summarizes the advantages offered by the photon as a means of examining the structural features of the atom.
TO	As a means of examining the structure of the atom, the photon offers several advantages.

Interesting Detail. Often an interesting detail will attract the readers' attention and arouse their curiosity. Readers of a **white paper** for a manufacturer of telescopes and scientific instruments, for example, may be persuaded to invest if they believe that the company is developing innovative, cutting-edge products.

▶ The rings of Saturn have puzzled astronomers ever since they were discovered by Galileo in 1610 using the first telescope. Recently, even more rings have been discovered....

  Our company's Scientific Instrument Division designs and manufactures research-quality, computer-controlled telescopes that promise to solve the puzzles of Saturn's rings by enabling scientists to use multicolor differential photometry to determine the rings' origins and compositions.

Definition. Although a definition can be useful as an opening, do not define something with which your audience is familiar or provide a definition that is obviously a contrived opening (such as "Webster defines *technology* as . . ."). A definition should be used as an opening only if it offers insight into what follows.

▶ *Risk* is often a loosely defined term. In this report, risk refers to a qualitative combination of the probability of an event and the severity of the consequences of that event. In fact, . . .

Anecdote. An anecdote can be used to attract and build interest in a subject that may otherwise be mundane; however, this strategy is best suited to longer documents and **presentations**.

▶ In his poem "The Calf Path" (1895), Sam Walter Foss tells of a wandering, wobbly calf trying to find its way home at night through the lonesome woods. It made a crooked path, which was taken up the next day by a lone dog. Then "a bellwether sheep pursued the trail over vale and steep, drawing behind him the flock, too, as all good bellwethers do." This forest path became a country lane that bent and turned and turned again. The lane became a village street, and at last the main street of a flourishing city. The poet ends by saying, "A hundred thousand men were led by a calf near three centuries dead."

Many companies today follow a "calf path" because they react to events rather than planning. . . .

Quotation. Occasionally, you can use a quotation to stimulate interest in your subject. To be effective, however, the quotation must be pertinent—not some loosely related remark selected from a book of quotations.

▶ Deborah Andrews predicted that "technical communicators in the twenty-first century must reach audiences and collaborate across borders of culture, language, and technology" [B3]. One way we are accomplishing that goal is to make sure our training includes cross-cultural experiences that provide . . .

Forecast. Sometimes you can use a forecast of a new development or trend to gain the audience's attention and interest.

▶ In the not-too-distant future, we may be able to use a handheld medical diagnostic device similar to those in science fiction to assess the complete physical condition of accident victims. This project and others are now being developed at The Seldi Group, Inc.

Persuasive Hook. While all opening strategies contain persuasive elements, the hook uses **persuasion** most overtly. A **brochure** touting the newest innovation in tax-preparation software might address readers as follows:

▶ Welcome to the newest way to do your taxes! TaxPro EZ ends the headache of last-minute tax preparation with its unique Web-Link feature.

Full-Scale Introductions

The purpose of a full-scale introduction is to give readers enough general information about the subject to enable them to understand the details in the body of the document. An introduction should accomplish any or all of the following:

- *State the subject.* Provide background information, such as definition, history, or theory, to provide context for your readers.
- *State the purpose.* Make your readers aware of why the document exists and whether the material provides a new perspective or clarifies an existing perspective.
- *State the scope.* Tell readers the amount of detail you plan to cover.
- *Preview the development of the subject.* Especially in a longer document, outline how you plan to develop the subject. Providing such information allows readers to anticipate how the subject will be presented and helps them evaluate your conclusions or recommendations.

Consider writing an opening or introduction last. Many writers find that it is only when they have drafted the body of the document that they have a full enough perspective on the subject to introduce it adequately.

Manuals and Specifications

You may need to write one kind of introduction for reports, academic papers, or **trade journal articles** and a different kind for **manuals** or **specifications**. When writing an introduction for a manual or set of specifications, identify the topic and its primary purpose or function in the first sentence or two. Be specific, but do not go into elaborate detail. Your introduction sets the stage for the entire document, and it should provide readers with a broad frame of reference and an understanding of the overall topic. Then the reader is ready for technical details in the body of the document.

How technical your introduction should be depends on your readers: What are their technical backgrounds? What kind of information are they seeking in the manual or specification? The topic should be introduced with a specific audience in mind—a computer user, for example, has different interests in an application program and a different technical vocabulary than a programmer. Whether you need to provide explanations or definitions of terminology will depend on your intended audience. The following example is written for readers who understand such terms as "constructor" and "software modules."

> ▶ The System Constructor is a program that can be used to create operating systems for a specific range of microcomputer systems. The constructor selects requested operating software modules from an existing file of software modules and combines those modules with a previously compiled application program to create a functional operating system designed for a specific hardware configuration. It selects the requested software modules, establishes the necessary linkage between the modules, and generates the control tables for the system according to parameters specified at run time.

You may encounter a dilemma that is common in technical writing: Although you cannot explain topic A until you have explained topic B, you cannot explain topic B before explaining topic A. The solution is to explain both topics in broad, general terms in the introduction, as in the following paragraph that introduces "source units" and "destination units":

> ▶ The NEAT/3 programming language, which treats all peripheral units as file storage units, allows your program to perform data input or output operations depending on the specific unit. Peripheral units from which your program can only input data are referred to as *source units*; those to which your program can only output data are referred to as *destination units*.

Then, when you need to write a detailed explanation of topic A (*source units*), you will be able to do so because your reader will know just enough about both topics A and B to be able to understand your detailed explanation.

investigative reports

An investigative **report** offers a precise analysis of a workplace problem or an issue in response to a request or need for information. The investigative report shown in Figure I–10 evaluates whether a company should adopt a program called *Basic English* to train and prepare documentation for non–English-speaking readers. See also **laboratory reports**.

Memo

To: Noreen Rinaldo, Training Manager
From: Charles Lapinski, Senior Instructor *CL*
Date: February 10, 2009
Subject: Adler's Basic English Program

As requested, I have investigated Adler Medical Instruments' (AMI's) Basic English Program to determine whether we might adopt a similar program.

The purpose of AMI's program is to teach medical technologists outside the United States who do not speak or read English to understand procedures written in a special 800-word vocabulary called *Basic English*. This program eliminates the need for AMI to translate its documentation into a number of different languages. The Basic English Program does not attempt to teach the medical technologists to be fluent in English but, rather, to recognize the 800 basic words that appear in Adler's documentation.

Course Analysis

The course teaches technologists a basic medical vocabulary in English; it does not provide training in medical terminology. Students must already know, in their own language, the meaning of medical vocabulary (e.g., the meaning of the word *hemostat*). Students must also have basic knowledge of their specialty, must be able to identify a part in an illustrated parts book, must have used AMI products for at least one year, and must be able to read and write in their own language.

Students are given an instruction manual, an illustrated book of equipment with parts and their English names, and pocket references containing the 800 words of the Basic English vocabulary plus the English names of parts. Students can write the corresponding word in their language beside the English word and then use the pocket reference as a bilingual dictionary. The course consists of 30 two-hour lessons, each lesson introducing approximately 27 words. No effort is made to teach pronunciation; the course teaches only recognition of the 800 words, which include 450 nouns; 70 verbs; 180 adjectives and adverbs; and 100 articles, prepositions, conjunctions, and pronouns.

Course Success

The 800-word vocabulary enables the writers of documentation to provide medical technologists with any information that might be required because the subject areas are strictly limited to usage, troubleshooting, safety, and operation of AMI medical equipment. All nonessential words (*apple*, *father*, *mountain*, and so on) are eliminated, as are most synonyms (for example, *under* appears, but *beneath* does not).

Conclusions and Recommendations

AMI's program appears to be quite successful, and a similar approach could also be appropriate for us. I see two possible ways in which we could use some or all of the elements of AMI's program: (1) in the preparation of our student manuals or (2) as AMI uses the program.

I think it would be unnecessary to use the Basic English methods in the preparation of manuals for *all* of our students. Most of our students are English speakers to whom an unrestricted vocabulary presents no problem.

As for our initiating a program similar to AMI's, we could create our own version of the Basic English vocabulary and write our instructional materials in it. Because our product lines are much broader than AMI's, however, we would need to create illustrated parts books for each of the different product lines.

FIGURE I–10. Investigative Report

Open an investigative report with a statement of its primary and (if any) secondary **purposes**, then define the **scope** of your investigation. If the report includes a survey of opinions, for example, indicate the number of people surveyed and other identifying information, such as income categories and occupations. (See also **questionnaires**.) Include any information that is pertinent in defining the extent of the investigation. Then report your findings and discuss their significance with your **conclusions**.

Sometimes the person requesting the investigative report may ask you to make recommendations as a result of your findings. In that case, the report may be referred to as a *recommendation report*. See also **feasibility reports** and **trouble reports**.

italics

Italics is a style of type used to denote **emphasis** and to distinguish foreign expressions, book titles, and certain other elements. *This sentence is printed in italics.* Italic type is often signaled by underlining in manuscripts submitted for publication or where italic font is not available (see also **e-mail**). You may need to italicize words that require special emphasis in a sentence. ("Contrary to projections, sales have *not* improved.") Do not overuse italics for emphasis, however. ("*This* will hurt *you* more than *me*.")

Foreign Words and Phrases

Foreign words and phrases are italicized (*bonjour*, *guten tag*, the sign said "*Se habla español*"). Foreign words that have been fully assimilated into English need not be italicized (cliché, etiquette, vis-à-vis, de facto, résumé). When in doubt about whether to italicize a word, consult a current dictionary. See also **foreign words in English**.

Titles

Italicize the **titles** of separately published documents, such as books, periodicals, newspapers, pamphlets, brochures, legal cases, movies, and television programs.

▶ The book *Turning Workplace Conflicts into Collaboration* was reviewed in the *New York Times*.

Abbreviations of such titles are italicized if their spelled-out forms would be italicized.

▶ The *NYT* is one of the nation's oldest newspapers.

Italicize the titles of compact discs, videotapes, plays, long poems, paintings, sculptures, and long musical works.

CD-ROM	*Computer Security Tutorial on CD-ROM*
PLAY	Arthur Miller's *Death of a Salesman*
LONG POEM	T. S. Eliot's *The Wasteland*
MUSICAL WORK	Gershwin's *Porgy and Bess*

Use **quotation marks** for parts of publications, such as chapters of books and articles or sections within periodicals.

Proper Names

The names of ships, trains, and aircraft (but not the companies or governments that own them) are italicized (U.S. aircraft carrier *Independence*, U.S. space shuttle *Endeavour*). Craft that are known by model or serial designations are not italicized (DC-7, Boeing 747).

Words, Letters, and Figures

Words, letters, and figures discussed as such are italicized.

- The word *inflammable* is often misinterpreted.
- The *S* and *6* keys on my keyboard do not function.

Subheads

Subheads in a report are sometimes italicized.

- *Training Managers.* We are leading the way in developing first-line managers who not only are professionally competent but . . . .

See also **headings** and **layout and design**.

its / it's

Its is a possessive **pronoun** and does not use an **apostrophe**. *It's* is a **contraction** of *it is*.

- *It's* essential that the lab maintain *its* quality control.

See also **expletives** and **possessive case**.

jargon

Jargon is a highly specialized slang that is unique to an occupational or a professional group. Jargon is at first understood only by insiders; over time, it may become known more widely. For example, computer programmers adopted the term *debugging* to describe the discovery and correction of errors in software. If all your readers are members of a particular occupational group, jargon may provide an efficient means of communicating. However, if you have any doubt that your entire **audience** is part of such a group, avoid using jargon. See also **affectation**, **functional shift**, and **gobbledygook**.

job descriptions

Most large companies and many small ones use formal job descriptions to specify the duties of and requirements for many of the jobs in the firm.* Job descriptions fulfill several important functions: They provide information on which equitable salary scales can be based; they help management determine whether all functions within a company are adequately supported; and they let both prospective and current employees know exactly what is expected of them. Together, all the job descriptions in a firm present a picture of the organization's structure.

Sometimes middle managers are given the task of writing job descriptions for their employees. In many organizations, though, employees are required to draft their own job descriptions, which supervisors then check and approve.

Although job-description formats vary from organization to organization, they commonly contain the following sections:

- The *accountability section* identifies, by title only, the person to whom the employee reports.

---

*Job descriptions are sometimes called *position descriptions*, a term also used for formal announcements of openings for professional or administrative positions.

- The *scope of responsibilities* section provides an overview of the primary and secondary functions of the job and states, if applicable, who reports to the employee.
- The *specific duties* section gives a detailed account of the particular duties of the job as concisely as possible.
- The *personal requirements* section lists the required or preferred education, training, experience, and licensing for the job.

The job description shown in Figure J–1 is typical. It never mentions the person holding the job described; it focuses, instead, on the job and the qualifications required to fill the position.

WRITER'S CHECKLIST Writing Job Descriptions

- Before attempting to write your job description, list all the different tasks you do in a week or a month. Otherwise, you will almost certainly leave out some of your duties.
- Focus on content. Remember that you are describing your job, not yourself.
- List your duties in decreasing order of importance. Knowing how your various duties rank in importance makes it easier to set valid job qualifications.
- Begin each statement of a duty with a **verb** and be specific. Write "Orient new staff members to the department" rather than "New staff orientation."
- Review existing job descriptions that are considered well written.

job search

Whether you are applying for your first job or want to change careers entirely, begin by assessing your skills, interests, and abilities, perhaps through **brainstorming**.* Next, consider your career goals and values. For instance, do you prefer working independently or collaboratively? Do you enjoy public settings? Do you like meeting people? How important are career stability and location? What would you most like to be doing in the immediate future? in two years? in five years?

*A good source for stimulating your thinking is the most recent edition of *What Color Is Your Parachute? A Practical Manual for Job-Hunters & Career-Changers* by Richard Nelson Bolles, published by Ten Speed Press.

Manager, Technical Publications
Dakota Electrical Corporation

Accountability
Reports directly to the Vice President, Customer Service.

Scope of Responsibilities
The Manager of Technical Publications plans, coordinates, and supervises the design and development of technical publications and documentation required to support the sale, installation, and maintenance of Dakota products. The manager is responsible for the administration and morale of the staff. The supervisor for instruction manuals and the supervisor for parts manuals report directly to the manager.

Specific Duties
- Directs an organization currently comprising 20 people (including two supervisors), over 75 percent of whom are writing professionals and graphic artists
- Screens, selects, and hires qualified applicants for the department
- Prepares a formal orientation program to familiarize writing trainees with the production of reproducible copy and graphic arts
- Evaluates the performance of and determines the salary adjustments for all department employees
- Plans documentation to support new and existing products
- Subcontracts publications and acts as a purchasing agent when needed
- Offers editorial advice to supervisors
- Develops and manages an annual budget for the Technical Publications Department
- Cooperates with the Engineering, Parts, and Service Departments to provide the necessary repair and spare-parts manuals upon the introduction of new equipment
- Serves as a liaison among technical specialists, the publications staff, and field engineers
- Recommends new and appropriate uses for the department within the company
- Keeps up with new technologies in printing, typesetting, art, and graphics and uses them to the advantage of Dakota Electrical Corporation where applicable

Requirements
- B.A. in professional writing or equivalent
- Minimum of three years' professional writing experience and a general knowledge of graphics, production, and Web design
- Minimum of two years' management experience with a knowledge of the general principles of management
- Strong interpersonal skills

FIGURE J–1. Job Description

Once you have reflected and brainstormed about the job that is right for you, a number of sources can help you locate the job you want. Of course, you should not rely on any one of the following sources exclusively:

- Networking and informational interviews
- Campus career services
- Web resources
- Job advertisements
- Trade and professional journal listings
- Employment agencies (private, temporary, government)
- Letters of inquiry

Keep a file during your job search of dated job ads, copies of **application letters** and **résumés**, and the names of important contacts. This collection can serve as a future resource and reminder.

Networking and Informational Interviews

Networking involves communicating with people who might provide useful advice or may know of potential jobs in your interest areas. They may include people already working in your chosen field, contacts in professional organizations, professors, family members, or friends. Discussion groups and Web job-networking sites can be helpful in this process. Use your contacts to expand your network of contacts. Most career professionals estimate that between 60 and 80 percent of all open positions are filled through networking.

Informational interviews are appointments you schedule with working professionals who can give you "insider" views of an occupation or industry. These brief meetings (usually 20 to 30 minutes) also offer you the opportunity to learn about employment trends as well as leads for employment opportunities. Because you ask the questions, these interviews allow you to participate in an interview situation that is less stressful than the job interview itself. To make the most of informational interviews, prepare carefully and review both **interviewing for information** and **interviewing for a job**.

Campus Career Services

A visit to a college career-development center is another good way to begin your job search. Government, business, and industry recruiters often visit campus career offices to interview prospective employees; recruiters also keep career counselors aware of their companies' current employment needs and submit **job descriptions** to them. Not only can career counselors help you select a career, but they can also put you in touch with the best and most current resources—identifying where to begin your search and saving you time. Career-development centers often hold workshops on résumé preparation and offer other job-finding resources on their Web sites.

Web Resources

Using the Web can enhance your job search in a number of ways. First, you can consult sites that give advice about careers, job seeking, and résumé preparation. Second, you can learn about businesses and organizations that may hire employees in your area by visiting their Web sites. Such sites often list job openings, provide instructions for applicants, and offer other information, such as employee benefits. Third, you can learn about jobs in your field and post your résumé for prospective employers at employment databases, such as Monster.com or America's Job Bank. Fourth, you can post your résumé at your personal Web site. Although posting your résumé at an employment database will undoubtedly attract more potential employers, including your résumé at your own site has benefits. For example, you might provide a link to your site in e-mail correspondence or provide your Web site's URL in an inquiry letter to a prospective employer. If you use a personal Web site, however, it should contain only material that would be of interest to prospective employers, such as examples of your work, awards, and other professional items.

▶ **ETHICS NOTE** Be careful with the material that you post at such online social communities as MySpace, Facebook, and YouTube. Potential employers often conduct Web searches that could access content at such Web sites, so you need to be aware of how you present yourself in online public spaces. ◆

Employment specialists suggest that you spend time on the Web in the evening or early morning so that you can focus on in-person contacts during working hours.

WEB LINK	Finding a Job

For job-hunting tips, sample documents, and links to the sites mentioned in this entry, see bedfordstmartins.com/alredtech and select *Finding an Internship or Job* and *Links for Handbook Entries*.

Job Advertisements

Many employers advertise jobs in the classified sections of newspapers. Because job listings can differ, search in both the printed and Web editions of local and big-city newspapers under *employment* or *job market*.

Use the search options they provide or the general strategies for database searches discussed in the entry **research**.

A clinical medical technologist seeking a job, for example, might find the specialty listed under "Medical Technologist" or "Clinical Laboratory Technologist." Depending on a hospital's or pathologist's needs, the listing could be even more specific, such as "Blood Bank Technologist" or "Hematology Technologist."

As you read the ads, take notes on salary ranges, job locations, job duties and responsibilities, and even the terminology used in the ads to describe the work. A knowledge of keywords and key expressions that are generally used to describe a particular type of work can be helpful when you prepare your résumé and letters of application.

Trade and Professional Journal Listings

In many industries, associations publish periodicals of interest to people working in the industry. Such periodicals (print and online) often contain job listings. To learn about the trade or professional associations for your occupation, consult resources on the Web, such as Google's Directory of Professional Organizations or online resources offered by your library or campus career office. You may also consult the following references at a library: *Encyclopedia of Associations*, *Encyclopedia of Business Information Sources*, and *National Directory of Employment Services*.

Employment Agencies (Private, Temporary, Government)

Private employment agencies are profit-making organizations that are in business to help people find jobs—for a fee. Reputable agencies provide you with job leads, help you organize your job search, and supply information on companies doing the hiring. A staffing agency or temporary placement agency could match you with an appropriate temporary or permanent job in your field. Temporary work for an organization for which you might want to work permanently is an excellent way to build your network while continuing your job search.

Choose an employment or a temporary-placement agency carefully. Some are well established and reputable; others are not. Check with your local Better Business Bureau and your college career office before you sign an agreement with a private employment agency. Further, be sure you understand who is paying the agency's fee. Often the employer pays the agency's fee; however, if you have to pay, make sure you know exactly how much. As with any written agreement, read the fine print carefully.

Local, state, and federal government agencies also offer many free or low-cost employment services. Local government agencies are listed in telephone and Web directories under the name of your city, county, or state. For information on occupational trends, see the Occupational Outlook Handbook at *http://bls.gov/OCO*, and for information on jobs with the federal government, see the U.S. Office of Personnel Management Web site at *http://usajobs.opm.gov*. See also **salary negotiations**.

Letters of Inquiry

If you would like to work for a particular firm, write and ask whether it has any openings for people with your qualifications. Normally, you can send the letter to the department head, the director of human resources, or both; for a small firm, however, write to the head of the firm. Such letters work best if you have networked as described earlier in this entry.

journal articles (*see* trade journal articles)

K

kind of / sort of

The phrases *kind of* and *sort of* should be used only to refer to a class or type of things.

▶ He described a special *kind of* mirror called a quantum switch.

Do not use *kind of* or *sort of* to mean "rather," "somewhat," or "somehow." That usage can lead to vagueness; it is better to be specific.

VAGUE	It was *kind of* a bad year for the company.
SPECIFIC	The company's profits fell 10 percent last year.

know-how

The informal term *know-how*, meaning "special competence or knowledge," should be avoided in formal writing **style**.

▶ The applicant has impressive technical ~~know-how~~ *skill*.

L

laboratory reports

A laboratory report communicates information acquired from laboratory testing or a major investigation. It should begin by stating the reason that a laboratory investigation was conducted; it should also list the equipment and methods used during the test, the problems encountered, the results and conclusions reached, and any recommendations. The example in Figure L–1 shows sections from a typical laboratory report.

A laboratory report emphasizes the equipment and procedures used in the investigation because those two factors can be critical in determining the accuracy of the data and even replicating the procedure if necessary. Although this emphasis often requires the use of the passive voice, you should present the results of the laboratory investigation clearly and precisely. If your report requires **graphs** or **tables**, integrate them into your report as described in the entry **visuals**. For reporting routine or simple laboratory tests, see **test reports**.

lay / lie

Lay is a transitive **verb**—a verb that requires a direct object to complete its meaning—that means "place" or "put."

▶ We will *lay* the foundation one section at a time.

The past-tense form of *lay* is *laid*.

▶ We *laid* the first section of the foundation last month.

The perfect-tense form of *lay* is also *laid*.

▶ Since June, we *have laid* all but two sections of the foundation.

Lay is frequently confused with *lie*, which is an intransitive verb—a verb that does not require an object to complete its meaning—that means "recline" or "remain."

▶ A person in shock should *lie* down with legs slightly elevated.

PCB Exposure from Oil Combustion
Wayne County Firefighters Association

Submitted to:
Mr. Philip Landowe
President, Wayne County Firefighters Association
Wandell, IN 45602

Submitted by:
Analytical Laboratories, Incorporated
1220 Pfeiffer Parkway
Indianapolis, IN 46223
February 27, 2009

INTRODUCTION
Waste oil used to train firefighters was suspected by the Wayne County Firefighters Association of containing polychlorinated biphenyls (PCBs). According to information provided by Mr. Philip Landowe, President of the Association, it has been standard practice in training firefighters to burn 20–100 gallons of oil in a diked area of approximately 25–50 m^3. Firefighters would then extinguish the fire at close range. Exposure would last several minutes, and the exercise would be repeated two or three times each day for one week.

Oil samples were collected from three holding tanks near the training area in Englewood Park on November 14, 2008. To determine potential firefighter exposure to PCBs, bulk oil analyses were conducted on each of the samples. In addition, the oil was heated and burned to determine the degree to which PCB is volatized from the oil, thus increasing the potential for firefighter exposure via inhalation.

TESTING PROCEDURES
Bulk oil samples were diluted with hexane, put through a cleanup step, and analyzed in electron-capture gas chromatography. The oil from the underground tank that contained PCBs was then exposed to temperatures of 1008°C without ignition and 2008°C with ignition. Air was passed over the enclosed sample during heating, and volatized PCB was trapped in an absorbing medium. The absorbing medium was then extracted and analyzed for PCB released from the sample.

RESULTS
Bulk oil analyses are presented in Table 1. Only the sample from the underground tank contained detectable amounts of PCB. Aroclor 1260, containing 60 percent chlorine, was found to be present in this sample at 18 mg. Concentrations of 50 mg PCB in oil are considered hazardous. Stringent storage and disposal techniques are required for oil with PCB concentrations at these levels.

TABLE 1. Bulk Oil Analyses

Source	Sample #	PCB Content (mg/g)
Underground tank (11' deep)	6062	18*
Circle tank (3' deep)	6063a	<1
	6063b	<1
Square pool (3' deep)	6064a	<1
	6064b	<1

*Aroclor 1260 is the PCB type. This sample was taken for volatilization study.

FIGURE L–1. Laboratory Report

DISCUSSION AND CONCLUSIONS
At a concentration of 18 mg/g, 100 gallons of oil would contain approximately 5.5 g of PCB. Of the 5.5 g of PCB, about 0.3 g would be released to the atmosphere under the worst conditions.

The American Conference of Governmental Industrial Hygienists has established a threshold limit value (TLV)* of 0.5 mg/m^3 air for a PCB containing 54 percent C1 as a time-weighted average over an 8-hour work shift and has stipulated that exposure over a 15-minute period should not exceed 1 mg/m^3. The 0.3 g of released PCB would have to be diluted to 600 m^3 air to result in a concentration of 0.5 mg/m^3 or less. Because the combustion of oil lasted several minutes, a dilution to more than 600 m^3 is likely; thus, exposure would be less than 0.5 mg/m^3.

In summary, because exposure to this oil was limited and because PCB concentrations in the oil were low, it is unlikely that exposure from inhalation would be sufficient to cause adverse health effects. However, we cannot rule out the possibility that excessive exposure may have occurred under certain circumstances, based on factors such as excessive skin contact and the possibility that oil with a higher-level PCB concentration could have been used earlier. The practice of using this oil should be terminated.

*The safe average concentration that most individuals can be exposed to in an 8-hour day.

FIGURE L–1. Laboratory Report (*continued*)

The past-tense form of *lie* is *lay* (not *lied*). This form causes the confusion between *lie* and *lay*.

▶ The injured employee *lay* still for approximately five minutes.

The perfect-tense form of *lie* is *lain*.

▶ The injured employee *had lain* still for five minutes before the EMTs arrived.

layout and design

DIRECTORY
Typography 296
   Typeface and Type Size 296
   Type Style and Emphasis 297
Page-Design Elements 298
   Justification 298
   Headings 298

Headers and Footers 298
Lists 299
Columns 299
White Space 299
Color 299

layout and design

> Visuals 299
>   Icons 299
>   Captions 300
> Rules 300
> Page Layout and Thumbnails 300

The layout and design of a document can make even the most complex information accessible and give **readers** a favorable impression of the writer and the organization. To accomplish those goals, a design should help readers find information easily; offer a simple and uncluttered presentation; and highlight structure, hierarchy, and order. The design must also fit the **purpose** of the document and its **context**. For example, if clients are paying a high price for consulting services, they may expect a sophisticated, polished design; if employees inside an organization expect management to be frugal, they may accept—even expect—an economical and standard company design. See also **audience**.

Effective design is based on visual simplicity and harmony and can be achieved with careful selection of typography, page-design elements, and appropriate **visuals**, as well as thoughtful layout of text and visual components on a page.

Typography

Typography refers to the style and arrangement of type on a page. A complete set of all the letters, numbers, and symbols available in one typeface (or style) is called a *font*. The letters in a typeface have a number of distinctive characteristics, as shown in Figure L–2.

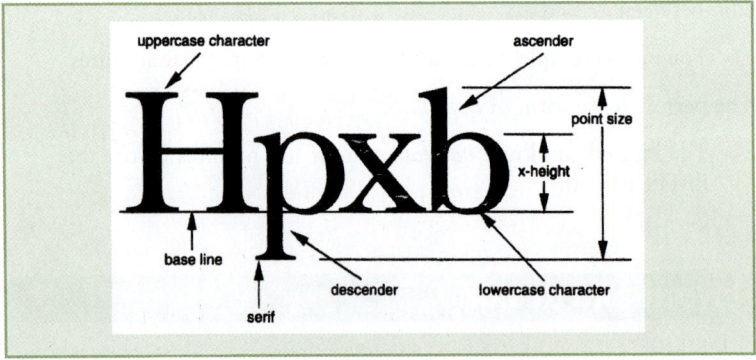

FIGURE L–2. Primary Components of Letter Characters

Typeface and Type Size. For most on-the-job writing, select a typeface primarily for its legibility. Avoid typefaces that may distract readers. Instead, choose popular typefaces with which readers are familiar, such as

Times Roman, Garamond, or Gill Sans. Avoid using more than two typefaces in the text of a document. For certain documents, however, such as **newsletters**, you may wish to use distinctively different typefaces for contrast among various elements such as headlines, **headings**, inset **quotations**, and sidebars. Experiment before making final decisions, keeping in mind your **audience**.

One way typefaces are characterized is by the presence or absence of serifs. Serif typefaces have projections, as shown in Figure L–2; sans serif styles do not. (*Sans* is French for "without.") The text of this book is set in Sabon, a serif typeface. Although sans serif type has a modern look, serif type is easier to read, especially in the smaller sizes. Sans serif, however, works well for headings (like the entry titles in this book) and for Web sites and other documents read on-screen.

Ideal font sizes for the main text of paper documents range from 10 to 12 points.* However, for some elements or documents, you may wish to select typeface sizes that are smaller (as in footnotes) or larger (as in headlines for **brochures**). See Figure L–3 for a comparison of type sizes in a serif typeface. Your readers and the distance from which they will read a document should help determine type size. For example, instructions that will rest on a table at which the reader stands require a larger typeface than a document that will be read up close. For **presentations** and **writing for the Web**, preview your document to see the effectiveness of your choice of point sizes and typefaces.

---

6 pt. This size might be used for dating a source.
8 pt. This size might be used for footnotes.
10 pt. This size might be used for figure captions.
12 pt. This size might be used for main text.
14 pt. This size might be used for headings.

FIGURE L–3. Type Sizes (6- to 14-Point Type)

Type Style and Emphasis. One method of achieving emphasis through typography is to use capital letters. HOWEVER, LONG STRETCHES OF ALL UPPERCASE LETTERS ARE DIFFICULT TO READ. (See also **e-mail**.) Use all uppercase letters only in short spans, such as in headings. Likewise, use italics sparingly because *continuous italic type reduces legibility and thus slows readers*. Of course, **italics** are useful if

*A *point* is a unit of type size equal to 0.01384 inch, or approximately $1/72$ of an inch.

your aim is to slow readers, as in cautions and warnings. **Boldface**, used in moderation, may be the best cuing device because it is visually different yet retains the customary shapes of letters and numbers.

Page-Design Elements

Thoughtfully used design elements can provide not only emphasis but also visual logic within a document by highlighting organization. Consistency and moderation are important—use the same technique to highlight a particular feature throughout your document and be careful not to overuse any single technique. The following typical elements can be used to make your document accessible and effective: justification, headings, <u>headers and footers</u>, <u>lists</u>, columns, white space, and color. Some of these elements are illustrated in Figure F–5 on page 201. For advice on design elements for Web pages, see <u>Web design</u>.

Justification. Left-justified (ragged-right) margins are generally easier to read than full-justified margins, especially for text using wide margins on 8½ × 11″ pages. Left justification is also better if full justification causes your word-processing or desktop-publishing software to insert irregular spaces between words, producing unwanted white space or unevenness in blocks of text. Full-justified text is more appropriate for publications aimed at a broad audience that expects a more formal, polished appearance. Full justification is also useful with narrow, multiple-column formats because the spaces between the columns (called *alleys*) need the definition that full justification provides.

Headings. Headings reveal the organization of a document and help readers decide which sections they need to read. You should provide typographic contrast between headings and the body text with either a different typeface or a different style (**bold**, *italic*, CAPS, and so on). Headings are often effective in boldface or in a sans serif typeface that contrasts with a body text in a serif typeface.

Headers and Footers. A header in a report, letter, or other document appears at the top of each page (as in this book), and a footer appears at the bottom of each page. Document pages may have headers or footers (or both) that include such elements as the topic or subtopic of a section, an identifying number, the date the document was written, the page number, and the document name. Keep your <u>headers and footers</u> concise because too much information in them can create visual clutter. However—at a minimum—a multipage document should include the page number in a header or footer. For more information on adding page numbers and laying out a page, see "Web Link: Designing Documents" on page 300. Headers are also important in <u>letters</u> and <u>memos</u>.

Lists. Vertically stacked words, phrases, and other items with numbers or bullets can effectively highlight such information as steps in sequence, materials or parts needed, key or concluding points, and recommendations. For further detail, see <u>lists</u>.

Columns. As you design pages, consider how columns may improve the readability of your document. A single-column format works well with larger typefaces, double-spacing, and left-justified margins. For smaller typefaces and single-spaced lines, the two-column structure keeps text columns narrow enough so that readers need not scan back and forth across the width of the entire page for every line. Avoid widows and orphans: a *widow* is a single word carried over to the top of a column or page; an *orphan* is a word on a line by itself at the end of a column.

White Space. White space visually frames information and breaks it into manageable chunks. For example, white space between paragraphs helps readers see the information in each paragraph as a unit. White space between sections can also serve as a visual cue to signal that one section is ending and another is beginning.

Color. Color and screening (shaded areas on a page) can distinguish one part of a document from another or unify a series of documents. They can set off sections within a document, highlight examples, or emphasize warnings. In tables, screening can highlight column titles or sets of data to which you want to draw the reader's attention.

Visuals

Readers notice visuals before they notice text, and they notice larger visuals before they notice smaller ones. Thus, the size of an illustration suggests its relative importance. For newsletter articles and publications aimed at wide audiences, consider especially the proportion of the visual to the text. Magazine designers often use the three-fifths rule: Page layout is more dramatic and appealing when the major element (**<u>photograph</u>**, **<u>drawing</u>**, or other visual) occupies three-fifths rather than one-half the available space. The same principle can be used to enhance the visual appeal of a **<u>report</u>**.

Visuals can be gathered in one place (for example, at the end of a report), but placing them in the text closer to their accompanying explanations makes them more effective. Using illustrations in the text also provides visual relief. For advice on the placement of visuals, see the *Writer's Checklist: Creating and Integrating Visuals* (pages 556–57).

Icons. Icons are pictorial representations used to describe such concepts as computer files, programs, or commands. Commonly used icons on the Web include the national flags to symbolize different language versions of a document. To be effective, icons must be simple and easily recognized or defined as well as culturally appropriate. See also **global graphics**.

Captions. Captions are titles that highlight or describe visuals, such as photographs. Captions often appear below or above figures and **tables** and are aligned with the visual to the left or are centered.

Rules. Rules are vertical or horizontal lines used to box or divide one area of the page from another. Rules and boxes, for example, set off visuals from surrounding explanations or highlight warning statements from the steps in **instructions**.

Page Layout and Thumbnails

Page layout involves combining typography, design elements, and visuals on a page to make a coherent whole. The flexibility of your design is affected by your design software, your method of printing the document, your budget, and whether your employer or client requires you to use a template.

Before you spend time positioning actual text and visuals on a page, especially for documents such as brochures, you may want to create a thumbnail sketch, in which blocks of simulated text and visuals indicate the placement of elements. You can go further by roughly assembling all the thumbnail pages to show the size, shape, form, and general style of a large document. Such a mock-up, called a *dummy*, allows you to see how a finished document will look.

WEB LINK | **Designing Documents**

Word-processing and desktop-publishing programs offer many options for improving the layout-and-design elements of your document. For a tutorial on these options, see *bedfordstmartins.com/alredtech* and select *Tutorials*, "Designing Documents with a Word Processor." For step-by-step instructions for setting margins, alignment, columns, and other design elements, select *Digital Tips*, "Laying Out a Page" and "Creating Styles and Templates."

---

lend / loan

Both *lend* and *loan* can be used as **verbs**, but *lend* is more common. ("You can *lend* [or *loan*] them the money if you wish.") Unlike *lend*, *loan* can be a **noun**. ("The bank approved our *loan*.")

letters

Business letters—normally written for those outside an organization—are often the most appropriate choice for formal communications with professional associates or customers. Letters printed on organizational letterhead communicate formality, respect, and authority. See **correspondence** for advice on writing strategy and style. See also **selecting the medium**.

Although word-processing software includes templates for formatting business letters, the templates may not provide the appropriate dimensions and spacing you need. The following sections offer specific advice on formatting and etiquette for business letters.*

Common Letter Styles

If your employer requires a particular format, use it. Otherwise, follow the guidelines provided here, and review the examples shown in Figures L–4 and L–5.

The two most common formats for business letters are the full-block style shown in Figure L–4 and the modified-block style shown in Figure L–5. In the *full-block style*, the entire letter is aligned at the left margin. In the *modified-block style*, the return address, date, and complimentary closing begin at the center of the page and the other elements are aligned at the left margin. All other letter styles are variations of the full-block and modified-block styles.

To achieve a professional appearance, center the letter on the page vertically and horizontally. Although one-inch margins are the default standard in many word-processing programs, it is more important to establish a picture frame of blank space surrounding the page of text. When you use organizational letterhead stationery, consider the bottom of the letterhead as the top edge of the paper. The right margin should be approximately as wide as the left margin. To give a fuller appearance to very short letters, increase both margins to about an inch and a half. Use your full-page or print-preview feature to check for proportion.

Heading

Unless you are using letterhead stationery, place your full return address and the date in the heading. Because your name appears at the end of the letter, it need not be included in the heading. Spell out words such as *street*, *avenue*, *first*, and *west* rather than abbreviating them. You may either spell out the name of the state in full or use the standard Postal Service abbreviation available at *usps.com*. The date usually goes

---

*For additional details on letter formats and design, you may wish to consult a guide such as *The Gregg Reference Manual* (Tenth Edition) by William A. Sabin (New York: McGraw-Hill, 2004).

Letterhead

520 Niagara Street
Braintree, MA 02184

Phone: (781) 787-1175
Fax: (781) 787-1213
E-mail: info@evans.com

EVANS
and Associates
Transportation Engineers

Date: May 15, 2009

Inside address:
Mr. George W. Nagel
Director of Operations
Boston Transit Authority
57 West City Avenue
Boston, MA 02210

Salutation: Dear Mr. Nagel:

Body:

Enclosed is our final report evaluating the safety measures for the Boston Intercity Transit System.

We believe that the report covers the issues you raised in our last meeting and that you will be pleased with the results. However, if you have any further questions, we would be happy to meet with you again at your convenience.

We would also like to express our appreciation to Mr. L. K. Sullivan of your committee for his generous help during our trips to Boston.

Complimentary closing: Sincerely,

Signature: *Carolyn Brown*

Writer's signature block:
Carolyn Brown, Ph.D.
Director of Research
cbrown@evans.com

End notations:
CB/ls
Enclosure: Final Safety Report
cc: ITS Safety Committee Members

FIGURE L–4. Full-Block-Style Letter (with Letterhead)

3814 Oak Lane
Dedham, MA 02180
December 8, 2009

Dr. Carolyn Brown
Director of Research
Evans and Associates
Transportation Engineers
520 Niagara Street
Braintree, MA 02184

Dear Dr. Brown:

Thank you very much for allowing me to tour your testing facilities. The information I gained from the tour will be of great help to me in preparing the report for my class at Marshall Institute. The tour has also given me some insight into the work I may eventually do as a laboratory technician.

I especially appreciated the time and effort Vikram Singh spent in showing me your facilities. His comments and advice were most helpful.

Again, thank you.

Sincerely,

Leslie Warden

Leslie Warden
781-555-1212

FIGURE L–5. Modified-Block-Style Letter (Without Letterhead)

directly beneath the last line of the return address. Do not abbreviate the name of the month. Begin the heading about two inches from the top of the page. If you are using letterhead that gives the company address, enter only the date, three lines below the last line of the letterhead.

Inside Address

Include the recipient's full name, title, and address in the inside address, two to six lines below the date, depending on the length of the letter. The inside address should be aligned with the left margin, and the left margin should be at least one inch wide.

Salutation

Place the salutation, or greeting, two lines below the inside address and align it with the left margin. In most business letters, the salutation contains the recipient's personal title (such as *Mr.*, *Ms.*, *Dr.*) and last name, followed by a colon. If you are on a first-name basis with the recipient, use only the first name in the salutation.

Address women as *Ms.* unless they have expressed a preference for *Miss* or *Mrs.* However, professional titles (such as *Professor*, *Senator*, *Major*) take precedence over *Ms.* and similar courtesy titles.

When a person's first name could be either feminine or masculine, one solution is to use both the first and last names in the salutation (*Dear Pat Smith:*). Avoid "To Whom It May Concern" because it is impersonal and dated.

For multiple recipients, the following salutations are appropriate:

▶ Dear Professor Allen and Dr. Rivera: [two recipients]

▶ Dear Ms. Becham, Ms. Moore, and Mr. Stein: [three recipients]

▶ Dear Colleagues: [*Members*, or other suitable collective term]

If you are writing to a large company and do not know the name or title of the recipient, you may address a letter to an appropriate department or identify the subject in a subject line and use no salutation. (For information on creating subject lines, see page 110 of **correspondence**.)

▶ National Medical Supply Group
501 West National Avenue
Minneapolis, MN 55407

Attention: Customer Service Department
or
Subject: Defective Cardio-100 Stethoscopes

I am returning six stethoscopes with damaged diaphragms that . . .

In other circumstances in which you do not know the recipient's name, use a title appropriate to the context of the letter, such as *Dear Customer* or *Dear IT Professional*.

Body

The body of the letter should begin two lines below the salutation (or any element that precedes the body, such as a subject or an attention line). Single-space within and double-space between paragraphs, as shown in Figures L–4 and L–5. To provide a fuller appearance to a very short letter, you can increase the side margins or increase the font size. You can also insert extra space above the inside address, the typed (signature) name, and the initials of the person typing the letter—but do not exceed twice the recommended space for each of these elements.

Complimentary Closing

Type the complimentary closing two spaces below the body. Use a standard expression such as *Sincerely*, *Sincerely yours*, or *Yours truly*. (If the recipient is a friend as well as a business associate, you can use a less formal closing, such as *Best wishes* or *Best regards* or, simply, *Best*.) Capitalize only the initial letter of the first word, and follow the expression with a comma.

Writer's Signature Block

Type your full name four lines below and aligned with the complimentary closing. On the next line include your business title, if appropriate. The following lines may contain your individual contact information, such as a telephone number or an **e-mail** address, if not included in the letterhead or the body of your letter. Sign the letter in the space between the complimentary closing and your name.

End Notations

Business letters sometimes require additional information that is placed at the left margin, two spaces below the typed name and title of the writer in a long letter, four spaces below in a short letter.

Reference initials show the letter writer's initials in capital letters, followed by a slash mark (or colon), and then the initials of the person typing the letter in lowercase letters, as shown in Figure L–4. When the writer is also the person typing the letter, no initials are needed.

Enclosure notations indicate that the writer is sending material along with the letter (an invoice, an article, and so on). Note that you should mention the enclosure in the body of the letter. Enclosure notations may take several forms:

- Enclosure: Final Safety Report
- Enclosures (2)
- Enc. *or* Encs.

Copy notation ("cc:") tells the reader that a copy of the letter is being sent to the named recipient(s) (see Figure L–4). Use a blind-copy notation ("bcc:") when you do not want the addressee to know that a copy is being sent to someone else. A blind-copy notation appears only on the copy, not on the original ("bcc: Dr. Brenda Shelton").

Continuing Pages

If a letter requires a second page (or, in rare cases, more), always carry at least two lines of the body text over to that page. Use plain (nonletterhead) paper of quality equivalent to that of the letterhead stationery for the second page. It should have a header with the recipient's name, the page number, and the date. Place the header in the upper left-hand corner or across the page, as shown in Figure L–6.

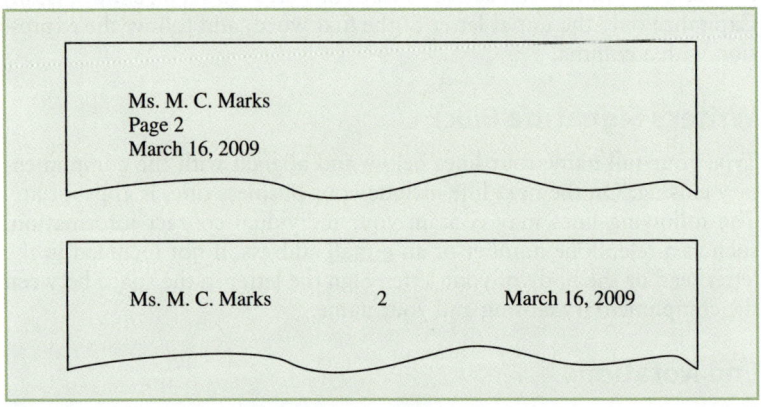

FIGURE L–6. Headers for the Second Page of a Letter

like / as

To avoid confusion between *like* and *as*, remember that *like* is a **preposition** and *as* (or *as if*) is a **conjunction**. Use *like* with a **noun** or **pronoun** that is not followed by a **verb**.

- The supervisor still behaves *like* a novice.

Use *as* before **clauses**, which contain verbs.

- He responded *as* we expected he would.
- The presentation seemed *as if* it would never end.

Like and *as* are used in **comparisons**: *Like* is used in constructions that omit the verb, and *as* is used when the verb is retained.

- He adapted to the new system *like* a duck to water.
- He adapted to the new system *as* a duck adapts to water.

listening

Effective listening enables the listener to understand the directions of an instructor, the message in a speaker's **presentation**, the goals of a manager, and the needs and wants of customers. Above all, it lays the foundation for cooperation. Productive communication occurs when both the speaker and the listener focus clearly on the content of the message and attempt to eliminate as much interference as possible.

Fallacies About Listening

Most people assume that because they can hear, they know how to listen. In fact, *hearing* is passive, whereas *listening* is active. Hearing voices in a crowd or a ringing telephone requires no analysis and no active involvement. We hear such sounds without choosing to listen to them—we have no choice but to hear them. Listening, however, requires actively focusing on a speaker, interpreting the message, and assessing its worth. Listening also requires that you consider the **context** of messages and the differences in meaning that may be the result of differences in the speaker's and the listener's occupation, education, culture, sex, race, or other factors. See also **biased language**, **connotation / denotation**, **English as a second language**, and **global communication**.

Active Listening

To become an active listener, you need to take the following steps:

Step 1: Make a Conscious Decision. The first step to active listening is simply making up your mind to do so. Active listening requires a conscious effort, something that does not come naturally. The well-known precept offers good advice: "Seek first to understand and *then* to be understood."*

Step 2: Define Your Purpose. Knowing why you are listening can go a long way toward managing the most common listening problems: drifting attention, formulating your response while the speaker is still talking, and interrupting the speaker. To help you define your purpose for listening, ask yourself these questions:

- What kind of information do I hope to get from this exchange, and how will I use it?
- What kind of message do I want to send while I am listening? (Do I want to portray understanding, determination, flexibility, competence, or patience?)
- What factors—boredom, daydreaming, anger, impatience—might interfere with listening during the interaction? How can I keep these factors from placing a barrier between the speaker and me?

Step 3: Take Specific Actions. Becoming an active listener requires a willingness to become a responder rather than a reactor. A *responder* is a listener who slows down the communication to be certain that he or she is accurately receiving the message sent by the speaker. A *reactor* simply says the first thing that comes to mind, without checking to make sure that he or she accurately understands the message. Take the following actions to help you become a responder and not a reactor.

- Make a conscious effort to be impartial when evaluating a message. For example, do not dismiss a message because you dislike the speaker or are distracted by the speaker's appearance, mannerisms, or accent.
- Slow down the communication by asking for more information or by **paraphrasing** the message received before you offer your thoughts. Paraphrasing lets the speaker know you are listening, gives the speaker an opportunity to clear up any misunderstanding, and keeps you focused.
- Listen with empathy by putting yourself in the speaker's position. When people feel they are being listened to empathetically, they tend to respond with appreciation and cooperation, thereby improving the communication.
- Take notes, when possible, to help you stay focused on what a speaker is saying. **Note-taking** not only communicates your atten-

---

*Stephen R. Covey, *The 7 Habits of Highly Effective People: Powerful Lessons in Personal Change*, 15th ed. (New York: Free Press, 2004).

tiveness to the speaker but also reinforces the message and helps you remember it.

Step 4: Adapt to the Situation. The requirements of active listening differ from one situation to another. For example, when you are listening to a lecture, you may be listening only for specific information. However, if you are on a team project that depends on everyone's contribution, you need to listen at the highest level so that you can gather information as well as pick up on nuances the other speakers may be communicating. See also **collaborative writing**.

lists

Vertically stacked lists of words, **phrases**, and other items that are often highlighted with bullets, numbers, or letters can save readers time by allowing them to see at a glance specific items or key points. As shown in Figure L–7, lists also help readers by breaking up complex statements and by focusing on such information as steps in sequence, materials or parts needed, questions or concluding points, and recommendations.

As Figure L–7 also shows, you should provide **context** by introducing a list, preferably with a complete sentence followed by a **colon** or without punctuation if the introductory statement is a fragment. Ensure **coherence** by following the list with some reference to the list or the statement that introduced the list.

Before we agree to hold the software engineering conference at the Brent Hotel, we need to make sure the hotel can provide the following resources:

- Business center with high-quality Internet access and printing services
- Main exhibit area that can accommodate thirty 8-foot-by-15-foot booths
- Eight meeting rooms, each with a podium or table and seating for 25 people
- High-speed Internet access and digital projection in each room
- Ballroom dining facilities for 250 people with a dais for four speakers

To confirm that the Brent Hotel is our best choice, we should tour the facilities during our stay in Kansas City.

FIGURE L–7. Bulleted List in a Paragraph

> **WRITER'S CHECKLIST** Using Lists
>
> Follow the practices of your organization or use the guidelines below for consistency and formatting.
>
> **CONSISTENCY**
> - Do not overuse lists or include too many items, as in full pages of lists or **presentation** slides that are dense with lists.
> - List only comparable items, such as tasks or equipment, that are balanced in importance (as in Figure L–7).
> - Begin each listed item in the same way—whether with **nouns**, **verbs**, or other **parts of speech**—and maintain **parallel structure** throughout.
>
> **FORMATTING**
> - Capitalize the first word in each listed item, unless doing so is visually awkward.
> - Use **periods** or other ending **punctuation** when the listed items are complete sentences.
> - Avoid **commas** or **semicolons** following items and do not use the **conjunction** *and* before the last item in a list.
> - Use bullets (round, square, arrow) when you do not wish to indicate rank or sequence.
> - List bulleted items in a logical order, keeping your **audience** and **purpose** in mind. See also **methods of development** and **persuasion**.
> - Use numbers to indicate sequence or rank.
> - Follow each number with a period and start the item with a capital letter.
> - When lists need subdivisions, use letters with numbers. See also **outlining**.

literature reviews

A literature review is a summary of the relevant literature (printed and electronic) available on a particular subject over a specified period of time. For example, a literature review might describe significant material published in the past five years on a technique for improving emergency medical diagnostic procedures, or it might describe all **reports** written in the past ten years on efforts to improve safety procedures at a particular company. A literature review tells readers what has been published on a particular subject and gives them an idea of what material they should read in full.

Some **trade journal articles** or theses begin with a brief literature review to bring the reader up to date on current research in the field. The writer then uses the review as background for his or her own discussion of the subject. Figure L–8 shows a literature review that serves as an **introduction** for an article that surveys the use of visual communication in the workplace. A literature review also could be a whole document in itself. Fully developed literature reviews are good starting points for detailed **research**.

To prepare a literature review, you must first research published material on your topic. Because your readers may begin their research based on your literature review, carefully and accurately cite all bibliographic information. As you review each source, note the **scope** of the

---

Our field's growing awareness of the centrality of visual communication to the work of professional writers has given rise to much of the visually oriented research in pedagogy, rhetoric, and related areas within professional communication (see, for example, Brasseur, 1997; Kostelnick, 1994). Complementing this work, educators have also been actively developing courses and programs to help prepare students for the expanded professional roles they will likely assume. It is not surprising, therefore, that a 2003 survey of members of the Association of Teachers of Technical Writing (ATTW) ascertained that teachers of professional writing give visual communication top billing as a topic of interest (Dayton & Bernhardt, 2004). However, while visual communication has established a firm foothold in professional communication research and pedagogy (Allen & Benninghoff, 2004), program coordinators and faculty more generally lack current data about the role of visual communication in the workplace. Such data could help to determine whether academic perceptions of the importance of visual communication are well founded, and, in turn, whether academic programs are meeting student needs in terms of workplace preparation. Such data could also provide concrete evidence in arguments about curricular design and enrich the information educators are able to offer students about the workplace life of the professional writer. As professional communication programs continue to mature and proliferate, such information will become even more useful.

Source: Brumberger, Eva. "Visual Communication in the Workplace: A Survey of Practice." *Technical Communication Quarterly* 16, no. 4 (Autumn 2007): 369–95.

FIGURE L–8. Literature Review

book or article and judge its value to the reader. Save all printouts of computer-assisted searches—you may want to incorporate the sources in a bibliography. See also **documenting sources**.

Begin a literature review by defining the area to be covered and the types of works to be reviewed. For example, a literature review may be limited to articles and reports and not include any books. You can arrange your discussion chronologically, beginning with a description of the earliest relevant literature and progressing to the most recent (or vice versa). You can also subdivide the topic, discussing works in various subcategories of the topic.

Related to literature reviews are annotated **bibliographies**, which also give readers information about published material. Unlike literature reviews, however, an annotated bibliography cites each bibliographic item and then describes it in a single block of text. The description (or *annotation*) may include the **purpose** of the article or book, its scope, the main topics covered, its historical importance, and anything else the writer feels the reader should know.

logic errors

Logic is essential to convincing an **audience** that your conclusions are valid. Errors in logic can undermine the point you are trying to communicate and your credibility. Some typical logic errors are described in this entry. See also **persuasion**.

▶ ETHICS NOTE Many of the following errors in logic, when used to mislead **readers**, are unethical as well as illogical. See also **ethics in writing**. ◆

Lack of Reason

When a statement is contrary to the reader's common sense, that statement is not reasonable. If, for example, you stated, "New York is a small town," your reader might immediately question your statement. However, if you stated, "Although New York's population is over eight million, it is composed of neighborhoods that function as small towns," your reader could probably accept the statement as reasonable.

Sweeping Generalizations

Sweeping generalizations are statements that are too broad or all-inclusive to be supportable; they generally enlarge an observation about a small group to refer to an entire population. A flat statement such as "Management is never concerned about employees" ignores any evidence that management could be concerned for its employees. Using such generalizations weakens your credibility.

Non Sequiturs

A non sequitur is a statement that does not logically follow a previous statement.

> ▶ I cleared off my desk, and the report is due today.

The missing link in these statements is that the writer cleared his or her desk to make space for materials to help finish the report that is due today. In your own writing, be careful that you do not allow gaps in logic to produce non sequiturs.

False Cause

A false cause (also called *post hoc, ergo propter hoc*) refers to the logical fallacy that because one event followed another event, the first somehow caused the second.

> ▶ I didn't bring my umbrella today. No wonder it is now raining.
>
> ▶ Because we now have our board meetings at the Education Center, our management turnover rate has declined.

Such errors in reasoning can happen when the writer hastily concludes that two events are related without examining whether a causal connection between them, in fact, exists.

Biased or Suppressed Evidence

A conclusion reached as a result of biased or suppressed evidence—self-serving data, questionable sources, purposely omitted or incomplete facts—is both illogical and unethical. Suppose you are preparing a report on the acceptance of a new policy among employees. If you distribute **questionnaires** only to those who think the policy is effective, the resulting evidence will be biased. Intentionally ignoring relevant data that might not support your position not only produces inaccurate results but also is unethical.

Fact Versus Opinion

Distinguish between fact and opinion. Facts include verifiable data or statements, whereas opinions are personal conclusions that may or may not be based on facts. For example, it is verifiable that distilled water boils at 100°C; that it tastes better or worse than tap water is an opinion. Distinguish the facts from your opinions in your writing so that your readers can draw their own conclusions.

Loaded Arguments

When you include an opinion in a statement and then reach conclusions that are based on that statement, you are loading the argument. Consider the following opening for a memo:

> ▶ I have several suggestions to improve the poorly written policy manual. First, we should change . . .

Unless everyone agrees that the manual is poorly written, readers may reject a writer's entire message because they disagree with this loaded premise. Do not load arguments in your writing; conclusions reached with loaded statements are weak and can produce negative reactions in readers who detect the loading.

WEB LINK	Understanding an Argument

Dr. Frank Edler of Metropolitan Community College in Omaha, Nebraska, offers a tutorial in recognizing the logical components of an argument and thinking critically. See *bedfordstmartins.com/alredtech* and select *Links for Handbook Entries*.

loose / lose

Loose is an **adjective** meaning "not fastened" or "unrestrained." ("He found a *loose* wire.") *Lose* is a **verb** meaning "be deprived of" or "fail to win." ("I hope we do not *lose* the contract.")

malapropisms

A malapropism is a word that sounds similar to the one intended but is ludicrously wrong in the **context**.

    INCORRECT   Our employees are less *sedimentary* now that we have a fitness center.

    CORRECT   Our employees are less *sedentary* now that we have a fitness center.

Intentional malapropisms are sometimes used in humorous writing; unintentional malapropisms can confuse readers and embarrass a writer. See also **figures of speech**.

manuals

Manuals (printed or electronic) help customers and technical specialists use and maintain products. These manuals are often written by professional technical writers, although in smaller companies, engineers or technicians may write them. See also **instructions** and review the entries **process explanation** and **technical writing style**.

Types of Manuals

User Manuals. User manuals are aimed at skilled or unskilled users of equipment and provide instructions for the setup, operation, and maintenance of a product. User manuals also typically include safety precautions and troubleshooting charts and guides.

Tutorials. Tutorials are self-study guides for users of a product or system. Either packaged with user manuals or provided electronically, tutorials guide novice users through the operation of a product or system.

Training Manuals. Training manuals are used to prepare individuals for some procedure or skill, such as operating a respirator, flying an airplane, or processing an insurance claim. Training manuals may be printed or delivered in electronic or online forms.

Operators' Manuals. Written for skilled operators of construction, manufacturing, computer, or military equipment, operators' manuals contain essential instructions and safety warnings. They are often published in a convenient format that allows operators to use them at a work site.

Service Manuals. Service manuals help trained technicians repair equipment or systems, usually at the customer's location. Such manuals often contain troubleshooting guides for locating technical problems.

Special-Purpose Manuals. Some users need manuals that fulfill special purposes; these include programmer reference manuals, overhaul manuals, handling and setup manuals, and safety manuals.

Designing and Writing Effective Manuals

Identify and Write for Your Audience. Will you be writing for novice users, intermediate users, or experts? Or will you be instructing a combination of users with different levels of technical knowledge and experience? Identify your **audience** before you begin writing. Depending on your audience, you will make the following decisions:

- Which details to include (fewer for experts, more for novices)
- What level of technical vocabulary to use (necessary technical terminology for experts, plain language for novices)
- Whether to include a summary list of steps (experts will probably prefer using this summary list and not the entire manual; once novices and intermediate users gain more skill, they will also prefer referring just to this summary list)

Design your manual so that **readers** can use the equipment, software, or machinery while they are also reading your instructions. See also **layout and design**.

- Provide **headings** and subheadings using words that readers will find familiar so that they can easily locate particular sections and instructions.
- Write with precision and accuracy so that readers can perform the procedures easily.
- Use clearly drawn and labeled **visuals**, in addition to written text, to show readers exactly what equipment, online screens, or other items in front of them should look like.

Provide an Overview. An overview at the beginning of a manual should explain the overall purpose of the procedure, how the procedure can be useful to the reader, and any cautions or warnings the reader should know about before starting. If readers know the purpose of the procedure and its specific workplace applications, they will be more likely to pay close attention to the steps of the procedure and apply them appropriately in the future.

Create Major Sections. Divide any procedure into separate goals, create major sections to cover those goals, and state them in the section headings. If your manual has chapters, you can divide each chapter into specific subsections. In a manual for students about designing a Web page, an introductory chapter might include these subsections: (1) obtaining Web space, (2) viewing the new space, (3) using a Web authoring tool.

Use Headings to Indicate the Goals of Actions. Within each section, you can use headings to describe *why* readers need to follow each step or each related set of steps. The conventional way to indicate a goal in a heading is to use the infinitive form of a **verb** ("*To scan* the document") or the gerund form ("*Scanning* the document"). Whichever verb form you choose for headings, use it consistently throughout your manual each time you want to indicate a goal. The following example shows how a heading can indicate the goal of a set of actions with the gerund form of the verb ("Scanning").

▶ **Scanning Documents**
    Action 1   Place the document in the feeder
    Action 2   Check the box below the window for single-page items
    Action 3   Click "SCAN" to open the "Scan Manager"

Use the Imperative Verb Form to Indicate Actions. The previous example shows the conventional way to indicate an action by using the imperative form of verbs ("*Place* the document"; "*Check* the box"). Use the imperative form consistently each time you want to indicate an action. In the example about scanning documents, the manual writer designated actions by placing subheads (Action 1, Action 2, and so on) in the left margin and imperative verbs ("Place," "Check," and "Click") in the instructions to the right of the subheads.

Use Simple and Direct Verbs. Simple and direct verbs are most meaningful, especially for novice readers. Avoid **jargon** and terms known only to intermediate readers or experts, unless you know that they constitute the entire reading audience.

POOR VERB CHOICE	BETTER VERB CHOICE
Attempt	Try
Depress, Hit	Press
Discontinue	Stop
Display	Show
Employ	Use
Enumerate	Count
Execute	Do
Observe	Watch
Segregate	Divide

Indicate the Response of Actions. When appropriate, indicate the expected response of an action to reassure readers that they are performing the procedure correctly (for example, "A blinking light will appear" or "You will see a red triangle"). Use the form you choose consistently throughout a set of instructions to designate a response. In the following example, you can see how the response is indented farther to the right than the action, begins with a subhead ("Response"), and is indicated by a full sentence.

▶ Action 3   Click "SCAN" to open the "Scan Manager"

<u>Response</u>   The "Scan Manager" will open and ask you to choose "preview," "scan," or "help."

Checking for Usability. To test the accuracy and **clarity** of your instructions, at a minimum ask someone who is not familiar with the subject to use the manual to spot missing steps or confusing passages. In most cases, you should conduct systematic **usability testing** on manual drafts so that you can detect errors and other problems readers might encounter.

WRITER'S CHECKLIST Preparing Manuals

- Determine the best medium for your manual: Web, online document, CD-ROM, spiral binding, loose-leaf binding, and so on.
- Pay attention to **organization** and **outlining** because complex products and systems need well-organized manuals to be useful to readers.
- Use a consistent format for each part of a manual: headings, subheadings, goals, actions, responses, cautions, warnings, and tips.
- Use standardized symbols, as described in **global graphics**, especially for international readers or where regulations require them.
- Use visuals—such as screen shots, schematics, exploded-view **drawings**, **flowcharts**, **photographs**, and **tables**—placing them where they would most benefit readers.
- Include **indexes** to help readers find information.

Writer's Checklist: Preparing Manuals (continued)

- Use warning statements and standard symbols for potential dangers, as described on pages 260–61.
- Have manuals reviewed by your peers as well as by technical experts and other specialists to ensure that the manuals are helpful, accurate, and appropriate. See also **revision** and **proofreading**.

maps

Maps are often used to show specific geographic areas and features (roads, mountains, rivers, and the like). They can also illustrate geographic distributions of populations, climate patterns, corporate branch offices, and so forth. The map in Figure M–1, from an environmental assessment, shows the overlapping geographic areas served by three electric utilities in Missouri, Iowa, and Illinois. Note that the map contains a figure number and title, scale of distances, key (or legend), compass, and distinctive highlighting for emphasis. Maps are often used in **reports**, **proposals**, **brochures**, **environmental impact statements**, and other documents in which readers need to know the location or geographic orientation of buildings and other facilities.

WRITER'S CHECKLIST Creating and Using Maps

- Follow the general guidelines discussed in **visuals** for placement of maps.
- Label each map clearly, and assign each map a figure number if it is one of a number of illustrations.
- Clearly identify all boundaries in the map. Eliminate unnecessary boundaries.
- Eliminate unnecessary information that may clutter a map. For example, if the purpose of the map is to show population centers, do not include mountain elevations, rivers, or other physical features.
- Include a scale of miles/kilometers or feet/meters to give your readers an indication of the map's proportions.
- Indicate which direction is north with an arrow or a compass symbol.
- Emphasize key features by using color, shading, dots, crosshatching, or other appropriate symbols.
- Include a key, or legend, that explains what the different colors, shadings, or symbols represent.

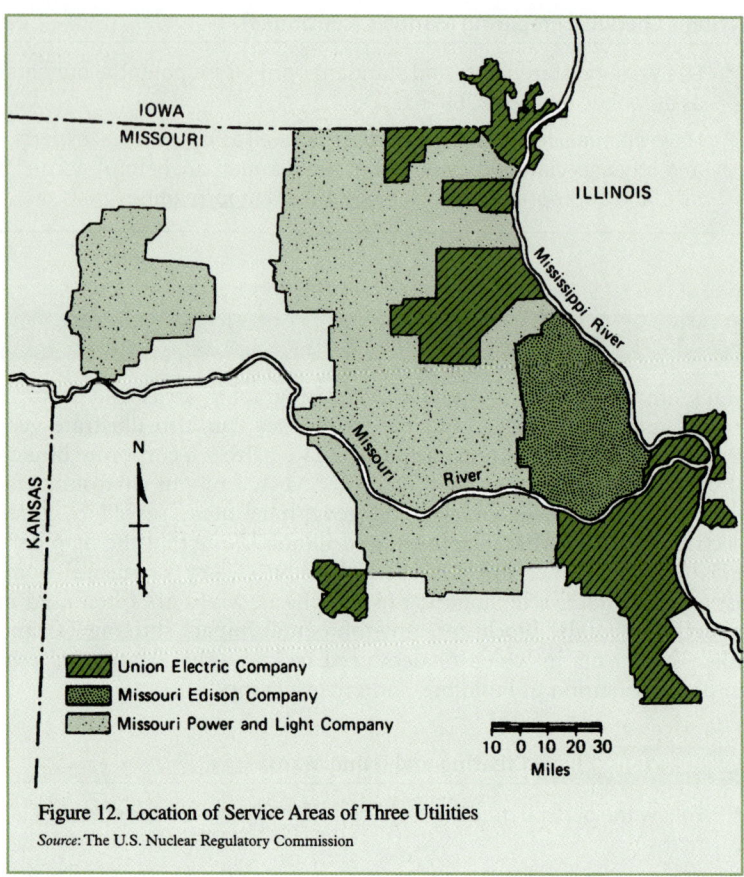

FIGURE M–1. Map

> **WEB LINK** Maps and Mapping Information
>
> The University of Texas Libraries offer a useful site with links to online maps and other types of mapping and cartographic resources. See *bedfordstmartins.com/alredtech* and select *Links for Handbook Entries*.

mathematical equations

You can accurately prepare material with mathematical equations and make it easy to read by following consistent standards throughout a document. Mathematicians often use scientific software packages to prepare and format research articles and reports with mathematical equations. Some scientific journals, in fact, require documents to be written with software packages such as LaTeX, Scientific Workplace, AMSTeX, or similar versions of TeX (*www.latex-project.org/*). Unless you need to follow such specific styles or specifications, the guidelines in this entry should serve you well.

Formatting Equations

Set short and simple equations, such as $x(y) = y^2 + 3y + 2$, as part of the running text rather than displaying them on separate lines, as long as an equation does not appear at the beginning of a sentence. If a document contains multiple equations, identify them with numbers, as the following example shows:

$$x(y) = y^2 + 3y + 2 \tag{1}$$

Number displayed equations consecutively throughout the work. Place the equation number, in parentheses, at the right margin of the same line as the equation (or of the first line if the equation runs longer than one line). Leave at least four spaces between the equation and the equation number. Refer to displayed equations by number, for example, as "Equation 1" or "Eq. 1."

Positioning Displayed Equations

Equations that are set off from the text need to be surrounded by space. Triple-space between displayed equations and normal text. Double-space between one equation and another and between the lines of multi-line equations. Count space above the equation from the uppermost character in the equation; count space below from the lowermost character.

Type displayed equations either at the left margin or indented five spaces from the left margin, depending on their length. When a series of short equations is displayed in sequence, align them on the equal signs.

$$p(x,y) = \sin(x + y) \tag{2}$$
$$p(x,y) = \sin x \cos y + \cos x \sin y \tag{3}$$
$$p(x_0, y_0) = \sin x_0 \cos y_0 + \cos x_0 \sin y_0 \tag{4}$$
$$q(x,y) = \cos(x + y) \tag{5}$$
$$= \cos x \cos y - \sin x \sin y$$
$$q(x_0, y_0) = \cos x_0 \cos y_0 - \sin x_0 \sin y_0 \tag{6}$$

Break an equation that requires two lines at the equal sign, carrying the equal sign over to the second portion of the equation.

$$\int_0^1 (f_n - \tfrac{n}{r} f_n)^2 \, r \, dr + 2n \int_0^1 f_n f_n dr \qquad (7)$$
$$= \int_0^1 (f_n - \tfrac{n}{r} f_n)^2 \, r \, dr + n f_n^2(1)$$

If you cannot break an equation at the equal sign, break it at a plus or minus sign that is not in **parentheses** or **brackets**. Bring the plus or minus sign to the next line of the equation, which should be positioned to end near the right margin of the equation.

$$\emptyset(x, y, z) = (x^2 + y^2 + z^2)^{1/2} (x - y + z)(x + y - z)^2 \qquad (8)$$
$$- [f(x, y, z) - 3x^2]$$

The next best place to break an equation is between parentheses or brackets that indicate multiplication of two major elements.

For equations that require more than two lines, start the first line at the left margin, end the last line at the right margin (or four spaces to the left of the equation's number), and center intermediate lines between the margins. Whenever possible, break equations at operational signs, parentheses, or brackets.

Omit punctuation after displayed equations, even when they end a sentence and even when a key list defining terms follows (for example, P = pressure, psf; V = volume, cu ft; T = temperature, °C). Punctuation may be used before an equation, however, depending on the grammatical construction.

The term $(n)_r$ may be written in a more familiar way by using the following algebraic device:

$$(n)_r = \frac{(n)(n-1)(n-2)\ldots(n-r+1)(n-r)(n-r-1)\ldots 3\cdot 2\cdot 1}{(n-r)(n-r-1)\ldots 3\cdot 2\cdot 1} \qquad (9)$$
$$= \frac{n!}{(n-r)!}$$

maybe / may be

Maybe (one word) is an **adverb** meaning "perhaps." ("*Maybe* the legal staff can resolve this issue.") *May be* (two words) is a **verb** phrase. ("It *may be* necessary to hire a specialist.")

media / medium

Media is the plural of *medium* and should always be used with a plural **verb**.

- Many communication *media are* available today.
- The Internet *is* a multifaceted *medium*.

meetings

Meetings allow people to share information and collaborate to produce better results than exchanges of **e-mail** messages or other means would allow. Like a **presentation**, a successful meeting requires planning and preparation. See also **selecting the medium**.

Planning a Meeting

As you plan a meeting, determine the focus of the meeting, decide who should attend, and choose the best time and place to hold it. Prepare an agenda for the meeting and determine who should take the minutes. See **minutes of meetings**.

Determine the Purpose of the Meeting.   The first step in planning a meeting is to focus on the desired outcome. Ask yourself the following question to help you determine the **purpose** of the meeting: What should participants know, believe, do, or be able to do as a result of attending the meeting?

Once you have your desired outcome in focus, use the information to write a purpose statement for the meeting that answers the questions *what* and *why*.

- The purpose of this meeting is to gather ideas from the sales force [*what*] in order to create a successful sales campaign for our new scanner [*why*].

Decide Who Should Attend.   Schedule a meeting for a time when all or most of the key people can be present. If a meeting must be held without some key participants, ask those people for their contributions prior to the meeting or invite them to participate by speakerphone, videoconference, or such remote methods as described in *Digital Tip: Conducting Meetings from Remote Locations*. Of course, the meeting minutes should be distributed to everyone, including significant non-attendees.

Choose the Meeting Time.   The time of day and the length of the meeting can influence its outcome. Consider the following when you are planning a meeting:

- Monday morning is often used to prepare for the coming week's work.
- Friday afternoon is often focused on completing the current week's tasks.
- Long meetings may need to include breaks to allow participants to respond to messages and refresh themselves.
- Meetings held during the last 15 minutes of the day will be quick, but few people will remember what happened.
- Remote participants may need consideration for their time zones.

Choose the Meeting Location. Having a meeting at your own location can give you an advantage: You feel more comfortable, which, along with your guests' newness to their surroundings, may give you an edge. Holding the meeting on someone else's premises, however, can signal cooperation. For balance, especially when people are meeting for the first time or are discussing sensitive issues, meet at a neutral site where no one gains an advantage and attendees may feel freer to participate.

DIGITAL TIP

Conducting Meetings from Remote Locations

When participants cannot meet face to face, consider holding a videoconference or Web conference using groupware or whiteboard software. Groupware allows meeting participants at remote locations to perform collaborative activities, such as sharing documents, participating in real-time discussions, or commenting on documents. Whiteboard software allows participants to write on an interactive LCD panel as though they were in a room with a whiteboard. Writings are then displayed on each participant's screen and can be saved or printed. For links to Web sites with information on groupware and whiteboard software, see *bedfordstmartins.com/alredtech* and select *Links for Handbook Entries.*

Establish the Agenda. A tool for focusing the group, the agenda is an outline of what the meeting will address. Always prepare an agenda for a meeting, even if it is only an informal list of main topics. Ideally, the agenda should be distributed to attendees a day or two before the meeting. For a longer meeting in which participants are required to make a presentation, try to distribute the agenda a week or more in advance.

The agenda should list the attendees, the meeting time and place, and the topics you plan to discuss. If the meeting includes presentations, list the time allotted for each speaker. Finally, indicate an approximate length for the meeting so that participants can plan the rest of their day. Figure M–2 shows a typical agenda.

Design Meeting Agenda

Purpose: To get creative ideas for the CZX software
Date: Monday, May 11, 2009
Place: Conference Room E
Time: 9:30 a.m.–11:00 a.m.
Attendees: New Products Manager, Software Engineering Manager and Designers, Technical Publications Manager, Technical Training Manager

Topic	Presenter	Time
CZX Software	Bob Arbuckle	9:30–9:45
The Campaign	Maria Lopez	9:45–10:00
The Design Strategy	Mary Winifred	10:00–10:15
Discussion	Led by Dave Grimes	10:15–11:00

FIGURE M–2. Meeting Agenda

If the agenda is distributed in advance of the meeting, it should be accompanied by a **cover letter** or message informing people of the following:

- The purpose of the meeting
- The date and place of the meeting
- The meeting start and stop times
- The names of the people invited
- Instructions on how to prepare

Figure M–3 shows a cover message to accompany an agenda.

Assign the Minute-Taking. Delegate the minute-taking to someone other than the leader. The minute-taker should record major decisions made and tasks assigned. To avoid misunderstandings, the minute-taker needs to record each assignment, the person responsible for it, and the date on which it is due.

For a standing committee, it is best to rotate the responsibility of taking minutes. See also **note-taking**.

> From: "E. Lauter" <elauter@mmsoftware.com>
> To: "R. Arbuckle" <arbuckle@mmsoftware.com>,
> "D. Grimes" <dgrimes@mmsoftware.com>,
> "M. Lopez" <mlopez@mmsoftware.com>,
> "M. Winifred" <mwinifred@mmsoftware.com>,
> "Design" <design_all@mmsoftware.com>
> Sent: Tues, 05 May 2009 13:30:12 EST
> Subject: Design Meeting (May 11 at 9:30 a.m.)
> Attachments: Meeting Agenda.doc (29 KB)
>
> **Purpose of the Meeting**
>
> The purpose of this meeting is to get your ideas for our new CZX software.
>
> **Date, Time, and Location**
>
> Date: Monday, May 11, 2009
> Time: 9:30 a.m.–11:00 a.m.
> Place: Conference Room E (go to the ground floor, take a right off the elevator, third door on the left)
>
> **Attendees**
>
> Those addressed above
>
> **Meeting Preparation**
>
> Everyone should be prepared to offer suggestions on the following topics:
>
> - Features of the new software
> - Techniques for designing the software
> - Customer profile of potential buyers
> - FAQs—questions customers may ask
>
> **Agenda**
>
> Please see the attached document.

FIGURE M–3. E-mail to Accompany an Agenda

Conducting the Meeting

Assign someone to write on a board or project a computer image of information that needs to be viewed by everyone present.

During the meeting, keep to your agenda; however, create a productive environment by allowing room for differing views and fostering an environment in which participants listen respectfully to one another.

- Consider the feelings, thoughts, ideas, and needs of others—do not let your own agenda blind you to other points of view.

- Help other participants feel valued and respected by **listening** to them and responding to what they say.
- Respond positively to the comments of others whenever possible.
- Consider communication styles and approaches that are different from your own, particularly those from other cultures. See also **global communication**.

Deal with Conflict.   Despite your best efforts, conflict is inevitable. However, conflict is potentially valuable; when managed positively, it can stimulate creative thinking by challenging complacency and showing ways to achieve goals more efficiently or economically. See **collaborative writing**.

Members of any group are likely to vary in their personalities and attitudes, and you may encounter people who approach meetings differently. Consider the following tactics for the interruptive, negative, rambling, overly quiet, and territorial personality types.

- The *interruptive person* rarely lets anyone finish a sentence and may intimidate the group's quieter members. Tell that person in a firm but nonhostile tone to let the others finish in the interest of getting everyone's input. By addressing the issue directly, you signal to the group the importance of putting common goals first.
- The *negative person* has difficulty accepting change and often considers a new idea or project from a negative point of view. Such negativity, if left unchecked, can demoralize the group and suppress enthusiasm for new ideas. If the negative person brings up a valid point, however, ask for the group's suggestions to remedy the issue being raised. If the negative person's reactions are not valid or are outside the agenda, state the necessity of staying focused on the agenda and perhaps recommend a separate meeting to address those issues.
- The *rambling person* cannot collect his or her thoughts quickly enough to verbalize them succinctly. Restate or clarify this person's ideas. Try to strike a balance between providing your own interpretation and drawing out the person's intended meaning.
- The *overly quiet person* may be timid or may just be deep in thought. Ask for this person's thoughts, being careful not to embarrass the person. In some cases, you can have a quiet person jot down his or her thoughts and give them to you later.
- The *territorial person* fiercely defends his or her group against real or perceived threats and may refuse to cooperate with members of other departments, companies, and so on. Point out that although such concerns may be valid, everyone is working toward the same overall goal and that goal should take precedence.

Close the Meeting. Just before closing the meeting, review all decisions and assignments. Paraphrase each to help the group focus on what they have agreed to do and to ensure that the minutes will be complete and accurate. Now is the time to raise questions and clarify any misunderstandings. Set a date by which everyone at the meeting can expect to receive copies of the minutes. Finally, thank everyone for participating, and close the meeting on a positive note.

> **WRITER'S CHECKLIST** **Planning and Conducting Meetings**
>
> - Develop a purpose statement for the meeting to focus your planning.
> - Invite only those essential to fulfilling the purpose of the meeting.
> - Select a time and place convenient to all those attending.
> - Create an agenda and distribute it a day or two before the meeting.
> - Assign someone to take meeting minutes.
> - Ensure that the minutes record key decisions; assignments; due dates; and the date, time, and location of any follow-up meeting.
> - Follow the agenda to keep everyone focused.
> - Respect the views of others and how they are expressed.
> - Use the strategies in this entry for handling conflict and attendees whose style of expression may prevent getting everyone's best thinking.
> - Close the meeting by reviewing key decisions and assignments.

memos

Memos are documents that use a standard form (*To:*, *From:*, *Date:*, *Subject:*) whether sent on paper or as attachments to **e-mail** messages. They are used within organizations to report results, instruct employees, announce policies, disseminate information, and delegate responsibilities.

Even in organizations where e-mail messages have largely taken the function of memos, a printed or an attached memo with organizational letterhead can communicate with formality and authority in addition to offering the full range of word-processing features. Paper memos are also useful in manufacturing and service industries, as well as in other businesses where employees do not have easy access to e-mail. For a discussion of writing strategies for memos, e-mail, and **letters**, see **correspondence**. See also **selecting the medium**.

Memo Format

The memo shown in Figure M–4 illustrates that the memo format can be used not only for routine correspondence but also for short reports, proposals, and other internal documents.

Although memo formats and conventions vary, the format (*To:*, *From:*, *Date:*, *Subject:*) in Figure M–4 is typical. As this example also illustrates, the use of **headings** and **lists** often fosters **clarity** and provides emphasis in memos. For a discussion of subject lines, see **correspondence**.

As with e-mail, be alert to the practices of addressing and distributing memos in your organization. Consider who should receive or needs to be copied on a memo and in what order—senior managers, for example, take precedence over junior managers. If rank does not apply, alphabetizing recipients by last name is safe.

Some organizations ask writers to initial or sign printed memos to verify that the writer accepts responsibility for a memo's contents. Electronic copies of memos should not include simulated initials.

Additional Pages

When memos require more than one page, use a second-page header and always carry at least two lines of the body text over to that page. The header should include either the recipient's name or an abbreviated subject line (if there are too many names to fit), the page number, and the date. Place the header in the upper left-hand corner or across the page, as shown in Figure M–5.

> **WEB LINK** **Writing Memos**
>
> For links to articles about when to write memos and tips for following organization protocol, see *bedfordstmartins.com/alredtech* and select *Links for Handbook Entries*.

methods of development

A logical method of development satisfies the readers' need for shape and structure in a document, whether it is an **e-mail**, a **report**, or a Web page. It helps you as a writer move smoothly and logically from the **introduction** to a **conclusion**.

Choose the method or, as is often the case, a combination of methods that best suits your subject, **audience**, and **purpose**. Following are

Professional Publishing Services

MEMORANDUM

TO: Barbara Smith, Publications Manager
FROM: Hannah Kaufman, Vice President   *HK*
DATE: April 14, 2009
SUBJECT: Schedule for ACM Electronics Brochures

ACM Electronics has asked us to prepare a comprehensive set of brochures for its Milwaukee office by August 10, 2009. We have worked with similar firms in the past, so this job should be relatively easy to prepare. My guess is that the job will take nearly two months. Ted Harris has requested time and cost estimates for the project. Fred Moore in production will prepare the cost estimates, and I would like you to prepare a tentative schedule for the project.

Additional Personnel
In preparing the schedule, check the status of the following:
- Production schedule for all staff writers
- Availability of freelance writers
- Availability of dependable graphic designers

Ordinarily, we would not need to depend on outside personnel; however, because our bid for the *Wall Street Journal* special project is still under consideration, we could be short of staff in June and July. Further, we have to consider vacations that have already been approved.

Time Estimates
Please give me time estimates by April 17. A successful job done on time will give us a good chance to obtain the contract to do ACM Electronics' annual report for its stockholders' meeting this fall.

I have enclosed several brochures that may be helpful.

cc: Ted Harris, President
      Fred Moore, Production Editor

Enclosures: Sample Brochures

FIGURE M–4. Typical Memo Format (Printed with Sender's Handwritten Initials)

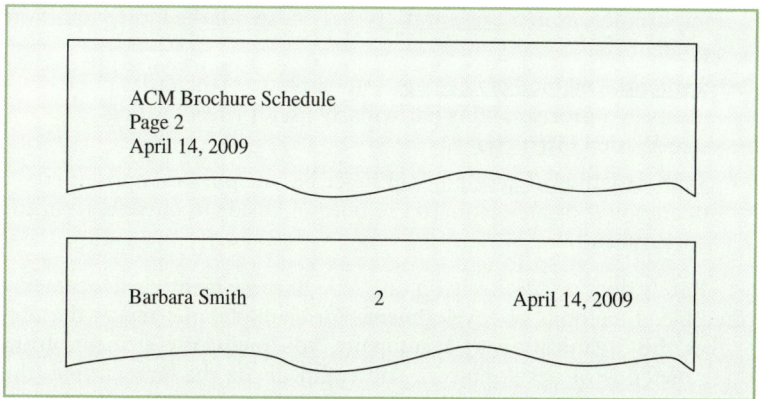

FIGURE M–5. Headers for the Second Page of Memos

the most common methods, each of which is discussed in further detail in its own entry.

- **Cause-and-effect method of development** begins with either the cause or the effect of an event. This approach can be used to develop a report that offers a solution to a problem, beginning with the problem and moving on to the solution or vice versa.
- **Chronological method of development** emphasizes the time element of a sequence, as in a **trouble report** that traces events as they occurred in time.
- **Comparison method of development** is useful when writing about a new topic that is in many ways similar to another topic that is more familiar to your readers.
- **Definition method of development** extends definitions with additional details, examples, comparisons, or other explanatory devices. See also **defining terms**.
- **Division-and-classification method of development** either separates a whole into component parts and discusses each part separately (*division*) or groups parts into categories that clarify the relationship of the parts (*classification*).
- **General and specific methods of development** proceed either from general information to specific details or from specific information to a general conclusion.
- **Order-of-importance method of development** presents information in either decreasing order of importance, as in a **proposal** that begins with the most important point, or increasing order of

importance, as in a **presentation** that ends with the most important point.

- **Sequential method of development** emphasizes the order of elements in a process and is particularly useful when writing step-by-step **instructions**.
- **Spatial method of development** describes the physical appearance of an object or area from top to bottom, inside to outside, front to back, and so on.

Rarely does a writer rely on only one of these methods. Documents often blend methods of development. For example, in a report that describes the organization of a company, you might use elements from three methods of development. You could divide the larger topic (the company) into operations (division and classification), arrange the operations according to what you see as their impact within the company (order of importance), and present their manufacturing operations in the order they occur (sequential). As this example illustrates, when outlining a document, you may base your major division on one primary method of development appropriate to your purpose and then subordinate other methods to it.

minutes of meetings

Organizations and committees that keep official records of their **meetings** refer to such records as *minutes*. Because minutes are often used to settle disputes, they must be accurate, complete, and clear. When approved, minutes of meetings are official and can be used as evidence in legal proceedings. An example of minutes is shown in Figure M–6.

Keep your minutes brief and to the point. Except for recording formally presented motions, which must be transcribed word for word, summarize what occurs and paraphrase discussions. To keep the minutes concise, follow a set format, and use **headings** for each major point discussed. See also **note-taking**.

Avoid abstractions and generalities; always be specific. Refer to everyone in the same way—a lack of consistency in titles or names may suggest a deference to one person at the expense of another. Avoid **adjectives** and **adverbs** that suggest good or bad qualities, as in "Mr. Sturgess's *capable* assistant read the *comprehensive* report to the subcommittee." Minutes should be objective and impartial.

If a member of the committee is to follow up on something and report back to the committee at its next meeting, clearly state the person's name and the responsibility he or she has accepted.

minutes of meetings

NORTH TAMPA MEDICAL CENTER
Minutes of the Monthly Meeting
Medical Audit Committee

DATE: June 23, 2009

PRESENT: G. Miller (Chair), C. Bloom, J. Dades, K. Gilley, D. Ingoglia (Secretary), S. Ramirez

ABSENT: D. Rowan, C. Tsien, C. Voronski, R. Fautier, R. Wolf

Dr. Gail Miller called the meeting to order at 12:45 p.m. Dr. David Ingoglia made a motion that the June 2, 2009, minutes be approved as distributed. The motion was seconded and passed.

The committee discussed and took action on the following topics.

(1) TOPIC: Meeting Time

<u>Discussion</u>: The most convenient time for the committee to meet.
<u>Action taken</u>: The committee decided to meet on the fourth Tuesday of every month, at 12:30 p.m.

FIGURE M–6. Minutes of a Meeting

WRITER'S CHECKLIST   Preparing Minutes of Meetings

Include the following in meeting minutes:

- The name of the group or committee holding the meeting
- The topic of the meeting
- The kind of meeting (a regular meeting or a special meeting called to discuss a specific subject or problem)
- The number of members present and, for committees or boards of ten or fewer members, the names of those present and absent
- The place, time, and date of the meeting
- A statement that the chair and the secretary were present or the names of any substitutes
- A statement that the minutes of the previous meeting were approved or revised
- A list of any reports that were read and approved

Writer's Checklist: Preparing Minutes of Meetings (continued)

- All the main motions that were made, with statements as to whether they were carried, defeated, or tabled (vote postponed), and the names of those who made and seconded the motions (motions that were withdrawn are not mentioned)
- A full description of resolutions that were adopted and a simple statement of any that were rejected
- A record of all ballots with the number of votes cast for and against resolutions
- The time the meeting was adjourned (officially ended) and the place, time, and date of the next meeting
- The recording secretary's signature and typed name and, if desired, the signature of the chairperson

mixed constructions

A mixed construction is a sentence in which the elements do not sensibly fit together. The problem may be a **grammar** error, a **logic error**, or both.

▶ Because the copier wouldn't start, ~~explains why~~ we called a technician.

The original sentence mixes a subordinate **clause** (*Because the copier wouldn't start*) with a **verb** (*explains*) that attempts to incorrectly use the subordinate clause as its subject. The revision correctly uses the **pronoun** *we* as the subject of the main clause. See also **sentence construction**.

modifiers

Modifiers are words, phrases, or clauses that expand, limit, or make otherwise more specific the meaning of other elements in a sentence. Although we can create sentences without modifiers, we often need the detail and clarification they provide.

WITHOUT MODIFIERS	Production decreased.
WITH MODIFIERS	*Glucose* production decreased *rapidly*.

Most modifiers function as **adjectives** or **adverbs**. Adjectives describe qualities or impose boundaries on the words they modify.

▶ *noisy* machinery, *ten* files, *this* printer, *a* workstation

An adverb modifies an adjective, another adverb, a **verb**, or an entire **clause**.

▶ Under test conditions, the brake pad showed *much* less wear than it did under actual conditions.
[The adverb *much* modifies the adjective *less*.]
▶ The redesigned brake pad lasted *much* longer.
[The adverb *much* modifies another adverb, *longer*.]
▶ The wrecking ball hit the wall of the building *hard*.
[The adverb *hard* modifies the verb *hit*.]
▶ *Surprisingly*, the motor failed even after all the durability and performance tests it had passed.
[The adverb *surprisingly* modifies an entire clause.]

Adverbs are **intensifiers** when they increase the impact of adjectives (*very* fine, *too* high) or adverbs (*very* slowly, *rather* quickly). Be cautious using intensifiers; their overuse can lead to vagueness and a resulting lack of precision.

Stacked (Jammed) Modifiers

Stacked (or jammed) modifiers are strings of modifiers preceding **nouns** that make writing unclear or difficult to read.

▶ Your *staffing-level authorization reassessment* plan should result in a major improvement.

The noun *plan* is preceded by three long modifiers, a string that forces the reader to slow down to interpret its meaning. Stacked modifiers often result from a tendency to overuse **buzzwords** or **jargon**. See how breaking up the stacked modifiers makes the example easier to read.

▶ Your plan for reassessing the staffing-level authorizations should result in a major improvement.

Misplaced Modifiers

A modifier is misplaced when it modifies the wrong word or phrase. A misplaced modifier can cause **ambiguity**.

▶ We *almost* lost all of the parts.
[The parts were *almost* lost but were not.]

▶ We lost *almost* all of the parts.
[Most of the parts were in fact lost.]

To avoid ambiguity, place modifiers as close as possible to the words they are intended to modify. Likewise, place phrases near the words they modify. Note the two meanings possible when the phrase is shifted in the following sentences:

▶ The equipment *without the accessories* sold the best.
[Different types of equipment were available, some with and some without accessories.]

▶ The equipment sold the best *without the accessories*.
[One type of equipment was available, and the accessories were optional.]

Place clauses as close as possible to the words they modify.

REMOTE	We sent the brochure to several local firms *that had four-color art*.
CLOSE	We sent the brochure *that had four-color art* to several local firms.

Squinting Modifiers

A squinting modifier is one that can be interpreted as modifying either of two sentence elements simultaneously, thereby confusing readers about which is intended. See also **dangling modifiers**.

▶ We agreed *on the next day* to make the adjustments.
[Did they agree *to make the adjustments* on the next day?
Or *on the next day*, did they agree to make the adjustments?]

A squinting modifier can sometimes be corrected simply by changing its position, but often it is better to rewrite the sentence.

▶ We agreed that *on the next day* we would make the adjustments.
[The adjustments were to be made *on the next day*.]

▶ *On the next day*, we agreed that we would make the adjustments.
[The agreement was made *on the next day*.]

mood

The grammatical term *mood* refers to the **verb** functions that indicate whether the verb is intended to make a statement, ask a question, give a command, or express a hypothetical possibility.

The *indicative mood* states a fact, gives an opinion, or asks a question.

- The setting *is* correct.
- *Is* the setting correct?

The *imperative mood* expresses a command, suggestion, request, or plea. In the imperative mood, the implied subject *you* is not expressed. (*Install* the system today.)

The *subjunctive mood* expresses something that is contrary to fact, conditional, hypothetical, or purely imaginative; it can also express a wish, a doubt, or a possibility. In the subjunctive mood, *were* is used instead of *was* in clauses that speculate about the present or future, and the base form (*be*) is used following certain verbs, such as *propose*, *request*, or *insist*. See also progressive **tense**.

- If we *were* to finish the tests today, we would be ahead of schedule.
- The research director insisted that she [I, you, we, they] *be* the project leader.

The most common use of the subjunctive mood is to express clearly that the writer considers a condition to be contrary to fact. If the condition is not considered to be contrary to fact, use the indicative mood.

> **ESL TIPS** for Determining Mood
>
> In written and especially in spoken English, the tendency increasingly is to use the indicative mood where the subjunctive traditionally has been used. Note the differences between traditional and contemporary usage in the following examples.
>
> *Traditional (formal) use of the subjunctive mood*
> - I wish he *were* here now.
> - If I *were* going to the conference, I would room with him.
> - I requested that she *show* up on time.
>
> *Contemporary (informal) use of the indicative mood*
> - I wish he *was* here now.
> - If I *was* going to the conference, I would room with him.
> - I requested that she *shows* up on time.
>
> In professional writing, it is better to use the more traditional expressions.

SUBJUNCTIVE	If I *were* president of the firm, I would change several hiring policies.
INDICATIVE	Although I *am* president of the firm, I don't control every aspect of its policies.

Ms. / Miss / Mrs.

Ms. is widely used in business and public life to address or refer to a woman, especially if her marital status is either unknown or irrelevant to the **context**. Traditionally, *Miss* is used to refer to an unmarried woman, and *Mrs.* is used to refer to a married woman. Some women may indicate a preference for *Ms.*, *Miss*, or *Mrs.*, which you should honor. If a woman has an academic or a professional title, use the appropriate form of address (*Doctor*, *Professor*, *Captain*) instead of *Ms.*, *Miss*, or *Mrs.* See also **biased language**.

mutual / common

Common is used when two or more persons (or things) share something or possess it jointly.

- ▶ We have a *common* desire to make the program succeed.
- ▶ The fore and aft guidance assemblies have a *common* power source.

Mutual may also mean "shared" (*mutual* friend, of *mutual* benefit), but it usually implies something given and received reciprocally and is used with reference to only two persons or parties.

- ▶ Melek respects Roth, and from my observations the respect is *mutual*.
 [Roth also respects Melek.]

N

narration

Narration is the presentation of a series of events in a prescribed (often chronological) sequence. Much narrative writing explains how something happened: a laboratory study, a site visit, an accident, the decisions in an important meeting. See also **chronological method of developoment**, **trip reports**, and **trouble reports**.

Effective narration rests on two key writing techniques: the careful, accurate sequencing of events and a consistent **point of view** on the part of the narrator. Narrative sequence and essential shifts in the sequence are signaled in three ways: chronology (clock and calendar time), transitional words pertaining to time (*before*, *after*, *next*, *first*, *while*, *then*), and verb tenses that indicate whether something has happened (past **tense**) or is under way (present tense). The point of view indicates the writer's relation to the information being narrated as reflected in the use of **person**. Narration usually expresses a first- or third-person point of view. First-person narration indicates that the writer is a participant, and third-person narration indicates that the writer is writing about what happened to someone or something else.

The narrative shown in Figure N–1 reconstructs the chronology of an early-morning accident of Chicago Transit Authority Green Line train run 2. This train struck two signal maintainers who were working near a tower on the section of the Chicago Loop that is above the intersection of Lake and Wells streets. The Loop is elevated, and one maintainer fell from the structure, landing on a parked car and then the street. The investigators needed to "tell the story" in detail so that any lessons learned could be used to improve safety. To do that, they recount and sequence events as precisely as possible. The verb tenses throughout indicate past action: *approached*, *continued*, *heard*, *removed*, *stopped*.

Although narration often exists in combination with other forms of discourse (**description**, **exposition**, **persuasion**), avoid interrupting a narrative with lengthy explanations or analyses. Explain only what is necessary for readers to follow the events. See **audience**.

The Accident*

On the morning of the accident, two night-shift signal maintainers were repairing a switch at tower 18. Between 4:00 and 4:30 a.m., two day-shift maintainers joined them.[1] As the two crews conferred about the progress of the repair, Green Line train run 1 approached the tower. A trainee was operating the train, and a train operator/line instructor[2] was observing. Both crew members on the train later stated that they had not heard the control center's radioed advisory that workers were on the track structure at tower 18. The line instructor said that as the train approached the tower with a *proceed* (green) signal, he observed wayside maintenance personnel from about 150 feet away and told the trainee to stop the train, which he did. One of the maintainers gave the train a hand signal to proceed, and the train continued on its way. Shortly after the train left, the night-shift maintainers also left.

The day-shift maintainers continued to work. Just before the accident, they removed a defective part and started to install a replacement. According to both men, they were squatting over the switch machine. One was facing the center of the track and attaching wires, while the other was facing the Loop with his back to the normal direction of train movements. He was shining a flashlight on the work area.

The accident train approached the tower on the *proceed* signal. One maintainer later said that he remembered being hit by the train, while the other said that he was hit by "something." A train operator/line instructor was operating the train, and a trainee was observing. Both later said that they had not seen any wayside workers. They said that they had heard noise that the student described as a "thump" in the vicinity of the accident and caught a "glimpse" of something.

Both maintainers later stated that they had not seen or heard the train as it approached. After being struck, one of the maintainers fell from the structure. The other fell to the deck of the platform on the outside of the structure. He used his radio to tell the control center that he and another maintainer had been "hit by the train." Emergency medical personnel were dispatched to the scene, and an ambulance took both men to a local hospital.

In the meantime, the accident train continued past the tower and stopped at the next station, Clark and Lake, where the crew members inspected the train from the platform and found no damage. They continued on their way until they heard the radio report that workers had been struck by a train. They stopped their train at the next station and reported to a supervisor.

According to the operator of the accident train, nothing had distracted her from her duties, and she had been facing forward and watching the track before the train arrived at tower 18. The trainee supported her account. Both crew members said that they had not heard the control center's radioed advisory that workers were on the track structure at tower 18.

---

[1] All times referred to in this report are central standard time.
[2] Line instructors are working train operators who provide on-the-job training to operator trainees.

**Source*: National Transportation Safety Board, "Railroad Accident Brief: Chicago Transit Authority, DCA-02-FR-005, Chicago, Illinois, February 26, 2002." www.ntsb.gov/publictn/2003/RAB0304.htm

FIGURE N–1. Narration from an Accident Report

nature

Nature, when used to mean "kind" or "sort," is vague. Avoid this usage in your writing. Say exactly what you mean.

▶ The ~~nature of~~ the contract caused the problem.
      ^exclusionary clause in

needless to say

The phrase *needless to say* sometimes occurs in speech and writing. Eliminate the phrase or replace it with a more descriptive **word choice**.

▶ ~~Needless to say,~~ staff reductions have decreased customer loyalty.
    ^Service logs indicate that

newsletter articles

If your organization publishes a **newsletter**, you may be asked to contribute an article on a subject in your area of expertise. In fact, an article is a good way to promote your work or your department.

Before you begin to write, consider the traditional *who, what, where, when,* and *why* of journalism ("*Who* did it? *What* was done? *Where* was it done? *When* was it done? *Why* was it done?") and then add *how*, which may be of as much interest to your colleagues as any of the five *w*'s. Next, determine whether the company has an official policy or position on your subject. If it does, adhere to it as you prepare your article. If there is no company policy, determine your management's attitude toward your subject.

Gather several fairly recent issues of the newsletter and study the **style** and **tone** of the writing and the approach used for various kinds of subjects. Understand those perspectives before you begin to work on your own article. Ask yourself the following questions about your subject: What is its significance to the organization? What is its significance to my coworkers? The answers to those questions should help you establish the style, tone, and approach for your article and also heavily influence your conclusion. See also **context**.

Research for a newsletter article frequently consists of **interviewing for information**. Interview key personnel concerned with your subject. Get all available information and all points of view. Be sure to give maximum credit to the maximum number of people by quoting statements from those involved in projects and naming those who have developed initiatives. See also **quotations**.

Figure N–2 shows an article written for *Connection*, a newsletter produced by Ken Cook Company and distributed to current and prospective clients. This article describes how the company developed a print-on-demand technology called media①off™. Notice that the sidebar at the right of the page ("Tech Tools") uses visual elements to draw

FIGURE N–2. Newsletter Article

readers to the article as well as giving them a sense of its content. By describing this system, the article aims to demonstrate Ken Cook Company's commitment "to providing solutions that meet customer needs and keep pace with the latest breakthroughs in technology."

> **WRITER'S CHECKLIST** Writing Newsletter Articles
>
> Writing a newsletter article requires a more journalistic approach than writing a **report**. Because newsletters are not required reading, you do not have a captive **audience**.
>
> - Write an intriguing **title** to catch the audience's attention; **rhetorical questions** often work well.
> - Include as many eye-catching **photographs** or **visuals** as appropriate to entice your audience to read the lead **paragraph** of your **introduction**. See also **layout and design**.
> - Fashion a lead, or first paragraph, that will encourage further reading. The first paragraph generally makes the **transition** from the title to the body of the article.
> - Offer a well-developed presentation of your subject to hold the readers' interest all the way to the end of the article.
> - Write a **conclusion** that emphasizes the significance of your subject to your audience and stresses the points you want your readers to retain. See **emphasis**.
> - Follow the steps listed in the Checklist of the Writing Process on pages xxiii–xxiv as you prepare your newsletter article.

newsletters

Newsletters are publications that are designed to inform and to create and sustain interest and membership in an organization. They can also be used to sell products and services. The two main types of newsletters are organizational newsletters and subscription newsletters.

Types of Newsletters

Organizational newsletters like the one shown in Figure N–3 are sent to employees, clients, or members of an association to keep them informed about issues regarding their company or group, such as the development of new products or policies or the accomplishments of individuals or teams. Stories in organizational newsletters can both enhance the image and foster pride among employees of the organization's products

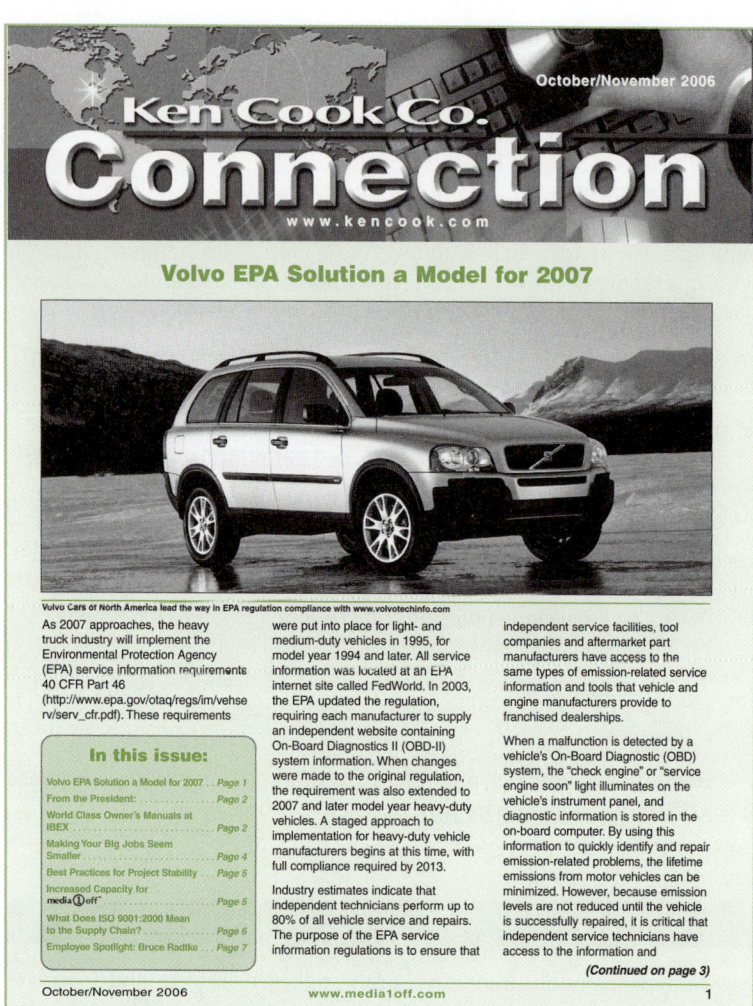

FIGURE N–3. Company Newsletter (Front Page)

or services. For example, Figure N–3 shows how Ken Cook Company links the quality of its publications with the ability of service personnel to meet Environmental Protection Agency (EPA) standards for an auto manufacturer.

Subscription newsletters are designed to attract and build a readership interested in buying specific products or services or in learning more about a specific subject. Subscribers are buying information, and they expect value for their money. For example, a person with experience in the stock market could create a financial newsletter and charge subscribers a monthly fee for the investing advice in that newsletter; a person who collects movie memorabilia could create an online newsletter that includes stories about ways to find and sell rare movie posters.

Developing Newsletters

Before you begin to develop a newsletter, decide on its specific **purpose** and the specific **audience** you will be targeting; then make sure the newsletter's appearance and editorial choices create a sense of identification among the readership. Newsletters often involve **collaborative writing** in which different individuals work on design, content, and project management. See also **persuasion**. If you are asked to contribute an article to a newsletter, see **newsletter articles**.

You will need to acquire a mailing list (names and addresses of your readers) and decide on the most strategic way to get the newsletter to these readers (whether through postal mail, interoffice mail, e-mail, or Web posting). Because it can be time-consuming and technically problematic to send out hundreds or thousands of online newsletters by yourself, you may also need to subscribe to a online list-hosting service.*

Your **research** should include trade journals, business and technology magazines, the Web, and other sources to find specific angles for the articles that will appeal to your select audience. Attempt to provide content that your readers will not find elsewhere, for example, by interviewing and profiling customers, association members, or employees. Check your facts meticulously—newsletter readers are often specialists in their fields. Because newsletters are often distributed to branches and clients abroad, see **global communication** and **global graphics**. See also **interviewing for information**.

As shown in Figures N–2 and N–3, a newsletter's format should be simple and consistent, yet visually appealing to your readership. Use the active **voice** and a conversational **tone**. Use **headings** and bullets to break up the text and make the newsletter easy to read. Keep your sentences simple and paragraphs short. See **conciseness** and **layout and design**.

Using word-processing, desktop-publishing, or Web-development software, create newspaper columns and one or two **visuals** per page that complement the text. On the front page, identify the organization, include the date, volume and issue numbers, and a contents box. For Web newsletters, follow the principles of good **Web design**. See also **layout and design**, **photographs** and **writing for the Web**.

---

*E-mail list-hosting services have their own servers and provide commercial delivery of premium e-mail that often contains graphic and other digital forms used for advertising.

nominalizations

A nominalization is a **noun** form of a **verb** that is often combined with vague and general (or "weak") verbs like *make, do, give, perform, provide*. Avoid nominalizations when you can use specific verbs that communicate the same idea more directly and concisely.

▶ The tests will ~~give an indication~~ *indicate* if the virus is present.

▶ The staff should ~~perform an evaluation of~~ *evaluate* the new software.

If you use nominalizations solely to make your writing sound more formal, the result will be **affectation**. You may occasionally have an appropriate use for a nominalization. For example, you might use a nominalization to slow the **pace** of your writing. See also **conciseness**, **technical writing style**, and **voice**.

none

None can be either a singular or a plural **pronoun**, depending on the context. See also **agreement**.

▶ *None* of the material *has* been ordered.
  [Always use a singular **verb** with a singular **noun**, in this case, "material."]

▶ *None* of the clients *has* been called yet.
  [Use a singular verb even with a plural noun (*clients*) if the intended emphasis is on the idea of *not one*.]

▶ *None* of the clients *have* been called yet.
  [Use a plural verb if you intend *none* to refer to all clients.]

For **emphasis**, substitute *no one* or *not one* for *none* and use a singular verb.

▶ We paid the retail price for three of the machines, ~~none~~ *not one* of which was worth the money.

nor / or

Nor always follows *neither* in sentences with continuing negation. ("They will *neither* support *nor* approve the plan.") Likewise, *or* follows *either* in sentences. ("The firm will accept *either* a short-term *or* a long-term loan.")

Two or more singular subjects joined by *or* or *nor* usually take a singular **verb**. However, when one subject is singular and one is plural, the verb agrees with the subject nearer to it. See also **conjunctions** and **parallel structure**.

SINGULAR	Neither the *architect* nor the *client was* happy with the design.
PLURAL	Neither the *architect* nor the *clients were* happy with the design.
SINGULAR	Neither the *architects* nor the *client was* happy with the design.

note-taking

The purpose of note-taking is to summarize and record information you extract during **research**. (For taking notes at a meeting, see **minutes of meetings**.) The challenge in taking notes is to condense someone else's thoughts into your own words without distorting the original thinking. As you extract information, let your knowledge of the **audience** and the **purpose** of your writing guide you.

▶ **ETHICS NOTE** Resist copying your source word for word as you take notes; instead, paraphrase the author's idea or concept. If you only change a few words from a source and incorporate that text into your document, you will be guilty of **plagiarism**. See also **paraphrasing**. ◆

On occasion, when an expert source states something that is especially precise, striking, or noteworthy or that reinforces your point, you can justifiably quote the source directly and incorporate it into your document. If you use a direct quote, enclose the material in **quotation marks** in your notes. In your finished writing, provide the source of your quotation. Normally, you will rarely need to quote anything longer than a paragraph. See also **documenting sources** and **quotations**.

When taking notes on abstract ideas, as opposed to factual data, do not sacrifice **clarity** for brevity — notes expressing concepts can lose their meaning if they are too brief. The critical test is whether you can understand the note a week later and recall the significant ideas of the passage. Consider the information in the following paragraph:

> Long before the existence of bacteria was suspected, techniques were in use for combating their influence in, for instance, the decomposition of meat. Salt and heat were known to be effective, and these do in fact kill bacteria or prevent them from multiplying. Salt acts by the osmotic effect of extracting water from the bacterial cell fluid. Bacteria are less easily destroyed by osmotic action

than are animal cells because their cell walls are constructed in a totally different way, which makes them much less permeable.

The paragraph says essentially three things:

1. Before the discovery of bacteria, salt and heat were used to combat the effects of bacteria.
2. Salt kills bacteria by extracting water from their cells by osmosis, hence its use in curing meat.
3. Bacteria are less affected by the osmotic effect of salt than are animal cells, because bacterial cell walls are less permeable.

If your readers' needs and your objective involve tracing the origin of the bacterial theory of disease, you might want to note that salt was traditionally used to kill bacteria long before people realized what caused meat to spoil. It might not be necessary to your topic to say anything about the relative permeability of bacterial cell walls.

You should record notes in a way you find efficient. Some find various shareware note and index programs useful. However, jotting notes on 3 × 5-inch index cards is often more flexible, and the cards are especially useful for **outlining** complex and long-term projects.*

WRITER'S CHECKLIST Taking Notes

- Ask yourself the following questions: What information do I need to fulfill my purpose? What are the needs of my audience?
- Record only the most important ideas and concepts. Be sure to record all vital names, dates, and definitions.
- When in doubt about whether to take a note, consider the difficulty of finding the source again should you want it later.
- Use quotations when sources state something that is precise, striking, or noteworthy or that succinctly reinforces a point you are making.
- Ensure proper credit. Record the author; title; publisher; place; page number; URL; and date of publication, posting, or retrieval. (On subsequent notes from the same source, include only the author and page number.)
- Use your own shorthand and record notes in a way that you find efficient, whether in an electronic document or on index cards.
- Photocopy or download pages and highlight passages that you intend to quote.
- Check your notes for accuracy against the original material before moving on to another source.

---

*Peter Walsh, in *How to Organize Just About Everything* (New York: Free Press, 2004), points out that index cards, despite their image, are often best for organizing a writing task.

nouns

DIRECTORY
Types of Nouns 349
Noun Functions 349
Collective Nouns 350
Plural Nouns 350

A noun is a **part of speech** that names a person, place, thing, concept, action, or quality.

Types of Nouns

The two basic types of nouns are proper nouns and common nouns. *Proper nouns*, which are capitalized, name specific people, places, and things (H. G. Wells, Boston, United Nations, Nobel Prize). See also **capitalization**.

Common nouns, which are not capitalized unless they begin sentences, name general classes or categories of persons, places, things, concepts, actions, and qualities (writer, city, organization, award). Common nouns include collective nouns, concrete nouns, abstract nouns, count nouns, and mass nouns.

Collective nouns are common nouns that indicate a group or collection. They are plural in meaning but singular in form (audience, jury, brigade, staff, committee). (See the subsection Collective Nouns on page 350 for advice on using singular or plural forms with collective nouns.)

Concrete nouns are common nouns used to identify those things that can be discerned by the five senses (paper, keyboard, glue, nail, grease).

Abstract nouns are common nouns that name ideas, qualities, or concepts that cannot be discerned by the five senses (loyalty, pride, valor, peace, devotion).

Count nouns are concrete nouns that identify things that can be separated into countable units (desks, envelopes, printers, pencils, books).

Mass nouns are concrete nouns that identify things that cannot be separated into countable units (water, air, electricity, oil, cement). See also **English as a second language**.

Noun Functions

Nouns function as subjects of **verbs**, direct and indirect objects of verbs and **prepositions**, subjective and objective **complements**, or **appositives** (shown below).

▶ The *metal* failed during the test. [subject]

▶ The bricklayer cemented the *blocks* efficiently. [direct object of a verb]

▶ The state presented our *department* a safety award. [indirect object]

▶ The event occurred within the *year*. [object of a preposition]

▶ A dynamo is a *generator*. [subjective complement]

▶ The regional manager was appointed *chairperson*. [objective complement]

▶ Philip Garcia, the *treasurer*, gave his report last. [appositive]

Words normally used as nouns can also be used as **adjectives** and **adverbs**.

▶ It is *company* policy. [adjective]

▶ He went *home*. [adverb]

Collective Nouns

When a collective noun refers to a group as a whole, it takes a singular verb and pronoun.

▶ The staff *was* divided on the issue and could not reach *its* decision until May 15.

When a collective noun refers to individuals within a group, it takes a plural verb and pronoun.

▶ The staff *returned* to *their* offices after the conference.

A better way to emphasize the individuals on the staff would be to use the phrase *the staff members*.

▶ The staff members *returned* to *their* offices after the conference.

Treat organization names and titles as singular.

▶ LRM Associates *has* grown 30 percent in the last three years; *it* will move to a new facility in January.

Plural Nouns

Most nouns form the plural by adding -*s* (dolphin/dolphins, pencil/pencils). Nouns ending in *ch*, *s*, *sh*, *x*, and *z* form the plural by adding -*es*.

▶ search/searches, glass/glasses, wish/wishes, six/sixes, buzz/buzzes

Nouns that end in a consonant plus *y* form the plural by changing the *y* to -*ies* (delivery/deliveries). Some nouns ending in *o* add -*es* to form the plural, but others add only -*s* (tomato/tomatoes, dynamo/dynamos).

Some nouns ending in *f* or *fe* add *-s* to form the plural; others change the *f* or *fe* to *ves*.

▶ cliff/cliffs, fife/fifes, hoof/hooves, knife/knives

Some nouns require an internal change to form the plural.

▶ woman/women, man/men, mouse/mice, goose/geese

Some nouns do not change in the plural form.

▶ many *fish*, several *deer*, fifty *sheep*

Some nouns remain in the plural form whether singular or plural.

▶ headquarters, means, series, crossroads

Hyphenated and open compound nouns form the plural in the main word.

▶ sons-in-law, high schools, editors in chief

Compound nouns written as one word add *-s* to the end (two *tablespoonfuls*).

If you are unsure of the proper usage, check a dictionary. See **possessive case** for a discussion of how nouns form possessives.

number (grammar)

Number is the grammatical property of **nouns**, **pronouns**, and **verbs** that signifies whether one thing (singular) or more than one (plural) is being referred to. (See also **agreement**.) Nouns normally form the plural by simply adding *-s* or *-es* to their singular forms.

▶ *Partners* in successful *businesses* are not always personal friends.

Some nouns require an internal change to form the plural.

▶ woman/women, man/men, goose/geese, mouse/mice

All pronouns except *you* change internally to form the plural.

▶ I/we, he/they, she/they, it/they

By adding *-s* or *-es*, most verbs show the singular of the third **person**, present **tense**, indicative **mood**.

▶ he *stands*, she *works*, it *goes*

The verb *be* normally changes form to indicate the plural.

SINGULAR I *am* ready to begin work.
PLURAL We *are* ready to begin work.

See also **agreement**.

> ## numbers
>
> **DIRECTORY**
> Numerals or Words 352
> Plurals 353
> Measurements 353
> Fractions 353
> Money 353
> Time 354
> Dates 354
> Addresses 354
> Documents 355

The standards for using numbers vary; however, unless you are following an organizational or a professional style manual, observe the following guidelines.

Numerals or Words

Write numbers from zero through ten as words, and write numbers above ten as numerals.

▶ I rehearsed my presentation *three* times.

▶ The association added *150* new members.

Spell out numbers that begin a sentence, however, even if they would otherwise be written as numerals.

▶ *One hundred and fifty* new members joined the association.

If spelling out such a number seems awkward, rewrite the sentence so that the number does not appear at the beginning ("We added *150* new members").

Spell out approximate and round numbers.

▶ We've had *over a thousand* requests this month.

In most writing, spell out small ordinal numbers, which express degree or sequence (first, second; *but* 27th, 42nd), when they are single words (our nineteenth year), or when they modify a century (the twenty-first century). However, avoid ordinal numbers in <u>dates</u> (use March 30 or 30 March, not March 30th).

Plurals

Indicate the plural of numerals by adding -*s* (7s, the late 1990s). Form the plural of a written number (like any noun) by adding -*s* or -*es* or by dropping the *y* and adding -*ies* (elevens, sixes, twenties). See also <u>apostrophes</u>.

Measurements

Express units of measurement as numerals (3 miles, 45 cubic feet, 9 meters). When numbers run together in the same phrase, write one as a numeral and the other as a word.

▶ The order was for ~~12~~ *twelve* 6-foot tables.

Generally give percentages as numerals and write out the word *percent*. (Approximately *85 percent* of the land has been sold.) However, in a <u>table</u>, use a numeral followed by the percent symbol (85%).

Fractions

Express fractions as numerals when they are written with whole numbers (27½ inches, 4¼ miles). Spell out fractions when they are expressed without a whole number (one-fourth, seven-eighths). Always write decimal numbers as numerals (5.21 meters).

Money

In general, use numerals to express exact or approximate amounts of money.

▶ We need to charge *$28.95* per unit.
▶ The new system costs *$60,000*.

Use words to express indefinite amounts of money.

▶ The printing system may cost *several thousand dollars*.

Use numerals and words for rounded amounts of money over one million dollars.

▶ The contract is worth *$6.8 million*.

Use numerals for more complex or exact amounts.

▶ The corporation paid *$2,452,500* in taxes last year.

For amounts under a dollar, ordinarily use numerals and the word *cents* ("The pens cost *50 cents* each"), unless other numerals that require dollar signs appear in the same sentence.

▶ The business-card holders cost *$10.49* each, the pens cost *$.50* each, and the pencil-cup holders cost *$6.49* each.

> **ESL TIPS** for Punctuating Numbers
>
> Some rules for punctuating numbers in English are summarized as follows.
>
> Use a comma to separate numbers with four or more digits into groups of three, starting from the right (*5,289,112,001* atoms).
>
> Do not use a comma in years, house numbers, ZIP Codes, and page numbers.
>
> ▶ June *2009*
> ▶ *92401* East Alameda Drive
> ▶ The ZIP Code is *91601*.
> ▶ Page *1204*
>
> Use a period to represent the decimal point (*4.2* percent, *$3,742,097.43*). See also global communication and global graphics.

Time

Divide hours and minutes with **colons** when *a.m.* or *p.m.* follows (7:30 a.m., 11:30 p.m.). Do not use colons with the 24-hour system (0730, 2330). Spelled-out time is not followed by *a.m.* or *p.m.* (seven o'clock in the evening).

Dates

In the United States dates are usually written in a month-day-year sequence (August 11, 2009). Never use the strictly numerical form for dates (8/11/09) because the date is not immediately clear, especially in **international correspondence**.

Addresses

Spell out numbered streets from one to ten unless space is at a premium (East Tenth Street). Write building numbers as numerals. The only exception is the building number *one* (One East Monument Street). Write highway numbers as numerals (U.S. 40, Ohio 271, I-94).

Documents

Page numbers are written as numerals in manuscripts (page 37). Chapter and volume numbers may appear as numerals or words (Chapter 2 or Chapter Two, Volume 1 or Volume One), but be consistent. Express figure and table numbers as numerals (Figure 4, Table 3).

Do not follow a word representing a number with a numeral in parentheses that represents the same number. Doing so is redundant.

▶ Send five ~~(5)~~ copies of the report.

O

objects

Objects are **nouns** or noun equivalents: **pronouns**, **verbals**, and noun **phrases** or **clauses**. (See also **complements** and **verbs**.) The three kinds of objects are direct objects, indirect objects, and objects of **prepositions**.

A *direct object* answers the question *what?* or *whom?* about a verb and its subject.

▶ We sent a *full report*. [We sent *what*?]

▶ Michelle e-mailed the *client*. [Michelle e-mailed *whom*?]

An *indirect object* is a noun or noun equivalent that occurs with a direct object after certain kinds of transitive verbs, such as *give*, *wish*, *cause*, and *tell*. The indirect object answers the question *to whom or what?* or *for whom or what?* The indirect object always precedes the direct object.

▶ We sent the *general manager* a full report.
[*Report* is the direct object; the indirect object, *general manager*, answers the question, "We sent a full report *to whom?*"]

The *object of a preposition* is a noun or pronoun that is introduced by a preposition, forming a prepositional phrase.

▶ At the *meeting*, the district managers approved the contract.
[*Meeting* is the object, and *at the meeting* is the prepositional phrase.]

OK / okay

The expression *okay* (also spelled *OK*) is common in informal writing, but it should be avoided in most technical writing.

▶ Mr. Sturgess ~~gave his okay to~~ *approved* the project.

on / onto / upon

On is normally used as a **preposition** meaning "attached to" or "located at." ("Install the shelf *on* the north wall.") *On* also stresses a position of rest. ("The patient lay *on* the stretcher.") *Onto* implies movement to a position on or movement up and on. ("The commuters surged *onto* the platform.") *Upon* emphasizes movement or a condition. ("The report is due *upon* completion of the project.")

one

When used as an indefinite **pronoun**, *one* may help you avoid repeating a **noun**. ("We need a new plan, not an old *one*.") *One* is often redundant in phrases in which it restates the noun, and it may take the proper emphasis away from the **adjective**.

▶ The training program was not a~~ unique one.~~ *unique.*

One can also be used in place of a noun or personal pronoun in a statement. ("*One* cannot ignore *one's* physical condition.") Using *one* in that way is formal and impersonal; in any but the most formal writing, you should address your reader directly and personally as *you*. ("*You* cannot ignore *your* physical condition.") See also **point of view**.

one of those . . . who

A dependent **clause** beginning with *who* or *that* and preceded by *one of those* takes a plural **verb**.

▶ She is *one of those* managers *who are* concerned about their writing.

▶ This is *one of those* policies *that make* no sense when you examine them closely.

In those two examples, *who* and *that* refer to plural antecedents (*managers* and *policies*) and thus take plural verbs (*are* and *make*). See also **agreement**.

If the phrase *one of those* is preceded by *the only*, however, the verb should be singular.

- She is *the only one of those* managers *who is* concerned about her writing.
  [The verb is singular because its subject, *who*, refers to a singular antecedent, *one*. If the sentence were reversed, it would read, "Of those managers, she is *the only one who is* concerned about her writing."]
- This is *the only one of those* policies *that makes* no sense when you examine it closely.
  [If the sentence were reversed, it would read, "Of those policies, this is *the only one that makes* no sense when you examine it closely."]

only

The word *only* should be placed immediately before the word or phrase it modifies. See also **modifiers**.

- We ~~only~~ lack ^*only*^ financial backing.

Be careful with the placement of *only* because it can change the meaning of a sentence.

- *Only* he said that he was tired.
  [He alone said that he was tired.]
- He *only* said that he was tired.
  [He actually was not tired, although he said he was.]
- He said *only* that he was tired.
  [He said nothing except that he was tired.]
- He said that he was *only* tired.
  [He said that he was nothing except tired.]

order-of-importance method of development

The order-of-importance **method of development** is a particularly effective and common organizing strategy. This method can use one of two ordering strategies—decreasing order (Figure O–1), which is often best for written documents, and increasing order (Figure O–2), which is especially effective for oral **presentations**.

Memorandum

To: Tawana Shaw, Director, Human Resources Department
From: Frank W. Nemitz, Chief, Product Marketing   *FWN*
Date: November 20, 2009
Subject: Selection of Manager of the Technical Writing Department

I have assessed the candidates for Manager of the Technical Writing Department as you requested, and the following are my evaluations.

Top-Ranked Candidate

The most qualified candidate is Michelle Bryant, acting manager of the department. In her 12 years in the department, Ms. Bryant has gained wide experience in all facets of its operations. She has maintained a consistently high production record and has demonstrated the skills and knowledge required for the supervisory duties she is now handling. She has continually been rated "outstanding" in all categories in her job-performance appraisals. However, her supervisory experience is limited to her present three-month tenure as acting manager of the department, and she lacks the college degree required by the job description.

Second-Ranked Candidate

Michael Bastick, Graphics Coordinator, my second choice, also has strong potential for the position. An able administrator, he has been with the company for seven years. Further, he is enrolled in a management-training course at Metro State University's downtown campus. I have ranked him second because he lacks supervisory experience and because his most recent work as a technical writer has been limited.

Third-Ranked Candidate

Jane Fine, my third-ranked candidate, has shown herself to be an exceptionally skilled writer in her three years with the Public Relations Department. Despite her obvious potential, she does not yet have the breadth of experience in technical writing that is required to manage the Technical Writing Department. Jane Fine also lacks on-the-job supervisory experience.

FIGURE O–1. Decreasing Order-of-Importance Method of Development

To: Sun-Hee Kim <kim@appliedsciences.com>
From: Harry Mathews <mathews@appliedsciences.com>
Sent: Friday, May 22, 2009 9:29 AM
Subject: Recruiting Qualified Electronics Technicians

As our company continues to expand, and with the planned opening of the Lakeland Facility late next year, we need to increase and refocus our recruiting program to keep our company staffed with qualified electronics technicians. Below is my analysis of possible recruiting options.

Technical School Recruitment
Over the past three years, we have relied on our in-house internship program and on local and regional technical school graduates to fill these positions. Although our in-house internship program provides a qualified pool of employees, technical school enrollments in the area have in the past provided candidates who are already trained. Each year, however, fewer technical school graduates are being produced, and even the most vigorous Career Day recruiting has yielded disappointing results.

Military Veteran Recruitment
In the past, we relied heavily on the recruitment of skilled veterans from all branches of the military. This source of qualified applicants all but disappeared when the military offered attractive reenlistment bonuses for skilled technicians in uniform. As a result, we need to become aggressive in our attempts to reach this group through advertising. I would like to meet with you soon to discuss the details of a more dynamic recruiting program for skilled technicians leaving the military.

I am certain that with the right recruitment campaign, we can find the skilled employees essential to our expanding role in electronics products and consulting.

FIGURE O–2. Increasing Order-of-Importance Method of Development

Decreasing Order

Decreasing order begins with the most important fact or point, then moves to the next most important, and so on, ending with the least important. This order is especially appropriate for a **memo** or other **correspondence** addressed to a busy decision-maker (see Figure O–1), who may be able to reach a decision after considering only the most important points. In a **report** addressed to various **readers**, some of whom may be interested in only the major points and others who may need all the information, decreasing order may be ideal for your **purpose**.

The advantages of decreasing order are that (1) it gets the reader's attention immediately by presenting the most important point first, (2) it makes a strong initial impression, and (3) it ensures that even the most hurried reader will not miss the most important point.

Increasing Order

Increasing order begins from the least important point or fact, then progresses to the next more important, and builds finally to the most important or strongest point.

Increasing order of importance is effective for writing in which (1) you want to save your strongest points until the end or (2) you need to build the ideas point by point to an important **conclusion** (see Figure O–2). Many oral presentations benefit especially from increasing order because it leaves the **audience** with the strongest points freshest in their minds. The disadvantage of increasing order, especially for written documents, is that it begins weakly, and the reader or listener may become impatient or distracted before reaching your main point. In the example given in Figure O–2, the writer begins with the least productive source of applicants and builds up to the most productive source.

organization

Organization is essential to the success of any writing project, from a **formal report** to a **Web design** or an effective **presentation**. Good organization is achieved by **outlining** and by using a logical and an appropriate **method of development** that suits your subject, your **audience**, and your **purpose**.

During the organization stage of the writing process, you must consider a **layout and design** that will be helpful to your reader and a **format** appropriate to your subject and purpose. If you intend to include **visuals** with your writing, consider them as you create your outline, especially if they need to be prepared by someone else while you are writing and revising the draft. See also "Five Steps to Successful Writing."

organizational charts

An organizational chart shows how the various components of an organization are related to one another. This type of **visual** is useful when you want to give **readers** an overview of an organization or to display the lines of authority within it, as in Figure O–3.

The title of each organizational component (office, section, division) is placed in a separate box. The boxes are then linked to a central authority. If readers need the information, include the name of the person and position title in each box.

As with all visuals, place the organizational chart as close as possible to but not preceding the text that refers to it.

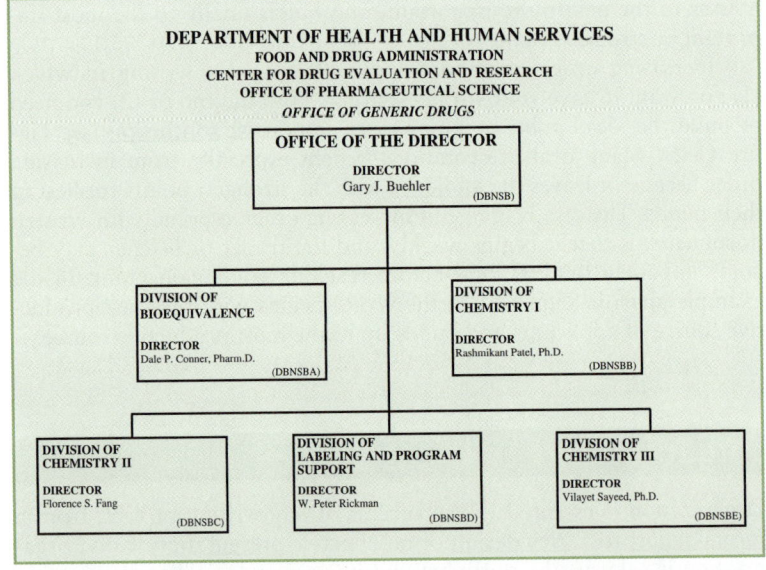

FIGURE O–3. Organizational Chart

outlining

An outline is the skeleton of the document you are going to write; at the least, it should list the main topics and subtopics of your subject in a logical **method of development**.

Advantages of Outlining

An outline provides structure to your writing by ensuring that it has a beginning (**introduction**), a middle (main body), and an end (**conclusion**). Using an outline offers many other benefits.

- Larger and more complex subjects are easier to handle because an outline breaks them into manageable parts.
- Like a road map, an outline indicates a starting point and keeps you moving logically so that you do not lose your way before you arrive at your conclusion.
- Parts of an outline are easily moved around so that you can select the most effective arrangement of your ideas.
- Creating a good outline frees you from concerns of **organization** while you are **writing a draft**.
- An outline enables you to provide **coherence** and **transition** so that one part flows smoothly into the next without omitting important details.
- **Logic errors** are much easier to detect and correct in an outline than in a draft.
- An outline helps with **collaborative writing** because it enables a team to refine a project's **scope**, divide responsibilities, and maintain focus.

Types of Outlines

Two types of outlines are most common: short topic outlines and lengthy sentence outlines. A *topic outline* consists of short phrases arranged to reflect your primary method of development. A topic outline is especially useful for short documents such as **e-mails**, **letters**, or **memos**. See also **correspondence**.

For a large writing project, create a topic outline first, and then use it as a basis for creating a sentence outline. A *sentence outline* summarizes each idea in a complete sentence that may become the topic sentence for a paragraph in the rough draft. If most of your notes can be shaped into topic sentences for paragraphs in your rough draft, you can be relatively sure that your document will be well organized. See also **note-taking** and **research**.

Creating an Outline

When you are outlining large and complex subjects with many pieces of information, the first step is to group related notes into categories. Sort the notes by major and minor division headings. Use an appropriate method of development to arrange items and label them with Roman numerals. For example, the major divisions for this discussion of outlining could be as follows:

I. Advantages of outlining
II. Types of outlines
III. Creating an outline

The second step is to establish your minor divisions within each major division. Arrange your minor points using a method of development under their major division and label them with capital letters.

II. Types of outlines
   A. Topic outlines   ] Division and Classification
   B. Sentence outlines
III. Creating an outline
   A. Establish major and minor divisions.
   B. Sort notes by major and minor divisions.] Sequential
   C. Complete the sentence outline.

You will sometimes need more than two levels of headings. If your subject is complicated, you may need three or four levels of headings to better organize all of your ideas in proper relationship to one another. In that event, use the following numbering scheme:

I. First-level heading
   A. Second-level heading
      1. Third-level heading
         a. Fourth-level heading

The third step is to mark each of your notes with the appropriate Roman numeral and capital letter. Arrange the notes logically within each minor heading, and mark each with the appropriate, sequential Arabic number. As you do, make sure your organization is logical and your headings have **parallel structure**. For example, all the second-level headings under "III. Creating an outline" are complete sentences in the active **voice**.

   Treat **visuals** as an integral part of your outline, and plan approximately where each should appear. Either include a rough sketch of the visual or write "illustration of . . ." at each place. As with other information in an outline, freely move, delete, or add visuals as needed.

   The outline samples shown earlier use a combination of numbers and letters to differentiate the various levels of information. You could also use a decimal numbering system, such as the following, for your outline.

1. FIRST-LEVEL HEADING
    1.1 Second-level heading
    1.2 Second-level heading
        1.2.1 Third-level heading
        1.2.2 Third-level heading
            1.2.2.1 Fourth-level heading
            1.2.2.2 Fourth-level heading
    1.3 Second-level heading
2. FIRST-LEVEL HEADING

This system should not go beyond the fourth level because the numbers get too cumbersome beyond that point. In many documents, such as **manuals**, the decimal numbering system is carried over from the outline to the final version of the document for ease of cross-referencing sections.

Create your draft by converting your notes into complete sentences and **paragraphs**. If you have a sentence outline, the most difficult part of the writing job is over. However, whether you have a topic or a sentence outline, remember that an outline is flexible; it may need to change as you write the draft, but it should always be your point of departure and return.

DIGITAL TIP

Creating an Outline

Using the outline feature of your word-processing software allows you to format your outline automatically — fill in, rearrange, and update your outline, as well as create an alphanumeric or a decimal numbering outlining style. For step-by-step instructions, see *bedfordstmartins.com/alredtech* and select *Digital Tips*, "Creating an Outline."

---

outside [of]

In the phrase *outside of*, the word *of* is redundant.

▶ Place the rack outside ~~of~~ the incubator.

Do not use *outside of* to mean "aside from" or "except for."

    Except for
▶ ~~Outside of~~ his frequent absences, Jim has a good work record.

over [with]

In the expression *over with*, the word *with* is redundant; such words as *completed* or *finished* often better express the thought.

▶ You may open the test chamber when the experiment is over ~~with.~~

▶ You may open the test chamber when the experiment is ~~over with.~~ *completed.*

P

pace

Pace is the speed at which you present ideas to the reader. Your goal should be to achieve a pace that fits your **audience**, **purpose**, and **context**. The more knowledgeable the reader is about the subject, the faster your pace can be. Be careful, though, not to lose control of the pace. In the first version of the following passage, facts are piled on top of each other at a rapid pace. In the second version, the same facts are presented in two more easily assimilated sentences. In addition, the second version achieves a different and more desirable **emphasis**.

> RAPID   The hospital's generator produces 110 volts at 60 hertz and is powered by a 90-horsepower engine. It is designed to operate under normal conditions of temperature and humidity, for use under emergency conditions, and may be phased with other units of the same type to produce additional power when needed.
>
> CONTROLLED   The hospital's generator, which is powered by a 90-horsepower engine, produces 110 volts at 60 hertz under normal conditions of temperature and humidity. Designed especially for use under emergency conditions, this generator may be phased with other units of the same type to produce additional power when needed.

paragraphs

A paragraph performs three functions: (1) It develops the unit of thought stated in the topic sentence; (2) it provides a logical break in the material; and (3) it creates a visual break on the page, which signals a new topic.

Topic Sentence

A topic sentence states the paragraph's main idea; the rest of the paragraph supports and develops that statement with related details. The topic sentence is often the first sentence because it tells the reader what the paragraph is about.

> *The arithmetic of searching for oil is stark.* For all the scientific methods of detection, the only way the oil driller can actually know for sure that there is oil in the ground is to drill a well. The average cost of drilling an oil well is at least $2,000,000 and drilling a single well may cost over $18,000,000. And once the well is drilled, the odds against its containing any oil at all are eight to one!

The topic sentence is usually most effective early in the paragraph, but a paragraph can lead to the topic sentence, which is sometimes done to achieve **emphasis**.

> Energy does far more than simply make our daily lives more comfortable and convenient. Suppose you wanted to stop—and reverse—the economic progress of this nation. What would be the surest and quickest way to do it? Find a way to cut off the nation's oil resources! The economy would plummet into the abyss of national economic ruin. *Our economy, in short, is energy-based.*
> —*The Baker World* (Los Angeles: Baker Oil Tools)

On rare occasions, the topic sentence may logically fall in the middle of a paragraph.

> . . . [It] is time to insist that science does not progress by carefully designed steps called "experiments," each of which has a well-defined beginning and end. *Science is a continuous and often a disorderly and accidental process.* We shall not do the young psychologist any favor if we agree to reconstruct our practices to fit the pattern demanded by current scientific methodology.
> —B. F. Skinner, "A Case History in Scientific Method"

Paragraph Length

A paragraph should be just long enough to deal adequately with the subject of its topic sentence. A new paragraph should begin whenever the subject changes significantly. A series of short, undeveloped paragraphs can indicate poor **organization** and sacrifice unity by breaking a single idea into several pieces. A series of long paragraphs, however, can fail to provide the reader with manageable subdivisions of thought. Paragraph length should aid the reader's understanding of ideas.

Occasionally, a one-sentence paragraph is acceptable if it is used as a **transition** between longer paragraphs or as a one-sentence **introduction** or **conclusion** in **correspondence**.

Writing Paragraphs

Careful paragraphing reflects the writer's logical organization and helps the reader follow the writer's thoughts. A good working outline makes it easy to group ideas into appropriate paragraphs. (See also **outlining**.) The following partial topic outline plots the course of the subsequent paragraphs:

TOPIC OUTLINE (PARTIAL)
I. Advantages of Chicago as location for new facility
   A. Transport infrastructure
      1. Rail
      2. Air
      3. Truck
      4. Sea (except in winter)
   B. Labor supply
      1. Engineering and scientific personnel
         a. Similar companies in area
         b. Major universities
      2. Technical and manufacturing personnel
         a. Community college programs
         b. Custom programs

RESULTING PARAGRAPHS

Probably the greatest advantage of Chicago as a location for our new facility is its excellent transport facilities. The city is served by three major railroads. Both domestic and international air-cargo service are available at O'Hare International Airport; Midway Airport's convenient location adds flexibility for domestic air-cargo service. Chicago is a major hub of the trucking industry, and most of the nation's large freight carriers have terminals there. Finally, except in the winter months when the Great Lakes are frozen, Chicago is a seaport, accessible through the St. Lawrence Seaway.

Chicago's second advantage is its abundant labor force. An ample supply of engineering and scientific staff is assured not only by the presence of many companies engaged in activities similar to ours but also by the presence of several major universities in the metropolitan area. Similarly, technicians and manufacturing personnel are in abundant supply. The colleges in the Chicago City College system, as well as half a dozen other two-year colleges in the outlying areas, produce graduates with associate's degrees in a wide variety of technical specialties appropriate to our needs.

Moreover, three of the outlying colleges have expressed an interest in developing off-campus courses attuned specifically to our requirements.

Paragraph Unity and Coherence

A good paragraph has <u>unity</u> and <u>coherence</u> as well as adequate development. *Unity* is singleness of purpose, based on a topic sentence that states the core idea of the paragraph. When every sentence in the paragraph develops the core idea, the paragraph has unity. *Coherence* is holding to one point of view, one attitude, one tense; it is the joining of sentences into a logical pattern. A careful choice of transitional words ties ideas together and thus contributes to coherence in a paragraph. Notice how the boldfaced italicized words tie together the ideas in the following paragraph.

> **TOPIC SENTENCE** *Over the past several months, I have heard complaints about the Merit Award Program.* ***Specifically****, many employees feel that this program should be linked to annual **salary increases**. They believe that **salary increases** would provide a much better incentive than the current $500 to $700 cash awards for exceptional service.* ***In addition****, these **employees believe** that their supervisors consider the cash awards a satisfactory alternative to salary increases. Although I don't think this practice is widespread, the fact that the **employees believe** that it is justifies a reevaluation of the Merit Award Program.*

Simple enumeration (*first, second, then, next,* and so on) also provides effective transition within paragraphs. Notice how the boldfaced italicized words and phrases give coherence to the following paragraph.

> ▶ Most adjustable office chairs have nylon tubes that hold metal spindle rods. To keep the chair operational, lubricate the spindle rods occasionally. ***First,*** *loosen the set screw in the adjustable bell.* ***Then*** *lift the chair from the base.* ***Next****, apply the lubricant to the spindle rod and the nylon washer.* ***When you have finished****, replace the chair and tighten the set screw.*

parallel structure

Parallel structure requires that sentence elements that are alike in function be alike in grammatical form as well. This structure achieves an economy of words, clarifies meaning, expresses the equality of the ideas, and achieves **emphasis**. Parallel structure assists **readers** because it allows them to anticipate the meaning of a sentence element on the basis of its construction.

Parallel structure can be achieved with words, **phrases**, or **clauses**.

▶ If you want to benefit from the jobs training program, you must be *punctual*, *courteous*, and *conscientious*. [parallel words]

▶ If you want to benefit from the jobs training program, you must recognize the importance *of punctuality*, *of courtesy*, and *of conscientiousness*. [parallel phrases]

▶ If you want to benefit from the jobs training program, *you must arrive punctually*, *you must behave courteously*, and *you must study conscientiously*. [parallel clauses]

Correlative **conjunctions** (*either . . . or*, *neither . . . nor*, *not only . . . but also*) should always join elements that use parallel structure. Both parts of the pairs should be followed immediately by the same grammatical form: two similar words, two similar phrases, or two similar clauses.

▶ Viruses carry either *DNA* or *RNA*, never both. [parallel words]

▶ Clearly, neither *serological tests* nor *virus isolation studies* alone would have been adequate. [parallel phrases]

▶ Either *we must increase our production efficiency* or *we must decrease our production goals*. [parallel clauses]

To make a parallel construction clear and effective, it is often best to repeat an article, a **pronoun**, a helping **verb**, a **preposition**, a subordinating conjunction, or the mark of an infinitive (*to*).

▶ The association has *a* mission statement and *a* code of ethics.

▶ The driver *must* check the gauge regularly and *must* act quickly when the indicator falls below the red line.

Parallel structure is especially important in creating **lists**, outlines, **tables of contents**, and **headings** because it lets readers know the relative value of each item in a table of contents and each heading in the body of a document. See also **outlining**.

Faulty Parallelism

Faulty parallelism results when joined elements are intended to serve equal grammatical functions but do not have equal grammatical form.

Faulty parallelism sometimes occurs because a writer tries to compare items that are not comparable.

| NOT PARALLEL | The company offers special college training to help nonexempt employees move into professional careers like engineering management, software development, service technicians, and sales trainees. [Notice faulty comparison of occupations—*engineering management* and *software development*—to people—*service technicians* and *sales trainees*.] |

To avoid faulty parallelism, make certain that each element in a series is similar in form and structure to all others in the same series.

| PARALLEL | The company offers special college training to help nonexempt employees move into professional careers like *engineering management*, *software development*, *technical services*, and *sales*. |

paraphrasing

Paraphrasing is restating or rewriting in your own words the essential ideas of another writer. The following example is an original passage explaining the concept of *object blur*. The paraphrased version restates the essential information of the passage in a form appropriate for a **report**.

| ORIGINAL | One of the major visual cues used by pilots in maintaining precision ground reference during low-level flight is that of object blur. We are acquainted with the object-blur phenomenon experienced when driving an automobile. Objects in the foreground appear to be rushing toward us, while objects in the background appear to recede slightly.
—Wesley E. Woodson and Donald W. Conover, *Human Engineering Guide for Equipment Designers* |
| PARAPHRASED | Object blur refers to the phenomenon by which observers in a moving vehicle report that foreground objects appear to rush at them, while background objects appear to recede slightly (Woodson & Conover, 1964). |

◆ ETHICS NOTE Because paraphrasing does not quote a source word for word, **quotation marks** are not used. However, paraphrased material should be credited because the *ideas* are taken from someone else. See also **ethics in writing**, **note-taking**, **plagiarism**, and **quotations**. ✦

parentheses

Parentheses are used to enclose explanatory or digressive words, phrases, or sentences. Material in parentheses often clarifies or defines the preceding text without altering its meaning.

▶ She severely bruised her tibia (or shinbone) in the accident.

Parenthetical information may not be essential to a sentence (in fact, parentheses deemphasize the enclosed material), but it may be helpful to some readers.

Parenthetical material does not affect the punctuation of a sentence, and any punctuation (such as a **comma** or **period**) should appear following the closing parenthesis.

▶ She could not fully extend her knee because of a torn meniscus (or cartilage), and she suffered pain from a severely bruised tibia (or shinbone).

When a complete sentence within parentheses stands independently, the ending punctuation is placed inside the final parenthesis.

▶ The project director listed the problems her staff faced. (This was the third time she had complained to the board.)

For some constructions, however, you should consider using **subordination** rather than parentheses.

▶ The early tests showed little damage ~~(the attending physician was pleased)~~ *, which pleased the attending physician,* but later scans revealed abdominal trauma.

Parentheses also are used to enclose numerals or letters that indicate sequence.

▶ The following sections deal with (1) preparation, (2) research, (3) organization, (4) writing, and (5) revision.

Do not follow spelled-out **numbers** with numerals in parentheses representing the same numbers.

▶ Send five ~~(5)~~ copies of the report.

Use **brackets** to set off a parenthetical item that is already within parentheses.

▶ We should be sure to give Emanuel Foose (and his brother Emilio [1912–1982]) credit for his part in founding the institute.

See also **documenting sources** and **quotations**.

parts of speech

The term *parts of speech* describes the class of words to which a particular word belongs, according to its function in a sentence.

PART OF SPEECH	FUNCTION
noun, **pronoun**	naming/referring
verb	acting/asserting
adjective, **adverb**	describing/modifying
conjunction, **preposition**	joining/linking
interjection	exclaiming

Many words can function as more than one part of speech. See also **functional shift**.

party

In legal language, *party* refers to an individual, a group, or an organization. ("The injured *party* sued my client.") The term is inappropriate in all but legal writing; when you are referring to a person, use the word *person*.

▶ The ~~party~~ *person* whose file you requested is here now.

Party is, of course, appropriate when it refers to a group. ("Jim arranged a tour of the facility for the members of our *party*.")

per

When *per* is used to mean "for each," "by means of," "through," or "on account of," it is appropriate (*per* annum, *per* capita, *per* diem, *per* head). When used to mean "according to" (*per* your request, *per* your order), the expression is **jargon** and should be avoided.

▶ As ~~per our discussion,~~ *we discussed,* I will send revised instructions.

percent / percentage

Percent is normally used instead of the symbol % ("only 15 *percent*"), except in **tables**, where space is at a premium. *Percentage*, which is never used with **numbers**, indicates a general size ("only a small *percentage*").

periods

A period is a mark of **punctuation** that usually indicates the end of a declarative or an imperative sentence. Periods also link when used as leaders (as in rows of periods in **tables of contents**) and indicate omissions when used as **ellipses**. Periods are also used to end questions that are actually polite requests, or instructions to which an affirmative response is assumed. ("Will you call me as soon as he arrives.") See also **sentence construction**.

Periods in Quotations

Use a **comma**, not a period, after a declarative sentence that is quoted in the context of another sentence.

▶ "There is every chance of success," she stated.

A period is placed inside **quotation marks**. See also **quotations**.

▶ He stated clearly, "My vote is yes."

Periods with Parentheses

Place a period outside the final parenthesis when a parenthetical element ends a sentence.

▶ The institute was founded by Harry Denman (1902–1972).

Place a period inside the final parenthesis when a complete sentence stands independently within **parentheses**.

▶ The project director listed the problems her staff faced. (This was the third time she had complained to the board.)

Other Uses of Periods

Use periods following the numerals in a numbered **list** and following complete sentences in a list.

▶ 1. Enter your name and PIN.
2. Enter your address with ZIP Code.
3. Enter your home telephone number.

When a sentence ends with an abbreviation that ends with a period, do not add another period.

▶ Please meet me at 3:30 p.m.

Periods are also used after initials in names (Wilma T. Grant, J. P. Morgan), as decimal points with **numbers** (27.3 degrees Celsius, $540.26, 6.9 percent), and with certain **abbreviations** (Ms., Dr., Inc.).

Period Faults

The incorrect use of a period is sometimes referred to as a *period fault*. When a period is inserted prematurely, the result is a **sentence fragment**.

FRAGMENT	After a long day at the office during which we finished the quarterly report. We left hurriedly for home.
SENTENCE	After a long day at the office, during which we finished the quarterly report, we left hurriedly for home.

When two independent clauses are joined without any punctuation, the result is a *fused*, or *run-on*, *sentence*. Adding a period between the clauses is one way to correct a run-on sentence.

RUN-ON	Bill was late for ten days in a row Ms. Sturgess had to dismiss him.
CORRECT	Bill was late for ten days in a row. Ms. Sturgess had to dismiss him.

Other options are to add a comma and a coordinating **conjunction** (*and*, *but*, *for*, *or*, *nor*, *so*, *yet*) between the clauses, to add a **semicolon**, or to add a semicolon with a conjunctive **adverb**, such as *therefore* or *however*.

P

person

Person refers to the form of a personal **pronoun** that indicates whether the pronoun represents the speaker, the person spoken to, or the person or thing spoken about. A pronoun representing the speaker is in the *first* person. ("*I* could not find the answer in the manual.") A pronoun that represents the person or people spoken to is in the *second* person. ("*You* will be a good manager.") A pronoun that represents the person or people spoken about is in the *third* person. ("*They* received the news quietly.") The following list shows first-, second-, and third-person pronouns. See also **case**, **number**, and **one**.

PERSON	SINGULAR	PLURAL
First	I, me, my, mine	we, us, our, ours
Second	you, your, yours	you, your, yours
Third	he, him, his, she, her, hers, it, its	they, them, their, theirs

persuasion

Persuasive writing attempts to convince an **audience** to adopt the writer's point of view or take a particular action. Workplace writing often uses persuasion to reinforce ideas that readers already have, to convince readers to change their current ideas, or to lobby for a particular suggestion or policy (as in Figure P–1 on page 378). You may find yourself advocating for safer working conditions, justifying the expense of a new program, or writing a **proposal** for a large purchase. See also **context** and **purpose**.

In persuasive writing, the way you present your ideas is as important as the ideas themselves. You must support your appeal with logic and a sound presentation of facts, statistics, and examples. See also **logic errors**.

A writer also gains credibility, and thus persuasiveness, through the readers' impressions of the document's appearance. For this reason, consider carefully a document's **layout and design**. See also **résumés**.

▶ ETHICS NOTE Avoid ambiguity. Do not wander from your main point and, above all, never make false claims. You should also acknowledge any real or potentially conflicting opinions; doing so allows you to anticipate and overcome objections and builds your credibility. See also **ethics in writing**. ◆

The **memo** shown in Figure P–1 was written to persuade the engineering staff to accept and participate in a change to a new computer system. Notice that not everything in this memo is presented in a positive light. Change brings disruption, and the writer acknowledges that fact.

A persuasive technique that places the focus on your reader's interest and perspective is discussed in the entry **"you" viewpoint**. See also **correspondence**.

phenomenon / phenomena

A *phenomenon* is an observable thing, fact, or occurrence ("a natural *phenomenon*"). Its plural form is *phenomena*.

photographs

Photographs are often the best way to show the appearance of an object, record an event, or demonstrate the development of a phenomenon over a period of time. Photographs, however, cannot depict the internal

Interoffice Memo

TO: Engineering Staff
FROM: Harold Kawenski, MIS Administrator *HK*
DATE: April 21, 2009
SUBJECT: Plans for the Changeover to the New Computer System

As you all know, the merger with Datacom has resulted in dramatic growth in our workload—a 30-percent increase in our customer support services during the last several months. To cope with this expansion, we will soon install the NRT/R4 server and QCS Enterprise software with Web-based applications.

Let me briefly describe the benefits of this system and the ways we plan to help you cope with the changeover.

The QCS system will help us access up-to-date marketing and product information when we need it. This system will speed processing dramatically and give us access to all relevant company-wide databases. Because we anticipate that our workload will increase another 30 percent in the next several months, we need to get the QCS system online and working smoothly as soon as possible.

The changeover to this system, unfortunately, will cause some disruption at first. We will need to transfer many of our legacy programs and software applications to the new system. In addition, all of us will need to learn to navigate in the R4 and QCS environments. Once we have made these adjustments, however, I believe we will welcome the changes.

To help everyone cope with the changeover, we will offer training sessions that will begin next week. I have attached a sign-up form with specific class times. We will also provide a technical support hotline at extension 4040, which will be available during business hours; e-mail support at qcs-support@conco.com; and online help documentation.

I would like to urge you to help us make a smooth transition to the QCS system. Please e-mail me with your suggestions or questions about the impact of the changeover on your department. I look forward to working with you to make this system a success.

Enclosure: Training Session Schedule

FIGURE P–1. Persuasive Memo

workings of a mechanism or below-the-surface details of objects or structures. Such details are better represented in **drawings**. Photographs are also effective in catching the readers' attention and adding personal relevance to such documents as **brochures** and **newsletters**. See also **readers**.

An effective photograph shows important details and indicates the relative size of the subject by including a familiar object—such as a ruler or a person—near the subject being photographed.

Figure P–2 shows a photograph from an interactive Web presentation for a pilot training program. This photograph is one in a series that simulates a pilot's "walk-around"—a procedure in which pilots visually examine an aircraft in a 360-degree safety inspection prior to take-off. In this photo, the stair steps are lowered to show the relative size of the aircraft.

For **reports**, treat photographs as you do other **visuals**, giving them figure numbers, call-outs (labels) to identify key features, and captions, if needed. Position the figure number and caption so that readers can view them and the photograph from the same orientation.

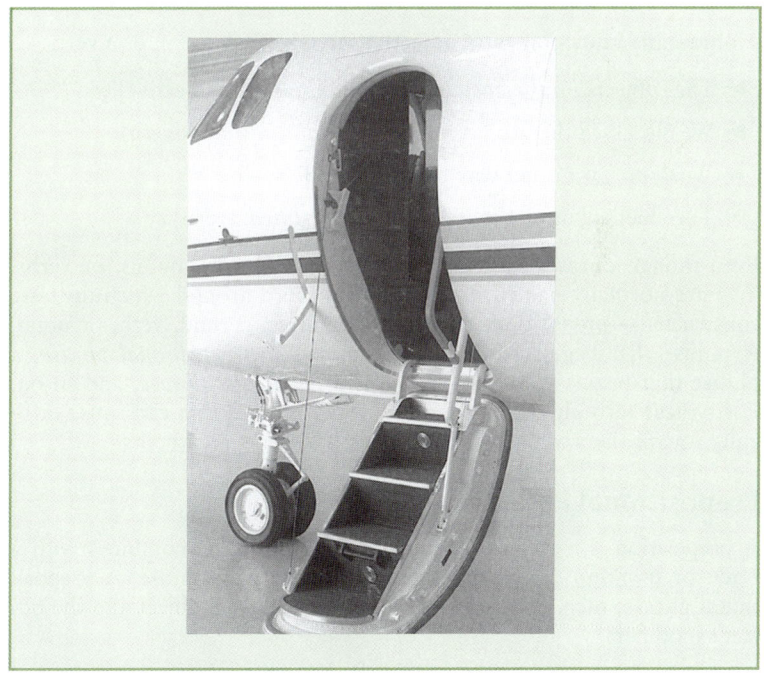

FIGURE P–2. Photo (of Aircraft Door). Photo courtesy of Ken Cook Company.

> ⊕ **ETHICS NOTE** Be careful to avoid **plagiarism** by appropriately **documenting sources** for photographs and to obtain **copyright** permission if you plan to publish photographs that you do not take yourself. ✦

phrases

DIRECTORY

Prepositional Phrases 380
Participial Phrases 381
   Dangling Participial Phrases 382
   Misplaced Participial Phrases 382
Infinitive Phrases 382
Gerund Phrases 383
Verb Phrases 383
Noun Phrases 383

A phrase is a meaningful group of words that cannot make a complete statement because it does not contain both a subject and a predicate, as **clauses** do. Phrases, which are based on **nouns**, nonfinite **verb** forms, or verb combinations, provide context within a clause or sentence in which they appear. See also **sentence construction**.

▶ She reassured her staff *by her calm confidence*. [phrase]

A phrase may function as an **adjective**, an **adverb**, a noun, or a verb.

▶ The subjects *on the agenda* were all discussed. [adjective]

▶ We discussed the project *with great enthusiasm*. [adverb]

▶ *Working hard* is her way of life. [noun]

▶ The chief engineer *should have been notified*. [verb]

Even though phrases function as adjectives, adverbs, nouns, or verbs, they are normally named for the kind of word around which they are constructed—**preposition**, participle, infinitive, gerund, verb, or noun. A phrase that begins with a preposition is a *prepositional phrase*, a phrase that begins with a participle is a *participial phrase*, and so on. For typical verb phrases and prepositional phrases that can cause difficulty for speakers of **English as a second language**, see **idioms**.

Prepositional Phrases

A preposition is a word that shows relationship and combines with a noun or **pronoun** (its **object**) to form a modifying phrase. A prepositional phrase, then, consists of a preposition plus its object and the object's modifiers.

▶ *After the meeting*, the district managers adjourned *to the cafeteria*.

Prepositional phrases, because they normally modify nouns or verbs, usually function as adverbs or adjectives. A prepositional phrase may function as an adverb of motion ("Turn the dial four degrees *to the left*") or an adverb of manner ("Answer customers' questions *in a courteous fashion*"). A prepositional phrase may function as an adverb of place and may appear in different places in the sentence.

▶ *In home and office computer systems*, security is essential.

▶ Security is essential *in home and office computer systems*.

Prepositional phrases may function as adjectives; when they do, they follow the nouns they modify.

▶ Food waste *with a high protein content* can be processed into animal food.

Be careful when you use prepositional phrases because separating a prepositional phrase from the noun it modifies can cause **ambiguity**.

AMBIGUOUS	*The woman* standing by the security guard *in the gray suit* is our division manager.
CLEAR	*The woman in the gray suit* who is standing by the security guard is our division manager.

Watch as well for the overuse of prepositional phrases where **modifiers** would be more economical.

OVERUSED	The man *with gray hair in the blue suit with pinstripes* is the former president *of the company*.
ECONOMICAL	The *gray-haired* man in the *blue pin-striped* suit is the former *company* president.

Participial Phrases

A participle is any form of a verb that is used as an adjective. A participial phrase consists of a participle plus its object and its modifiers.

▶ The division *having the largest number of patents* will work with NASA.

The relationship between a participial phrase and the rest of the sentence must be clear to the reader. For that reason, every sentence containing a participial phrase must have a noun or pronoun that the participial phrase modifies; if it does not, the result is a dangling participial phrase.

Dangling Participial Phrases. A dangling participial phrase occurs when the noun or pronoun that the participial phrase is meant to modify is not stated but only implied in the sentence. See also **dangling modifiers**.

DANGLING	*Being unhappy with the job*, his efficiency suffered. [His efficiency was not unhappy with the job; what the participial phrase really modifies—*he*—is not stated but merely implied.]
CORRECT	*Being unhappy with the job*, he grew less efficient. [In this version, what that participial phrase modifies—*he*—is explicitly stated.]

Misplaced Participial Phrases. A participial phrase is misplaced when it is too far from the noun or pronoun it is meant to modify and so appears to modify something else. Such an error can make the writer look ridiculous.

MISPLACED	We saw a large warehouse *driving down the highway*.
CORRECT	*Driving down the highway*, we saw a large warehouse.

Infinitive Phrases

An infinitive is the basic form of a verb (*go*, *run*, *talk*) without the restrictions imposed by **person** and **number**. An infinitive is generally preceded by the word *to* (which is usually a preposition but in this use is called the *sign*, or *mark*, of the infinitive). An infinitive phrase consists of the word *to* plus an infinitive and any objects or modifiers.

▶ *To succeed in this field*, you must be willing *to assume responsibility*.

Do not confuse a prepositional phrase beginning with *to* with an infinitive phrase. In an infinitive phrase, *to* is followed by a verb; in a prepositional phrase, *to* is followed by a noun or pronoun.

PREPOSITIONAL PHRASE	We went *to the building site*.
INFINITIVE PHRASE	Our firm tries *to provide a comprehensive training program*.

The implied subject of an introductory infinitive phrase should be the same as the subject of the sentence. If it is not, the phrase is a dangling modifier. In the following example, the implied subject of the infinitive is *you* or *one*, not *practice*.

▶ To learn a new language, ~~practice is needed.~~ *you must practice.*

Gerund Phrases

A gerund is a **verbal** ending in *-ing* that is used as a noun. A gerund phrase consists of a gerund plus any objects or modifiers and always functions as a noun.

SUBJECT   *Preparing a grant proposal* is a difficult task.
DIRECT OBJECT She liked *chairing the committee.*

Verb Phrases

A verb phrase consists of a main verb and its helping verb.

▶ He *is* [helping verb] *working* [main verb] hard this summer.

Words can appear between the helping verb and the main verb of a verb phrase. ("He *is* always *working*.") The main verb is always the last verb in a verb phrase.

Questions often begin with a verb phrase. ("*Will* he *verify* the results?") The adverb *not* may be appended to a helping verb in a verb phrase. ("He *did not work* today.")

Noun Phrases

A noun phrase consists of a noun and its modifiers. ("Have *the two new employees* fill out *these forms*.")

plagiarism

Plagiarism is the use of someone else's unique ideas without acknowledgment, or the use of someone else's exact words without **quotation marks** and appropriate credit. Plagiarism is considered to be the theft of someone else's creative and intellectual property and is not accepted in business, science, journalism, academia, or any other field. See also **ethics in writing** and **research**.

Citing Sources

Quoting a passage—including cutting and pasting a passage from an Internet source into your work—is permissible only if you enclose the passage in quotation marks and properly cite the source. For detailed guidance on quoting correctly, see **quotations**. If you intend to publish, reproduce, or distribute material that includes quotations from published works, including Web sites, you may need to obtain written permission from the **copyright** holders of those works.

Even Web sites that grant permission to copy, distribute, or modify material under the "copyleft"* principle, such as *Wikipedia*, nonetheless caution that you must give appropriate credit to the source from which material is taken (see *http://en.wikipedia.org/wiki/Wikipedia: Citing_Wikipedia*).

Paraphrasing the words and ideas of another *also requires that you cite your source*, even though you do not enclose paraphrased ideas or materials in quotation marks. (See also **documenting sources**.) Paraphrasing a passage without citing the source is permissible only when the information paraphrased is common knowledge.

Common Knowledge

Common knowledge generally refers to information that is widely known and readily available in handbooks, manuals, atlases, and other references. For example, the "law of supply and demand" is common knowledge and is found in virtually every economics and business textbook.

Common knowledge also refers to information within a specific field that is generally known and understood by most others in that field—even though it is not widely known by those outside the field. For examples, see the Web Link at the end of this entry.

An indication that something is common knowledge is whether it is repeated in multiple sources without citation. The best advice is, when in doubt, cite the source.

➽ **ETHICS NOTE** In the workplace, employees often borrow material freely from in-house manuals, reports, and other company documents. Using or **repurposing** such material is neither plagiarism nor a violation of copyright. For information on the use of public domain and government material, see **copyright**. ✦

WEB LINK	Avoiding Plagiarism

For a tutorial on using sources correctly, see *bedfordstmartins.com/alredtech* and select *Tutorials*, "The St. Martin's Tutorial on Avoiding Plagiarism." For links to other helpful resources, select *Links for Handbook Entries*.

---

*"Copyleft" is a play on the word *copyright* and is the effort to free materials from many of the restrictions of copyright. See *http://en.wikipedia.org/wiki/Copyleft*.

point of view

Point of view is the writer's relation to the information presented, as reflected in the use of grammatical **person**. The writer usually expresses point of view in first-, second-, or third-person personal **pronouns**. Use of first person indicates that the writer is a participant or an observer. Use of second or third person indicates that the writer is giving directions, **instructions**, or advice, or writing about other people or something impersonal.

FIRST PERSON	*I* scrolled down to find the settings option.
SECOND PERSON	*You* need to scroll down to find the settings option. [*You* is explicitly stated.]
	Scroll down to find the settings option. [*You* is understood in such an instruction.]
THIRD PERSON	*He* scrolled down to find the settings option.

Consider the following sentence, revised from an impersonal to a more personal point of view. Although the meaning of the sentence does not change, the revision indicates that people are involved in the communication.

▶ ~~It is regrettable~~ *I regret* that the equipment shipped on Friday ~~is unacceptable.~~ *we cannot accept*.

Many people think they should avoid the pronoun *I* in technical writing. Such practice, however, often leads to awkward sentences, with people referring to themselves in the third person as *one* or as *the writer* instead of as *I*.

▶ ~~One~~ *I* can only conclude that the absorption rate is too fast.

However, do not use the personal point of view when an impersonal point of view would be more appropriate or more effective because you need to emphasize the subject matter over the writer or the reader. In the following example, it does not help to personalize the situation; in fact, the impersonal version may be more tactful.

PERSONAL	I received objections to my proposal from several of your managers.
IMPERSONAL	Several managers have raised objections to the proposal.

Whether you adopt a personal or an impersonal point of view depends on the **purpose** and the **audience** of the document. For example, in an informal **e-mail** to an associate, you would most likely adopt a personal point of view. However, in a **report** to a large group, you would probably emphasize the subject by using an impersonal point of view.

◆ ETHICS NOTE In company **correspondence**, use of the pronoun *we* may be interpreted as reflecting company policy, whereas *I* clearly reflects personal opinion. Which pronoun to use should be decided according to whether you are speaking for yourself (*I*) or for the company (*we*).

▶ *I* understand your frustration with the approval process, but *we* must meet the new safety regulations. ✦

> **ESL TIPS** for Stating an Opinion
>
> In some cultures, stating an opinion in writing is considered impolite or unnecessary, but in the United States, readers expect to see a writer's opinion stated clearly and explicitly. The opinion should be followed by specific examples to help the reader understand the writer's point of view.

positive writing

Presenting positive information as though it were negative is confusing to **readers**.

NEGATIVE   If the error does *not* involve data transmission, the backup function will *not* be used.

In this sentence, the reader must reverse two negatives to understand the exception that is being stated. (See also **double negatives**.) The following sentence presents the exception in a positive and straightforward manner.

POSITIVE   The backup function is used only when the error involves data transmission.

◆ ETHICS NOTE Negative facts or conclusions, however, should be stated negatively; stating a negative fact or conclusion positively is deceptive because it can mislead the reader.

DECEPTIVE   In the first quarter of this year, employee exposure to airborne lead averaged within 10 percent of acceptable state health standards.

ACCURATE   In the first quarter of this year, employee exposure to airborne lead averaged 10 percent below acceptable state health standards.

See also **ethics in writing**. ✦

Even if what you are saying is negative, do not state it more negatively than necessary.

NEGATIVE   We are withholding your shipment because we have not received your payment.

POSITIVE   We will forward your shipment as soon as we receive your payment.

See also **correspondence** and **"you" viewpoint**.

possessive case

A **noun** or **pronoun** is in the possessive case when it represents a person, place, or thing that possesses something. Possession is generally expressed with an **apostrophe** and an *s* ("the *report's* title"), with a prepositional **phrase** using *of* ("the title *of the report*"), or with the possessive form of a pronoun ("*our* report").

Practices vary for some possessive forms, but the following guidelines are widely used. Above all, be consistent.

Singular Nouns

Most singular nouns show the possessive case with *'s*.

▶ a *manager's* office   the *witness's* testimony
   an *employee's* paycheck   the *bus's* schedule
   the *company's* stock value

When pronunciation with *'s* is difficult or when a multisyllable noun ends in a *z* sound, you may use only an apostrophe.

▶ *New Orleans'* convention hotels

Plural Nouns

Plural nouns that end in *-s* or *-es* show the possessive case with only an apostrophe.

▶ the *managers'* reports   the *witnesses'* testimony
   the *employees'* paychecks   the *buses'* schedules
   the *companies'* joint project

Plural nouns that do not end in *-s* show the possessive with *'s*.

▶ *children's* clothing, *women's* resources, *men's* room

Apostrophes are not always used in official names ("*Consumers* Union") or for words that may appear to be possessive nouns but function as **adjectives** ("a *computer peripherals* supplier").

Compound Nouns

Compound words form the possessive with *'s* following the final letter.

▶ the *attorney general's* decision, the *editor-in-chief's* desk, the *pipeline's* diameter

Plurals of some compound expressions are often best expressed with a prepositional phrase ("presentations *of the editors in chief*").

Coordinate Nouns

Coordinate nouns show joint possession with *'s* following the last noun.

▶ *Fischer and Goulet's* partnership was the foundation of their business.

Coordinate nouns show individual possession with *'s* following each noun.

▶ The difference between *Barker's* and *Washburne's* test results was not statistically significant.

Possessive Pronouns

The possessive pronouns (*its*, *whose*, *his*, *her*, *our*, *your*, *my*, *their*) are also used to show possession and do not require apostrophes. ("Even good systems have *their* flaws.") Only the possessive form of a pronoun should be used with a gerund (a noun formed from an *-ing* **verb**).

▶ The safety officer insisted on *our* wearing protective clothing. [*Wearing* is the gerund.]

Possessive pronouns are also used to replace nouns. ("The responsibility was *theirs*.") See also **its / it's**.

Indefinite Pronouns

Some indefinite pronouns (*all*, *any*, *each*, *few*, *most*, *none*, *some*) form the possessive case with the **preposition** *of*.

▶ We tested both packages and found the bacteria on the surface *of each*.

Other indefinite pronouns (*everyone*, *someone*, *anyone*, *no one*), however, use *'s*.

▶ *Everyone's* contribution is welcome.

prefixes

A prefix is a letter or group of letters placed in front of a root word that changes the meaning of the root word. When a prefix ends with a vowel and the root word begins with a vowel, the prefix is often separated from the root word with a **hyphen** (*re-enter*, *pro-active*, *anti-inflammatory*). Some words with the double vowel are written without a hyphen (*cooperate*) and others with or without a hyphen (*re-elect* or *reelect*).

Prefixes, such as *neo-* (derived from a Greek word meaning "new"), are often hyphenated when used with a proper **noun** (*neo-Keynesian*). Such prefixes are not normally hyphenated when used with common nouns, unless the base word begins with the same vowel (*neonatal*, *neo-orthodoxy*).

A hyphen may be necessary to clarify the meaning of a prefix; for example, *reform* means "correct" or "improve," and *re-form* means "change the shape of." When in doubt, check a current **dictionary**.

preparation

The preparation stage of the writing process is essential. By determining the needs of your **audience**, your **purpose**, the **context**, and the **scope** of coverage, you understand the information you will need to gather during **research**. See also **collaborative writing** and "Five Steps to Successful Writing."

WRITER'S CHECKLIST Preparing to Write

- Determine who your readers are and learn certain key facts about them—their knowledge, attitudes, and needs relative to your subject.
- Determine the document's primary purpose: What exactly do you want your readers to know, to believe, or to be able to do when they have finished reading your document?
- Consider the context of your message and how it should affect your writing.
- Establish the scope of your document—the type and amount of detail you must include—not only by understanding your readers' needs and purpose but also by considering any external constraints, such as

Writer's Checklist: Preparing to Write (continued)

word limits for **trade journal articles** or the space limitations of Web pages. See also **writing for the Web**.

▶ Select the medium appropriate to your readers and purpose. See also **selecting the medium**.

prepositions

A preposition is a word that links a **noun** or **pronoun** to another sentence element by expressing such relationships as direction (*to, into, across, toward*), location (*at, in, on, under, over, beside, among, by, between, through*), time (*before, after, during, until, since*), or position (*for, against, with*). Together, the preposition, its **object** (the noun or pronoun), and the object's **modifiers** form a prepositional **phrase** that acts as a modifier.

▶ Answer help-line questions *in a courteous manner*.
[The prepositional phrase *in a courteous manner* modifies the **verb** *answer*.]

The object of a preposition (the word or phrase following the preposition) is always in the objective **case**. When the object is a compound, both nouns and pronouns should be in the objective case. For example, the phrase "between you and *me*" is frequently and incorrectly written as "between you and *I*." *Me* is the objective form of the pronoun, and *I* is the subjective form.

Many words that function as prepositions also function as **adverbs**. If the word takes an object and functions as a connective, it is a preposition; if it has no object and functions as a modifier, it is an adverb.

PREPOSITIONS	The manager sat *behind* the desk *in* her office.
ADVERBS	The customer lagged *behind*; then he came *in* and sat down.

Certain verbs, adverbs, and adjectives are normally used with certain prepositions (interested *in*, aware *of*, equated *with*, adhere *to*, capable *of*, object *to*, infer *from*). See also **idioms**.

Prepositions at the End of a Sentence

A preposition at the end of a sentence can be an indication that the sentence is awkwardly constructed.

▶ ~~The~~ branch office ~~is where she was at.~~ ^*She was at the* ^.

However, if a preposition falls naturally at the end of a sentence, leave it there. ("I don't remember which file name I saved it *under*.")

Prepositions in Titles

Capitalize prepositions in **titles** when they are the first or last words, or when they contain five or more letters (unless you are following a style that recommends otherwise). See also **capitalization**.

▶ The newspaper column "*In* My Opinion" included a review of the article "New Concerns *About* Distance Education."

Preposition Errors

Do not use redundant prepositions, such as "off *of*," "in back *of*," "inside *of*," and "at *about*."

| EXACT | The client will arrive at ~~about~~ four o'clock. |
| APPROXIMATE | The client will arrive ~~at~~ about four o'clock. |

Avoid unnecessarily adding the preposition *up* to verbs.

▶ Call ~~up and~~ *to* see if he is in his office.

Do not omit necessary prepositions.

▶ He was oblivious *to* and not distracted by the view from his office window.

See also **conciseness** and **English as a second language**.

presentations

DIRECTORY

Determining Your Purpose   392
Analyzing Your Audience   392
Gathering Information   392
Structuring the Presentation   393
   The Introduction   393
   The Body   394
   The Closing   394
   Transitions   395
Using Visuals   395
   Flip Charts   396
   Whiteboard or Chalkboard   396
   Overhead Transparencies   396
   Presentation Software   396
Writer's Checklist: Using Visuals in a Presentation   398
Delivering a Presentation   398
   Practice   398
   Delivery Techniques That Work   399
   Presentation Anxiety   400
Writer's Checklist: Preparing for and Delivering a Presentation   400

The steps required to prepare an effective presentation parallel the steps you follow to write a document. As with writing a document, determine your **purpose** and analyze your **audience**. Then gather the facts that will support your point of view or proposal and logically organize that information. Presentations do, however, differ from written documents in a number of important ways. They are intended for listeners, not readers. Because you are speaking, your manner of delivery, the way you organize the material, and your supporting **visuals** require as much attention as your content.

Determining Your Purpose

Every presentation is given for a purpose, even if it is only to share information. To determine the primary purpose of your presentation, use the following question as a guide: What do I want the audience to know, to believe, or to do when I have finished the presentation? Based on the answer to that question, write a purpose statement that answers the *what?* and *why?* questions.

> ▶ The purpose of my presentation is to convince my company's chief information officer of the need to improve the appearance, content, and customer use of our company's Web site [*what*] so that she will be persuaded to allocate additional funds for site-development work in the next fiscal year [*why*].

Analyzing Your Audience

Once you have determined the desired end result of the presentation, analyze your audience so that you can tailor your presentation to their needs. Ask yourself these questions about your audience:

- What is your audience's level of experience or knowledge about your topic?
- What is the general educational level and age of your audience?
- What is your audience's attitude toward the topic you are speaking about, and—based on that attitude—what concerns, fears, or objections might your audience have?
- Do any subgroups in the audience have different concerns or needs?
- What questions might your audience ask about this topic?

Gathering Information

Once you have focused the presentation, you need to find the facts and arguments that support your point of view or the action you propose.

As you gather information, keep in mind that you should give the audience only what will accomplish your goals; too much detail will overwhelm them, and too little will not adequately inform your listeners or support your recommendations. For detailed guidance about gathering information, see **research**.

Structuring the Presentation

When structuring the presentation, focus on your audience. Listeners are freshest at the outset and refocus their attention near the end. Take advantage of that pattern. Give your audience a brief overview of your presentation at the beginning, use the body to develop your ideas, and end with a summary of what you covered and, if appropriate, a call to action. See also **methods of development**.

The Introduction. Include in the **introduction** an opening that focuses your audience's attention, as in the following examples:

- [*Definition of a problem*] "You have to write an important report, but you'd like to incorporate lengthy handwritten notes from several meetings you attended. Your scanner will not read these notes, and you will have to type many pages. You groan because that seems an incredible waste of time. Have I got a solution for you!"

- [*An attention-getting statement*] "As many as 50 million Americans have high blood pressure."

- [*A rhetorical question*] "Would you be interested in a full-sized computer keyboard that is waterproof and noiseless, and can be rolled up like a rubber mat?"

- [*A personal experience*] "As I sat at my computer one morning, deleting my eighth spam message of the day, I decided that it was time to take action to eliminate this time-waster."

- [*An appropriate quotation*] "According to researchers at the Massachusetts Institute of Technology, 'Garlic and its cousin, the onion, confer major health benefits—including fighting cancer, infections, and heart disease.'"

Following your opening, use the introduction to set the stage for your audience by providing an overview of the presentation, which can include general or background information that will be needed to understand the detailed information in the body of your presentation. It can also show how you have organized the material.

- This presentation analyzes three high-volume, on-demand printers for us to consider purchasing. Based on a comparison of all three,

I will recommend the one I believe best meets our needs. To do so, I'll discuss the following five points:

1. Why we need a high-volume printer [*the problem*]
2. The basics of on-demand technology [*general information*]
3. The criteria I used to compare the three printer models [*comparison*]
4. The printer models I compared and why [*possible solutions*]
5. The printer I propose we buy [*proposed solution*]

The Body. If your goal is **persuasion**, present the evidence that will persuade the audience to agree with your conclusions and act on them. If you are discussing a problem, demonstrate that it exists and offer a solution or range of possible solutions. For example, if your introduction stated that the problem is low profits, high costs, outdated technology, or high employee absenteeism, you could use the following approach.

1. Prove your point.
   - Strategically organize the facts and data you need.
   - Present the information using easy-to-understand visuals.
2. Offer solutions.
   - Increase profits by lowering production costs.
   - Cut overhead to reduce costs, or abolish specific programs or product lines.
   - Replace outdated technology, or upgrade existing technology.
   - Offer employees more flexibility in their work schedules or other incentives.
3. Anticipate questions ("How much will it cost?") and objections ("We're too busy now—when would we have time to learn the new software?") and incorporate the answers into your presentation.

The Closing. Fulfill the goals of your presentation in the closing. If your purpose is to motivate the listeners to take action, ask them to do what you want them to do; if your purpose is to get your audience to think about something, summarize what you want them to think about. Many presenters make the mistake of not actually closing—they simply quit talking, shuffle papers, and then walk away.

Because your closing is what your audience is most likely to remember, use that time to be strong and persuasive. Consider the following typical closing.

> ▶ Based on all the data, I believe that the Worthington TechLine 5510 Production Printer best suits our needs. It produces 40 pages per minute more than its closest competitor and provides modular

systems that can be upgraded to support new applications. The Worthington is also compatible with our current computer network, and staff training at our site is included with our purchase. Although the initial cost is higher than that for the other two models, the additional capabilities, compatibility with most standard environments, lower maintenance costs, and strong customer support services make it a better value.

I recommend we allocate the funds necessary for this printer by the fifteenth of this month in order to be well prepared for the production of next quarter's customer publications.

This closing brings the presentation full circle and asks the audience to fulfill the purpose of the presentation—exactly what a **conclusion** should do.

Transitions. Planned **transitions** should appear between the introduction and the body, between major points in the body, and between the body and the closing. Transitions are simply a sentence or two to let the audience know that you are moving from one topic to the next. They also prevent a choppy presentation and provide the audience with assurance that you know where you are going and how to get there.

▶ Before getting into the specifics of each printer I compared, I'd like to present the benefits of networked, on-demand printers in general. That information will provide you with the background you'll need to compare the differences among the printers and their capabilities discussed in this presentation.

It is also a good idea to pause for a moment after you have delivered a transition between topics to let your listeners shift gears with you. Remember, they do not know your plan.

Using Visuals

Well-planned **visuals** not only add interest and emphasis to your presentation but also clarify and simplify your message because they communicate clearly, quickly, and vividly. Charts, graphs, and illustrations can greatly increase audience understanding and retention of information, especially for complex issues and technical information that could otherwise be misunderstood or overlooked.

▶ **ETHICS NOTE** Be sure to provide credit for any visual taken from a print or an online source. You can include a citation either on an individual visual (such as a slide) or in a list of references or works cited that you distribute to your audience. For information on citing visuals from print or Web sources, see **documenting sources**. ◆

You can create and present the visual components of your presentation by using a variety of media—flip charts, whiteboard or chalkboard, overhead transparencies, slides, or computer presentation software. See also **layout and design**.

Flip Charts. Flip charts are ideal for smaller groups in a conference room or classroom and are also ideal for **brainstorming** with your audience.

Whiteboard or Chalkboard. The whiteboard or chalkboard common to classrooms is convenient for creating sketches and for jotting notes during your presentation. If your presentation requires extensive notes or complex **drawings**, however, prepare handouts on which the audience can jot notes and which they can keep for future reference.

Overhead Transparencies. With transparencies you can create a series of overlays to explain a complex device or system, adding (or removing) the overlays one at a time. You can also lay a sheet of paper over a list of items on a transparency, uncovering one item at a time as you discuss it, to focus audience attention on each point in the sequence.

Presentation Software. Presentation software, such as Microsoft PowerPoint, Corel Presentations, and OpenOffice.org Impress, lets you create your presentation on your computer. You can develop charts and graphs with data from spreadsheet software or locate visuals on the Web, and then import those files into your presentation. This software also offers standard templates and other features that help you design effective visuals and integrated text. Enhancements include a selection of typefaces, highlighting devices, background textures and colors, and clip-art images. Images can also be printed out for use as overhead transparencies or handouts. However, avoid using too many enhancements, which may distract viewers from your message. Figure P–3 shows well-balanced slides for a presentation based on the sample "formal report" in Figure F–5.

Rehearse your presentation using your electronic slides, and practice your transitions from slide to slide. Also practice loading your pre-

> **WEB LINK** Preparing Presentation Slides
>
> For a helpful tutorial on creating effective slides, see *bedfordstmartins.com/alredtech* and select *Tutorials*, "Preparing Presentation Slides." For links to additional information and tutorials for using presentation software, see *bedfordstmartins.com/alredtech* and select *Links for Handbook Entries*.

FIGURE P–3. Slides for a Presentation

sentation and anticipate any technical difficulties that might arise. Should you encounter a technical snag during the presentation, stay calm and give yourself time to solve the problem. If you cannot solve the problem, move on without the technology. As a backup, carry a printout of your electronic presentation as well as an extra electronic copy on a storage medium.

> **WRITER'S CHECKLIST** Using Visuals in a Presentation
>
> - Use text sparingly in visuals. Use bulleted or numbered **lists**, keeping them in **parallel structure** and with balanced content. Use numbers if the sequence is important and bullets if it is not.
> - Limit the number of bulleted or numbered items to no more than five or six per visual.
> - Limit each visual to no more than 40 to 45 words. Any more will clutter the visual and force you to use a smaller font that could impair the audience's ability to read it.
> - Make your visuals consistent in type style, size, and spacing.
> - Use a type size visible to members of the audience at the back of the room. Type should be boldface and no smaller than 30 points. For headings, 45- or 50-point type works even better.
> - Use graphs and charts to show data trends. Use only one or two illustrations per visual to avoid clutter and confusion.
> - Make the contrast between your text and the background sharp. Use light backgrounds with dark lettering and avoid textured or decorated backgrounds.
> - Use no more than 12 visuals per presentation. Any more will tax the audience's concentration.
> - Match your delivery of the content to your visuals. Do not put one set of words or images on the screen and talk about the previous visual or, even worse, the next one.
> - Do not read the text on your visual word for word. Your audience can read the visuals; they look to you to provide the key points in detail.

Delivering a Presentation

Once you have outlined and drafted your presentation and prepared your visuals, you are ready to practice your presentation and delivery techniques.

Practice. Familiarize yourself with the sequence of the material—major topics, notes, and visuals—in your outline. Once you feel comfortable with the content, you are ready to practice the presentation.

PRACTICE ON YOUR FEET AND OUT LOUD. Try to practice in the room where you will give the presentation. Practicing on-site helps you get the feel of the room: the lighting, the arrangement of the chairs, the position of electrical outlets and switches, and so forth. Practice out loud to gauge the length of your presentation, to uncover problems such as awkward transitions, and to eliminate verbal tics (such as "um," "you know," and "like").

PRACTICE WITH YOUR VISUALS AND TEXT. Integrate your visuals into your practice sessions to help your presentation go more smoothly. Operate the equipment (computer, slide projector, or overhead projector) until you are comfortable with it. Decide if you want to use a remote control or wireless mouse or if you want to have someone else advance your slides. Even if things go wrong, being prepared and practiced will give you the confidence and poise to continue.

Delivery Techniques That Work. Your delivery is both audible and visual. In addition to your words and message, your nonverbal communication affects your audience. Be animated—your words have impact and staying power when they are delivered with physical and vocal animation. If you want listeners to share your point of view, show enthusiasm for your topic. The most common delivery techniques include making eye contact; using movement and gestures; and varying voice inflection, projection, and pace.

EYE CONTACT. The best way to establish rapport with your audience is through eye contact. In a large audience, directly address those people who seem most responsive to you in different parts of the room. Doing that helps you establish rapport with your listeners by holding their attention and gives you important visual cues that let you know how your message is being received. Do the listeners seem engaged and actively listening? Based on your observations, you may need to adjust the pace of your presentation.

MOVEMENT. Animate the presentation with physical movement. Take a step or two to one side after you have been talking for a minute or so. That type of movement is most effective at transitional points in your presentation between major topics or after pauses or emphases. Too much movement, however, can be distracting, so try not to pace.

Another way to integrate movement into your presentation is to walk to the screen and point to the visual as you discuss it. Touch the screen with the pointer and then turn back to the audience before beginning to speak (remember the three *t*'s: touch, turn, and talk).

GESTURES. Gestures both animate your presentation and help communicate your message. Most people gesture naturally when they talk; nervousness, however, can inhibit gesturing during a presentation. Keep one hand free and use that hand to gesture.

VOICE. Your voice can be an effective tool in communicating your sincerity, enthusiasm, and command of your topic. Use it to your advantage to project your credibility. *Vocal inflection* is the rise and fall of your voice at different times, such as the way your voice naturally rises at the end of a question ("You want it *when?*"). A conversational delivery and eye contact promote the feeling among members of the audience that you are addressing them directly. Use vocal inflection to highlight differences between key and subordinate points in your presentation.

PROJECTION. Most presenters think they are speaking louder than they are. Remember that your presentation is ineffective for anyone in the audience who cannot hear you. If listeners must strain to hear you, they may give up trying to listen. Correct projection problems by practicing out loud with someone listening from the back of the room.

PACE. Be aware of the speed at which you deliver your presentation. If you speak too fast, your words will run together, making it difficult for your audience to follow. If you speak too slowly, your listeners will become impatient and distracted.

Presentation Anxiety. Everyone experiences nervousness before a presentation. Survey after survey reveals that for most people dread of public speaking ranks among their top five fears. Instead of letting fear inhibit you, focus on channeling your nervous energy into a helpful stimulant. Practice will help you, but the best way to master anxiety is to know your topic thoroughly — knowing what you are going to say and how you are going to say it will help you gain confidence and reduce anxiety as you become immersed in your subject.

> **WRITER'S CHECKLIST** Preparing for and Delivering a Presentation
>
> ✔ Familiarize yourself with the surroundings at the location of the presentation ahead of time.
> ✔ Practice your presentation with visuals at the location and in front of listeners if possible.
> ✔ Prepare a set of notes that will trigger your memory during the presentation.
> ✔ Make as much eye contact as possible with your audience to establish rapport and maximize opportunities for audience feedback.
> ✔ Animate your delivery by integrating movement, gestures, and vocal inflection into your presentation. However, keep your movements and speech patterns natural.

Writer's Checklist: Preparing for and Delivering a Presentation (continued)

- Speak loudly and slowly enough to be heard and understood.
- Do not read the text on your visuals word for word; explain the key points in detail.

For information and tips on communicating with cross-cultural audiences, see **global communication**, **global graphics**, and **international correspondence**.

principal / principle

Principal, meaning "an amount of money on which interest is earned or paid" or "a chief official in a school or court proceeding," is sometimes confused with *principle*, which means "a basic truth or belief."

- ▶ The bank will pay 3.5 percent on the *principal*.
- ▶ He sent a letter to the *principal* of the high school.
- ▶ She is a person of unwavering *principles*.

Principal is also an adjective, meaning "main" or "primary." ("My *principal* objection is that it will be too expensive.")

process explanation

A process explanation may describe the steps in a process, an operation, or a procedure, such as the steps necessary to design and manufacture a product. The **introduction** often presents a brief overview of the process or lets **readers** know why it is important for them to become familiar with the process you are explaining. Be sure to define terms that readers might not understand and provide **visuals** to clarify the process. See also **defining terms** and **instructions**.

In describing a process, use transitional words and phrases to create unity within **paragraphs**, and select **headings** to mark the **transition** from one step to the next. Figure T–3 describes both the process and the procedure of "Monocular Recording." Visuals, like the one used in Figure T–3, can help convey your message.

progress and activity reports

Progress reports provide details on the tasks completed for major workplace projects, whereas *activity reports* focus on the ongoing work of individual employees. Both are sometimes called *status reports*. Although many organizations use standardized forms for these **reports**, the content and structure shown in Figures P–4 and P–5 are typical.

Progress Reports

A progress report provides information to decision-makers about the status of a project—whether it is on schedule and within budget. Progress reports are often submitted by a contracting company to a client company, as shown in Figure P–4. They are used mainly for projects that involve many steps over a period of time and are issued at regular intervals to describe what has been done and what remains to be done. Progress reports help projects run smoothly by helping managers assign work, adjust schedules, allocate budgets, and order supplies and equipment. All progress reports for a particular project should have the same **format**.

The **introduction** to the first progress report should identify the project, methods used, necessary materials, expenditures, and completion date. Subsequent reports summarize the progress achieved since the preceding report and list the steps that remain to be taken. The body of the progress report should describe the project's status, including details such as schedules and costs, a statement of the work completed, and perhaps an estimate of future progress. The report ends with **conclusions** and recommendations about changes in the schedule, materials, techniques, and other information important to the project.

Activity Reports

Within an organization, employees often submit activity reports to managers on the status of ongoing projects. Managers may combine the activity reports of several individuals or teams into larger activity reports and, in turn, submit those larger reports to their own managers. The activity report shown in Figure P–5 was submitted by a manager (Wayne Tribinski) who supervises 11 employees; the reader of the report (Kathryn Hunter) is Tribinski's manager.

Because the activity report is issued periodically (usually monthly) and contains material familiar to its **readers**, it normally needs no introduction or conclusion, although it may need a brief opening to provide **context**. Although the format varies from company to company, the following sections are typical: Current Projects, Current Problems, Plans for the Next Period, and Current Staffing Level (for managers).

Hobard Construction Company

9032 Salem Avenue
Lubbock, TX 79409

www.hobardcc.com
(808) 769-0832
Fax: (808) 769-5327

August 14, 2009

Walter M. Wazuski
County Administrator
109 Grand Avenue
Manchester, NH 03103

Dear Mr. Wazuski:

Subject: Progress Report 8 for July 1–July 29, 2009

The renovation of the County Courthouse is progressing on schedule and within budget. Although the cost of certain materials is higher than our original bid indicated, we expect to complete the project without exceeding the estimated costs because the speed with which the project is being completed will reduce overall labor expenses.

Costs
Materials used to date have cost $78,600, and labor costs have been $193,000 (including some subcontracted plumbing). Our estimate for the remainder of the materials is $59,000; remaining labor costs should not exceed $64,000.

Work Completed
As of July 29, we had finished the installation of the circuit-breaker panels and meters, the level-one service outlets, and all the subfloor wiring. The upgrading of the courtroom, the upgrading of the records-storage room, and the replacement of the air-conditioning units are in the preliminary stages.

Work Scheduled
We have scheduled the upgrading of the courtroom to take place from August 31 to October 7, the upgrading of the records-storage room from October 12 to November 18, and the replacement of the air-conditioning units from November 23 to December 16. We see no difficulty in having the job finished by the scheduled date of December 23.

Sincerely yours,

Tran Nuguélen
ntran@hobardcc.com

FIGURE P–4. Progress Report

INTEROFFICE MEMO

Date: June 8, 2009
To: Kathryn Hunter, Director of IT
From: Wayne Tribinski, Manager, Applications Programs   *WT*
Subject: Activity Report for May 2009

We are dealing with the following projects and problems, as of May 31.

Projects
1. For the *Software Training Mailing Campaign*, we anticipate producing a set of labels for mailing software training information to customers by June 12.
2. The *Search Project* is on hold until the PL/I training has been completed, probably by the end of June.
3. The project to provide a database for the *Information Management System* has been expanded in scope to provide a database for all training activities. We are rescheduling the project to take the new scope into account.

Problems

The *Information Management System* has been delayed. The original schedule was based on the assumption that a systems analyst who was familiar with the system would work on this project. Instead, the project was assigned to a newly hired systems analyst who was inexperienced and required much more learning time than expected.

Bill Michaels, whose activity report is attached, is correcting a problem in the *CNG Software*. This correction may take a week.

Plans for Next Month
- Complete the *Software Training Mailing Campaign*.
- Resume the *Search Project*.
- Restart the project to provide a database on information management with a schedule that reflects its new scope.
- Write a report to justify the addition of two software developers to my department.
- Congratulate publicly the recipients of Meritorious Achievement Awards: Bill Thomasson and Nancy O'Rourke.

Current Staffing Level

Current staff: 11
Open requisitions: 0

Attachment

FIGURE P–5. Activity Report

pronoun reference

A <u>pronoun</u> should refer clearly to a specific antecedent. Avoid vague and uncertain references.

> We got the account, which was a big one, after we wrote the proposal. ~~It was a big one.~~

For <u>coherence</u>, place pronouns as close as possible to their antecedents—distance increases the likelihood of <u>ambiguity</u>.

> The office building next to City Hall, praised for its architectural design, is ~~is praised for its architectural design~~.

A general (or broad) reference or one that has no real antecedent is a problem that often occurs when the word *this* is used by itself.

> He deals with personnel problems in his work. This experience helps him in his personal life.

Another common problem is a hidden reference, which has only an implied antecedent.

> A high-lipid, low-carbohydrate diet is "ketogenic" because it favors ~~their~~ the formation of ketone bodies.

Do not repeat an antecedent in parentheses following the pronoun. If you feel you must identify the pronoun's antecedent in that way, rewrite the sentence.

AWKWARD	The senior partner first met Bob Evans when he (Evans) was a trainee.
IMPROVED	Bob Evans was a trainee when the senior partner first met him.

For advice on avoiding pronoun-reference problems with gender, see <u>biased language</u>.

pronouns ·

DIRECTORY

Case 407	Number 409
Gender 409	Person 409

A pronoun is a word that is used as a substitute for a **noun** (the noun for which a pronoun substitutes is called the *antecedent*). Using pronouns in place of nouns relieves the monotony of repeating the same noun over and over. See also **pronoun reference**.

Personal pronouns refer to the person or people speaking (*I, me, my, mine*; *we, us, our, ours*); the person or people spoken to (*you, your, yours*); or the person, people, or thing(s) spoken of (*he, him, his*; *she, her, hers*; *it, its*; *they, them, their, theirs*). See also **person** and **point of view**.

▶ If *their* figures are correct, *ours* must be in error.

Demonstrative pronouns (*this, these, that, those*) indicate or point out the thing being referred to.

▶ *This* is my desk. *These* are my coworkers. *That* will be a difficult job. *Those* are incorrect figures.

Relative pronouns (*who, whom, which, that*) perform a dual function: (1) They take the place of nouns and (2) they connect and establish the relationship between a dependent **clause** and its main clause.

▶ The department manager decided *who* would be hired.

Interrogative pronouns (*who, whom, what, which*) are used to ask questions.

▶ *What* is the trouble?

Indefinite pronouns specify a class or group of persons or things rather than a particular person or thing (*all, another, any, anyone, anything, both, each, either, everybody, few, many, most, much, neither, nobody, none, several, some, such*).

▶ Not *everyone* liked the new procedures; *some* even refused to follow them.

A *reflexive pronoun*, which always ends with the suffix *-self* or *-selves*, indicates that the subject of the sentence acts upon itself. See also **sentence construction**.

▶ The electrician accidentally shocked *herself*.

The reflexive pronouns are *myself, yourself, himself, herself, itself, oneself, ourselves, yourselves,* and *themselves. Myself* is not a substitute for *I* or *me* as a personal pronoun.

▶ Victor and ~~myself~~ *I* completed the report on time.

▶ The assignment was given to Ingrid and ~~myself~~ *me*.

Intensive pronouns are identical in form to the reflexive pronouns, but they perform a different function: Intensive pronouns emphasize their antecedents.

▶ I *myself* asked the same question.

Reciprocal pronouns (*one another*, *each other*) indicate the relationship of one item to another. *Each other* is commonly used when referring to two persons or things and *one another* when referring to more than two.

▶ Lashell and Kara work well with *each other*.

▶ The crew members work well with *one another*.

Case

Pronouns have forms to show the subjective, objective, and possessive cases.

SINGULAR	SUBJECTIVE	OBJECTIVE	POSSESSIVE
First person	I	me	my, mine
Second person	you	you	your, yours
Third person	he, she, it	him, her, it	his, her, hers, its

PLURAL	SUBJECTIVE	OBJECTIVE	POSSESSIVE
First person	we	us	our, ours
Second person	you	you	your, yours
Third person	they	them	their, theirs

> **ESL TIPS** for Using Possessive Pronouns
>
> In many languages, possessive pronouns agree in number and gender with the nouns they modify. In English, however, possessive pronouns agree in number and gender with their antecedents. Check your writing carefully for agreement between a possessive pronoun and the word, phrase, or clause to which it refers.
>
> ▶ The *woman* brought *her* brother a cup of soup.
>
> ▶ Robert sent *his* mother flowers on Mother's Day.

A pronoun that functions as the subject of a clause or sentence is in the subjective **case** (*I*, *we*, *he*, *she*, *it*, *you*, *they*, *who*). The subjective case is also used when the pronoun follows a linking **verb**.

- *She* is my boss.
- My boss is *she*.

A pronoun that functions as the object of a verb or **preposition** is in the objective case (*me, us, him, her, it, you, them, whom*).

- Ms. Davis hired Tom and *me*. [object of verb]
- Between *you* and *me*, she's wrong. [object of preposition]

A pronoun that expresses ownership is in the **possessive case** (*my, mine, our, ours, his, her, hers, its, your, yours, their, theirs, whose*).

- He took *his* notes with him on the business trip.
- We took *our* notes with us on the business trip.

A pronoun **appositive** takes the case of its antecedent.

- Two systems analysts, Joe and *I*, were selected to represent the company.
  [*Joe and I* is in apposition to the subject, *two systems analysts*, and must therefore be in the subjective case.]
- The manager selected two representatives — Joe and *me*.
  [*Joe and me* is in apposition to *two representatives*, which is the object of the verb, *selected*, and therefore must be in the objective case.]

If you have difficulty determining the case of a compound pronoun, try using the pronoun singly.

- In his letter, Eldon mentioned *him* and *me*.
  In his letter, Eldon mentioned *him*.
  In his letter, Eldon mentioned *me*.
- *They* and *we* must discuss the terms of the merger.
  They must discuss the terms of the merger.
  We must discuss the terms of the merger.

When a pronoun modifies a noun, try it without the noun to determine its case.

- [*We/Us*] pilots fly our own planes.
  We fly our own planes.
  [You would not write, "*Us* fly our own planes."]

▶ He addressed his remarks directly to [*we/us*] technicians.
He addressed his remarks directly to *us*.
[You would not write, "He addressed his remarks directly to *we*."]

Gender

A pronoun must agree in gender with its antecedent. A problem sometimes occurs because the masculine pronoun has traditionally been used to refer to both sexes. To avoid the sexual bias implied in such usage, use *he or she* or the plural form of the pronoun, *they*.

▶ ~~Each~~ *All* may stay or go as ~~he chooses.~~ *they choose.*

As in this example, when the singular pronoun (*he*) changes to the plural (*they*), the singular indefinite pronoun (*each*) must also change to its plural form (*all*). See also **biased language**.

Number

Number is a frequent problem with only a few indefinite pronouns (*each*, *either*, *neither*, and those ending with *-body* or *-one*, such as *anybody*, *anyone*, *everybody*, *everyone*, *nobody*, *no one*, *somebody*, *someone*) that are normally singular and so require singular verbs and are referred to by singular pronouns.

▶ As *each member arrives* for the meeting, please hand *him or her* a copy of the confidential report. *Everyone* must return the copy before *he or she* leaves. *Everybody* on the committee *understands* that *neither* of our major competitors *is* aware of the new process we have developed.

Person

Third-person personal pronouns usually have antecedents.

▶ Gina presented the report to the members of the board of directors. *She* [Gina] first summarized *it* [the report] for *them* [the directors] and then asked for questions.

First- and second-person personal pronouns do not normally require antecedents.

▶ *I* like my job.
▶ *You* were there at the time.
▶ *We* all worked hard on the project.

proofreaders' marks

Publishers have established symbols called *proofreaders' marks* that writers and editors use to communicate in the production of publications. Familiarity with those symbols makes it easy for you to communicate your changes to others. Figure P–6 lists standard proofreaders' marks. For using Comment and Track Changes in word-processing programs, see *Digital Tip: Incorporating Tracked Changes* on page 489.

MARK/SYMBOL	MEANING	EXAMPLE	CORRECTED TYPE
	Delete	the ~~manager's~~ report	the report
∧	Insert	the ∧ report	the manager's report
dots (stet)	Let stand	the manager's report	the manager's report
≡ (cap)	Capitalize	the monday meeting	the Monday meeting
/ (lc)	Lowercase	the Monday Meeting	the Monday meeting
∽ (tr)	Transpose	the cover lettre	the cover letter
⌒	Close space	a loud speaker	a loudspeaker
#	Insert space	a loudspeaker	a loud speaker
¶	Paragraph	...report. The meeting...	...report. The meeting...
⌒	Run in with previous line or paragraph	...report. The meeting...	...report. The meeting...
___ (ital)	Italicize	the New York Times	the *New York Times*
∼∼ (bf)	Boldface	Use boldface sparingly.	Use **boldface** sparingly.
⊙	Insert period	I wrote the e-mail	I wrote the e-mail.
⌃	Insert comma	However we cannot...	However, we cannot...
=	Insert hyphen	clear cut decision	clear-cut decision
1/M	Insert em dash	Our goal productivity	Our goal—productivity
⌄ or :/	Insert colon	We need the following	We need the following:
⌄ or ;/	Insert semicolon	we finished we achieved	we finished; we achieved
⌄ ⌄	Insert quotation marks	He said, I agree.	He said, "I agree."
⌄	Insert apostrophe	the managers report	the manager's report

FIGURE P–6. Proofreaders' Marks

proofreading

Proofreading is essential whether you are writing a brief **e-mail** or a **résumé**. Grammar checkers and spell checkers are important aids to proofreading, but they can make writers overconfident. If a typographical error results in a legitimate English word (for example, *coarse* instead of *course*), the spell checker will not flag the misspelling. You may find some of the tactics discussed in **revision** useful when proofreading; in fact, you may find passages during proofreading that will require further revision.

Whether the material you proofread is your own writing or that of someone else, consider proofreading in several stages. Although you need to tailor the stages to the specific document and to your own problem areas, the following *Writer's Checklist* should provide a useful starting point for proofreading.

WRITER'S CHECKLIST   Proofreading in Stages

FIRST-STAGE REVIEW
- Appropriate **format**, as for **reports** or **correspondence**
- Consistent style, including **headings**, terminology, spacing, fonts
- Correct numbering of figures and **tables**

SECOND-STAGE REVIEW
- Specific **grammar** and **usage** problems
- Appropriate **punctuation**
- Correct **abbreviations** and **capitalization**
- Correct **spelling** (especially names and places)
- Complete Web or e-mail addresses
- Accurate data in tables and **lists**
- Cut-and-paste errors; for example, a result of moved or deleted text and **numbers**

FINAL-STAGE REVIEW
- Survey of your overall goals: **audience** needs and **purpose**
- Appearance of the document (see **layout and design**)
- Review by a trusted colleague, especially for crucial documents (see **collaborative writing**)

Consider using standard proofreaders' marks for proofreading someone else's document.

DIGITAL TIP

Proofreading for Format Consistency

Viewing whole pages on-screen is an effective way to check formatting, spacing, and typographical consistency, as well as the general appearance of documents. For comparing layouts, view multiple pages or "tile" separate documents side by side. For instructions, see *bedfordstmartins.com/alredtech* and select *Digital Tips*, "Proofreading for Format Consistency."

proposals

DIRECTORY
Proposal Contexts and Strategies   412
  Audience and Purpose   413
  Solicited and Unsolicited Proposals   413
  Internal and External Proposals   414
  Project Management   414
Writer's Checklist: Writing Persuasive Proposals   417
Proposal Forms   418
  Informal Proposal Structure   418
  Formal Proposal Structure   418
Proposal Types   419
  Sales Proposals   420
  Grant and Research Proposals   421

A proposal is a document written to persuade **readers** that what is proposed will benefit them by solving a problem or fulfilling a need. When you write a proposal, therefore, you must convince readers that they need what you are proposing, that it is practical and appropriate, and that you are the right person or organization to provide the proposed product or service. See also **persuasion** and **"you" viewpoint**.

Proposal Contexts and Strategies

In a proposal, support your assertions with relevant facts, statistics, and examples. Your supporting evidence must lead logically to your proposed plan of action or solution. Cite relevant sources of information that provide strong credibility to your argument. Avoid ambiguity, do not wander from your main point, and never make false claims. See **ethics in writing**.

Audience and Purpose. Whether you send a proposal inside or outside your organization, readers will evaluate your plan on how well you answer their questions about what you are proposing to do, how you plan to do it, how much it will cost, and how it will benefit them.

Because proposals often require more than one level of approval, take into account all the readers in your **audience**. Consider especially their levels of technical knowledge of the subject. For example, if your primary reader is an expert on your subject but a supervisor who must also approve the proposal is not, provide an **executive summary** written in nontechnical language for the supervisor. You might also include a **glossary** of terms used in the body of the proposal or an **appendix** that explains highly detailed information in nontechnical language. If your primary reader is not an expert but a supervisor is, write the proposal with the nonexpert in mind and include an appendix that contains the technical details.

Writing a persuasive and even complex proposal can be simplified by composing a concise statement of **purpose**—the exact problem or opportunity that your proposal is designed to address and how you plan to persuade your readers to accept what you propose. Composing a purpose statement before outlining and writing your proposal will also help you and any collaborators understand the direction, **scope**, and goals of your proposal.

Solicited and Unsolicited Proposals. *Solicited proposals* are prepared in response to a request for goods or services. Such proposals usually follow the format prescribed by the procuring organization or agency, which issues a **request for proposals** (RFP) or an invitation for bids (IFB).

An RFP often defines a need or problem and allows those who respond to propose possible solutions. The procuring organization generally distributes an RFP to several predetermined vendors. The RFP usually outlines the specific requirements for the ideal solution. For example, if an organization needs an accounting system, it may require the proposed system to create customized reports. The RFP also may contain specific formatting requirements, such as page length, font type and size, margin widths, **headings**, numbering systems, sections, and appendix items. When responding to RFPs, you should follow their requirements exactly—proposals that do not provide the required information or do not follow the required format may be considered "nonresponsive" and immediately rejected.

In contrast to an RFP, an IFB is commonly issued by federal, state, and local government agencies to solicit bids on clearly defined products or services. An IFB is restrictive, binding the bidder to produce an item or a service that meets the exact requirements of the organization issuing the IFB. The goods or services are defined in the IFB by references to performance standards stated in technical **specifications**.

Bidders must be prepared to prove that their product will meet all requirements of the specifications. The procuring organization generally publishes its IFB on its Web site or in a specialized venue, such as Federal Business Opportunities at *www.fedbizopps.gov*. Like RFPs, IFBs usually have specific format requirements; proposals that do not follow the required format can be rejected without review.

Unsolicited proposals are submitted to a company or department without a prior request for a proposal. Companies or departments often operate for years with a problem they have never recognized (unnecessarily high maintenance costs, for example, or poor inventory-control methods). Many unsolicited proposals are preceded by an inquiry from a salesperson to determine potential interest and need. If you receive a positive response, you would conduct a detailed study of the prospective customer's needs to determine whether you can be of help and, if so, exactly how. You would then prepare your proposal on the basis of your study.

Internal and External Proposals. *Internal proposals*, which can be either solicited or unsolicited, are written by employees of an organization for decision-makers inside that organization. The level of formality of internal proposals often depends on the frequency with which they are written and the degree of change proposed. Routine proposals are typically informal and involve small spending requests, requests for permission to hire new employees or increase salaries, and requests to attend conferences or purchase new equipment. Special-purpose proposals are usually more formal and involve requests to commit relatively large sums of money. They have various names, but a common designation is a *capital appropriations request* or a *capital appropriations proposal*. Figure P–7 shows a special-purpose internal proposal.

External proposals are prepared for clients, customers, or other decision-makers outside a company or an organization. They are either solicited or unsolicited. External proposals are almost always written as formal proposals.

Project Management. Proposal writers are often faced with writing high-quality, persuasive proposals under tight organizational deadlines. Dividing the task into manageable parts is the key to accomplishing your goals, especially when proposals involve substantial **collaborative writing**. For example, you might set deadlines for completing various proposal sections or stages of the writing process.* See also "Five Steps to Successful Writing."

*For help with project management, a good source is JoAnn T. Hackos, *Information Development: Managing Your Documentation Projects, Portfolio, and People* (Indianapolis, IN: Wiley, 2007).

ABO, Inc.
Interoffice Memo

To: Joan Marlow, Director, Human Resources Division

From: Leslie Galusha, Chief *LG*
Employee Benefits Department

Date: June 15, 2009

Subject: Employee Fitness and Health-Care Costs

Health-care and workers'-compensation insurance costs at ABO, Inc., have risen 100 percent over the last five years. In 2004, costs were $5,675 per employee per year; in 2009, they have reached $11,560 per employee per year. This doubling of costs mirrors a national trend, with health-care costs anticipated to continue to rise at the same rate for the next ten years. Controlling these escalating expenses will be essential. They are eating into ABO's profit margin because the company currently pays 70 percent of the costs for employee coverage.

Healthy employees bring direct financial benefits to companies in the form of lower employee insurance costs, lower absenteeism rates, and reduced turnover. Regular physical exercise promotes fit, healthy people by reducing the risk of coronary heart disease, diabetes, osteoporosis, hypertension, and stress-related problems. I propose that to promote regular, vigorous physical exercise for our employees, ABO implement a health-care program that focuses on employee fitness....

Problem of Health-Care Costs
The U.S. Department of Health and Human Services recently estimated that health-care costs in the United States will triple by the year 2020. Corporate expenses for health care are rising at such a fast rate that, if unchecked, in seven years they will significantly erode corporate profits.

Researchers agree that people who do not participate in a regular and vigorous exercise program incur double the health-care costs and are hospitalized 30 percent more days than people who exercise regularly. Nonexercisers are also 41 percent more likely to submit medical claims over $10,000 at some point during their careers than are those who exercise regularly.

FIGURE P–7. Special-Purpose Internal Proposal (Introduction)

Joan Marlow   2   June 15, 2009

My study of Tenneco, Inc., found that the average health-care claim for unfit men was $2,006 per illness compared with an average claim of $862 for those who exercised regularly. For women, the average claim for those who were unfit was $2,535, more than double the average claim of $1,039 for women who exercised. In addition, Control Data Corporation found that each non-exerciser cost the company an extra $515 a year in health-care expenses.

These figures are further supported by data from independent studies. A model created by the National Institutes of Health (NIH) estimates that the average white-collar company could save $596,000 annually in medical costs (per 1,000 employees) just by promoting wellness. NIH researchers estimated that for every $1 a firm invests in a health-care program, it saves up to $3.75 in health-care costs. Another NIH study of 667 insurance-company employees showed savings of $2.65 million over a five-year period. The same study also showed a 400 percent drop in absentee rates after the company implemented a company-wide fitness program.

Possible Solutions for ABO
The benefits of regular, vigorous physical activity for employees and companies are compelling. To achieve these benefits at ABO, I propose that we choose from one of two possible options: Build in-house fitness centers at our warehouse facilities, or offer employees several options for membership at a national fitness club. The following analysis compares . . .

Conclusion and Recommendation
I recommend that ABO, Inc., participate in the corporate membership program at AeroFitness Clubs, Inc., by subsidizing employee memberships. By subsidizing memberships, ABO shows its commitment to the importance of a fit workforce. Club membership allows employees at all five ABO warehouses to participate in the program. The more employees who participate, the greater the long-term savings in ABO's health-care costs. Building and equipping fitness centers at all five warehouse sites would require an initial investment of nearly $2.5 million. These facilities would, in addition, occupy valuable floor space—on average, 4,000 square feet at each warehouse. Therefore, this option would be very costly.

FIGURE P–7. Special-Purpose Internal Proposal (*continued*) (Body)

Joan Marlow				3				June 15, 2009

Enrolling employees in the corporate program at AeroFitness would allow them to receive a one-month free trial membership. Those interested in continuing could then join the club and pay half of the one-time membership fee of $900 and receive a 30-percent discount on the $600 yearly fee. The other half of the membership fee ($450) would be paid for by ABO. If employees leave the company, they would have the option of purchasing ABO's share of the membership to continue at AeroFitness or selling their half of the membership to another ABO employee wishing to join AeroFitness.

Implementing this program will help ABO, Inc., reduce its health-care costs while building stronger employee relations by offering employees a desirable benefit. If this proposal is adopted, I have some additional thoughts about publicizing the program to encourage employee participation. I look forward to discussing the details of this proposal with you and answering any questions you may have.

FIGURE P–7. Special-Purpose Internal Proposal (*continued*) (Conclusion)

WRITER'S CHECKLIST Writing Persuasive Proposals

- Analyze your audience carefully to determine how to best meet your readers' needs or requirements.
- Write a concise purpose statement at the outset to clarify your proposal's goals.
- Divide the writing task into manageable segments and develop a timeline for completing tasks.
- Review the descriptions of proposal contexts, structure, and types in this entry.
- Focus on the proposal's benefits to readers and anticipate their questions or objections.
- Incorporate evidence to support the claims of your proposal.
- Select an appropriate, visually appealing format. See **layout and design**.
- Use a confident, positive **tone** throughout the proposal.

Proposal Forms

Proposals are written within a specific **context**. Understanding the context, as described in that entry, will help you determine the most appropriate writing strategy as well as the proposal's length, formality, and structure.

Informal Proposal Structure. Informal proposals are relatively short (about five pages or fewer) and typically consist of an **introduction**, a body, and a **conclusion**.

   INTRODUCTION.   The introduction should define the purpose and scope of your proposal as well as the problem you propose to address or solve. You may also include any relevant background or context that will help readers appreciate the benefits of what you will propose in the body.

   BODY.   The body should offer the details of your plan to address or solve the problem and explain (1) what service or product you are offering; (2) how you will perform the work and what special materials you may use; (3) the schedule you plan to follow that designates when each phase of the project will be completed; and (4) if appropriate, a breakdown of project costs.

   CONCLUSION.   The conclusion should persuasively resell your proposal by emphasizing the benefits of your plan, solution, product, or service over any competing ideas or projects. You may also need to include details about the time period during which the proposal is valid. Effective conclusions show confidence in your proposal, your appreciation for the opportunity to submit the proposal, and your willingness to provide further information, as well as encouraging your reader to act on your proposal.

Formal Proposal Structure.   Proposals longer than five pages are often called *formal proposals* and typically include front matter and back matter. The number of sections in a proposal depends on the audience, the purpose, and the scope of the proposal, or on the specific requirements outlined in an RFP or IFB. If you are responding to an RFP or IFB, follow the proposal organization and format requirements exactly as stated; otherwise your proposal may be considered noncompliant and may be rejected without review (see Solicited and Unsolicited Proposals on pages 413–14). If you are not responding to an RFP or IFB, sections of formal proposals can often be grouped into front matter, body, and back matter. (For more information on many of the following components, see **formal reports**.)

FRONT MATTER

- *Cover Letter or Letter of Transmittal.* In the **cover letter**, express appreciation for the opportunity to submit your proposal, any help from the customer (or decision-maker), and any prior positive associations with the customer. Then summarize the proposal's recommendations and express confidence that they will satisfy the customer's or decision-maker's needs.
- *Title Page.* Include the **title** of the proposal, the date, the name and logo of the organization to which it is being submitted, and your company name and logo.
- *Table of Contents.* Include a **table of contents** in longer proposals to guide readers to important sections, which should be listed according to beginning page numbers.
- *List of Figures.* If your proposal has six or more figures, include a list of figures with captions as well as figure and page numbers. See **visuals**.

BODY

- *Executive Summary.* Briefly summarize the proposal's highlights in persuasive, nontechnical language for decision-makers.
- *Introduction.* (See the sections *Sales Proposals* and *Grant and Research Proposals* for content typically included in this section.)
- *Body.* (See the sections *Sales Proposals* and *Grant and Research Proposals* for content typically included in this section.)
- *Conclusion.* (See the sections *Sales Proposals* and *Grant and Research Proposals* for content typically included in this section.)

BACK MATTER

- *Appendixes.* Provide résumés of key personnel or material of interest to some readers, such as statistical analyses, organizational charts, and workflow diagrams.
- *Bibliography.* List sources consulted in preparing the proposal, such as research studies, specifications, and standards. See **bibliographies** and **documenting sources**.
- *Glossary.* If your proposal contains terms that will be unfamiliar to your intended audience, list and define them in the glossary.

Proposal Types

Proposals are written for many specific purposes, but the two most common types of proposals are *sales proposals* and *grant and research proposals*.

Sales Proposals. A sales proposal is a company's offer to provide specific goods or services to a potential buyer within a specified period of time and for a specified price. A sales proposal must demonstrate above all that the prospective customer's purchase of the seller's products or services will solve a problem, improve operations, or offer other substantial benefits.

Sales proposals vary greatly in size and sophistication — from several pages written by one person, to dozens of pages written collaboratively by several people, to hundreds of pages written by a team of professional proposal writers. A short sales proposal might bid for the construction of a single home, a moderate-length proposal might bid for the installation of a computer network, and a large proposal might bid for the construction of a multimillion-dollar water-purification system or shopping center.

ETHICS NOTE Always keep in mind that, once submitted, a sales proposal is a legally binding document that promises to offer goods or services within a specified time and for a specified price. ◆

Your first task in writing a sales proposal is to find out exactly what your prospective customer needs. Then determine whether your organization can satisfy that customer's needs. If appropriate, compare your company's strengths with those of competing firms, determine your advantages over them, and emphasize those advantages in your proposal.

If you are creating a sales proposal in response to an RFP or IFB, be sure to comply with information and section requirements exactly as stated — do not include irrelevant or extraneous information or you may risk having your proposal rejected. If you are not responding to an RFP or IFB, your sales proposal may include some or all of the following sections (keep in mind that formal sales proposals will also include front matter and back matter):

- *Introduction*. Explain the reasons for the proposal, emphasizing reader benefits.
- *Background or Problem*. Describe the problem or opportunity your proposal addresses. To make your proposal more persuasive, your problem statement should illustrate how your proposal will benefit your client's organization.
- *Product Description*. If your proposal offers products as well as services, include a general description of the products and any technical specifications.
- *Detailed Solutions (Rationale)*. Explain in a detailed section that will be read by technical specialists exactly how you plan to do what you are proposing.

- *Cost Analysis*. Itemize the estimated costs of all the products and services you are offering.
- *Delivery Schedule or Work Plan*. Outline how you will accomplish the work and show a timetable for each phase of the project.
- *Staffing*. Summarize the expertise (education, experience, and certifications) of key personnel who will work on the project. Include their résumés in an appendix.
- *Site Preparation*. If your recommendations include modifying the customer's physical facilities, include a site-preparation description that details the required modifications.
- *Training Requirements*. If the products and services you are proposing require training the customer's employees, specify the required training and its cost.
- *Statement of Responsibilities*. To prevent misunderstandings about what your and your customer's responsibilities will be, state those responsibilities.
- *Organizational Sales Pitch*. Describe your company, its history, and its present position in the industry. An organizational sales pitch is designed to sell your company and its general capability in the field. It promotes the company and concludes the proposal on a positive, persuasive note.
- *Authorization Request and Deadline*. Close with a request for approval and a deadline that explains how long the proposed prices are valid.
- *Conclusion*. Include a persuasive conclusion that summarizes the proposal's key points and stresses your company's strong points.

Figure P–8 shows sections from a major sales proposal (shown in full at the following Web Link).

> **WEB LINK** **Sample Sales Proposals**
>
> For a complete and an annotated version of Figure P–8, as well as additional sample proposals and the RFP to which Figure P–8 responded, see *bedfordstmartins.com/alredtech* and select *Model Documents Gallery*.

Grant and Research Proposals. Grant proposals request funds or material goods to support a specific project or cause. Grants are not loans and usually do not have to be repaid. Many government and private agencies solicit and fund research or grant proposals. A scientist, for example, may write a grant to the National Institutes of Health to request a specific sum of money to conduct research on a new cancer drug.

The Waters Corporation
17 North Waterloo Blvd.
Tampa, Florida 33607
Phone: (813) 919-1213 Fax: (813) 919-4411
www.waters.com

September 2, 2009

Mr. John Yeung, General Manager
Cookson's Retail Stores, Inc.
101 Longuer Street
Savannah, Georgia 31499

Dear Mr. Yeung:

The Waters Corporation appreciates the opportunity to respond to Cookson's Request for Proposals dated July 20, 2009. We would like to thank Mr. Becklight, Director of your Management Information Systems Department, for his invaluable contributions to the study of your operations that we conducted before preparing our proposal.

It has been Waters's privilege to provide Cookson's with retail systems and equipment since your first store opened many years ago. Therefore, we have become very familiar with your requirements as they have evolved during the expansion you have experienced since that time. Waters's close working relationship with Cookson's has resulted in a clear understanding of Cookson's philosophy and needs.

Our proposal describes a Waters Interactive Terminal/Retail Processor System designed to meet Cookson's network and processing needs. It will provide all of your required capabilities, from the point-of-sale operational requirements at the store terminals to the host processor. The system uses the proven Retail III modular software, with its point-of-sale applications, and the superior Interactive Terminal, with its advanced capabilities and design. This system is easily installed without extensive customer reprogramming.

FIGURE P–8. Sales Proposal (Cover Letter)

Mr. J. Yeung
Page 2
September 2, 2009

The Waters Interactive Terminal/Retail Processor System, which is compatible with much of Cookson's present equipment, not only will answer your present requirements but will provide the flexibility to add new features and products in the future. The system's unique hardware modularity, efficient microprocessor design, and flexible programming capability greatly reduce the risk of obsolescence.

Thank you for the opportunity to present this proposal. You may be sure that we will use all the resources available to the Waters Corporation to ensure the successful implementation of the new system.

Sincerely yours,

Janet A. Curtain

Janet A. Curtain
Executive Account Manager
General Merchandise Systems
(jcurtain@waters.com)

Enclosure: Proposal

FIGURE P–8. Sales Proposal (*continued*) (Cover Letter)

The Waters Proposal   September 2, 2009

EXECUTIVE SUMMARY

The Waters 319 Interactive Terminal/615 Retail Processor System will provide your management with the tools necessary to manage people and equipment more profitably with procedures that will yield more cost-effective business controls for Cookson's.

The equipment and applications proposed for Cookson's were selected through the combined effort of Waters and Cookson's Management Information Systems Director, Mr. Becklight. The architecture of the system will respond to your current requirements and allow for future expansion.

The features and hardware in the system were determined from data acquired through the comprehensive survey we conducted at your stores in July of this year. The total of 71 Interactive Terminals proposed to service your four store locations is based on the number of terminals currently in use and on the average number of transactions processed during normal and peak periods. The planned remodeling of all four stores was also considered, and the suggested terminal placement has been incorporated into the working floor plan. The proposed equipment configuration and software applications have been simulated to determine system performance based on the volumes and anticipated growth rates of the Cookson's stores.

The information from the survey was also used in the cost justification, which was checked and verified by your controller, Mr. Deitering. The cost-effectiveness of the Waters Interactive Terminal/Retail Processor System is apparent. Expected savings, such as the projected 46 percent reduction in sales audit expenses, are realistic projections based on Waters's experience with other installations of this type.

Waters has a proven track record of success in the manufacture, installation, and servicing of retail business information systems stretching over decades. We believe that the system we propose will extend and strengthen our successful and long-term partnership with Cookson's.

– 1 –

FIGURE P–8. Sales Proposal (*continued*) (Executive Summary)

The Waters Proposal   September 2, 2009

GENERAL SYSTEM DESCRIPTION

The point-of-sale system that Waters is proposing for Cookson's includes two primary Waters products. These are the 319 Interactive Terminal and the 615 Retail Processor.

Waters 319 Interactive Terminal
The primary component in the proposed retail system is the Interactive Terminal. It contains a full microprocessor, which gives it the flexibility that Cookson's has been looking for.

The 319 Interactive Terminal provides you with freedom in sequencing a transaction. You are not limited to a preset list of available steps or transactions. The terminal program can be adapted to provide unique transaction sets, each designed with a logical sequence of entry and processing to accomplish required tasks. In addition to sales transactions recorded on the selling floor, specialized transactions such as theater-ticket sales and payments can be designed for your customer-service area.

The 319 Interactive Terminal also functions as a credit authorization device, either by using its own floor limits or by transmitting a credit inquiry to the 615 Retail Processor for authorization.

Data-collection formats have been simplified so that transaction editing and formatting are much more easily accomplished. The IS manager has already been provided with documentation on these formats and has outlined all data-processing efforts that will be necessary to transmit the data to your current systems. These projections have been considered in the cost justification.

Waters 615 Retail Processor
The Waters 615 Retail Processor is a minicomputer system designed to support the Waters family of retail terminals. The . . .

[*The proposal next describes the Waters 615 Retail Processor before moving on to the detailed solution section.*]

FIGURE P–8. Sales Proposal (*continued*) (General Description of Products)

The Waters Proposal   September 2, 2009

PAYROLL APPLICATION

Current Procedure
Your current system of reporting time requires each hourly employee to sign a time sheet; the time sheet is reviewed by the department manager and sent to the Payroll Department on Friday evening. Because the week ends on Saturday, the employee must show the scheduled hours for Saturday and not the actual hours; therefore, the department manager must adjust the reported hours on the time sheet for employees who do not report on the scheduled Saturday or who do not work the number of hours scheduled.

The Payroll Department employs a supervisor and three full-time clerks. To meet deadlines caused by an unbalanced workflow, an additional part-time clerk is used for 20 to 30 hours per week. The average wage for this clerk is $9.00 per hour.

Advantage of Waters's System
The 319 Interactive Terminal can be programmed for entry of payroll data for each employee on Monday morning by department managers, with the data reflecting actual hours worked. This system would eliminate the need for manual batching, controlling, and data input. The Payroll Department estimates conservatively that this work consumes 40 hours per week.

Hours per week	40
Average wage (part-time clerk)	× 9.00
Weekly payroll cost	$360.00
Annual savings	$18,720

Elimination of the manual tasks of tabulating, batching, and controlling can save 0.25 hourly unit. Improved workflow resulting from timely data in the system without data-input processing will allow more efficient use of clerical hours. This would reduce payroll by the 0.50 hourly unit currently required to meet weekly check disbursement.

Eliminate manual tasks	0.25
Improve workflow	0.75
40-hour unit reduction	1.00
Hours per week	40
Average wage (full-time clerk)	11.00
Savings per week	$440.00
Annual savings	$22,880

TOTAL SAVINGS: $41,600

– 5 –

FIGURE P–8. Sales Proposal (*continued*) (Detailed Solution)

The Waters Proposal   September 2, 2009

COST ANALYSIS

This section of our proposal provides detailed cost information for the Waters 319 Interactive Terminal and the Waters 615 Retail Processor. It then multiplies these major elements by the quantities required at each of your four locations.

319 Interactive Terminal

Equipment	Price	Maint. (1 yr.)
Terminal	$2,895	$167
Journal Printer	425	38
Receipt Printer	425	38
Forms Printer	525	38
Software	220	—
TOTALS	$4,490	$281

[*The cost section goes on to describe other costs to install the system before summarizing the costs.*]

The following table summarizes all costs.

Location	Hardware	Maint. (1 yr.)	Software
Store No. 1	$72,190	$4,975	$3,520
Store No. 2	89,190	6,099	4,400
Store No. 3	76,380	5,256	3,740
Store No. 4	80,650	5,537	3,960
Data Center	63,360	6,679	12,480
Subtotals	$381,770	$28,546	$28,100

TOTAL $438,416

DELIVERY SCHEDULE

Waters is normally able to deliver 319 Interactive Terminals and 615 Retail Processors within 30 days of the date of the contract. This can vary depending on the rate and size of incoming orders.

All the software recommended in this proposal is available for immediate delivery. We do not anticipate any difficulty in meeting your tentative delivery schedule.

– 8 –

FIGURE P–8. Sales Proposal (*continued*) (Cost Analysis and Delivery Schedule)

The Waters Proposal   September 2, 2009
───────────────────────

SITE PREPARATION

Waters will work closely with Cookson's to ensure that each site is properly prepared prior to system installation. You will receive a copy of Waters's installation and wiring procedures manual, which lists the physical dimensions, service clearance, and weight of the system components in addition to the power, logic, communications-cable, and environmental requirements. Cookson's is responsible for all building alterations and electrical facility changes, including the purchase and installation of communications cables, connecting blocks, and receptacles.

Wiring
For the purpose of future site considerations, Waters's in-house wiring specifications for the system call for two twisted-pair wires and twenty-two shielded gauges. The length of communications wires must not exceed 2,500 feet.

As a guide for the power supply, we suggest that Cookson's consider the following:

1. The branch circuit (limited to 20 amps) should service no equipment other than 319 Interactive Terminals.
2. Each 20-amp branch circuit should support a maximum of three Interactive Terminals.
3. Each branch circuit must have three equal-size conductors—one hot leg, one neutral, and one insulated isolated ground.
4. Hubbell IG 5362 duplex outlets or the equivalent should be used to supply power to each terminal.
5. Computer-room wiring will have to be upgraded to support the 615 Retail Processor.

–9–

FIGURE P–8. Sales Proposal (*continued*) (Site-Preparation Section)

The Waters Proposal	September 2, 2009

TRAINING

To ensure a successful installation, Waters offers the following training course for your operators.

Interactive Terminal/Retail Processor Operations
Course number: 8256
Length: three days
Tuition: $500.00

This course provides the student with the skills, knowledge, and practice required to operate an Interactive Terminal/Retail Processor System. Online, clustered, and stand-alone environments are covered.

We recommend that students have a department-store background and that they have some knowledge of the system configuration with which they will be working.

FIGURE P–8. Sales Proposal (*continued*) (Training Section)

The Waters Proposal   September 2, 2009
_____

RESPONSIBILITIES

On the basis of its years of experience in installing information-processing systems, Waters believes that a successful installation requires a clear understanding of certain responsibilities.

Waters's Responsibilities
Generally, it is Waters's responsibility to provide its users with needed assistance during the installation so that live processing can begin as soon thereafter as is practical. The following items describe our specific responsibilities:

- Provide operations documentation for each application that you acquire from Waters.
- Provide forms and other supplies as ordered.
- Provide specifications and technical guidance for proper site planning and installation.
- Provide adviser assistance in the conversion from your present system to the new system.

Cookson's Responsibilities
Cookson's will be responsible for the suggested improvements described earlier, as well as the following:

- Identify an installation coordinator and system operator.
- Provide supervisors and clerical personnel to perform conversion to the system.
- Establish reasonable time schedules for implementation.
- Ensure that the physical site requirements are met.
- Provide personnel to be trained as operators and ensure that other employees are trained as necessary.
- Assume the responsibility for implementing and operating the system.

FIGURE P–8. Sales Proposal (*continued*) (Statement of Responsibilities)

The Waters Proposal   September 2, 2009

DESCRIPTION OF VENDOR

The Waters Corporation develops, manufactures, markets, installs, and services total business information-processing systems for selected markets. These markets are primarily in the retail, financial, commercial, industrial, health-care, education, and government sectors.

The Waters total-system concept encompasses one of the broadest hardware and software product lines in the industry. Waters computers range from small business systems to powerful general-purpose processors. Waters computers are supported by a complete spectrum of terminals, peripherals, and data-communication networks, as well as an extensive library of software products. Supplemental services and products include data centers, field service, systems engineering, and educational centers.

The Waters Corporation was founded in 1934 and presently has approximately 26,500 employees. The Waters headquarters is located at 17 North Waterloo Boulevard, Tampa, Florida, with district offices throughout the United States and Canada.

WHY WATERS?

Strong Commitment to the Retail Industry
Waters's commitment to the retail industry is stronger than ever. We are continually striving to provide leadership in the design and implementation of new retail systems and applications that will ensure our users of a logical growth pattern.

Dynamic Research and Development
Over the years, Waters has spent increasingly large sums on research-and-development efforts to ensure the availability of products and systems for the future. In 2008, our research-and-development expenditure for advanced systems design and technological innovations reached the $70-million level.

Leading Point-of-Sale Vendor
Waters is a leading point-of-sale vendor, having installed over 150,000 units. The knowledge and experience that Waters has gained over the years from these installations ensure well-coordinated and effective systems implementations.

FIGURE P–8. Sales Proposal (*continued*) (Vendor Description and Organizational Sales Pitch)

The Waters Proposal   September 2, 2009
_____

CONCLUSION

Waters welcomes the opportunity to submit this proposal to Cookson's. The Waters Corporation is confident that we have offered the right solution at a competitive price. Based on the hands-on analysis we conducted, our proposal takes into account your current and projected workloads and your plans to expand your facilities and operations. Our proposal will also, we believe, afford Cookson's future cost-avoidance measures in employee time and in enhanced accounting features.

Waters has a proven track record of success in the manufacture, installation, and servicing of retail business information systems stretching over many decades. We also have a demonstrated record of success in our past business associations with Cookson's. We believe that the system we propose will extend and strengthen this partnership.

Should you require additional information about any facet of this proposal, please contact Janet A. Curtain, who will personally arrange to meet with you or arrange for Waters's technical staff to meet with you or send you the information you need.

We look forward to your decision and to continued success in our working relationship with Cookson's.

– 13 –

FIGURE P–8. Sales Proposal (*continued*) (Conclusion)

Or the president of Habitat for Humanity may write a grant to a lumber company asking for a donation of lumber to help construct new housing for disadvantaged families.

Research proposals request approval to conduct **research** to investigate a problem or possible improvements to a product or an operation. Because their purpose is to gain approval to conduct research, they do not focus on particular solutions or ultimate results. For example, an engineer may submit a research proposal to a manager for permission to research a new method that improves cement strength for bridges. Similarly, students often submit research proposals to request approval of their research plans for term projects, such as formal reports, or thesis projects.

Grant and research proposals are persuasive when they clearly define your research goals, your plan for achieving those goals, and your qualifications to perform the research. The proposal typically includes the following key components:

- *Introduction.* Explain the reasons for and the benefits of the proposal. What can readers expect as a result of the proposed research, and what is the value of your potential findings?
- *Background.* Describe the problem your research will address so that readers are confident that you understand the problem completely. Illustrate how both your primary audience and others will benefit from the results of your proposed research.
- *Research Plan.* Discuss in detail your plan for conducting the research. First, focus on your research objectives—what specifically you plan to investigate. Then, focus on your research methods—how you plan to achieve your objectives (through interviewing? on the Web? through other sources?).
- *Work Schedule.* Outline realistic deadlines for specific research tasks that will help you achieve your objectives and meet the final deadline.
- *Qualifications.* Summarize the expertise of those who will conduct the research. You might also include their **résumés** in an appendix.
- *Budget.* Provide a list of projected costs for your research project, as appropriate, including costs of all resources needed to carry out your research plan.
- *Conclusion.* Remind the reader of the benefits from your research and any specific products that will result, such as a formal report. Close with a request for approval by a specific date and offer to answer any of the readers' questions.

pseudo- / quasi-

As a **prefix**, *pseudo-*, meaning "false or counterfeit," is joined to the root word without a **hyphen** unless the root word begins with a capital letter (*pseudo*science, *pseudo*-Newtonian). *Pseudo-* is sometimes confused with *quasi-*, meaning "somewhat" or "partial." Unlike *semi-*, *quasi-* means "resembling something" rather than "half." *Quasi-* is usually hyphenated in combinations (*quasi*-scientific theories). See also **bi- / semi-**.

punctuation

Punctuation helps **readers** understand the meaning and relationships of words, phrases, clauses, and sentences. Marks of punctuation link, separate, enclose, indicate omissions, terminate, and classify. Most punctuation marks can perform more than one function. See also **sentence construction**.

The use of punctuation is determined by grammatical conventions and the writer's intention. Understanding punctuation is essential for writers because it enables them to communicate with **clarity** and precision. See also **grammar**.

Detailed information on each mark of punctuation is given in its own entry. The following are the 13 marks of punctuation.

apostrophe	'	**parentheses**	( )
brackets	[ ]	**period**	.
colon	:	**question mark**	?
comma	,	**quotation marks**	" "
dash	—	**semicolon**	;
exclamation mark	!	**slash**	/
hyphen	-		

See also **abbreviations**, **capitalization**, **contractions**, **dates**, **ellipses**, **italics**, and **numbers**.

WEB LINK Practicing Punctuation

For online exercises on punctuation, see *bedfordstmartins.com/alredtech* and select *Exercise Central*.

purpose

What do you want your **readers** to know, to believe, or to do when they have read your document? When you answer that question about your **audience**, you have determined the primary purpose, or objective, of your document. Be careful not to state a purpose too broadly. A statement of purpose such as "to explain continuing-education standards" is too general to be helpful during the writing process. In contrast, "to explain to Critical-Care Nurse Society (CCNS) members how to determine if a continuing-education course meets CCNS professional standards" is a specific purpose that will help you focus on what you need your document to accomplish. Often the **context** will help you focus your purpose.

The writer's primary purpose is often more complex than simply "to explain" something, as shown in the previous paragraph. To fully understand this complexity, you need to ask yourself not only *why* you are writing the document but *what* you want to influence your reader to believe or to do after reading it. Suppose a writer for a **newsletter** has been assigned to write an article about cardiopulmonary resuscitation (CPR). In answer to the question *what?* the writer could state the purpose as "to emphasize the importance of CPR." To the question *why?* the writer might respond, "to encourage employees to sign up for evening CPR classes." Putting the answers to the two questions together, the writer's purpose might be stated as, "To write a document that will emphasize the importance of CPR and encourage employees to sign up for evening CPR classes." Note that the primary purpose of this document is to persuade the readers of the importance of CPR, and the secondary goal is to motivate them to register for a class. Secondary goals often involve such abstract notions as to motivate, to persuade, to reassure, or to inspire your reader. See also **persuasion**.

If you answer the questions *what?* and *why?* and put the answers into writing as a stated purpose that includes both primary and secondary goals, you will simplify your writing task and more likely achieve your purpose. For a **collaborative writing** project, it is especially important to collectively write a statement of your purpose to ensure that the document achieves its goals. Do not lose sight of that purpose as you become engrossed in the other steps of the writing process. See also "Five Steps to Successful Writing."

question marks

The question mark (?) most often ends a sentence that is a direct question or request.

- Where did you put the specifications? [direct question]
- Will you e-mail me if your shipment does not arrive by June 10? [request]

Use a question mark to end a statement that has an interrogative meaning—a statement that is declarative in form but asks a question.

- The lab report is finished? [question in declarative form]

Question marks may follow a series of separate items within an interrogative sentence.

- Do you remember the date of the contract? Its terms? Whether you signed it?

Use a question mark to end an interrogative clause within a declarative sentence.

- It was not until July (or was it August?) that we submitted the report.

Retain the question mark in a title that is being cited, even though the sentence in which it appears has not ended.

- *Can Quality Be Controlled?* is the title of her book.

Never use a question mark to end a sentence that is an indirect question.

- He asked me where I put the specifications.

When a question is a polite request or an instruction to which an affirmative response is assumed, a question mark is not necessary.

- Will you call me as soon as he arrives. [polite request]

When used with **quotations**, the placement of the question mark is important. When the writer is asking a question, the question mark belongs outside the quotation marks.

▶ Did she actually say, "I don't think the project should continue"?

If the quotation itself is a question, the question mark goes inside the **quotation marks**.

▶ She asked, "Do we have enough funding?"

If both cases apply — the writer is asking a question and the quotation itself is a question — use a single question mark inside the quotation marks.

▶ Did she ask, "Do we have enough funding?"

questionnaires

A questionnaire — a series of questions on a particular topic sent out to a number of people — serves the same function as an interview but does so on paper, as an **e-mail** attachment, or as an online form. As you prepare a questionnaire, therefore, keep in mind your **purpose** and your intended **audience**. See also **interviewing for information** and **research**.

Questionnaires have several advantages over the personal interview as well as several disadvantages.

ADVANTAGES
- A questionnaire allows you to gather information from more people more quickly than you could by conducting personal interviews.
- It enables you to obtain responses from people who are difficult to reach or who are in various geographic locations.
- Those responding to a questionnaire have more time to think through their answers than when faced with the pressure of composing thoughtful and complete answers to an interviewer.
- The questionnaire may yield more objective data because it reduces the possibility that the interviewer's tone of voice or facial expressions might influence an answer.
- The cost of distributing and tabulating a questionnaire is lower than the cost of conducting numerous personal interviews.

DISADVANTAGES

- The results of a questionnaire may be slanted in favor of those people who have strong opinions on a subject because they are more likely to respond than those with only moderate views.
- The questionnaire does not allow specific follow-up questions to answers; at best, a questionnaire can be designed to let one question lead logically to another.
- Distributing questionnaires and waiting for replies may take considerably longer than conducting personal interviews.

The sample cover memo and questionnaire in Figure Q–1 were sent to employees in a large organization who had participated in a six-month program of flexible working hours.

Selecting the Recipients

Selecting the proper recipients for your questionnaire is crucial if you are to gather representative and usable data. If you wanted to survey the opinions of large groups in the general population—for example, all medical technologists working in private laboratories or all independent garage owners—your task would not be easy. Because you cannot include everybody in your survey, you need to choose a representative cross section. For example, you would want to include enough people from around the country, respondents of both genders, and people with different educational backgrounds. Only then could you make a generalized statement based on your findings from the sample. (The best sources of information on sampling techniques are market research and statistics texts.)

WEB LINK	Online Surveys

Surveymonkey.com offers help for designing online surveys as well as collecting and analyzing the results. For links to this site and more, see *bedfordstmartins.com/alredtech* and select *Links for Handbook Entries.*

Preparing the Questions

A key goal in designing the questionnaire is to keep it as brief as possible. The longer a questionnaire is, the less likely the recipient will be to complete and return it. Also, the questions should be easy to understand. A confusing question will yield confusing results, whereas a carefully worded question will be easy to answer. Ideally, recipients should be able to answer most questions with a "yes" or "no" or by checking or circling a choice among several options. Such answers are easy to

Luxwear Products Corporation
MEMO

To:	All Company Employees
From:	Nelson Barrett, Director *NB* Human Resources Department
Date:	October 19, 2009
Subject:	Review of Flexible Working Hours Program

Please complete and return the questionnaire enclosed regarding Luxwear's trial program of flexible working hours. Your answers will help us decide whether we should make the program permanent.

Return the completed questionnaire to Ken Rose, Mail Code 12B, by October 28. Your signature on the questionnaire is not necessary. All responses will be confidential and given serious consideration. Feel free to raise additional issues pertaining to the program.

If you want to discuss any item in the questionnaire, call Pam Peters in the Human Resources Department at extension 8812 or e-mail at pp1@lpc.com.

Enclosure: Questionnaire

FIGURE Q–1. Questionnaire (Cover Memo)

Flexible Working Hours Program
Questionnaire

1. What kind of position do you occupy?

   ☐ Supervisory
   ☐ Nonsupervisory

2. Indicate to the nearest quarter of an hour your starting time under flextime.

   ☐ 7:00 a.m.   ☐ 8:15 a.m.
   ☐ 7:15 a.m.   ☐ 8:30 a.m.
   ☐ 7:30 a.m.   ☐ 8:45 a.m.
   ☐ 7:45 a.m.   ☐ 9:00 a.m.
   ☐ 8:00 a.m.   ☐ Other (specify) _____

3. Where do you live?

   ☐ Talbot County   ☐ Greene County
   ☐ Montgomery County   ☐ Other (specify) _____

4. How do you usually travel to work?

   ☐ Drive alone   ☐ Walk
   ☐ Bus   ☐ Car pool
   ☐ Train   ☐ Motorcycle
   ☐ Bicycle   ☐ Other (specify) _____

5. Has flextime affected your commuting time?

   ☐ Increase: Approximate number of minutes _____
   ☐ Decrease: Approximate number of minutes _____
   ☐ No change

6. If you drive alone or in a car pool, has flextime increased or decreased the amount of time it takes you to find a parking space?

   ☐ Increased   ☐ Decreased   ☐ No change

7. Has flextime had an effect on your productivity?

   a. Quality of work
      ☐ Increased   ☐ Decreased   ☐ No change

   b. Accuracy of work
      ☐ Increased   ☐ Decreased   ☐ No change

   c. Quiet time for uninterrupted work
      ☐ Increased   ☐ Decreased   ☐ No change

FIGURE Q–1. Questionnaire (*continued*)

8. Have you had difficulty getting in touch with coworkers who are on different work schedules from yours?

   ☐ Yes   ☐ No

9. Have you had trouble scheduling meetings within flexible starting and quitting times?

   ☐ Yes   ☐ No

10. Has flextime affected the way you feel about your job?

    ☐ Yes   ☐ No

    If yes, please answer (a) or (b):

    a. Feel better about job
       ☐ Slightly   ☐ Considerably

    b. Feel worse about job
       ☐ Slightly   ☐ Considerably

11. How important is it for you to have flexibility in your working hours?

    ☐ Very   ☐ Not very   ☐ Somewhat   ☐ Not at all

12. Has flextime allowed you more time to be with your family?

    ☐ Yes   ☐ No

13. If you are responsible for the care of a young child or children, has flextime made it easier or more difficult for you to obtain babysitting or day-care services?

    ☐ Easier   ☐ More difficult   ☐ No change

14. Do you recommend that the flextime program be made permanent?

    ☐ Yes   ☐ No

15. Please describe below or on an attached sheet any major changes you recommend for the program.

Thank you for your assistance.

FIGURE Q–1. Questionnaire (*continued*)

tabulate and require minimum effort on the part of the respondent, thus increasing your chances of obtaining a response. See also **forms**.

▶ Do you recommend that the flextime program be made permanent?

☐ Yes   ☐ No   ☐ No opinion

If you need more information than such questions produce, provide an appropriate range of answers, as in the following example:

▶ How many hours of overtime would you be willing to work each week?

☐ 4 hours   ☐ 8 hours   ☐ Over 10 hours
☐ 6 hours   ☐ 10 hours   ☐ No overtime

Questions should be neutral; they should not be worded in such a way as to lead respondents to give a particular answer, which can result in inaccurate or skewed data.

SLANTED	Would you prefer the freedom of a four-day workweek?
NEUTRAL	Would you choose to work a four-day workweek, ten hours a day, with every Friday off?

WRITER'S CHECKLIST Designing a Questionnaire

- Prepare a **cover letter** (or a memo or an e-mail) explaining who you are, the questionnaire's purpose, the date by which you need a response, and how and where to send the completed questionnaire.
- Include a stamped, self-addressed envelope if you are using regular mail.
- Construct as many questions as possible for which the recipient does not have to compose an answer.
- Include a section on the questionnaire for additional comments where the recipient may clarify his or her overall attitude toward the subject.
- Include questions about the respondent's age, gender, education, occupation, and so on, only if such information will be of value in interpreting the answers.
- State whether the information provided as well as the recipient's identity will be kept confidential.
- Include your contact information (mailing address, phone number, and e-mail address).

Writer's Checklist: Designing a Questionnaire (continued)

- Consider offering some tangible appreciation to those who answer the questionnaire by a specific date, such as a copy of the results or, for a customer questionnaire, a gift certificate.

quid pro quo

Quid pro quo, which is Latin for "one thing for another," suggests mutual cooperation or "tit for tat" in a relationship between two groups or individuals. The term may be appropriate to business and legal contexts if you are sure your **readers** understand its meaning. ("Before approving the plan, we insisted on a fair *quid pro quo*.") See also **foreign words in English**.

quotation marks

Quotation marks (" ") are used to enclose a direct quotation of spoken or written words. Quotation marks have other special uses, but they should not be used for **emphasis**.

Direct Quotations

Enclose in quotation marks anything that is quoted word for word (a direct quotation) from speech or written material.

▶ She said clearly, "I want the progress report by three o'clock."

Do not enclose indirect quotations—usually introduced by the word *that*—in quotation marks. Indirect quotations are paraphrases of a writer's or speaker's words or ideas. See also **paraphrasing**.

▶ She said that she wanted the progress report by three o'clock.

ETHICS NOTE When you use quotation marks to indicate that you are quoting, do not make any changes or omissions in the quoted material unless you clearly indicate what you have done. For further information on incorporating quoted material and inserting comments, see **plagiarism** and **quotations**. ✦

Use single quotation marks (' ') to enclose a quotation that appears within a quotation.

▶ John said, "Jane told me that she was going to 'stay with the project if it takes all year.'"

Words and Phrases

Use quotation marks to set off special words or terms only to point out that the term is used in context for a unique or special purpose (that is, in the sense of the term *so-called*).

▶ A remarkable chain of events caused the sinking of the "unsinkable" *Titanic* on its maiden voyage.

Slang, colloquial expressions, and attempts at humor, although infrequent in workplace writing, should seldom be set off by quotation marks.

▶ Our first six months amounted to a ~~"shakedown cruise."~~ *shakedown cruise.*

Titles of Works

Use quotation marks to enclose **titles** of reports, short stories, articles, essays, single episodes of radio and television programs, and short musical works (including songs). However, do not use quotation marks for titles of books and periodicals, which should appear in **italics**.

▶ His report, "Effects of Government Regulations on Motorcycle Safety," cited the article "No-Fault Insurance and Motorcycles," published in *American Motorcyclist* magazine.

Use quotation marks for parts of publications, such as chapters of books and articles or sections within periodicals.

▶ "Bad Writing," an article by Barbara Wallraff, appeared in the "On Language" column of the *New York Times*.

Some titles, by convention, are not set off by quotation marks, underlining, or italics, although they are capitalized.

▶ Professional Writing [college course title], the Bible, the Constitution, Lincoln's Gettysburg Address, the Lands' End Catalog

Punctuation

Commas and **periods** always go inside closing quotation marks.

▶ "Reading *Computer World* gives me the insider's view," he says, adding, "It's like a conversation with the top experts."

Semicolons and **colons** always go outside closing quotation marks.

▶ He said, "I will pay the full amount"; this statement surprised us.

All other punctuation follows the logic of the context: If the punctuation is part of the material quoted, it goes inside the quotation marks; if the punctuation is not part of the material quoted, it goes outside the quotation marks.

quotations

Using direct and indirect quotations is an effective way to make or support a point. However, avoid the temptation to overquote during the **note-taking** phase of your **research**; concentrate on summarizing what you read.

▶ **ETHICS NOTE** When you use a quotation (or an idea of another writer), cite your source properly. If you do not, you will be guilty of **plagiarism**. For specific details on citation systems, see **documenting sources**. ✦

Direct Quotations

A direct quotation is a word-for-word copy of the text of an original source. Choose direct quotations (which can be of a word, a phrase, a sentence, or, occasionally, a paragraph) carefully and use them sparingly. Enclose direct quotations in **quotation marks** and separate them from the rest of the sentence by a **comma** or **colon**. Use the initial capital letter of a quotation if the quoted material originally began with a capital letter.

▶ The Dean of the Medical College stated, "We must attract more students to careers in medicine in order to meet the health-care needs of older Americans."

When dividing a quotation, set off the material that interrupts the quotation with commas, and use quotation marks around each part of the quotation.

▶ "We must attract more students to careers in medicine," he said in a recent interview, "in order to meet the health-care needs of older Americans."

Indirect Quotations

An indirect quotation is a paraphrased version of an original text. It is usually introduced by the word *that* and is not set off from the rest of the sentence by punctuation marks. See also **paraphrasing**.

▶ In an interview, he said *that* recruiting students to medical schools is essential to meeting the health-care demands of older Americans.

Deletions or Omissions

Deletions or omissions from quoted material are indicated by three **ellipsis** points (...) within a sentence and a period plus three ellipsis points (....) at the end of a sentence.

> ORIGINAL "Their touchstones, which to an outsider might appear as arbitrary and simplistic, were developed through highly specific research using both internal and outside expertise, distilled from a complex historical process of examining rhetorical variables, and documented as operational guidelines rather than narratives of processes or of theoretical models."
>
> WITH ELLIPSES "Their touchstones ... were developed through highly specific research using both internal and outside expertise, distilled from a complex historical process of examining rhetorical variables, and documented as operational guidelines...."

When a quoted passage begins in the middle of a sentence rather than at the beginning, ellipsis points are not necessary; the fact that the first letter of the quoted material is not capitalized tells the reader that the quotation begins in midsentence.

▶ Rivero goes on to conclude that "the Research and Development Center is essential to keeping high-tech companies in the state."

Inserting Material into Quotations

When it is necessary to insert a clarifying comment within quoted material, use **brackets**.

▶ "The industry is an integrated system that serves an extensive [geographic] area, with divisions existing as islands within the larger system's sphere of influence."

When quoted material contains an obvious error or might be questioned in some other way, insert the expression *sic* (Latin for "thus") in italic type and enclose it in brackets ([*sic*]) following the questionable material to indicate that the writer has quoted the material *exactly as it appeared in the original*.

▶ The company considers the Baker Foundation to be a "guilt-edged [*sic*] investment."

Incorporating Quotations into Text

Quote word for word only when a source with particular expertise states something that is especially precise, striking, or noteworthy, or that may reinforce a point you are making. Quotations must also logically, grammatically, and syntactically match the rest of the sentence and surrounding text. Notice in Figure Q–2 that the quotation blends with the content of the surrounding text, which uses **transition** to introduce and comment on the quotation.

After reviewing a large number of works on technical communication, Alred sees differences in the focus of practitioners and educators:

> Some works, for example, are essential to the immediate needs of professional technical communicators who face specific and demanding workplace tasks under tight deadlines. Other works are crucial to educators who must be concerned with the long-range implications of research or theoretical insights as they prepare students for professional careers in technical communication. (p. 586)

The aims of practitioners and educators, however, are not as distinct as this statement by itself might suggest. For example, many practicing technical communicators, as they design Web pages or produce documentation, see themselves as educators. Specifically, in preparing material for users of systems, practitioners must . . .

FIGURE Q–2. Long Quotation (APA Style)

Depending on the citation system, the style of incorporating quotations varies. For examples of three different styles, see **documenting sources**. Figure Q–2 shows APA style for a long quotation. At the end of the document, the following entry would appear in the APA-style list of references as the source of the quotation in Figure Q–2.

▶ Alred, G. J. (2003). Essential works on technical communication. *Technical Communication*, 50, 585–616.

Do not rely too heavily on the use of quotations in the final version of your document. Generally, avoid quoting anything that is longer than one paragraph.

R

raise / rise

Both *raise* and *rise* mean "move to a higher position." However, *raise* is a transitive **verb** and always takes an **object** ("*raise* crops"), whereas *rise* is an intransitive verb and never takes an object ("heat *rises*").

re

Re (and its variant form, *in re*) is business and legal **jargon** meaning "in reference to" or "in the case of." Although *re* is sometimes used in **memos** and **e-mails**, *subject* is a preferable term.

readers

The first rule of effective writing is to *help your readers*. If you overlook this commitment, your writing will not achieve its **purpose**, either for you or for your business or organization. For meeting the needs of both individual and multiple readers, see **audience**.

really

Really is an **adverb** meaning "actually" or "in fact." Although both *really* and *actually* are often used as **intensifiers** for **emphasis** or sarcasm in speech, avoid such use in writing.

▶ Did he ~~really~~ finish the report on time?

reason is [because]

Replace the redundant phrase *the reason is because* with *the reason is that* or simply *because*. See also **conciseness**.

reference letters

Writing a reference letter (or letter of recommendation) can range from completing an admission form for a prospective student to composing a detailed description of professional accomplishments and personal characteristics for someone seeking employment. In Figure R–1, a former employer has written a letter for someone who is seeking an advanced position as a researcher.

To write an effective letter of recommendation, you must be familiar enough with the applicant's abilities and performance to offer an evaluation, and you must keep in mind the following:

- Identify yourself by name, title or position, employer, and address.
- Respond directly to the inquiry, carefully addressing the specific questions asked.
- Address specifically the applicant's skills, abilities, knowledge, and character, guided by the person's **résumé** when possible.
- Communicate truthfully and without embellishment.

You could begin, as in Figure R–1, by stating the circumstances of your acquaintance and how long you have known the person for whom you are writing the letter. Mention, with as much substantiation as possible, one or two outstanding characteristics of the applicant. Organize the details in your letter using the decreasing **order-of-importance method of development**. Conclude with a brief summary of the applicant's qualifications and a clear statement of recommendation. See **correspondence** for letter format and general advice.

▶ ETHICS NOTE When you are asked to serve as a reference or to supply a letter of reference, be aware that applicants have a legal right to examine what you have written about them, unless they sign a waiver. ◆

refusal letters

A refusal delivers a negative message (or bad news) in the form of a **letter**, a **memo**, or an **e-mail**. The ideal refusal says "no" in such a way that you not only avoid antagonizing your reader but also maintain goodwill. See also **audience**.

When the stakes are high, you must convince your reader *before* you present the bad news that your reasons for refusing are logical or understandable. (See also **correspondence**.) Stating a negative message in your opening may cause your reader to react too quickly and dismiss

City of Springfield
Legislative Reference Bureau
200 East Main Street
Springfield, AK 99501

(414) 224-5555

January 12, 2009

Mr. Phillip Lester
Human Resources Director
Thompson Enterprises
201 State Street
St. Louis, MO 63102

Dear Mr. Lester:

[How long writer has known applicant and the circumstances]

As Kerry Hawkin's former employer, I am happy to have the opportunity to recommend her. I've known Kerry for five years, first as an intern in our office and for two years as a full-time research associate. This past year she has remained in contact as she completed graduate school.

Our office is the official research arm for the Springfield City Council, so we approved Kerry as an intern not only because of her outstanding grades but also because the university internship coordinator reported that she possessed excellent research skills. Kerry

[Outstanding characteristics of applicant]

proved her worth in our office as an intern and was offered a full-time position as a research associate to work on projects under my supervision. I found Kerry not only to be a careful researcher but also to be able to complete her assignments on schedule within tight deadlines. The material provided in her well-written reports unfailingly met the requirements for my work and more. During her time in our office, Kerry also proved herself to be a most valued colleague.

I strongly recommend Kerry for her ability to work independently, to organize her time efficiently, and to write clearly and articulately. I regret that we have no position to offer her at this time. Please do not hesitate to let me know if I can provide further information.

Sincerely yours,

Michelle Paul

[Recommendation and summary of qualifications]

Michelle Paul, Manager
mpaul@springfield.ar.gov

FIGURE R–1. Reference Letter

your explanation. The following pattern, used in the message shown in Figure R–2, is an effective way to handle this problem:

1. Open with **context** for the message (often called a "buffer").
2. Review the facts or details leading to the refusal or bad news.
3. Give the negative message based on the facts or details.
4. Close by establishing or reestablishing a positive relationship.

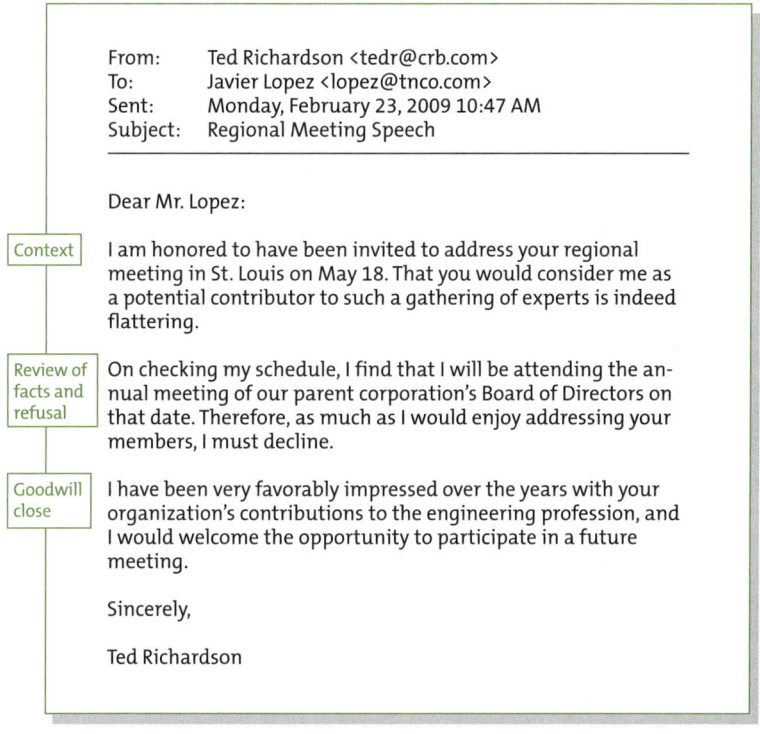

FIGURE R–2. Refusal Letter (Sent as E-mail)

Your opening should provide an appropriate context and establish a professional **tone**, for example by expressing appreciation for a reader's time, effort, or interest.

▶ The Screening Procedures Committee appreciates the time and effort you spent on your proposal for a new security-clearance procedure.

Next, review the circumstances of the situation sympathetically by placing yourself in the reader's position. Clearly detail the reasons you cannot do what the reader wants—even though you have not yet said you cannot do it. A good explanation, as shown in the following example, should ideally detail the reasons for your refusal so thoroughly that the reader will accept the negative message as a logical conclusion.

> We reviewed the potential effects of implementing your proposed security-clearance procedure company-wide. We not only asked the Security Systems Department to review the data but also surveyed industry practices, sought the views of senior management, and submitted the idea to our legal staff. As a result of this process, we have reached the following conclusions:
> - The cost savings you project are correct only if the procedure could be required throughout the company.
> - The components of your procedure are legal, but most are not widely accepted by our industry.
> - Based on our survey, some components could alienate employees who would perceive them as violating an individual's rights.
> - Enforcing company-wide use would prove costly and impractical.

Do not belabor the negative message—state your refusal quickly, clearly, and as positively as possible.

> For those reasons, the committee recommends that divisions continue their current security-screening procedures.

Close your message in a way that reestablishes goodwill—do not repeat the bad news (avoid writing "Again, we are sorry we cannot use your idea"). Ideally, provide an alternative, as in the following:

> Because some components of your procedure may apply in certain circumstances, we would like to feature your ideas in the next issue of *The Guardian*. I have asked the editor to contact you next week. On behalf of the committee, thank you for the thoughtful proposal.

If such an option is not possible or reasonable, offer a friendly remark that anticipates a positive future relationship, assure the reader of your high opinion of his or her product or service, or merely wish the reader success.

For responding to a complaint, see **adjustment letters**. For refusing a job offer, see **acceptance / refusal letters**.

regarding / with regard to

In regards to and *with regards to* are incorrect **idioms** for *in regard to* and *with regard to*. Both *as regards* and *regarding* are acceptable variants.

▶ In ~~regards~~ ^regard^ to your last question, I think a meeting is a good idea.

▶ ~~With regards to~~ ^Regarding^ your last question, I think a meeting is a good idea.

regardless

Always use *regardless* instead of the nonstandard *irregardless*, which expresses a double negative. The prefix *ir-* renders the base word negative, but *regardless* is already negative, meaning "unmindful."

repetition

The deliberate use of repetition to build a sustained effect or to emphasize a feeling or an idea can be a powerful device. See also **emphasis**.

▶ Similarly, atoms *come and go* in a molecule, but the molecule *remains*; molecules *come and go* in a cell, but the cell *remains*; cells *come and go* in a body, but the body *remains*; persons *come and go* in an organization, but the organization *remains*.
 —Kenneth Boulding, *Beyond Economics*

Repeating keywords from a previous sentence or paragraph can also be used effectively to achieve **transition**.

▶ For many years, *oil* has been a major industrial energy source. However, *oil* supplies are limited, and other sources of energy must be developed.

Be consistent in the word or phrase you use to refer to something. In technical writing, it is generally better to repeat a word or use a clear **pronoun reference** (so readers know that you mean the same thing) than to use **synonyms** to avoid repetition. See also **affectation**.

SYNONYMS	Several recent *analyses* support our conclusion. These *studies* cast doubt on the feasibility of long-range forecasting. The *reports*, however, are strictly theoretical.
CONSISTENT TERMS	Several recent *studies* support our conclusion. These *studies* cast doubt on the feasibility of long-range forecasting. *They* are, however, strictly theoretical.

Purposeless repetition, however, makes a sentence awkward and hides its key ideas. See also **conciseness**.

▶ She said that the customer ~~said that he~~ was canceling the order.

reports

A report is an organized presentation of factual information, often aimed at multiple **audiences**, that may present the results of an investigation, a trip, or a research project. For any report—whether formal or informal—assessing the readers' needs is essential. Following is a list of report entries in this book:

feasibility reports	185	**progress and activity reports**	402
formal reports	195	**test reports**	527
investigative reports	281	**trip reports**	539
laboratory reports	293	**trouble reports**	541

Formal reports often present the results of long-term projects or those that involve multiple participants. (See also **collaborative writing**.) Such projects may be done either for your own organization or as a contractual requirement for another organization. Formal reports generally follow a precise format and include such elements as **abstracts** and **executive summaries**. See also **proposals**.

Informal and short reports normally run from a few paragraphs to a few pages and ordinarily include only an **introduction**, a body, a **conclusion**, and (if necessary) recommendations. Because of their brevity, informal reports are customarily written as **correspondence**, including **letters** (if sent outside your organization), **memos**, and **e-mails**.

The introduction announces the subject of the report, states its **purpose**, and gives any essential background information. It may also summarize the conclusions, findings, or recommendations made in the report. The body presents a clearly organized account of the report's subject—the results of a test carried out, the status of a project, and other details readers may need. The amount of detail to include depends on your reader's knowledge and the complexity of the subject.

The conclusion summarizes your findings and interprets their significance for readers. In some reports, a final, separate section gives recommendations; in others, the conclusions and the recommendations are combined into one section. This final section makes suggestions for a course of action based on the data you have presented. See also **persuasion**.

repurposing

Repurposing is the copying or converting of existing content, such as written text and **visuals**, from one document or medium into another for a different **purpose**.* For example, if you are preparing a promotional **brochure**, you may be able to reuse material from a product **description** that is currently published on your organization's Web site. The brochure may then be printed or even placed on the Web site where it can be downloaded. See also **selecting the medium**.

In the workplace, this process saves time because content that often requires substantial effort to develop need not be re-created for each new application. The process of repurposing may be as simple as copying and pasting content from one document into another or as complex as distributing content and updates through a system of **content management**.

Content can be repurposed exactly as it is written only if it fits the **scope**, **audience**, and purpose of the new document. If the content alters these areas, you must adapt the content to fit its new **context**, as described in the following sections.

Repurpose for the Context

Staying focused on the purpose of your new document is critical, especially when repurposing content between different media. If you are writing a sales **proposal**, for example, and you only need to describe the **specifications** for a product, it may be useful to repurpose the specification list from your organization's Web site. However, the purpose of the Web-site content may be *to inform* customers about your products, whereas the purpose of a proposal is *to persuade* customers to buy your products. To effectively use the repurposed content in your proposal, you may need to adapt the **tense**, **voice**, **tone**, **grammar**, and **point of view** to make the repurposed content more persuasive and fit within the context of a sales proposal.

---

*The reuse of standard texts or content in technical publications is often referred to as "single-source publishing" or simply "single sourcing." See *www.stcsig.org/ss/*. Traditionally, such reuse of standard texts has been referred to as "boilerplate."

Repurpose for the Medium

The proper **style** and **format** of content written for a specific medium, such as a brochure or fact sheet, may not work as effectively when repurposed into a different medium, such as a Web site. Solid blocks of text may be easy to read in a brochure, but Web readers often need blocks of text to be separated into bulleted **lists** or very short paragraphs because readers process information differently when reading on a computer monitor. Adapt the **layout and design** of the repurposed content as appropriate to accommodate your reader's needs for the medium. See also **writing for the Web**.

◘ ETHICS NOTE In the workplace, repurposing content within an organization does not violate **copyright** because an organization owns the information it creates and can share it across the company. Likewise, a writer in an organization may use and repurpose material in the public domain and, with some limitations, content that is licensed under Creative Commons at *http://creativecommons.org/about/*. See also **blogs** and *Digital Tip: Wikis for Collaborative Documents* on page 72.

In the classroom, of course, the use of content or someone else's unique ideas without acknowledgment or the use of someone else's exact words without **quotation marks** and appropriate credit is **plagiarism**. ✦

requests for proposals

When a company or another organization needs a task performed or a service provided outside its staff expertise or capabilities, it seeks help from vendors. The company seeking help usually makes its requirements known by issuing a request for proposals (RFP) for vendors to evaluate and decide whether to bid on the project.

Although RFPs lay out an organization's requirements, they usually do not specify how these requirements are to be met.* Potential vendors prepare **proposals** to describe their solutions for meeting an RFP's requirements.

RFP Structure

Like proposals, RFPs vary in length and structure. The scope of information and even the terminology used will vary with the task or service

---

*A related document is an invitation for bids (IFB). An IFB specifically defines the quantity, type, and specifications for an item that an organization intends to purchase.

being solicited. The **purpose** and **context** of RFPs also vary because each involves unique legal, budgetary, confidentiality, and administrative considerations. The following sections describe typical components for RFPs.

Information About Your Company. One section should describe your company's mission, goals, size, facility locations, and position in the marketplace, as well as provide your company contact information. This section parallels the "Description of Vendor" section in proposals (see Figure P–8, page 431).

Project Description. Another section, sometimes called "Scope of Work" or "Workscope Description," describes the deliverable (the project or service) you need, with a detailed list of requirements. Requirements can be described in a summary statement, as shown in the following examples:

▶ The project deliverable will be a questionnaire to be mailed to approximately 400 adults regarding their consumer experiences with various health-care providers. . . .

▶ The deliverable will be a networked online forms package for approximately 1,200 employees, enabling information to be completed online, stored in a database, routed via e-mail, and printed on paper. The package shall be integrated into the company's current network environment at its headquarters. . . .

The project description may also indicate whether the deliverable must be created new or obtained commercially and adapted to the project's needs, as for software development. This section can also specify other details, such as a project-management plan and any warranty or liability requirements.

Delivery Schedule. For complex projects lasting a month or longer, specify the time allotted for the project and a proposed schedule of tasks.

Proposal Description. An RFP may provide format requirements; number of copies expected; due date; and where, to whom, and how (registered mail, courier service, etc.) it should be sent. If an electronic file is required, some firms include an electronic template.

Vendor Qualifications. Most RFPs request a summary of the vendor's experience, professional certifications, number of employees, years in business, quality-control procedures, awards and honors, and the résumés of the principal employees assigned to the project.

Proposal-Evaluation Criteria. The proposal-evaluation section informs vendors of the firm's selection criteria. They may select a firm solely on cost, cost and vendor past performance, or technical expertise. Some firms grade proposals with a point system on a scale of 100 or 1,000 total points, as in the following sample:

▶ Proposal-Evaluation Criteria

CRITERIA	POINTS
Administrative	50
Technical	700
Management	100
Price	100
Presentation and demonstration	50
Total	1,000

Appendixes. Some RFPs include one or more appendixes or "Attachments" for sample forms and questionnaires, a sample contract, a company's server and workstation configurations, workflow-analysis diagrams, dates and times when vendors can visit the work site before finalizing their proposals, and other essentials too detailed for the body of the RFP.

ETHICS NOTE If your RFP contains any company-confidential information, you could include a legally binding nondisclosure statement for vendors to sign before you send the full RFP. Many companies also include a guarantee not to open proposals before the due date and never to disclose information in one vendor's proposal to a competing vendor. ◆

WEB LINK **Sample Request for Proposals**

To view the RFP that accompanies the sample proposal in Figure P–8, see *bedfordstmartins.com/alredtech* and select *Model Documents Gallery.*

research

DIRECTORY
Primary Research   459
Secondary Research   460
Library Research Strategies   460
  Online Catalogs (Locating Books)   461
  Online Databases and Indexes (Locating Articles)   461
  Reference Works   462
Internet Research Strategies   463
  Search Engines   464
Writer's Checklist: Using Search Engines and Keywords   464
  Web Subject Directories   465
Evaluating Sources   465
Writer's Checklist: Evaluating Print and Online Sources   466

Research is the process of investigation—the discovery of information. To be focused, research must be preceded by **preparation**, especially consideration of your **audience**, **purpose**, and **scope**. During your research, effective **note-taking** is essential for a coherent **organization** that strategically integrates your own ideas, supporting facts, and any well-selected **quotations** into an effective final document. See also **documenting sources**, **outlining**, and "Five Steps to Successful Writing."

In an academic setting, your preparatory resources include conversations with your peers, instructors, and especially a reference librarian. On the job, your main resources are your own knowledge and experience and that of your colleagues. In the workplace, the most important sources of information may also include **test reports**, **laboratory reports**, and the like. In this setting, begin by **brainstorming** with colleagues about what sources will be most useful to your topic and how you can find them.

Primary Research

Primary research is the gathering of raw data from such sources as first-hand experience, interviews, direct observation, surveys and **questionnaires**, experiments, **meetings**, and the like. In fact, direct observation and interaction are the only ways to obtain certain kinds of information, such as about human and animal behavior, certain natural phenomena, and the operation of systems and equipment. You can also conduct primary research on the Internet by participating in discussion groups and newsgroups and by using **e-mail** to request information from specific audiences. See also **interviewing for information** and **listening**.

▸ **ETHICS NOTE** If you are planning research that involves observation, choose your sites and times carefully, and be sure to obtain permission in advance. During your observations, remain as unobtrusive as possible, and keep accurate, complete records that indicate date, time of day, duration of the observation, and so on. Save interpretations of your observations for future analysis. Be aware that observation can be valuable research, but it may also be time-consuming, complicated, and expensive. You may also inadvertently influence the subjects you are observing. ✦

Secondary Research

Secondary research is the gathering of information that has been previously analyzed, assessed, evaluated, compiled, or otherwise organized into accessible form. Sources include books and articles as well as **reports**, Web documents, audio and video recordings, podcasts, **correspondence**, **minutes of meetings**, **brochures**, and so forth. The following two sections—Library Research Strategies and Internet Research Strategies—provide methods for finding secondary sources.

As you seek information, keep in mind that in most cases the more recent the information, the better. Recently published periodicals and newspapers—as well as academic (.edu), organizational (.org), and government (.gov) Web sites—can be good sources of current information and can include published interviews, articles, papers, and conference proceedings. See *Writer's Checklist: Evaluating Print and Online Sources* on page 466.

When a resource seems useful, read it carefully and take notes that include any additional questions about your topic. Some of your questions may eventually be answered in other sources; those that remain unanswered can guide you to further primary research. For example, you may discover that you need to interview an expert. Not only can someone skilled in a field answer many of your questions, but he or she can also suggest further sources of information. See **paraphrasing** and **plagiarism**.

Library Research Strategies

The library provides organized paths into scholarship and the Internet as well as specialized resources—such as licensed online databases, indexes, catalogs, and directories—that can help you find current, reliable information. The first step in using library resources, either in an academic institution or in a workplace, is to develop a search strategy appropriate to the information needed for your topic. You may want to begin by asking a reference librarian for help (in person, by phone, by live chat, or by e-mail) to find the best print or online resources for your topic—a brief conversation can focus your research and save you time. In addition, use your library's homepage for access to its catalogs, databases of articles, subject directories to the Web, and more.

Your search strategy depends on the kind of information you are seeking. For example, if you need the latest data offered by government research, check the Web, as described later in this entry. Likewise, if you need a current scholarly article on a topic, search an online database (such as EBSCOhost's Academic Search Premier) subscribed to by your library. For an overview of a subject, you might turn to an encyclopedia; for historical background, your best resources are books, journals, and primary documents (such as a technical specification).

Online Catalogs (Locating Books). An online catalog—accessed through library research terminals and through the Internet—allows you to search a library's licensed holdings, indicates an item's location and availability, and may allow you to arrange an interlibrary loan.

You can search a library's online catalog by author, title, keyword, or subject. The most common ways of searching for a specific topic are by subject or by keyword. If your search turns up too many results, you can usually narrow it by using the "limit search" or "advanced search" option.

Online Databases and Indexes (Locating Articles). Most libraries subscribe to online databases, such as the following, which are often available through a library's Web site:

- *EBSCOhost's Academic Search Premier*: the largest academic multidisciplinary database, which provides full text for nearly 4,600 scholarly publications, including more than 3,500 peer-reviewed journals.
- *InfoTrac*: a large database that hosts general-interest and scholarly journals and resources in business, law, health, and other fields.
- *ProQuest*: a collection of databases, including newspapers and scholarly journals, in a wide range of subject areas.
- *FirstSearch*: a collection of specialized databases, such as WorldCat (library collections) and ArticleFirst.
- *Lexis/Nexis Academic*: a collection of databases containing news, business, legal, medical, and government information.

These databases, sometimes called *periodical indexes*, are excellent resources for articles published within the last 10 to 20 years. Many include descriptive abstracts and full texts of articles. To find older articles, you may need to consult a print index, such as the *Readers' Guide* to *Periodical Literature* and the *New York Times Index*, both of which have been digitized and may be available in some libraries.

To locate articles in a database, conduct a keyword search. If your search turns up too many results, refine your search by connecting two search terms with AND — "engineering AND employment" — or use other options offered by the database, such as a limited, a modified, or an advanced search.

Reference Works. In addition to articles, books, and Web sources, you may want to consult reference works such as encyclopedias, dictionaries, and atlases for a brief overview of your subject. Ask your reference librarian to recommend works and **bibliographies** that are most relevant to your topic. Many are available online and can be accessed through your library's homepage.

ENCYCLOPEDIAS. Encyclopedias are comprehensive, multivolume collections of articles arranged alphabetically. Some cover a wide range of subjects, while others — such as *The Encyclopedia of Careers and Vocational Guidance*, 13th ed., edited by William Hopke (Chicago: Ferguson, 2005) — focus on specific areas. The free online encyclopedia *Wikipedia* at *www.wikipedia.org* should be used only as a starting point for your research because users continuously update entries (with minimal editorial filtering).

DICTIONARIES. General **dictionaries** can be compact or comprehensive, unabridged publications. Specialized dictionaries define terms used in a particular field, such as business, computers, architecture, or consumer affairs, and offer detailed definitions of field-specific terms, usually written in straightforward language.

HANDBOOKS AND MANUALS. Handbooks and manuals are typically one-volume compilations of frequently used information in a particular field. They offer brief definitions of terms or concepts, standards for presenting information, procedures for documenting sources, and **visuals** such as **graphs** and **tables**.

BIBLIOGRAPHIES. Bibliographies list books, periodicals, and other research materials published in areas such as engineering, medicine, the humanities, and the social sciences.

GENERAL GUIDES. The annotated *Guide to Reference Books*, 12th ed., by Robert Balay (Chicago: American Library Association, 2008), can help you locate reference books, indexes, and other research materials. Check your library's homepage or ask a reference librarian if your library subscribes to a particular index and whether it is available online.

ATLASES. Atlases provide representations of the physical and political boundaries of countries, climate, population, or natural resources.

Microsoft® Encarta® World Atlas 2001. CD-ROM for Windows®. Microsoft Corporation, 2001. Allows users to access articles and media files on a wide range of topics.

Google Earth. http://earth.google.com/. Allows users to view virtually any location on earth, tilt and rotate site views for three-dimensional images, and save or share search results.

MapQuest. www.mapquest.com/. Allows users to search for specific locations and view maps and detailed directions to the site.

STATISTICAL SOURCES. Statistical sources are collections of numerical data. They are the best sources for such information as the U.S. gross domestic product, the consumer price index, or the demographic breakdown of the general population.

American Statistics Index. Washington, D.C.: Congressional Information Service, 1978–. Monthly, quarterly, and annual supplements.

U.S. Census Bureau. *Statistical Abstract of the United States.* Washington, D.C.: Government Printing Office, 1879–. Annual (*www.census.gov*).

Internet Research Strategies

The Internet provides access to a staggering amount of information. However, the information on the Web varies widely in its completeness and accuracy, so you need to evaluate Internet sources critically by following the advice in *Writer's Checklist: Evaluating Print and Online Sources* on page 466.

As comprehensive as search engines and directories may seem, keep in mind that none is complete or objective and they carry only a preselected range of content. Many, for example, do not index Adobe PDF files or Usenet newsgroups, and many cannot index databases and other non-HTML-based content. Search engines rank the sites they believe will be relevant to you based on a number of different strategies. Some search engines base relevance on how high on the given page your search term appears, on the number of appearances of your term, or on the number of other sites that link to the page. Almost all major search engines now sell high rankings to advertisers with the highest bids, so your results may not highlight the pages most relevant to your search. Your best strategy is to research how your favorite search engines work; nearly all provide detailed methodologies on their help pages.

Search Engines. A search engine locates information based on words or combinations of words that you specify. The software engine then lists for you the documents or files that contain one or more of these words in their titles, descriptions, or text. The following search engines are used widely:

Google	www.google.com
MSN	www.live.com
Yahoo!	www.yahoo.com

Also available are metasearch engines—tools that do not maintain an internal database but instead launch your query to multiple databases of various Web-based resources (other search engines or subject directories).

Dogpile	www.dogpile.com
Metacrawler	www.metacrawler.com

Many search engines allow advanced searches with options that provide more selective results. Although search engines vary in what and how they search, you can use some basic strategies, described in the following *Writer's Checklist*.

WRITER'S CHECKLIST Using Search Engines and Keywords

- Enter keywords and phrases that are as specific to your topic as possible. For example, if you are looking for information about *nuclear power* and enter only the term *nuclear*, the search will also yield listings for *nuclear* family, *nuclear* medicine, and hundreds of others not related to your topic.

- Use Boolean operators (AND, OR, NOT) to narrow your search. For example, if you are searching for information on breast cancer and are finding references to nothing but prostate cancer, try "breast AND cancer NOT prostate."

- Check any search tips available at the engine you use. For example, some engines allow you to use an advanced search that narrows your search by combining phrases with double quotation marks: "usability testing" will return only pages that have the full compound phrase.

- Try several search engines to get more varied results and remember that high rankings may be based on marketing by advertisers.

- Consider using a metasearch engine, such as Dogpile (*www.dogpile.com*) and Metacrawler (*www.metacrawler.com*), if you are interested in obtaining as many hits as possible. Be prepared to refine and narrow your search.

Web Subject Directories. A subject directory (also known as an index) organizes information by broad subject categories (natural sciences, entertainment, health, sports) and related subtopics (medicine, oceanography, physics). One such directory is *http://directory.google.com/*. A subject-directory search eventually produces a list of specific sites that contain information about the topics you request.

In addition to the subject directories offered by many search engines, the following directories will help you to conduct selective, scholarly research on the Web:

Infomine	*http://infomine.ucr.edu*
The Internet Public Library	*www.ipl.org*
The WWW Virtual Library	*www.vlib.org*

The Web includes numerous directories and sites devoted to specific subject areas. Following are some suggested resources for researching science and technology.

Pace University Library Internet Resources	*www.pace.edu/library/links/links.html*
CIO's Resource Centers	*www.cio.com/research*
IEEE Spectrum Online	*www.spectrum.ieee.org*
National Science Foundation	*www.nsf.gov*
LSU Libraries Federal Agencies Directory	*www.lib.lsu.edu/gov/fedgov.html*
FedStats	*www.fedstats.gov*

Some sites combine search engines with directories; for example, Google operates both a standard search engine and directory and a special contributor-generated directory referred to as an "Open Directory" (*http://dmoz.org*).

Evaluating Sources

The easiest way to ensure that information is valid is to obtain it from a reputable source. For Internet sources, be especially concerned about the validity of the information provided. Because anyone can publish on the Web, it is sometimes difficult to determine authorship of a document, and frequently a person's qualifications for speaking on a topic are absent or questionable. The Internet versions of established, reputable journals in medicine, management, engineering, computer software, and the like merit the same level of trust as the printed versions. Use the following domain abbreviations to help you determine an Internet site sponsor:

.aero	aerospace industry	.mil	U.S. military
.biz	business	.museum	museum
.com	company or individual	.name	individual

.coop	business cooperative	.net	network provider
.edu	college or university	.org	nonprofit organization
.gov	federal government	.int	international
.pro	professionals	.info	general use

As you move away from established, reputable sites, exercise more caution. Be especially wary of unmoderated, public Web forums. Collectively generated Web sites, such as *Craigslist* and *Wikipedia*, often make no guarantee of the validity of information on their sites (see *http://en .wikipedia.org/wiki/Wikipedia:General_disclaimer*).

WRITER'S CHECKLIST Evaluating Print and Online Sources

Keep in mind the following four criteria when evaluating Internet sources: authority, accuracy, bias, and currency.*

FOR ALL SOURCES

- Is the resource recent enough and relevant to your topic? Is it readily available?
- Who is the intended audience? The mainstream public? A small group of professionals?
- Who is the author(s)? Is the author(s) an authority on the subject?
- Does the author(s) provide enough supporting evidence and document sources so that you can verify the information's accuracy?
- Is the information presented in an objective, unbiased way? Are any biases made clear? Are opinions clearly labeled? Are viewpoints balanced, or are opposing opinions acknowledged?
- Are the language, tone, and style appropriate and cogent?

FOR A BOOK

- Does the preface or introduction indicate the author's or book's purpose?
- Does the table of contents relate to your topic? Does the index contain terms related to your topic?
- Are the chapters useful? Skim through one chapter that seems related to your topic — notice especially the introduction, headings, and closing.

FOR AN ARTICLE

- Is the publisher of the magazine or other periodical well known?
- What is the article's purpose? For a journal article, read the abstract; for a newspaper article, read the headline and lead sentences.

*For more details, see Leigh Ryan, *The Bedford Guide for Writing Tutors*, 3rd ed. (Boston: Bedford/St. Martin's, 2002).

Writer's Checklist: Evaluating Print and Online Sources (continued)

- Does the article contain informative diagrams or other visuals that indicate its scope?

FOR A WEB SITE

- Does a reputable group or organization sponsor or maintain the site?
- Are the purpose and scope of the site clearly stated? Check the "Mission Statement" or "About Us" pages. Are there any disclaimers?
- Is the site updated, thus current? Are the links functional and up to date?
- Is the documentation authoritative and credible? Check the links to other sources and cross-check facts at other reputable Web sites, such as academic ones.
- Is the site well designed? Is the material well written and error free?

> **WEB LINK** **Evaluating Online Sources**
>
> For a tutorial on evaluating information online, see *bedfordstmartins.com/alredtech* and select *Tutorials*, "Evaluating Online Sources." For links to additional related resources, select *Links for Handbook Entries*.

resignation letters

Resignation **letters** (or **memos**), as shown in Figures R–3 and R–4, should be as positive as possible, regardless of the reason you are leaving a job. You usually write a resignation letter to your supervisor or to an appropriate person in the Human Resources Department. Use the following guidelines.

- Start on a positive note, regardless of the circumstances under which you are leaving.
- Consider pointing out how you have benefited from working for the company or say something complimentary about the company.
- Comment on something positive about the people with whom you have been associated.
- Explain why you are leaving in an objective, factual tone.
- Avoid angry recriminations because your resignation will remain on file with the company and could haunt you in the future should you need references.

> 227 Kenwood Drive
> Austin, TX 78719
> January 6, 2009
>
> R.W. Johnson, Director of Metallurgy
> Hannibal Laboratories
> 1914 East 6th Street
> Austin, TX 78702
>
> Dear Mr. Johnson:
>
> [Positive opening] My three years at Hannibal Laboratories have been an invaluable period of learning and professional development. I arrived as a novice, and I believe that today I am a professional—primarily as a result of the personal attention and tutoring I have received from my superiors and the fine example set by both my superiors and my peers.
>
> [Reason for leaving] I believe, however, that the time has come for me to move on to a larger company that can give me an opportunity to continue my professional development. Therefore, I have accepted a position with Procter & Gamble, where I am scheduled to begin on January 26. Thus, my last day at Hannibal will be January 23. I will be happy to train my replacement during the next two weeks.
>
> [Positive closing] Many thanks for the experience I have gained and best wishes for the future.
>
> Sincerely,
>
> *J. L. Washburne*
>
> J. L. Washburne

FIGURE R–3. Resignation Letter (to Accept a Better Position)

Your letter or memo should give enough notice to allow your employer time to find a replacement. It might be no more than two weeks or it might be enough time to enable you to train your replacement. Some organizations may ask for a notice equivalent to the number of weeks of vacation you receive. Check the policy of your employer before you begin your letter.

The sample resignation letter in Figure R–3 is from an employee who is leaving to take a job offering greater opportunities. The memo

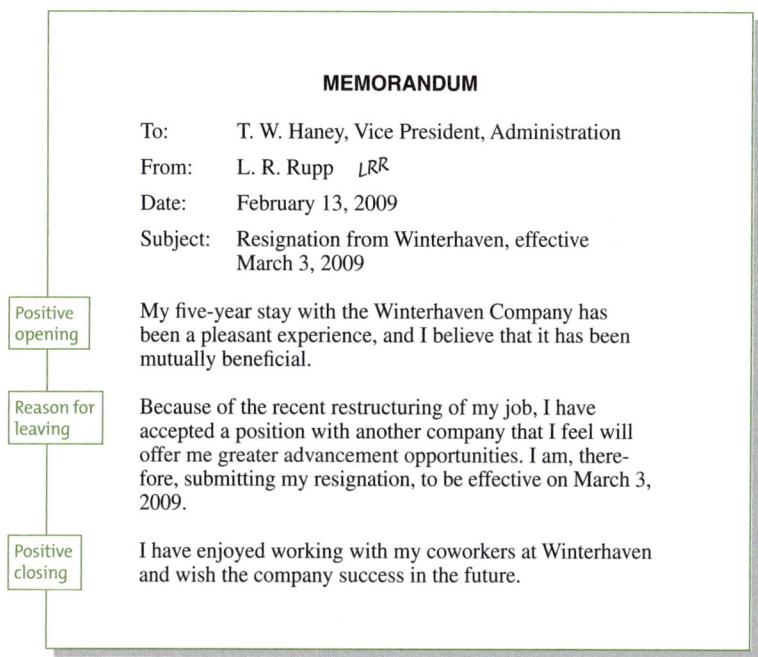

FIGURE R–4. Resignation Memo (Under Negative Conditions)

of resignation in Figure R–4 is written by an employee who is leaving because her position has been reclassified and her supervisor has not supported her advancement, although no personal conflict is mentioned. Notice that it opens and closes positively and that the reason for the resignation is stated without apparent anger or bitterness. For strategies concerning negative messages, see **correspondence** and **refusal letters**.

respective / respectively

Respective is an **adjective** that means "pertaining to two or more things regarded individually." ("The committee members returned to their *respective* offices.") *Respectively* is the adverb form of *respective*, meaning "singly, in the order designated."

▶ The first, second, and third engineering design award winners were Maria Juarez, Dan Wesp, and Simone Luce, *respectively*.

Respective and *respectively* are unnecessary if the meaning of individuality is clear.

> Each   member   a report.
> ▶ T̶h̶e̶ committee m̶e̶m̶b̶e̶r̶s̶ prepared t̶h̶e̶i̶r̶ ̶r̶e̶s̶p̶e̶c̶t̶i̶v̶e̶ ̶r̶e̶p̶o̶r̶t̶s̶.̶

restrictive and nonrestrictive elements

Modifying **phrases** and **clauses** may be either restrictive or nonrestrictive. A *nonrestrictive phrase or clause* provides additional information about what it modifies, but it does not restrict the meaning of what it modifies. A nonrestrictive phrase or clause can be removed without changing the essential meaning of the sentence. It is a parenthetical element that is set off by **commas** to show its loose relationship with the rest of the sentence.

> NONRESTRICTIVE   This instrument, *which is called a backscatter gauge*, fires beta particles at an object and counts the particles that bounce back.

A *restrictive phrase or clause* limits, or restricts, the meaning of what it modifies. If it were removed, the essential meaning of the sentence would change. Because a restrictive phrase or clause is essential to the meaning of the sentence, it is never set off by commas.

> RESTRICTIVE   All employees *wishing to donate blood* may take Thursday afternoon off.

Writers need to distinguish between nonrestrictive and restrictive elements. The same sentence can take on two entirely different meanings, depending on whether a modifying element is set off by commas (because it is nonrestrictive) or is not (because it is restrictive). A slip by the writer can not only mislead **readers** but also embarrass the writer.

> MISLEADING   He gave a poor performance evaluation to the staff members who protested to the Human Resources Department.
> [This suggests that he gave the poor evaluation because the staff members had protested.]
>
> ACCURATE   He gave a poor performance evaluation to the staff members, who protested to the Human Resources Department.
> [This suggests that the staff members protested because of the poor evaluation.]

Use *which* to introduce nonrestrictive clauses and *that* to introduce restrictive clauses.

NONRESTRICTIVE   After John left the restaurant, *which* is one of the finest in New York, he came directly to my office.
RESTRICTIVE   Companies *that* diversify usually succeed.

résumés

DIRECTORY
Sample Résumés   472
Analyzing Your Background   481
Organizing Your Résumé (Sections)   481
   Heading   482
   Job Objective   482
   Qualifications Summary   483
   Education   483
   Employment Experience   483
   Related Skills and Abilities   484
   Honors and Activities   485
   References and Portfolios   485
Salary and Advice for Returning Job Seekers   485
   Salary   485
   Returning Job Seekers   485
Electronic Résumés   486
   Web Résumés   486
   Scannable and Plain-Text Résumés   487
   E-mail–Attached Résumés   487

A résumé is the key tool of the **job search** that itemizes your qualifications and serves as a foundation for your **application letter** (often referred to as a *cover letter*). A résumé* should be limited to one page—or two pages if you have substantial experience. On the basis of the information in the résumé and application letter, prospective employers decide whether to ask you to come in for an interview. If you are invited to an interview, the interviewer can base specific questions on the contents of your résumé. See also **interviewing for a job**.

Because résumés affect a potential employer's first impression, make sure that yours is well organized, carefully designed, consistently

---

*A detailed résumé for someone in an academic and a scientific area is often called a *curriculum vitae* (also *vita* or *c.v.*). It may include education, publications, projects, grants, and awards as well as a full work history. Outside the United States, the term *curriculum vitae* is often used to mean *résumé*.

formatted, easy to read, and free of errors. Consider first an **organization** that highlights your strengths and fits your goals, as suggested by the examples shown in this entry. Experiment to determine a **layout and design** that is attractive and uncluttered. Consistency is especially important on a résumé. Be sure to use, for example, the same date formats (5/2009 or May 2009), punctuation, and spacing throughout. **Proofreading** is essential. Verify the accuracy of the information and have someone else review it. For hard-copy résumés, use a quality printer and high-grade paper.

▶ **ETHICS NOTE** Be truthful. The consequences of giving false information in your résumé are serious. In fact, the truthfulness of your résumé reflects not only on your own ethical stance but also on the integrity with which you would represent the organization. See also **ethics in writing**. ✦

Sample Résumés

The sample résumés in this entry are provided to stimulate your thinking about how to tailor your résumé to your own job search. Before you design and write your résumé, look at as many samples as possible and then organize and format your own to best suit your previous experience and your professional goals and to make the most persuasive case to your target employers. See also **persuasion**.

- Figure R–5 presents a conventional student résumé in which the student is seeking an entry-level position.
- Figure R–6 shows a résumé with a variation of the conventional headings to highlight professional credentials.
- Figure R–7 presents a student résumé with a format that is appropriately nonconventional because this student needs to demonstrate skills in graphic design for his potential audience. This résumé matches the letter in Figure A–9.
- Figure R–8 shows a résumé that focuses on the applicant's management experience. This résumé matches the letter in Figure A–10.
- Figure R–9 focuses on how the applicant advanced and was promoted within a single company.
- Figure R–10 illustrates how an applicant can organize a résumé by combining functional and chronological elements.
- Figure R–11 presents an electronic résumé in ASCII (American Standard Code for Information Interchange) format. Notice how this résumé emphasizes keywords so that potential employers searching a database for applicants will be able to find it easily.

Ana María López

Campus Address
148 University Drive
Bloomington, Indiana 47405
(812) 652-4781
aml@iu.edu

Home (after June 2009)
1436 West Schantz Avenue
Laurel, Pennsylvania 17322
(717) 399-2712
aml@yahoo.com

OBJECTIVE

Position as dental hygienist, with long-term goal of developing a community practice dental service.

EDUCATION

Bachelor of Science in Dental Hygiene, expected June 2009
Indiana University

Licensure: August 2008
Grade Point Average: 3.88 out of possible 4.0
Senior Honor Society
Minor: Management Information Systems

DENTAL EXPERIENCE

NORTHPOINT DENTAL ASSOCIATES, Bloomington, Indiana, 2008
Dental Assistant, Summer and Fall Quarters
    Developed office and laboratory management system.

RODRIGUEZ DENTAL ASSOCIATES, Bloomington, Indiana, 2007
Dental Assistant Intern
    Prepared patients for exams; processed X-rays; maintained patient treatment records.
Associate Editor, *Community Health Newsletter*, 2006–2007
    Wrote articles on good dental health practices; researched community health needs for editor; edited submissions.

COMPUTER SKILLS

Software: Microsoft Word, Excel, PowerPoint, Lotus
Hardware: Macintosh, IBM-PC
Medical: Magnus Patient Database System, Dentrix

REFERENCES

Available on request.

FIGURE R–5. Student Résumé (for an Entry-Level Position)

CHRIS RENAULT, RN

3785 Raleigh Court, #46 • Phoenix, AZ 67903 •
(555) 467-1115 • chris@resumepower.com

Qualifications

➤ Recent Honors Graduate of Approved Nursing Program
➤ Current Arizona Nursing Licensure and BLS Certification
➤ Presently Completing Clinical Nurse Internship Program

Education & Licensure

ARIZONA STATE UNIVERSITY
Tempe, AZ
Bachelor of Science in Nursing (BSN), 2009
Graduated summa cum laude (GPA: 4.0)

MOHAVE COMMUNITY COLLEGE
Kingman, AZ
Associate Degree in Nursing (AN), 2007
Graduated cum laude (GPA: 3.5)

Coursework Highlights: Family and Community Nursing, Health-Care Delivery Models, Health Assessment, Pathology, Microbiology, Nursing Research, Nursing of Older Adults, Health-Care Ethics

Arizona RN License, 2009
BLS Certification, 2009

Clinical Internship

CAMELBACK MEDICAL CENTER — Phoenix, AZ
Nurse Intern, 2008 to Present

- Accepted into new graduate RN training program and completing in-depth, eight-month rotation working under a trained preceptor.
- Gaining valuable clinical experience to assume the role of a professional nurse within an acute-care setting. Rotating through all medical center areas, including Postsurgical, Orthopedics, Pediatrics, Oncology, Emergency Department, Psychiatric Nursing, Cardiac Telemetry, and Critical Care.
- Developing speed and skill in the day-to-day functions of a staff nurse. Participating in patient assessment, treatment, medication disbursement, and surgical preparation as a member of the health-care team.
- Earned written commendations from preceptor for *"excellent ability to interact with patients and their families, showing a high degree of empathy, medical knowledge, and concern for quality and continuity of patient care."*

Community Involvement

Active Volunteer and Fundraising Coordinator, The American Cancer Society — Scottsdale, AZ, Chapter (2007 to Present)
Participant, Annual AIDS Walkathon (2004 to 2007) and "Find the Cure" Breast Cancer Awareness Marathon (2006, 2007)

FIGURE R–6. Résumé (Highlighting Professional Credentials). Prepared by Kim Isaacs, Advanced Career Systems, Inc.

Joshua S. Goodman
222 Morewood Avenue
Pittsburgh, PA 15212
cell: 412-555-1212
jgoodman@aol.com

OBJECTIVE

A position as a graphic designer with responsibilities in information design, packaging, and media presentations.

EDUCATION

**Carnegie Mellon University, Pittsburgh, Pennsylvania
BFA in Graphic Design — May 2009**

Graphic Design
Corporate Identity
Industrial Design
Graphic Imaging Processes
Color Theory
Computer Graphics
Typography
Serigraphy
Photography
Video Production

GRAPHIC DESIGN EXPERIENCE

**Assistant Designer • Dyer/Khan, Los Angeles, California
Summer 2007, Summer 2008**
Assistant Designer in a versatile design studio. Responsible for design, layout, comps, mechanicals, and project management.
Clients: Paramount Pictures, Mattel Electronics, and Motown Records.

**Photo Editor • Paramount Pictures Corporation, Los Angeles, California
Summer 2006**
Photo Editor for merchandising department. Established art files for movie and television properties. Edited images used in merchandising. Maintained archive and database.

**Production Assistant • Grafis, Los Angeles, California
Summer 2005**
Production assistant at fast-paced design firm. Assisted with comps, mechanicals, and miscellaneous studio work.
Clients: ABC Television, A&M Records, and Ortho Products Division.

COMPUTER SKILLS

XML, HTML, JavaScript, Forms, Macromedia Dreamweaver, Macromedia Flash Professional, Photoshop, Illustrator, Image Ready (Animated GIFs), CorelDRAW, DeepPaint, iGrafx Designer, MapEdit (Image Mapping), Scanning, Microsoft Access/Excel, QuarkXPress.

ACTIVITIES

Member, Pittsburgh Graphic Design Society; Member, The Design Group.

FIGURE R-7. Student Résumé (for a Graphic Design Job)

ROBERT MANDILLO
7761 Shalamar Drive
Dayton, Ohio 45424
(937) 255-4137
mand@juno.com

OBJECTIVE

A management position in the aerospace industry with responsibility for developing new designs and products.

MANAGEMENT EXPERIENCE

MANAGER, EXHIBIT DESIGN LAB — May 2002–Present
Wright-Patterson Air Force Base, Dayton, Ohio

Supervise 11 technicians in support of engineering exhibit design and production. Develop, evaluate, and improve materials and equipment for the design and construction of exhibits. Write specifications, negotiate with vendors, and initiate procurement activities for exhibit design support.

SUPERVISOR, GRAPHICS ILLUSTRATORS — June 1999–April 2002
Henderson Advertising Agency, Cincinnati, Ohio

Supervised five illustrators and four drafting mechanics after promotion from Graphics Technician. Analyzed and approved work-order requirements. Selected appropriate media and techniques for orders. Rendered illustrations in pencil and ink. Converted department to CAD system.

EDUCATION

MASTER OF BUSINESS ADMINISTRATION, 2008
University of Dayton, Ohio

BACHELOR OF SCIENCE IN MECHANICAL ENGINEERING TECHNOLOGY, 1989
Edison State College, Wooster, Ohio

ASSOCIATE'S DEGREE IN MECHANICAL DRAFTING, 1987
Wooster Community College, Wooster, Ohio

PROFESSIONAL AFFILIATION

National Association of Mechanical Engineers

REFERENCES / WEB SITE

References, letters of recommendation, and a portfolio of original designs and drawings available online at <www.juno.com/mand>.

FIGURE R–8. Résumé (Applicant with Management Experience)

CAROL ANN WALKER
1436 West Schantz Avenue
Laurel, Pennsylvania 17322
(717) 399-2712
caw@yahoo.com

FINANCIAL EXPERIENCE

KERFHEIMER CORPORATION, Philadelphia, Pennsylvania

Senior Financial Analyst, June 2004–Present
Report to Senior Vice President for Corporate Financial Planning. Develop manufacturing cost estimates totaling $30 million annually for mining and construction equipment with Department of Defense.

Financial Analyst, November 2001–June 2004
Developed $50-million funding estimates for major Department of Defense contracts for troop carriers and digging and earth-moving machines. Researched funding options, resulting in savings of $1.2 million.

FIRST BANK, INC., Bloomington, Indiana

Planning Analyst, September 1996–November 2001
Developed successful computer models for short- and long-range planning.

EDUCATION

Ph.D. in Finance: expected, June 2009
The Wharton School of the University of Pennsylvania

M.S. in Business Administration, 2000
University of Wisconsin–Milwaukee
"Executive Curriculum" for employees identified as promising by their employers.

B.S. in Business Administration (*magna cum laude*), 1996
Indiana University
Emphasis: Finance   Minor: Professional Writing

PUBLISHING AND MEMBERSHIP

Published "Developing Computer Models for Financial Planning," *Midwest Finance Journal* 34.2 (2006): 126–36.

Association for Corporate Financial Planning, Senior Member.

REFERENCES

References and a portfolio of financial plans are available on request.

FIGURE R–9. Advanced Résumé (Showing Promotion Within a Single Company)

CAROL ANN WALKER

1436 West Schantz Avenue • Laurel, PA 17322
(717) 399-2712 • caw@yahoo.com

Award-Winning Senior Financial Analyst

Astute senior analyst and corporate financial planner with 11 years of experience and proven success enhancing P&L scenarios by millions of dollars. Demonstrated ability to apply critical thinking and sound strategic/economic analysis to multidimensional business issues. Advanced computer skills include Hyperion, SQL, MS Office, and Crystal Reports.

Financial Analyst of the Year, 2007

Recipient of prestigious national award from the Association for Investment Management and Research (AIMR)

Areas of Expertise

- Financial Analysis & Planning
- Forecasting & Trend Projection
- Trend/Variance Analysis
- Comparative Analysis
- Expense Analysis
- Strategic Planning
- SEC & Financial Reporting
- Risk Assessment

Career Progression

KERFHEIMER CORPORATION—Philadelphia, PA   2001 to Present

Senior Financial Analyst, June 2004 to Present
Financial Analyst, November 2001 to June 2004

Rapidly promoted to lead team of 15 analysts in the management of financial/SEC reporting and analysis for publicly traded, $2.3-billion company. Develop financial/statistical models used to project and maximize corporate financial performance. Support nationwide sales team by providing financial metrics, trends, and forecasts.

Key Accomplishments:

- **Developed long-range funding requirements crucial to firm's subsequent capture of $1 billion** in government and military contracts.
- **Facilitated a 45% decrease in company's long-term debt** during several major building expansions through personally developed computer models for capital acquisition.
- **Jointly led large-scale systems conversion to Hyperion**, including personal upload of database in Essbase. Completed conversion without interrupting business operations.

FIGURE R–10. Advanced Résumé (Combining Functional and Chronological Elements). Prepared by Kim Isaacs, Advanced Career Systems, Inc.

———— CAROL ANN WALKER ————

Résumé • Page Two

Career Progression (*continued*)

FIRST BANK, INC.—Bloomington, IN　　　　　　1996 to 2001
Planning Analyst, September 1996 to November 2001
Compiled and distributed weekly, monthly, quarterly, and annual closings/financial reports, analyzing information for presentation to senior management. Prepared depreciation forecasts, actual-vs.-projected financial statements, key-matrix reports, tax-reporting packages, auditor packages, and balance-sheet reviews.

Key Accomplishments:

- **Devised strategies to acquire over $1 billion at 3% below market rate.**
- **Analyzed financial performance for consistency to plans and forecasts**, investigated trends and variances, and alerted senior management to areas requiring action.
- **Achieved an average 23% return on all personally recommended investments.** Applied critical thinking and sound financial and strategic analysis in all funding options research.

Education

THE WHARTON SCHOOL of the UNIVERSITY OF PENNSYLVANIA (Philadelphia)
Ph.D. in Finance Candidate, Expected June 2009

UNIVERSITY OF WISCONSIN–MILWAUKEE
M.S. in Business Administration, May 2000

INDIANA UNIVERSITY (Bloomington)
B.S. in Business Administration, Emphasis in Finance (*magna cum laude*), May 1996

Affiliations

- Association for Investment Management and Research (AIMR), Member, 2003 to Present
- Association for Corporate Financial Planning (ACFP), Senior Member, 2001 to Present

Portfolio of Financial Plans Available on Request
(717) 399-2712 • caw@yahoo.com

FIGURE R–10. Advanced Résumé (Combining Functional and Chronological Elements) (*continued*)

DAVID B. EDWARDS
6819 Locustview Drive
Topeka, Kansas 66614
(913) 233-1552
dedwards@cpu.fairview.edu

JOB OBJECTIVE
Programmer with writing, editing, and training responsibilities,
leading to a career in information design management.

KEYWORDS
Programmer, Operating Systems, Unipro, Newsletter, Graphics,
Listserv, Professional Writer, Editor, Trainer, Instructor,
Technical Writer, Tutor, Designer, Manager, Information Design.

EDUCATION
** Fairview Community College, Topeka, Kansas
** Associate's Degree, Computer Science, June 2006
** Dean's Honor List Award (six quarters)

RELEVANT COURSE WORK
** Operating Systems Design
** Database Management
** Introduction to Cybernetics
** Technical Writing

EMPLOYMENT EXPERIENCE
** Computer Consultant: September 2006 to Present
Fairview Community College Computer Center: Advised and trained novice
users; wrote and maintained Unipro operating system documentation.
** Tutor: January 2005 to June 2006
Fairview Community College: Assisted students in mathematics and
computer programming.

SKILLS AND ACTIVITIES
** Unipro Operating System: Thorough knowledge of word-processing,
text-editing, and file-formatting programs.
** Writing and Editing Skills: Experience in documenting computer
programs for beginning programmers and users.
** Fairview Community Microcomputer Users Group: Cofounder and
editor of monthly newsletter ("Compuclub"); listserv manager.

FURTHER INFORMATION
** References, college transcripts, a portfolio of computer programs,
and writing samples available on request.

FIGURE R–11. Electronic Résumé (in ASCII Format)

> **WEB LINK** Annotated Sample Résumés
>
> For more examples of résumés with helpful annotations, see *bedfordstmartins.com/alredtech* and select *Model Documents Gallery*.

Analyzing Your Background

In preparing to write your résumé, determine what kind of job you are seeking. Then ask yourself what information about you and your background would be most important to a prospective employer. List the following:

- Schools you attended, degrees you hold, your major field of study, academic honors you were awarded, your grade point average, particular academic projects that reflect your best work
- Jobs you have held, your principal and secondary duties in each job, when and how long you held each job, promotions, skills you developed in your jobs that potential employers value and seek in ideal job candidates, projects or accomplishments that reflect your important contributions
- Other experiences and skills you have developed that would be of value in the kind of job you are seeking; extracurricular activities that have contributed to your learning experience; leadership, interpersonal, and communication skills you have developed; any collaborative work you have performed; computer skills you have acquired

Use this information to brainstorm any further key details. Then, based on all the details, decide which to include in your résumé and how you can most effectively present your qualifications.

Organizing Your Résumé (Sections)

A number of different organizational patterns can be used effectively. The following sections are typical—which you choose should depend on your experience and goals, the employer's needs, and any standard practices in your profession.

- Heading (name and contact information)
- Job Objective
- Qualifications Summary
- Education
- Employment Experience

- Related Skills and Abilities
- Honors and Activities
- References and Portfolios

Whether you place "education" before "employment experience" depends on the job you are seeking and on which credentials would strengthen your résumé the most. If you are a recent graduate without much work experience, list education first. If you have years of job experience, including jobs directly related to the kind of position you are seeking, list employment experience first. In your education and employment sections, use a reverse chronological sequence: List the most recent experience first, the next most recent experience second, and so on.

Heading. At the top of your résumé, include your name, address, telephone number (home or cell), and a professional **e-mail** address.* Make sure that your name stands out on the page. If you have both a school address and a permanent home address, place your school address on the left side of the page and your permanent home address on the right side of the page. Place both underneath your name, as shown in Figure R–5. Indicate the dates you can be reached at each address (but do not date the résumé itself).

Job Objective. Some potential employers prefer to see a clear employment objective in résumés. An objective introduces the material in a résumé and helps the reader quickly understand your goal. If you decide to include an objective, use a heading such as "Objective," "Employment Objective," "Career Objective," or "Job Objective." State your immediate goal and, if you know that it will give you an advantage, the direction you hope your career will take. Try to write your objective in no more than three lines, and tailor it to the specific job for which you are applying, as illustrated in the following examples:

- A full-time computer-science position aimed at solving engineering problems and contributing to a management team.
- A position involving meeting the concerns of women, such as family planning, career counseling, or crisis management.
- Full-time management of a high-quality local restaurant.
- A summer research or programming position providing opportunities to use software-development and software-debugging skills.

*Do not use a clever or hobby-related e-mail address in employment correspondence; e-mail addresses that are based on your last name work well.

Qualifications Summary. You may wish to include a brief summary of your qualifications to persuade hiring managers to select you for an interview. Sometimes called a *summary statement* or *career summary*, a qualifications summary can include skills, achievements, experience, or personal qualities that make you especially well suited to the position. You may wish to give this section a heading such as "Profile," "Career Highlights," or simply "Qualifications." Or you may use a headline, as shown on the first page of the résumé in Figure R–10 ("Award-Winning Senior Financial Analyst").

Education. List the school(s) you have attended, the degree(s) you received and the dates you received them, your major field(s) of study, and any academic honors you have earned. Include your grade point average only if it is 3.0 or higher—or include your average in your major if that is more impressive. List courses only if they are unusually impressive or if your résumé is otherwise sparse (see Figures R–6 and R–7). Consider including the skills developed or projects completed in your course work. Mention your high school only if you want to call attention to special high school achievements, awards, projects, programs, internships, or study abroad.

Employment Experience. Organize your employment experience in reverse chronological order, starting with your most recent job and working backward under a single major heading called "Experience," "Employment," "Professional Experience," or the like. You could also organize your experience functionally by clustering similar types of jobs into one or several sections with specific headings such as "Management Experience" or "Major Accomplishments."

One type of arrangement might be more persuasive than the other, depending on the situation. For example, if you are applying for an accounting job but have no specific background in accounting, you would probably do best to list past and present jobs in chronological order, from most to least recent. If you are applying for a supervisory position and have had three supervisory jobs in addition to two nonsupervisory positions, you might choose to create a single section called "Supervisory Experience" and list only your three supervisory jobs. Or you could create two sections—"Supervisory Experience" and "Other Experience"—and include the three supervisory jobs in the first section and the nonsupervisory jobs in the second section.

The functional résumé groups work experience by types of workplace activities or skills rather than by jobs in chronological order. However, many employers are suspicious of functional résumés because they can be used to hide a poor work history, such as excessive job hopping or extended employment gaps. Functional elements can be combined with a chronological arrangement by using a qualifications summary or skills section, as shown in Figure R–10.

In general, follow these conventions when working on the "Experience" section of your résumé.

- Include jobs or internships when they relate directly to the position you are seeking. Although some applicants choose to omit internships and temporary or part-time jobs, including such experiences can make a résumé more persuasive if they have helped you develop specific related skills.
- Include extracurricular experiences, such as taking on a leadership position in a college organization or directing a community-service project, if they demonstrate that you have developed skills valued by potential employers.
- List military service as a job; give the dates served, the duty specialty, and the rank at discharge. Discuss military duties if they relate to the job you are seeking.
- For each job or experience, list both the job and company titles. Throughout each section, consistently begin with either the job or the company title, depending on which will likely be more impressive to potential employers.
- Under each job or experience, provide a concise description of your primary and secondary duties. If a job is not directly relevant, provide only a job title and a brief description of duties that helped you develop skills valued in the position you are seeking. For example, if you were a lifeguard and now seek a management position, focus on supervisory experience or even experience in averting disaster to highlight your management, decision-making, and crisis-control skills.
- Focus as much as possible on your achievements in your work history ("Increased employee retention rate by 16 percent by developing a training program"). Employers want to hire doers and achievers.
- Use action verbs (for example, "managed" rather than "as the manager") and state ideas succinctly, as shown in Figure R–8. Even though the résumé is about you, do not use "I" (for example, instead of "I was promoted to Section Leader," use "Promoted to Section Leader"). For electronic résumés that will be scanned for keywords, however, replace such verbs with nouns (instead of "managed" use *manager*, as in Figure R–11).

Related Skills and Abilities. Employers are interested in hiring applicants with a variety of skills or the ability to learn new ones quickly. Depending on the position, you might list in a skills section items such as fluency in foreign languages, writing and editing abilities, specialized technical knowledge, or computer skills (including knowledge of specific languages, software, and hardware).

Honors and Activities. If you have room on your résumé, list any honors and unique activities near the end. Include items such as student or community activities, professional or club memberships, awards received, and published works. Be selective: Do not duplicate information given in other categories, and include only information that supports your employment objective. Provide a heading for this section that fits its contents, such as "Activities," "Honors," "Professional Affiliations," or "Publications and Memberships."

References and Portfolios. Avoid listing references unless that is standard practice in your profession or your résumé is sparse. If you create a separate list of references for prospective employers, you can include a phrase such as "References available on request" to signal the end of a résumé, or write "Available on request" after the heading "References" as a design element to balance a page. Always seek permission from anyone you list as a reference.

A portfolio is a collection of samples in a binder or on the Web of your most impressive work and accomplishments. The portfolio can include successful documents you have written, articles, letters of praise from employers, and copies of awards and certificates. If you have developed a portfolio, you could also include the phrase "Portfolio available on request." If portfolios are standard in your profession, you might even include a small section that lists the contents of your portfolio.

Salary and Advice for Returning Job Seekers

Salary. Avoid listing the salary you desire in the résumé. On the one hand, you may price yourself out of a job you want if the salary you list is higher than a potential employer is willing to pay. On the other hand, if you list a low salary, you may not get the best possible offer. See **salary negotiations**.

Returning Job Seekers. If you are returning to the workplace after an absence, most career experts say that it is important to acknowledge the gap in your career. That is particularly true if, for example, you are reentering the workforce because you have devoted a full-time period to care for children or dependent adults. Do not undervalue such work. Although unpaid, it often provides experience that develops important time-management, problem-solving, organizational, and interpersonal skills. Although gaps in employment can be explained in the application letter, the following examples illustrate how you might reflect such experiences in a résumé. They would be especially appropriate for an applicant seeking employment in a field related to child or health care.

▶ **Primary Child-Care Provider, 2006 to 2008**
Provided full-time care to three preschool children at home. Instructed in beginning academic skills, time management, basics of nutrition, arts, and swimming. Organized activities, managed household, and served as neighborhood-watch captain.

▶ **Home Caregiver, 2006 to 2008**
Provided 60 hours per week in-home care to Alzheimer's patient. Coordinated medical care, developed exercise programs, completed and processed complex medical forms, administered medications, organized budget, and managed home environment.

If you have performed volunteer work during such a period, list that experience. Volunteer work often results in the same experience as does full-time, paid work, a fact that your résumé should reflect, as in the following example.

▶ **School Association Coordinator, 2006 to 2008**
Managed special activities of the Briarwood High School Parent-Teacher Association. Planned and coordinated meetings, scheduled events, and supervised fund-drive operations. Raised $70,000 toward refurbishing the school auditorium.

Electronic Résumés

In addition to the traditional paper résumé, you can post a Web-based résumé. You may also need to submit a scannable, plain-text résumé through e-mail or an online form to a potential employer to be included in an organization's database. As Internet and database technologies converge, remain current with the forms and protocols that employers prefer by reviewing popular job-search sites, such as HotJobs at *http://hotjobs.yahoo.com* and Monster.com at *www.monster.com*.

Web Résumés. If you plan to post your résumé on your own Web site, keep the following points in mind.

- Follow the general advice for **Web design**, such as viewing your résumé on several browsers to see how it looks.
- Just below your name, you may wish to provide a series of internal page links to such important categories as "experience" and "education."
- Consider building a multipage site for displaying a work portfolio, publications, reference letters, and other related materials.
- If privacy is an issue, include an e-mail link ("mailto") at the top of the résumé rather than your home address and phone number.

The disadvantage of posting a résumé at your own Web site is that you must attract the attention of employers on your own. For that reason, commercial services may be a better option because they can attract recruiters with their large databases.

Scannable and Plain-Text Résumés. A scannable résumé is normally mailed to an employer in paper form, scanned, and downloaded into a company's searchable database. Such a résumé can be well formatted, but it should not contain decorative fonts, underlining, shading, letters that touch each other, and other features that will not scan easily. Scan such a résumé yourself to make sure there are no problems.

Some employers request ASCII or plain-text résumés via e-mail, which can be added directly into the résumé database without scanning. The ASCII résumés also allow employers to read the file no matter what type of software they are using. You can copy and paste such a résumé directly into the body of the e-mail message.

> **DIGITAL TIP**
>
> Preparing an ASCII Résumé
>
> When preparing an ASCII document, proper formatting is critical. For example, you need to insert manual line breaks at 65 characters to prevent long, single lines when the documents are opened in various systems. Further, many word-processing elements such as bullets, underlining, and boldface are incompatible with ASCII, which is limited to letters, numbers, and basic punctuation. For more on this topic, see bedfordstmartins.com/alredtech and select *Digital Tips*, "Preparing an ASCII Résumé."

For résumés that will be downloaded into databases, it is better to use nouns than verbs to describe experience and skills (*designer* and *management* rather than *designed* and *managed*). You may also include a section in such a résumé titled "Keywords" (or perhaps give a descriptive name, such as "Areas of Expertise"). Keywords, also called *descriptors*, allow potential employers to search the database for qualified candidates. So be sure to use keywords that are the same as those used in the employer's descriptions of the jobs that best match your interests and qualifications. This section can follow the main heading or appear near the end of your résumé. Figure R–11 is an example of an electronic résumé that demonstrates the use of keywords.

E-mail–Attached Résumés. An employer may request, or you may prefer to submit, a résumé as an e-mail attachment to be printed out by the employer. If so, consider using a relatively plain design and sending the résumé as a rich text format (.rtf) document. Or, if precise design is important, send the résumé as a portable document format (PDF) file that will preserve the fonts, images, graphics, and layout. You can attach this file to an e-mail message that will then serve as your **application letter**. See *Digital Tip: Using PDF Files* on page 574.

revision

When you revise your draft, read and evaluate it primarily from the point of view of your **audience**. In fact, revising requires a different frame of mind than **writing a draft**. To achieve that frame of mind, experienced writers have developed the following tactics:

- Allow a "cooling period" between writing the draft and revision in order to evaluate the draft objectively.
- Print out your draft and mark up the paper copy; it is often difficult to revise on-screen.
- Read your draft aloud—often, hearing the text will enable you to spot problem areas that need improvement.
- Revise in passes by reading through your draft several times, each time searching for and correcting a different set of problems.

When you can no longer spot improvements, you may wish to give the draft to a colleague for review—especially for projects that are crucial for you or your organization as well as for collaborative projects, as described in **collaborative writing**.

WRITER'S CHECKLIST Revising Your Draft

- *Completeness.* Does the document achieve its primary **purpose**? Will it fulfill the readers' needs? Your writing should give readers exactly what they need but not overwhelm them.
- *Appropriate introduction and conclusion.* Check to see that your **introduction** frames the rest of the document and your **conclusion** ties the main ideas together. Both should account for revisions to the content of the document.
- *Accuracy.* Look for any inaccuracies that may have crept into your draft.
- *Unity and coherence.* Check to see that sentences and ideas are closely tied together (**coherence**) and contribute directly to the main idea expressed in the topic sentence of each **paragraph** (see **unity**). Provide **transitions** where they are missing and strengthen those that are weak.

Writer's Checklist: Revising Your Draft (continued)

- *Consistency.* Make sure that **layout and design**, **visuals**, and use of language are consistent. Do not call the same item by one term on one page and a different term on another page.
- *Conciseness.* Tighten your writing so that it says exactly what you mean. Prune unnecessary words, phrases, sentences, and even paragraphs. See **conciseness**.
- *Awkwardness.* Look for **awkwardness** in **sentence construction** — especially any **garbled sentences**.
- *Ethical writing.* Check for **ethics in writing** and eliminate **biased language**.
- *Active voice.* Use the active **voice** unless the passive voice is more appropriate.
- *Word choice.* Delete or replace **vague words** and unnecessary **intensifiers**. Check for **affectation** and unclear **pronoun references**. See also **word choice**.
- *Jargon.* If you have any doubt that all your readers will understand any **jargon** or special terms you have used, eliminate or define them.
- *Clichés.* Replace **clichés** with fresh **figures of speech** or direct statements.
- *Grammar.* Check your draft for grammatical errors. Use computer **grammar** checkers with caution. Because they are not always accurate, treat their recommendations only as *suggestions*.
- *Typographical errors.* Check your final draft for typographical errors both with your spell checker and with thorough **proofreading** because spell checkers do not catch all errors.
- *Wordy phrases.* Use the search-and-replace command to find and revise wordy phrases, such as *that is, there are, the fact that*, and *to be*, and unnecessary helping **verbs** such as *will*.

DIGITAL TIP

Incorporating Tracked Changes

When colleagues review your document, they can track changes and insert comments within the document itself. Tracking and commenting vary with types and versions of word-processing programs, but in most programs you can view the tracked changes on a single draft or review the multiple drafts of your reviewers' versions. For more on this topic, see *bedfordstmartins.com/alredtech* and select *Digital Tips*, "Tracked Changes."

rhetorical questions

A rhetorical question is a question to which a specific answer is neither needed nor expected. The question is often intended to make an **audience** think about the subject from a different perspective; the writer or speaker then answers the question in an article or a **presentation**. The answer to a rhetorical question such as "Is space exploration worth the cost?" may not be a simple yes or no; it might be a detailed explanation of the pros and cons of the value of space exploration.

The rhetorical question can be used as an effective **title** or opening, especially in **newsletter articles**, **brochures**, or even **blogs**. However, it should be used judiciously in other, more formal documents. A rhetorical question, for example, would not be appropriate for the title of a **report** or an **e-mail** addressed to a manager who needs to quickly understand the subject and purpose of the document or message.

run-on sentences

A run-on sentence, sometimes called a *fused sentence*, is two or more sentences without punctuation to separate them. The term is also sometimes applied to a pair of independent **clauses** separated by only a **comma**, although this variation is usually called a **comma splice**. See also **sentence construction** and **sentence faults**.

Run-on sentences can be corrected, as shown in the following examples, by (1) making two sentences, (2) joining the two clauses with a **semicolon** (if they are closely related), (3) joining the two clauses with a comma and a coordinating **conjunction**, or (4) subordinating one clause to the other.

▶ The client suggested several solutions ~~some~~ *. Some* are impractical.

▶ The client suggested several solutions *;* some are impractical.

▶ The client suggested several solutions *, but* some are impractical.

▶ The client suggested several solutions *, although* some are impractical.

salary negotiations

Salary negotiations usually take place either at the end of an interview or after a formal job offer has been made. If possible, delay discussing salary until after you receive a formal written job offer because you will have more negotiating leverage at that point. See also **job search**.

Before **interviewing for a job**, prepare for possible salary negotiations by researching the following:

- The current range of salaries for the work you hope to do at your level (entry? intermediate? advanced?) in your region of the country. Check trade journals and organizations in your field, or ask a reference librarian for help in finding this information. Job listings that include salary can also be helpful.

- Salaries made by last year's graduates from your college or university at your level and in your line of work. Your campus career-development office should have these figures.

- Salaries made by people you know at your level and in your line of work. Attend local organizational meetings in your field or contact officers of local organizations who might have this information or steer you to useful contacts.

If a potential employer requests your salary requirements with your **résumé**, consider your options carefully. If you provide a salary that is too high, the company might never interview you; if you provide a salary that is too low, you may have no opportunity later in the hiring process to negotiate for a higher salary. However, if you fail to follow the potential employer's directions and omit the requested information, an employer may disqualify you on principle. If you choose to provide salary requirements, always do so in a range (for example, $35,000 to $40,000). See also **application letters**.

If an interviewer asks your salary requirements toward the end of the job interview, you can try the following strategies to delay salary negotiations:

- Say something like, "I am sure that this company pays a fair salary for a person with my level of experience and qualifications" or "I am ready to consider your best offer."

- Indicate that you would like to learn more details about the position before discussing salary; point out that your primary goal is to work in a stimulating environment with growth potential, not to earn a specific salary.
- Express generally a strong interest in the position and the organization without referring to the salary.
- Reemphasize your unique qualifications (or combination of skills) for the job and what you can do for the company that other candidates cannot.

If the interviewer or company demands to know your salary requirements during a job interview, provide a wide salary range that you know would be reasonable for someone at your level in your line of work in that region of the country. For example, you could say, "I would hope for a salary somewhere between $28,000 and $38,000, but of course this is negotiable."

> **WEB LINK — Salary Information Resources**
>
> For links to Web sites offering useful resources for salary negotiations, see *bedfordstmartins.com/alredtech* and select *Links for Handbook Entries*.

Once salary negotiations begin, resist the temptation to immediately accept the first salary offer you receive. If you have done thorough **research**, you will know if the first salary offer is at the low, middle, or high end of the salary range for your level of experience in your line of work. If you have little or no experience and receive an offer for a salary at the low end of the range, you will realize that the offer probably is fair and reasonable. However, if you receive the same low-end offer but bring considerable experience to the job, you can negotiate for a higher salary that is more reasonable for someone with your background and credentials in your line of work in your region of the country.

Never say that you are unable to accept a salary below a particular figure. To keep negotiations going, simply indicate that you would have trouble accepting the first offer because it was lower than you had expected.

Remember that you are negotiating a package and not just a starting salary. For example, consider the value of a chance for an early promotion, thus a higher salary, within a few years or a particular job title or special responsibilities that would provide you with impressive chances for career growth. If the starting salary seems low, consider negotiating for some of the following possible job perks:

- Tuition reimbursement for continued education
- Payment of relocation costs
- Paid personal leave or paid vacations
- Paid personal or sick days
- Overtime potential and compensation
- Flexible hours and work-from-home options
- Health, dental, optical or eye care, disability, and life insurance
- Retirement plans, such as 401(k) and pension plans
- Profit sharing; investment or stock options
- Bonuses or cost-of-living adjustments
- Commuting or parking-cost reimbursement
- Family leave
- Child-care or elder-care benefits
- Fitness or wellness benefits
- Discounts on company products and services

You might find it most comfortable to respond to an initial offer in writing and then meet later with the potential employer for further negotiation. If possible, indicate all of your preferences and requirements at one time instead of continually asking for new and different benefits as you negotiate. Throughout this process, focus on what is most important to you (which might differ from what is most important to your friends) and on what you would find acceptable and comfortable.

scope

Scope is the depth and breadth of detail you include in a document as defined by your audience's needs, your **purpose**, and the **context**. (See also **audience**.) For example, if you write a **trip report** about a routine visit to a company facility, your readers may need to know only the basic details and any unusual findings. However, if you prepare a trip report about a visit to a division that has experienced problems and your purpose is to suggest ways to solve those problems, your report will contain many more details, observations, and even recommendations.

You should determine the scope of a document during the **preparation** stage of the writing process, even though you may refine it later. Defining your scope will expedite your **research** and can help determine team members' responsibilities in **collaborative writing**.

Your scope will also be affected by the type of document you are writing as well as the medium you select for your message. For example, government agencies often prescribe the general content and length for **proposals**, and some organizations set limits for the length of **memos** and **e-mails**. See **selecting the medium** and "Five Steps to Successful Writing."

selecting the medium

With so many media and forms of communication available, selecting the most appropriate can be challenging. Which electronic or paper medium is best, for example, depends on a wide range of factors related to your **audience**, your **purpose**, and the **context** of the communication.* Those factors include the following:

- the audience's preferences and expectations
- an individual's personal work style
- how widely information needs to be distributed
- what kind of record you need to keep
- the urgency of the communication
- the sensitivity or confidentiality required
- the technological resources available
- the organizational practices or regulations

As this partial list suggests, choosing the best medium may involve personal considerations, such as your own strengths as a communicator. If you need to collaborate with someone to solve a problem, for example, you may find e-mail exchanges less effective than a phone call or face-to-face meeting. Other considerations may depend more on the purpose or context of the communication. If you need precise wording or a record of a complex or sensitive message, for example, using a written medium is often essential.

The following descriptions of typical media and forms of communication and their functions will help you select the most appropriate for your needs.

Letters

Business **letters** with handwritten signatures are often the most appropriate choice for formal communications with professional associates or customers outside an organization. Letters are often used for job applications, for recommendations, and in other official and social contexts. Letters printed on organizational letterhead stationery communicate formality, respect, and authority.

Memos

Memos are appropriate for internal communication among members of the same organization; they use a standard header and are sent on paper or as attachments to e-mails. Many organizations use e-mail rather than

*Some refer to this process as selecting the best "channel" for communication.

memos for routine internal communications; however, organizations may use memos printed on organizational stationery when they need to communicate with the formality and authority of business letters. Memos may also be used in manufacturing or service industries, for example, where employees do not have easy access to e-mail. In such cases, memos can be used to instruct employees, announce policies, report results, disseminate information, and delegate responsibilities.

E-mail

<u>E-mail</u> (or *email*) functions in the workplace as a primary medium to communicate and share electronic files with colleagues, clients, and customers. Although e-mail messages may function as informal notes, they should follow the writing strategy and style described in <u>correspondence</u>. Because e-mail recipients can easily forward messages and attachments to others and because e-mail messages are subject to legal disclosure, e-mail requires writers to review their messages carefully before clicking the "Send" button.

Instant Messages

<u>Instant messaging</u> (IM) on a computer or cell phone may be an efficient way to communicate in real time with coworkers, suppliers, and customers—especially those at sites without access to e-mail. Instant messaging often uses online slang and such shortened spellings as "u" for *you* to save screen space on a cell phone. Instant messaging is obviously limited because recipients must be ready and willing to participate in an online conversation.

Telephone Calls

Telephone calls are best used for exchanges that require substantial interaction and the ability of participants to interpret each other's tone of voice. They are useful for discussing sensitive issues and resolving misunderstandings, although they do not provide the visual cues possible during face-to-face meetings. Cell (or *mobile*) phones are useful for communicating away from an office, but users should follow appropriate etiquette and organizational policies, such as speaking in an appropriate tone and switching to the vibrate mode during meetings.*

A teleconference, or conference call among three or more participants, is a less expensive alternative to face-to-face meetings requiring travel. Such conference calls work best when the person coordinating the call works from an agenda shared by all the participants and directs the discussion as if chairing a meeting. Participants can use the Web

---

*Based on LetsTalk.com's cell-phone etiquette survey and resulting guidelines at *www.letstalk.com/promo/unclecell/unclecell2.htm*.

during conference calls to share and view common documents. Conference calls in which decisions have been reached should be followed with written confirmation.

Voice-mail messages should be clear and brief. ("I got your package, so you don't need to call the distributor.") If the message is complicated or contains numerous details, use another medium, such as e-mail. If you want to discuss a subject at length, let the recipient know the subject so that he or she can prepare a response before returning your call. When you leave a message, give your name and contact information as well as the date and time of the call (if you are unsure whether the message will be time-stamped).

Faxes

A fax is most useful when the information—a drawing or signed contract, for example—must be viewed in its original form. Faxes are also useful when the recipient either does not have e-mail or prefers faxed documents. Fax machines in offices can be located in shared areas, so call the intended recipient before you send confidential or sensitive messages.

Meetings

In-person **meetings** are most appropriate for initial or early contacts with associates and clients with whom you intend to develop an important, long-term relationship or need to establish rapport. Meetings may also be best for brainstorming, negotiating, interviewing someone on a complex topic, solving a technical problem, or handling a controversial issue. For advice on how to record discussions and decisions, see **minutes of meetings**.

Videoconferences

Videoconferences are particularly useful for meetings when travel is impractical. Unlike telephone conference calls, videoconferences have the advantage of allowing participants to see as well as to hear one another. Videoconferences work best with participants who are at ease in front of the camera and when the facilities offer good production quality.

Web Sites

A public Internet or company intranet Web site is ideal for posting announcements or policies as well as for making available or exchanging documents and files with others. Your Web site can serve not only as a home base for resources but also as a place where ideas can be developed through, for example, discussion boards, **blogs**,* and wikis. See also **collaborative writing**, **writing for the Web**, **Web design**, and *Digital Tip: Wikis for Collaborative Documents* on page 72.

semicolons

The semicolon (;) links independent **clauses** or other sentence elements of equal weight and grammatical rank when they are not joined by a **comma** and a **conjunction**. The semicolon indicates a greater pause between clauses than does a comma but not as great a pause as a **period**.

Independent clauses joined by a semicolon should balance or contrast with each other, and the relationship between the two statements should be so clear that further explanation is not necessary.

▶ The new Web site was a success; every division reported increased online sales.

Do not use a semicolon between a dependent clause and its main clause.

▶ No one applied for the position; even though it was heavily advertised.

With Strong Connectives

In complicated sentences, a semicolon may be used before transitional words or **phrases** (*that is*, *for example*, *namely*) that introduce examples or further explanation. See also **transition**.

▶ The press understands Commissioner Curran's position on the issue; *that is*, local funds should not be used for the highway project.

A semicolon should also be used before conjunctive **adverbs** (*therefore*, *moreover*, *consequently*, *furthermore*, *indeed*, *in fact*, *however*) that connect independent clauses.

▶ The test results are not complete; *therefore*, I cannot make a recommendation.
[The semicolon in the example shows that *therefore* belongs to the second clause.]

*A *blog* (short for "Web log") functions like a journal or diary and often includes commentary, discussion, or news on a particular subject.

For Clarity in Long Sentences

Use a semicolon between two independent clauses connected by a coordinating conjunction (*and*, *but*, *for*, *or*, *nor*, *so*, *yet*) if the clauses are long and contain other **punctuation**.

▶ In most cases, these individuals are executives, bankers, or lawyers; *but* they do not, as the press seems to believe, simply push the button of their economic power to affect local politics.

A semicolon may also be used if any items in a series contain commas.

▶ Among those present were John Howard, president of the Omega Paper Company; Carol Delgado, president of Environex Corporation; and Larry Stanley, president of Stanley Papers.

Use **parentheses** or **dashes**, not semicolons, to enclose a parenthetical element that contains commas.

▶ All affected job classifications (receptionist, secretary, transcriptionist, and clerk) will be upgraded this month.

Use a **colon**, not a semicolon, as a mark of anticipation or enumeration.

▶ Three decontamination methods are under consideration: a zeolite-resin system, an evaporation system, and a filtration system.

The semicolon always appears outside closing **quotation marks**.

▶ The attorney said, "You must be accurate"; her client replied, "I will."

sentence construction

DIRECTORY

Subjects   499
Predicates   500
Sentence Types   500
  Structure   500
Intention   501
Stylistic Use   501
Constructing Effective Sentences   501

A sentence is the most fundamental and versatile tool available to writers. Sentences generally flow from a subject to a **verb** to any **objects**, **complements**, or **modifiers**, but they can be ordered in a variety of ways to achieve **emphasis**. When shifting word order for emphasis, however, be aware that word order can make a great difference in the meaning of a sentence.

- He was *only* the service technician.
- He was the *only* service technician.

The most basic components of sentences are subjects and predicates.

Subjects

The *subject* of a sentence is a **noun** or **pronoun** (and its modifiers) about which the predicate of the sentence makes a statement. Although a subject may appear anywhere in a sentence, it most often appears at the beginning. ("*The wiring* is defective.") Grammatically, a subject must agree with its verb in **number**.

- *These departments have* much in common.
- *This department has* several functions.

The subject is the actor in sentences using the active **voice**.

- *The Webmaster reported* an increase in site visits for May.

> **ESL TIPS** for Understanding the Subject of a Sentence
>
> In English, every sentence, except commands, must have an explicit subject.
>
> - *Paul* worked fast. ~~Established~~ *He established* the parameters for the project.
>
> In commands, the subject *you* is understood and is used only for emphasis.
>
> - (*You*) Show up at the airport at 6:30 tomorrow morning.
> - (*You*) Do your homework, young man. [parent to child]
>
> If you move the subject from its normal position (subject-verb-object), English often requires you to replace the subject with an expletive (*there, it*). In this construction, the verb agrees with the subject that follows it.
>
> - *There are* two files on the desk. [The subject is *files*.]
> - *It is* presumptuous for me to speak for Jim.
>   [The subject is *to speak for Jim*.]
>
> Time, distance, weather, temperature, and environmental expressions use *it* as their subject.
>
> - *It* is ten o'clock.
> - *It* is ten miles down the road.
> - *It* seldom snows in Florida.
> - *It* is very hot in Jorge's office.

A *compound subject* has two or more substantives (nouns or noun equivalents) as the subject of one verb.

▶ *The doctor* and *the nurse* agreed on a treatment plan.

Predicates

The *predicate* is the part of a sentence that makes an assertion about the subject and completes the thought of the sentence.

▶ Bill *has piloted the corporate jet.*

The *simple predicate* is the verb and any helping verbs (*has piloted*). The *complete predicate* is the verb and any modifiers, objects, or complements (*has piloted the corporate jet*). A *compound predicate* consists of two or more verbs with the same subject.

▶ The company *tried* but *did not succeed* in that field.

Such constructions help achieve **conciseness** in writing. A *predicate nominative* is a noun construction that follows a linking verb and renames the subject.

▶ She is my *attorney.* [noun]
▶ His excuse was *that he had been sick.* [noun clause]

Sentence Types

Sentences may be classified according to *structure* (simple, compound, complex, compound-complex); *intention* (declarative, interrogative, imperative, exclamatory); and *stylistic use* (loose, periodic, minor).

Structure. A *simple sentence* consists of one independent clause. At its most basic, a simple sentence contains only a subject and a predicate.

▶ The power [subject] failed [predicate].

A *compound sentence* consists of two or more independent clauses connected by a comma and a coordinating **conjunction**, by a **semicolon**, or by a semicolon and a conjunctive **adverb**.

▶ Drilling is the only way to collect samples of the layers of sediment below the ocean floor, *but* it is not the only way to gather information about these strata. [comma and coordinating conjunction]

▶ The chemical composition of seawater bears little resemblance to that of river water; the various elements are present in entirely different proportions. [semicolon]

▶ It was 500 miles to the site; *therefore,* we made arrangements to fly. [semicolon and conjunctive adverb]

A *complex sentence* contains one independent clause and at least one dependent clause that expresses a subordinate idea.

> The generator will shut off automatically [independent clause] if the temperature rises above a specified point [dependent clause].

A *compound-complex sentence* consists of two or more independent clauses plus at least one dependent clause.

> Productivity is central to controlling inflation [independent clause]; when productivity rises [dependent clause], employers can raise wages without raising prices [independent clause].

Intention. A *declarative sentence* conveys information or makes a factual statement. ("The motor powers the conveyor belt.") An *interrogative sentence* asks a direct question. ("Does the conveyor belt run constantly?") An *imperative sentence* issues a command. ("Restart in MS-DOS mode.") An *exclamatory sentence* is an emphatic expression of feeling, fact, or opinion. It is a declarative sentence that is stated with great feeling. ("The files were deleted!")

Stylistic Use. A *loose sentence* makes its major point at the beginning and then adds subordinate phrases and clauses that develop or modify that major point. A loose sentence could end at one or more points before it actually does end, as the periods in brackets illustrate in the following sentence:

> It went up[.], a great ball of fire about a mile in diameter[.], an elemental force freed from its bonds[.] after being chained for billions of years.

A *periodic sentence* delays its main ideas until the end by presenting subordinate ideas or modifiers first.

> During the last century, the attitude of the American citizen toward automation underwent a profound change.

A *minor sentence* is an incomplete sentence that makes sense in its context because the missing element is clearly implied by the preceding sentence.

> In view of these facts, is the service contract really useful? *Or economical?*

Constructing Effective Sentences

The subject-verb-object pattern is effective because it is most familiar to **readers**. In "The company increased profits," we know the subject (*company*) and the object (*profits*) by their positions relative to the verb (*increased*).

An *inverted sentence* places the elements in unexpected order, thus emphasizing the point by attracting the readers' attention.

- ▶ A better job I never had. [direct object-subject-verb]
- ▶ More optimistic I have never been. [subjective complement-subject-linking verb]
- ▶ A poor image we presented. [complement-subject-verb]

Use uncomplicated sentences to state complex ideas. If readers have to cope with a complicated sentence in addition to a complex idea, they are likely to become confused. Just as simpler sentences make complex ideas more digestible, a complex sentence construction makes a series of simple ideas more smooth and less choppy.

Avoid loading sentences with a number of thoughts carelessly tacked together. Such sentences are monotonous and hard to read because all the ideas seem to be of equal importance. Rather, distinguish the relative importance of sentence elements with **subordination**. See also **garbled sentences**.

LOADED   We started the program three years ago, only three members were on the staff, and each member was responsible for a separate state, but it was not an efficient operation.

IMPROVED   When we started the program three years ago, only three members were on the staff, each responsible for a separate state; however, that arrangement was not efficient.

Express coordinate or equivalent ideas in similar form. The structure of the sentence helps readers grasp the similarity of its components, as illustrated in **parallel structure**.

> **ESL TIPS** for Understanding the Requirements of a Sentence
>
> - A sentence must start with a capital letter.
> - A sentence must end with a period, a question mark, or an exclamation point.
> - A sentence must have a subject.
> - A sentence must have a verb.
> - A sentence must conform to subject-verb-object word order (or inverted word order for questions or emphasis).
> - A sentence must express an idea that can stand on its own (called the *main*, or *independent*, *clause*).

sentence faults

A number of problems can create sentence faults, including faulty **subordination**, **clauses** with no subjects, rambling sentences, omitted **verbs**, and illogical assertions.

Faulty subordination occurs when a grammatically subordinate element contains the main idea of the sentence or when a subordinate element is so long or detailed that it obscures the main idea. Both of the following sentences are logical, depending on what the writer intends as the main idea and as the subordinate element.

▶ Although the new filing system saves money, many of the staff are unhappy with it.
[If the main point is that *many of the staff are unhappy*, this sentence is correct.]

▶ The new filing system saves money, although many of the staff are unhappy with it.
[If the main point is that *the new filing system saves money*, this sentence is correct.]

In the following example, the subordinate element overwhelms the main point.

FAULTY	Because the noise level in the assembly area on a typical shift is as loud as a smoke detector's alarm ten feet away, employees often develop hearing problems.
IMPROVED	Employees in the assembly area often develop hearing problems because the noise level on a typical shift is as loud as a smoke detector's alarm ten feet away.

Missing subjects occur when writers inappropriately assume a subject that they do not state in the clause. See also **sentence fragments**.

| INCOMPLETE | Your application program can request to end the session after the next command.
[Your application program can request *who* or *what* to end the session?] |
|---|---|
| COMPLETE | Your application program can request *the host program* to end the session after the next command. |

Rambling sentences contain more information than the reader can comfortably absorb. The obvious remedy for a rambling sentence is to divide it into two or more sentences. When you do that, put the main message of the rambling sentence into the first of the revised sentences.

RAMBLING	The payment to which a subcontractor is entitled should be made promptly in order that in the event of a subsequent contractual dispute we, as general contractors, may not be held in default of our contract by virtue of nonpayment.
DIRECT	Pay subcontractors promptly. Then if a contractual dispute occurs, we cannot be held in default of our contract because of nonpayment.

Missing verbs produce some sentence faults.

▶ I never have and probably never will write the annual report.
              written

Faulty logic results when a predicate makes an illogical assertion about its subject. "Mr. Wilson's *job* is a sales representative" is not logical, but "*Mr. Wilson* is a sales representative" is logical. "Jim's *height* is six feet tall" is not logical, but "*Jim* is six feet tall" is logical. See also **logic errors**.

sentence fragments

A sentence fragment is an incomplete grammatical unit that is punctuated as a sentence.

FRAGMENT	And quit his job.
SENTENCE	He quit his job.

A sentence fragment lacks either a subject or a **verb** or is a subordinate **clause** or **phrase**. Sentence fragments are often introduced by relative **pronouns** (*who, whom, which, that*) or subordinating **conjunctions** (such as *although, because, if, when,* and *while*).

▶ The new manager instituted several new procedures. ~~Although~~ she didn't clear them with Human Resources.
                                                       , although

A sentence must contain a finite verb; **verbals** (nonfinite) do not function as verbs. The following sentence fragments use verbals (*providing, to work*) that cannot function as finite verbs.

FRAGMENT	*Providing* all employees with disability insurance.
SENTENCE	The company *provides* all employees with disability insurance.
FRAGMENT	*To work* a 40-hour week.
SENTENCE	Most of our employees *must work* a 40-hour week.

Explanatory phrases beginning with *such as*, *for example*, and similar terms often lead writers to create sentence fragments.

▶ The staff wants additional benefits. ~~For example,~~ *, such as* the use of company cars.

A hopelessly snarled fragment simply must be rewritten. To rewrite such a fragment, pull the main points out of the fragment, list them in the proper sequence, and then rewrite the sentence as illustrated in **garbled sentences**. See also **sentence construction** and **sentence faults**.

sentence variety

Sentences can vary in length, structure, and complexity. As you revise, vary your sentences so that they do not become tiresomely alike. See also **sentence construction**.

Sentence Length

A series of sentences of the same length is monotonous, so varying sentence length makes writing less tedious to the **reader**. For example, avoid stringing together a number of short independent **clauses**. Either connect them with subordinating connectives, thereby making some dependent clauses, or make some clauses into separate sentences. See also **subordination**.

STRING	The river is 63 miles long, and it averages 50 yards in width, and its depth averages 8 feet.
IMPROVED	The river, which is 63 miles long and averages 50 yards in width, has an average depth of 8 feet.
IMPROVED	The river is 63 miles long. It averages 50 yards in width and 8 feet in depth.

You can often effectively combine short sentences by converting **verbs** into **adjectives**.

▶ The digital shift indicator ~~failed. It~~ *failed* was pulled from the market.

Although too many short sentences make your writing sound choppy and immature, a short sentence can be effective following a long one.

▶ During the past two decades, many changes have occurred in American life—the extent, durability, and significance of which no one has yet measured. *No one can.*

In general, short sentences are good for emphatic, memorable statements. Long sentences are good for detailed explanations and support. Nothing is inherently wrong with a long sentence, or even with a complicated one, as long as its meaning is clear and direct. Sentence length becomes an element of style when varied for **emphasis** or contrast; a conspicuously short or long sentence can be used to good effect.

Word Order

When a series of sentences all begin in exactly the same way (usually with an **article** and a **noun**), the result is likely to be monotonous. You can make your sentences more interesting by occasionally starting with a modifying word, **phrase**, or clause.

▶ *To salvage the project*, she presented alternatives when existing policies failed to produce results. [modifying phrase]

However, overuse of this technique itself can be monotonous, so use it in moderation.

Inverted word order can be an effective way to achieve variety, but be careful not to create an awkward construction.

AWKWARD	Then occurred the event that gained us the contract.
EFFECTIVE	Never have sales been so good.

For variety, you can alter normal sentence order by inserting a phrase or clause.

▶ Titanium fills the gap, *both in weight and in strength*, between aluminum and steel.

The technique of inserting a phrase or clause is good for achieving emphasis, providing detail, breaking monotony, and regulating **pace**.

Loose and Periodic Sentences

A loose sentence makes its major point at the beginning and then adds subordinate phrases and clauses that develop or modify the point. A loose sentence could end at one or more points before it actually ends, as the periods in brackets illustrate in the following example:

▶ It went up[.], a great ball of fire about a mile in diameter[.], an elemental force freed from its bonds[.] after being chained for billions of years.

A periodic sentence delays its main idea until the end by presenting modifiers or subordinate ideas first, thus holding the readers' interest until the end.

▶ During the last century, the attitude of Americans toward technology underwent a profound change.

Experiment with shifts from loose sentences to periodic sentences in your own writing, especially during **revision**. Avoid the singsong monotony of a long series of loose sentences, particularly a series containing coordinate clauses joined by **conjunctions**. Subordinating some thoughts to others not only provides emphasis but also makes your sentences more interesting.

sequential method of development

The sequential, or step-by-step, **method of development** is especially effective for explaining a process or describing a mechanism in operation. (See **process explanation**.) It is also the logical method for writing **instructions**, as shown in Figure I–5 on page 261.

The main advantage of the sequential method of development is that it is easy to follow because the steps correspond to the process or operation being described. The disadvantages are that it can become monotonous and does not lend itself well to achieving **emphasis**.

Most methods of development have elements of sequence to a greater or lesser extent. The **chronological method of development**, for example, is also sequential: To describe a trip chronologically, from beginning to end, is also to describe it sequentially. The **cause-and-effect method of development** may contain certain elements of sequence. For example, a report of the causes leading to an accident (the effect) might describe those causes in the order they occurred (or their sequence).

service

When used as a **verb**, *service* means "keep up or maintain" as well as "repair." ("Our company will *service* your equipment.") If you mean "provide a more general benefit," use *serve*.

▶ Our company ~~services~~ *serves* the northwest area of the state.

set / sit

Sit is an intransitive **verb**; it does not, therefore, require an **object**. ("I *sit* by a window in the office.") Its past **tense** is *sat*. ("We *sat* around the conference table.") *Set* is usually a transitive verb, meaning "put or place," "establish," or "harden." Its past tense is *set*.

- Please *set* the supplies on the shelf.
- The jeweler *set* the stone carefully.
- Can we *set* a date for the meeting?
- The high temperature *sets* the epoxy quickly.

Set is occasionally an intransitive verb.

- The new adhesive *sets* in five minutes.

shall / will

Although traditionally *shall* was used to express the future **tense** with *I* and *we*, *will* is generally accepted with all persons. *Shall* is commonly used today only in questions requesting an opinion or a preference ("*Shall* we go?") rather than a prediction ("*Will* we go?"). It is also used in statements expressing determination ("I *shall* return!") or in formal regulations that express a requirement ("Applicants *shall* provide a proof of certification").

slashes

The slash (/)—called a variety of names, including *slant line*, *diagonal*, *virgule*, *bar*, and *solidus*—both separates and shows omission.

The slash can indicate alternatives or combinations.

- David's telephone numbers are (800) 549-2278/2235.
- Check the on/off switch before you leave.

The slash often indicates omitted words and letters.

- miles/hour (miles per hour); w/o (without)

In fractions and mathematical expressions, the slash separates the numerator from the denominator (3/4 for three-fourths; x/y for x over y).

Although the slash is used informally with **dates** (5/9/09), avoid this form in technical writing, especially in **international correspondence**.

The forward slash often separates items in URL (uniform resource locator) addresses for sites on the Internet (*bedfordstmartins.com/alredtech*). The backward slash is used to separate parts of file names (*c:\myfiles\reports\annual09.doc*).

so / so that / such

Avoid *so* as a substitute for *because*. See also **as / because / since**.

▶ ~~She~~ *Because she* reads faster, ~~so~~ she finished before I did.

Do not replace the phrase *so that* with *so* or *such that*.

▶ The report should be written ~~such that~~ *so that* it can be widely understood.

Such, an **adjective** meaning "of this or that kind," should never be used as a **pronoun**.

▶ Our company provides on-site child care, but I do not anticipate using ~~such~~ *it*.

some / somewhat

When *some* functions as an indefinite **pronoun** for a plural count **noun** or as an indefinite **adjective** modifying a plural count noun, use a plural **verb**.

▶ *Some* of us *are* prepared to work overtime.

▶ *Some* people *are* more productive than others.

Some is singular, however, when used with mass nouns.

▶ *Some* sand *has* trickled through the crack.

▶ Most of the water evaporated, but *some remains*.

When *some* is used as an adjective or a pronoun meaning "an undetermined quantity" or "certain unspecified persons," it should be replaced by the **adverb** *somewhat*, which means "to some extent."

▶ His writing has improved ~~some~~ *somewhat*.

some time / sometime / sometimes

Some time refers to a duration of time. ("We waited for *some time* before making the decision.") *Sometime* refers to an unknown or unspecified time. ("We will visit with you *sometime*.") *Sometimes* refers to occasional occurrences at unspecified times. ("He *sometimes* visits the branch offices.")

spatial method of development

The spatial **method of development** describes an object or a process according to the physical arrangement of its features. Depending on the subject, you describe its features from bottom to top, side to side, east to west, outside to inside, and so on. Descriptions of this kind rely mainly on dimension (height, width, length), direction (up, down, north, south), shape (rectangular, square, semicircular), and proportion (one-half, two-thirds). Features are described in relation to one another or to their surroundings, as illustrated in Figure S–1, which might be written by a home inspector or crime-scene investigator. The description in Figure S–1 relies on a bottom-to-top, clockwise (south to west to north to east) sequence, beginning with the front door. Such descriptions often benefit from **visuals**, such as **drawings**, that can provide overviews and details.

The spatial method of development might be used for descriptions of warehouse inventory, **proposals** for landscape work, construction-site **progress and activity reports**, and, in combination with a step-by-step sequence, many types of **instructions**.

specifications

A specification is a detailed and an exact statement that prescribes the materials, dimensions, and quality of something to be built, installed, or manufactured.

The two broad categories of specifications—industrial and government—both require precision. A specification must be written clearly and precisely and state *explicitly* what is needed. Because of the stringent requirements of specifications, careful **research** and **preparation** are especially important before you begin to write, as is careful **revision** after you have completed the draft. See also **clarity** and **ambiguity**.

Industrial Specifications

Industrial specifications are used, for example, in software development, in which there are no engineering drawings or other means of

DESCRIPTION

Interior of Two-Story, Six-Room House

Ground Floor

Front hall and stairwell. The front door faces south and opens into a hallway seven feet deep and ten feet wide. At the end of the hallway is a stairwell that begins on the right-hand (east) side of the hallway, rises five steps to a landing, and reverses direction at the left-hand (west) side of the hallway.

Dining room. To the left (west) of the hallway is the dining room, which measures 15 feet along its southern exposure and ten feet along its western exposure.

Kitchen. North of the dining room is the kitchen, which measures ten feet along its western exposure and 15 feet along its northern exposure.

Bathroom. East of the kitchen, along the northern side of the house, is a bathroom that measures ten feet (west to east) by five feet.

Living room. Parallel to the bathroom is a passageway the same size as the bathroom and leading from the kitchen to the living room. The living room (15 feet west to east by 20 feet north to south) occupies the entire eastern end of the floor.

Second Floor

Hallway. On the second floor, at the top of the stairs, is an L-shaped hallway, five feet wide. The base of the L, over the door, is 15 feet long. The vertical arm of the L is 13 feet long.

Southwest bedroom. To the west of the hall is the southwest bedroom, which measures ten feet along its southern exposure and eight feet along its western exposure.

Northwest bedroom. Directly to the north, over the kitchen, is the northwest bedroom, which measures 12 feet along its western exposure and ten feet along its northern exposure.

FIGURE S–1. Spatial Method of Development

documentation. An industrial specification is a permanent record that documents the item being developed so that it can be maintained by someone other than the person who designed it and provides detailed technical information about the item to all who need it (engineers, technical writers, technical instructors, etc.).

The industrial specification describes a planned project, a newly completed project, or an old project. The specification for each type of project must contain detailed technical descriptions of all aspects of the project: what was done and how it was done, as well as what is required to use the item, how it is used, what its function is, who would use it, and so on.

Government Specifications

Government agencies are required by law to contract for equipment strictly according to definitions provided in formal specifications. A government specification is a precise definition of exactly what the contractor is to provide. In addition to a technical description of the item to be purchased, the specification normally includes an estimated cost; an estimated delivery date; and standards for the design, manufacture, quality, testing, training of government employees, governing codes, inspection, and delivery of the item.

Government specifications contain details on the scope of the project; documents the contractor is required to furnish with the device; required product characteristics and functional performance of the device; required tests, test equipment, and test procedures; required preparations for delivery; notes; and **appendixes**. These specifications appear in **requests for proposals**, which prescribe the content and deadline for government **proposals** submitted by vendors bidding on a project.

spelling

Because spelling errors in your documents will damage your credibility, careful **proofreading** is essential. The use of a spell checker is crucial; however, it will not catch all mistakes, especially those in personal and company names. It cannot detect a spelling error if the error results in a valid word; for example, if you mean *to* but inadvertently type *too*, the spell checker will not detect the error. Likewise, spell checkers will not detect errors in the names of people, places, and organizations. If you are unsure about the spelling of a word, do not rely on guesswork or a spell checker—consult a standard **dictionary** or style guide.

strata / stratum

Strata is the plural form of *stratum*, meaning "a layer of material."

▶ The land's *strata* are exposed by erosion.

▶ Each *stratum* is clearly visible in the cliff.

style

A dictionary definition of *style* is "the way in which something is said or done, as distinguished from its substance." Writers' styles are determined by the way writers think and transfer their thoughts to paper—the way they use words, sentences, images, **figures of speech**, and so on.

A writer's style is the way his or her language functions in particular situations. For example, an **e-mail** to a friend would be relaxed, even chatty, in **tone**, whereas a job **application letter** would be more restrained and formal. Obviously, the style appropriate to one situation would not be appropriate to the other. In both situations, the **audience**, the **purpose**, and the **context** determine the manner or style the writer adopts. Beyond an individual's personal style, various kinds of writing have distinct stylistic traits, such as **technical writing style**.

Standard English can be divided into two broad categories of style—formal and informal—according to how it functions in certain situations. Understanding the distinction between formal and informal writing styles helps writers use the appropriate style. We must recognize, however, that no clear-cut line divides the two categories and that some writing may call for a combination of the two. See also **English, varieties of**.

Formal Writing Style

A formal writing style can perhaps best be defined by pointing to certain material that is clearly formal, such as scholarly and scientific articles in professional journals, lectures read at meetings of professional societies, and legal documents. Material written in a formal style is usually the work of a specialist writing to other specialists or writing that embodies laws or regulations. As a result, the vocabulary is specialized and precise. The writer's tone is impersonal and objective because the subject matter looms larger in the writing than does the author's personality. (See **point of view**.) A formal writing style does not use **contractions**, slang, or dialect. Because the material generally examines complex ideas, the **sentence construction** may be elaborate.

Formal writing need not be dull and lifeless. By using such techniques as the active **voice** whenever possible, **sentence variety**, and

subordination, a writer can make formal writing lively and interesting, especially if the subject matter is inherently interesting to **readers**. In the following example, a physicist uses a concept that is taken for granted (distance) to show how definitions are important in science.

> ▶ Distance is such a basic concept in our understanding of the world that it is easy to underestimate the depth of its subtlety. With the surprising effects that special and general relativity have had on our notions of space and time, and the new features arising from string theory, we are led to be a bit more careful even in our definition of distance. The most meaningful definitions in physics are those that are operational—that is, definitions that provide a means, at least in principle, for measuring whatever is being defined. After all, no matter how abstract a concept is, having an operational definition allows us to boil down its meaning to an experimental procedure for measuring its value.
> —Brian Greene, *The Elegant Universe: Superstrings, Hidden Dimensions, and the Quest for the Ultimate Theory*

Whether you should use a formal style in a particular instance depends on your readers and purpose. When writers attempt to force a formal style when it should not be used, their writing is likely to fall victim to **affectation**, **awkwardness**, and **gobbledygook**.

Informal Writing Style

An informal writing style is a relaxed and colloquial way of writing standard English. It is the style found in most personal e-mail and in some business **correspondence**, nonfiction books of general interest, and mass-circulation magazines. There is less distance between the writer and the reader because the tone is more personal than in a formal writing style. Consider the following passage, written in an informal style, from a nonfiction book on statistics.

> ▶ After a few months of reading risk statistics, I had a curious experience one morning, an epiphany of sorts. At the time, however, I felt more like Alice in Wonderland after taking a sip of the "Drink Me" potion. When I opened my eyes in bed and began to contemplate my day, I began to see it not in terms of what I had to accomplish but in terms of the risks that I would encounter. The world suddenly started looking different.
> —John F. Ross, *The Polar Bear Strategy: Reflections on Risk in Modern Life*

As the example illustrates, the vocabulary of an informal writing style is made up of generally familiar rather than unfamiliar words and expressions, although slang and dialect are usually avoided. An informal style approximates the cadence and structure of spoken English while conforming to the grammatical conventions of written English.

Writers who consciously attempt to create a distinctive style usually defeat their purpose. Attempting to impress readers with a flashy writing style can lead to affectation; attempting to impress them with scientific objectivity can produce a style that is dull and lifeless. Technical writing need be neither affected nor dull. It can and should be simple, clear, direct, even interesting—the key is to master basic writing skills and always to keep your readers in mind. What will be both informative and interesting to your readers? When that question is uppermost in your mind as you apply the steps of the writing process, you will achieve an interesting and informative writing style. See "Five Steps to Successful Writing."

WRITER'S CHECKLIST | **Developing an Effective Style**

- Use the active voice—not exclusively but as much as possible without becoming awkward or illogical.
- Use **parallel structure** whenever a sentence or **list** presents two or more thoughts of equal importance.
- Vary structures to avoid a monotonous style.
- Avoid stating positive thoughts in negative terms (write "40 percent responded" instead of "60 percent failed to respond"). See also **positive writing** and **ethics in writing**.
- Concentrate on achieving the proper balance between **emphasis** and subordination.

WEB LINK | **Style Guides**

Style guides, such as *The Chicago Manual of Style* and the Associated Press *Stylebook and Briefing on Media Law,* provide specific and sometimes varied advice for handling issues of usage, style, and formats for citations, correspondence, and documents. For a selected list of such style guides, see bedfordstmartins.com/alredtech and select *Links for Handbook Entries.*

subordination

Use subordination to show, by the structure of a sentence, the appropriate relationship between ideas of unequal importance. Subordination allows you to emphasize your main idea by putting less important ideas in subordinate **clauses** or **phrases**.

▶ Envirex Systems now employs 500 people. It was founded just three years ago.
[The two ideas are equally important.]

▶ Envirex Systems, *which now employs 500 people*, was founded just three years ago.
[The number of employees is subordinated; the founding date is emphasized.]

▶ Envirex Systems, *which was founded just three years ago*, now employs 500 people.
[The founding date is subordinated; the number of employees is emphasized.]

Effective subordination can be used to achieve **conciseness**, **emphasis**, and **sentence variety**. For example, consider the following sentences.

DEPENDENT CLAUSE	The research report, *which covered 86 pages*, was carefully documented.
PHRASE	The research report, *covering 86 pages*, was carefully documented.
SINGLE MODIFIER	The *86-page* research report was carefully documented.

Subordinating **conjunctions** (*because*, *if*, *while*, *when*, *although*) achieve subordination effectively.

▶ A buildup of deposits is impossible *because* the pipes are flushed with water every day.

You may use a coordinating conjunction (*and*, *but*, *for*, *nor*, *or*, *so*, *yet*) to concede that an opposite or balancing fact is true; however, a subordinating conjunction can often make the point more smoothly.

▶ *Although* their lab is well funded, ours is better equipped for DNA testing.

The relationship between a conditional statement and a statement of consequences is clearer if the condition is expressed as a subordinate clause.

▶ *Because* the tests were delayed, the surgery was postponed.

Relative **pronouns** (*who*, *whom*, *which*, *that*) can be used effectively to combine related ideas within sentences.

▶ The OnlinePro, *which* protects computers from malicious programs and e-mail attachments, makes your system "invisible" to hackers.

Avoid overlapping subordinate constructions that depend on the preceding construction. Overlapping can make the relationship between a relative pronoun and its antecedent less clear.

OVERLAPPING	Shock, *which* often accompanies severe injuries and infections, is a failure of the circulation, *which* is marked by a fall in blood pressure *that* initially affects the skin (*which* explains pallor) and later the vital organs such as the kidneys and brain.
CLEAR	Shock often accompanies severe injuries and infections. Marked by a fall in blood pressure, it is a failure of the circulation, initially to the skin (thus producing pallor) and later to the vital organs like the kidneys and the brain.

suffixes

A suffix is a letter or letters added to the end of a word to change its meaning in some way. Suffixes can change the part of speech of a word.

NO SUFFIX	The proposal was *thorough*. [**adjective**]
SUFFIX	The *thoroughness* is obvious. [**noun**]
SUFFIX	The proposal *thoroughly* described the problem. [**adverb**]

The suffix -*like* is sometimes added to nouns to make them into adjectives. The resulting compound word is hyphenated only if it is unusual or might not immediately be clear (childlike, lifelike, *but* dictionary-like, Einstein-like).

surveys (*see* questionnaires)

synonyms

A synonym is a word that means nearly the same thing as another word does (*seller*, *vendor*, *supplier*). The dictionary definitions of synonyms are similar, but the connotations may differ. For example, a *seller* may be the same thing as a *supplier*, but the term *supplier* does not suggest a retail transaction as strongly as *seller* does.

Do not try to impress your **readers** by finding fancy or obscure synonyms in a **thesaurus**; the result is likely to be **affectation**. See also **connotation / denotation** and **antonyms**.

syntax

Syntax refers to the way that words, phrases, and clauses are combined to form sentences. In English, the most common structure is the subject-verb-object pattern. For more information about the word order of sentences, see **sentence construction**, **sentence faults**, **sentence fragments**, and **sentence variety**.

tables of contents

A table of contents is typically included in a document longer than ten pages. It previews what the work contains and how it is organized, and it allows **readers** looking for specific information to locate sections by page number quickly and easily.

When creating a table of contents, use the major **headings** and subheadings of your document exactly as they appear in the text, as shown in the entry **formal reports**. (See the table of contents in Figure F–5 on page 204.) The table of contents is placed in the front matter following the title page and **abstract**, and precedes the list of tables or figures, the foreword, and the preface.

tables

A table organizes data, such as statistics, into parallel rows and columns that allow **readers** to make precise comparisons. Overall trends, however, are more easily conveyed in **graphs** and other **visuals**.

Table Elements

Tables typically include the elements shown in Figure T–1.

Table Number.   Table numbers should be placed above tables and assigned sequentially throughout the document.

Table Title.   The title (or *caption*), which is normally placed just above the table, should describe concisely what the table represents.

Box Head.   The box head contains the column headings, which should be brief but descriptive. Units of measurement should be either specified as part of the heading or enclosed in parentheses beneath it. Standard **abbreviations** and symbols are acceptable. Avoid vertical or diagonal lettering.

Table 1. Estimated Emissions from Electric Power Generation (tons per gigawatthour)

Fuel	Sulphur Dioxide	Nitrogen Oxides	Particulate Matter	Carbon Dioxide	Volatile Organic Compounds
Eastern coal	1.74	2.90	0.10	1,000	0.06
Western coal	0.81	2.20	0.06	1,039	0.09
Gas	0.003	0.57	0.02	640	0.05
Biomass	0.06	1.25	0.11	0*	0.61
Oil	0.51	0.63	0.02	840	0.03
Wind	0	0	0	0	0
Geothermal	0	0	0	0	0
Hydro	0	0	0	0	0
Solar	0	0	0	0	0
Nuclear	0	0	0	0	0

*Net emissions.
SOURCE: Department of Energy

Figure T–1. Elements of a Table

Stub. The stub, the left vertical column of a table, lists the items about which information is given in the body of the table.

Body. The body comprises the data below the column headings and to the right of the stub. Within the body, arrange columns so that the items to be compared appear in adjacent rows and columns. Align the numerical data in columns for ease of comparison, as shown in Figure T–1. Where no information exists for a specific item, substitute a row of dots or a dash to acknowledge the gap.

Rules. Rules are the lines (or *borders*) that separate the table into its various parts. Tables should include top and bottom borders. Tables often include right and left borders, although they may be open at the sides, as shown in Figure T–1. Generally, include a horizontal rule between the column headings and the body of the table. Separate the columns with vertical rules within a table only when they aid clarity.

Footnotes. Footnotes are used for explanations of individual items in the table. Symbols (such as * and †) or lowercase letters (sometimes in parentheses) rather than numbers are ordinarily used to key table footnotes because numbers might be mistaken for numerical data or could be confused with the numbering system for text footnotes. See also **documenting sources**.

Source Line. The source line identifies where the data originated. When a source line is appropriate, it appears below the table. Many organizations place the source line below the footnotes. See also **copyright** and **plagiarism**.

Continuing Tables. When a table must be divided so that it can be continued on another page, repeat the column headings and the table number and title on the new page with a "continued" label (for example, "Table 3. [title], *continued*").

Informal Tables

To list relatively few items that would be easier for the reader to grasp in tabular form, you can use an informal table, as long as you introduce it properly. Although informal tables do not need titles or table numbers to identify them, they do require column headings that accurately describe the information listed, as shown in Figure T–2.

The sound-intensity levels (decibels) for the three frequency bands (in hertz) were determined to be the following:

Frequency Band (Hz)	Decibels
600–1199	68
1200–2399	62
2400–4800	53

FIGURE T–2. Informal Table

technical manuals (*see* manuals)

technical writing style

The goal of technical writing is to enable **readers** to use a technology or understand a process or concept. Because the subject matter is more important than the writer's voice, technical writing style uses an objective, not a subjective, **tone**. The writing **style** is direct and utilitarian, emphasizing exactness and **clarity** rather than elegance or allusiveness. A technical writer uses figurative language only when a **figure of speech** would facilitate understanding.

Technical writing is often—but not always—aimed at readers who are not experts in the subject, such as consumers or employees learning to operate unfamiliar equipment. Because such **audiences** are inexperienced and the procedures described may involve hazardous material or equipment, clarity becomes an ethical as well as a stylistic concern. (See also **ethics in writing**.) Figure T–3 is an excerpt from a technical **manual** that instructs eye-care specialists about operating testing equipment. As the figure illustrates, <u>visuals</u> as well as <u>layout and design</u> enhance clarity, providing the reader with both text and visual information.

Technical writing style may use a technical vocabulary appropriate for the reader, as Figure T–3 shows (*monocular electrode*, *saccade*), but it avoids <u>affectation</u>. Good technical writing also avoids overusing the passive <u>voice</u>. See also <u>instructions</u>, <u>organization</u>, and <u>process explanation</u>.

MONOCULAR RECORDING

A monocular electrode configuration might be useful to measure the horizontal eye movement of each eye separately. This configuration is commonly used during the voluntary saccade tests.

1. Place electrodes at the inner canthi of both eyes (**A and B**, Figure 51).
2. Place electrodes at the outer canthi of both eyes (**C and D**).
3. Place the last electrode anywhere on the forehead (**E**).
4. Plug the electrodes into the jacks described in Figure 52 on the following page.

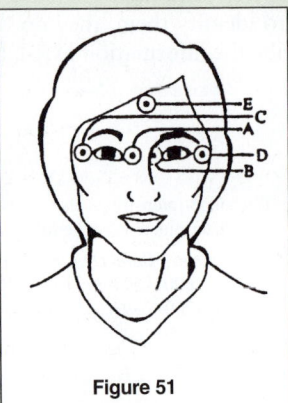

Figure 51

With this configuration, the Channel A display moves upward when the patient's left eye moves nasally and moves downward when the patient's left eye moves temporally. The Channel B display moves upward when the patient's right eye moves temporally and moves downward when the patient's right eye moves nasally.

NOTE: *During all testing, the Nystar Plus identifies and eliminates eye blinks without vertical electrode recordings.*

FIGURE T–3. Technical Writing Style

telegraphic style

Telegraphic style condenses writing by omitting <u>articles</u>, <u>pronouns</u>, <u>conjunctions</u>, and <u>transitions</u>. Although <u>conciseness</u> is important, especially in <u>instructions</u>, writers sometimes try to achieve conciseness by omitting necessary words. Telegraphic style forces <u>readers</u> to supply the missing words mentally, thus creating the potential for misunderstandings. Compare the following two passages and notice how much easier the revised version reads (the added words are italicized).

TELEGRAPHIC	Take following action when treating serious burns. Remove loose clothing on or near burn. Cover injury with clean dressing and wash area around burn. Secure dressing with tape. Separate fingers/toes with gauze/cloth to prevent sticking. Do not apply medication unless doctor prescribes.
CLEAR	Take *the* following action when treating *a* serious burn. Remove *any* loose clothing on or near *the* burn. Cover *the* injury with *a* clean dressing and wash *the* area around *the* burn. *Then* secure *the* dressing with tape. Separate *the* fingers *or* toes with gauze *or* cloth to prevent *them from* sticking *together*. Do not apply medication unless *a* doctor prescribes *it*.

Telegraphic style can also produce **ambiguity**, as the following example demonstrates.

AMBIGUOUS	Grasp knob and adjust lever before raising boom. [Does this sentence mean that the reader should *adjust the lever* or also *grasp an "adjust lever"*?]
CLEAR	Grasp *the* knob and adjust *the* lever before raising *the* boom.

Although you may save yourself work by writing telegraphically, your readers will have to work that much harder to decipher your meaning.

tenant / tenet

A *tenant* is one who holds or temporarily occupies a property owned by another person. ("The *tenant* was upset by the rent increase.") A *tenet* is an opinion or principle held by a person, an organization, or a system. ("Competition is a central *tenet* of capitalism.")

tense

DIRECTORY
Past Tense   524
Past Perfect Tense   524
Present Tense   525
Present Perfect Tense   525
Future Tense   525
Future Perfect Tense   525
Shift in Tense   526

Tense is the grammatical term for **verb** forms that indicate time distinctions. The six tenses in English are past, past perfect, present, present perfect, future, and future perfect. Each tense also has a corresponding progressive form.

TENSE	BASIC FORM	PROGRESSIVE FORM
Past	I began	I was beginning
Past perfect	I had begun	I had been beginning
Present	I begin	I am beginning
Present perfect	I have begun	I have been beginning
Future	I will begin	I will be beginning
Future perfect	I will have begun	I will have been beginning

Perfect tenses allow you to express a prior action or condition that continues in a present, past, or future time.

PRESENT PERFECT	I *have begun* to compile the survey results and will continue for the rest of the month.
PAST PERFECT	I *had begun* to read the manual when the lights went out.
FUTURE PERFECT	I *will have begun* this project by the time funds are allocated.

Progressive tenses allow you to describe some ongoing action or condition in the present, past, or future.

PRESENT PROGRESSIVE	I *am beginning* to be concerned that we will not meet the deadline.
PAST PROGRESSIVE	I *was beginning* to think we would not finish by the deadline.
FUTURE PROGRESSIVE	I *will be requesting* a leave of absence when this project is finished.

Past Tense

The simple past tense indicates that an action took place in its entirety in the past. The past tense is usually formed by adding *-d* or *-ed* to the root form of the verb. ("We *closed* the office early yesterday.")

Past Perfect Tense

The past perfect tense (also called *pluperfect*) indicates that one past event preceded another. It is formed by combining the helping verb *had* with the past-participle form of the main verb. ("He *had finished* by the time I arrived.")

Present Tense

The simple present tense represents action occurring in the present, without any indication of time duration. ("I *ride* the train.")

A general truth is always expressed in the present tense. ("Time *heals* all wounds.") The present tense can be used to present actions or conditions that have no time restrictions. ("Water *boils* at 212 degrees Fahrenheit.") Similarly, the present tense can be used to indicate habitual action. ("I *pass* the coffee shop every day.") The present tense is also used for the "historical present," as in newspaper headlines ("FDA *Approves* New Cancer Drug") or as in references to an author's opinion or a work's contents—even though it was written in the past.

▶ In her 1979 article, Carolyn Miller *argues* that technical writing possesses significant humanistic value.

Present Perfect Tense

The present perfect tense describes something from the recent past that has a bearing on the present—a period of time before the present but after the simple past. The present perfect tense is formed by combining a form of the helping verb *have* with the past-participle form of the main verb. ("We *have finished* the draft and can now revise it.")

Future Tense

The simple future tense indicates a time that will occur after the present. It uses the helping verb *will* (or *shall*) plus the main verb. ("I *will finish* the job tomorrow.") Do not use the future tense needlessly; doing so merely adds complexity.

▶ This system ~~will be~~ *is* explained on page 3.

▶ When you press this button, the hoist ~~will move~~ *moves* the plate into position.

Future Perfect Tense

The future perfect tense indicates action that will have been completed at the time of or before another future action. It combines *will have* and the past participle of the main verb. ("She *will have driven* 1,400 miles by the time she returns.")

Shift in Tense

Be consistent in your use of tense. The only legitimate shift in tense records a real change in time. Illogical shifts in tense will only confuse your **readers**.

> Before he installed the circuit, the technician ~~cleans~~ *cleaned* the contacts.

ESL TIPS for Using the Progressive Form

The progressive form of the verb is composed of two features: a form of the helping verb *be* and the *-ing* form of the base verb.

PRESENT PROGRESSIVE	I *am updating* the Web site.
PAST PROGRESSIVE	I *was updating* the Web site last week.
FUTURE PROGRESSIVE	I *will be updating* the Web site regularly.

The present progressive is used in three ways:

1. To refer to an action that is in progress at the moment of speaking or writing:
   > The technician *is repairing* the copier.
2. To highlight that a state or an action is not permanent:
   > The office temp *is helping* us for a few weeks.
3. To express future plans:
   > The summer intern *is leaving* to return to school this Friday.

The past progressive is used to refer to a continuing action or condition in the past, usually with specified limits.

> I *was failing* calculus until I got a tutor.

The future progressive is used to refer to a continuous action or condition in the future.

> We *will be monitoring* his condition all night.

Verbs that express mental activity (*believe, know, see,* and so on) are generally not used in the progressive.

> I ~~am believing~~ *believe* the defendant's testimony.

test reports

The test report differs from the more formal **laboratory report** in both size and **scope**. Considerably smaller, less formal, and more routine than the laboratory report, the test report can be presented as a **memo** or a **letter**, depending on whether its recipient is inside or outside the organization. Either way, the **report** should have a subject line at the beginning to identify the test being discussed. See also **correspondence**.

The opening of a test report should state the test's purpose, unless it is obvious from the subject line. The body of the report presents the data and, if relevant, describes procedures used to conduct the test. State the results of the test and, if necessary, interpret them. Conclude the report, if appropriate, with any recommendations made as a result of the test. Figure T–4 shows a report of tests required by a government agency to monitor asbestos in the air at a highway construction site.

that / which / who

The word *that* is often overused and can foster wordiness.

> ▶ ~~I think that when~~ *When* this project is finished, ~~that~~ *I think* you should publish the results.

However, include *that* in a sentence if it avoids **ambiguity** or improves the **pace**. See also **conciseness**.

> ▶ Some designers fail to appreciate *that* the workers who operate equipment constitute an important safety system.

Use *which*, not *that*, with nonrestrictive clauses (clauses that do not change the meaning of the basic sentence). See also **restrictive and nonrestrictive elements**.

NONRESTRICTIVE	After John left the law firm, *which* is the largest in the region, he started a private practice.
RESTRICTIVE	Companies *that* diversify usually succeed.

That and *which* should refer to animals and things; *who* should refer to people. See also **who / whom**.

Biospherics, Inc.

4928 Wyaconda Road
Rockville, MD 20852

Phone: 301-598-9011
Fax: 301-598-9570

September 11, 2009
Safety Committee
The Angle Company, Inc.
1869 Slauson Boulevard
Waynesville, VA 23927

Subject: Monitoring Airborne Asbestos at the Route 66 Site

On August 28, Biospherics, Inc., performed asbestos-in-air monitoring at your Route 66 construction site, near Front Royal, Virginia. Six people and three construction areas were monitored.

All monitoring and analyses were performed in accordance with "Occupational Exposure to Asbestos," U.S. Department of Health and Human Services, Public Health Service, National Institute for Occupational Safety and Health, 2008. Each worker or area was fitted with a battery-powered personal sampler pump operating at a flow rate of approximately two liters per minute. The airborne asbestos was collected on a 37 mm Millipore-type AA filter mounted in an open-face filter holder. Samples were collected over an 8-hour period.

In all cases, the workers and areas monitored were exposed to levels of asbestos fibers well below the standard set by OSHA. The highest exposure found was that of a driller exposed to 0.21 fibers per cubic centimeter. The driller's sample was analyzed by scanning electron microscopy followed by energy-dispersive X-ray techniques that identify the chemical nature of each fiber, to identify the fibers as asbestos or other fiber types. Results from these analyses show that the fibers pres-ent were tremolite asbestos. No nonasbestos fibers were found.

If you need more details, please let me know.

Yours truly,

Gary Geirelach

Gary Geirelach
Chemist
gg2@Bios.org

FIGURE T–4. Test Report

▶ Dr. Cynthia Winter, *who* recently joined the clinic, treated a dog *that* was severely burned.

there / their / they're

There is an **expletive** (a word that fills the position of another word, phrase, or clause) or an **adverb**.

    EXPLETIVE   *There* were more than 1,500 people at the conference.

    ADVERB   More than 1,500 people were *there*.

Their is the **possessive case** form of *they*. ("Managers check *their* e-mail regularly.") *They're* is a **contraction** of *they are*. ("Clients tell us *they're* pleased with our services.")

thesaurus

A thesaurus is a book or an electronic file of words and their **synonyms** and **antonyms**, arranged or retrievable by categories. Thoughtfully used, a thesaurus can help you with **word choice** during the **revision** phase of the writing process. However, the variety of words it offers may tempt you to choose an inappropriate word for the **context** or to use an obscure synonym. Use a thesaurus only to clarify or refine your meaning, not to impress your **readers**. (See also **affectation**.) Never use a word unless you are sure of its meanings; its **connotations** might be unknown to you and could mislead your readers.

titles

Titles are important because many **readers** decide whether to read documents—such as **reports**, **e-mails**, **memos**, and **newsletter articles**—based on their titles. This entry discusses both creating titles and referring to them in your writing. For advice on titles for figures and tables, see **visuals**.

Reports and Long Documents

Titles for reports, **proposals**, articles, and similar documents should state the document's topic, reflect its **tone**, and indicate its **scope** and **purpose** as in the following title of an academic article.

▶ "Effects of 60-Hertz Electric Fields on Embryo Chick Development, Growth, and Behavior"

Such titles should be concise but not so short that they are not specific. For example, the title "Electric Fields and Living Organisms" announces the topic and might be appropriate for a book, but it does not answer important questions that readers of an article would expect, such as "What is the relationship between electric fields and living organisms?" and "What aspects of the organisms are related to electric fields?"

Avoid titles with such redundancies as "Notes on," "Studies on," or "A Report on." However, works like annual reports or **feasibility reports** should be identified as such in the title because this information specifies the purpose and scope of the report. For titles of **progress and activity reports**, indicate the dates in a subtitle ("Quarterly Report on Hospital Admission Rates: January–March 2009"). Avoid using technical shorthand, such as chemical formulas, and other **abbreviations** in your title unless the work is addressed exclusively to specialists in the field. For multivolume publications, repeat the title on each volume and include the subtitle and number of each volume.

Titles should not use the sentence form, except for titles of articles in **newsletters** and magazines that ask a **rhetorical question**.

▶ "Is Online Learning Right for You?"

Memos, E-mail, and Internet Postings

Subject lines of **memos**, e-mail messages, and Internet postings function as titles and should concisely and accurately describe the topic of the message. Because recipients often use subject-line titles to prioritize and sort their **correspondence**, such titles must be specific.

| VAGUE | Subject: Tuition Reimbursement |
| SPECIFIC | Subject: Tuition Reimbursement for Time-Management Seminar |

Although the title in the subject line announces your topic, you should still provide an opening that provides **context** for the message.

Formatting Titles

Capitalization. Capitalize the initial letters of the first and last words of a title as well as all major words in the title. Do not capitalize articles (*a*, *an*, *the*), coordinating conjunctions (*and*, *but*), or short prepositions (*at*, *in*, *on*, *of*) unless they begin or end the title (*The Lives of a Cell*). Capitalize prepositions in titles if they contain five or more letters (*Between*, *Since*, *Until*, *After*).

Italics. Use **italics** or underlining when referring to titles of separately published works, such as books, periodicals, newspapers, pamphlets, brochures, legal cases, movies, television programs, and Web sites.

▶ *Turning Workplace Conflict into Collaboration* [book] by Joyce Richards was reviewed in the *New York Times* [newspaper].

▶ We will include a link to *The Weather Channel* (*www.weather.com*).

Abbreviations of such titles are italicized if their spelled-out forms would be italicized.

▶ *NEJM* for the *New England Journal of Medicine*

Italicize the titles of compact and digital video discs, videotapes, plays, long poems, paintings, sculptures, and long musical works.

CD-ROM	*Computer Security Tutorial* (CD-ROM edition)
PLAY	Arthur Miller's *Death of a Salesman*
LONG POEM	T. S. Eliot's *The Wasteland*
PAINTING	M. C. Escher's *Drawing Hands*
SCULPTURE	Auguste Rodin's *The Thinker*
MUSICAL WORK	Gershwin's *Porgy and Bess*

Quotation Marks. Use **quotation marks** when referring to parts of publications, such as chapters of books and articles or sections within periodicals.

▶ Her chapter titled "Effects of Government Regulations on Motorcycle Safety" in the book *Government Regulation and the Economy* was cited in a recent article, "No-Fault Insurance and Motorcycles," published in *American Motorcyclist* magazine.

Titles of reports, essays, short poems, short musical works (including songs), short stories, and single episodes of radio and television programs are also enclosed in quotation marks.

REPORT	"Analysis of Ethics Cases at CGF Corporation"
ESSAY	Ralph Waldo Emerson's "The American Scholar"
SHORT POEM	Robert Frost's "The Road Not Taken"
SONG	Bob Dylan's "Like a Rolling Stone"
SHORT STORY	Nathaniel Hawthorne's "The Birthmark"
TV PROGRAM	"In Depth: Ray Kurzweil" on C-Span [episode of *In Depth*]

Special Cases. Some titles, by convention, are not set off by quotation marks, underlining, or italics. Such titles follow standard practice for capitalization and the practice of the organization.

- Technical Writing [college course title], Old Testament, Magna Carta, the Constitution, Lincoln's Gettysburg Address, the Lands' End Catalog

For citing titles in references and works cited, see **documenting sources**.

to / too / two

To, *too*, and *two* are frequently confused because they sound alike. *To* is used as a **preposition** or to mark an infinitive. See **verbs**.

- Send the report *to* the district manager. [preposition]
- I do not wish *to* attend. [mark of the infinitive]

Too is an **adverb** meaning "excessively" or "also."

- The price was *too* high. [excessively]
- I, *too*, thought it was high. [also]

Two is a **number** (*two* buildings, *two* concepts).

tone

Tone is the attitude a writer expresses toward the subject and his or her readers. In workplace writing, tone may range widely—depending on the **purpose**, situation, **context**, **audience**, and even the medium of a communication. For example, in an **e-mail** message to be read only by an associate who is also a friend, your tone might be casual.

- Your proposal to Smith and Kline is super. We'll just need to hammer out the schedule. If we get the contract, I owe you lunch!

In a **memo** to your manager or superior, however, your tone might be more formal and respectful.

- I think your proposal to Smith and Kline is excellent. I have marked a couple of places where I'm concerned that we are committing ourselves to a schedule that we might not be able to keep. If I can help in any other way, please let me know.

In a message that serves as a **report** to numerous readers, the tone would be professional, without the more personal **style** that you would use with an individual reader.

▶ The Smith and Kline proposal appears complete and thorough, based on our department's evaluation. Several small revisions, however, would ensure that the company is not committing itself to an unrealistic schedule. These revisions are marked on the copy of the report attached to this message.

The **word choice**, the **introduction**, and even the **title** contribute to the overall tone of your document. For instance, a title such as "Ecological Consequences of Diminishing Water Resources in California" clearly sets a different tone from "What Happens When We've Drained California Dry?" The first title would be appropriate for a report; the second title would be more appropriate for a popular magazine or **newsletter article**. See also **correspondence** and **technical writing style**.

---

trade journal articles

DIRECTORY

Planning the Article   534
Gathering the Data   534
Organizing the Draft   535
Preparing Sections of the Article   535
   Abstract   535
   Introduction   535
Conclusion   536
Visuals and Tables   536
Headings   536
References   536
Preparing the Manuscript   536
Obtaining Publication Clearance   537

Trade journal articles are written for professional periodicals that aim to further the knowledge in a field among specialists in that field. (See **audience**.) Such periodicals, commonly known as trade journals (or professional or scholarly journals), are often the official publications of professional societies. *Technical Communication*, for example, is an official voice of the Society for Technical Communication. Other professional publications include *IEEE Transactions on Communications*, *Chemical Engineering*, *Nucleonics Week*, and hundreds of others. Professional staff people, such as engineers, scientists, educators, and legal professionals, regularly contribute articles to trade journals.

   Writing a trade journal article in your field can make your work more widely known, provide publicity for your employer, give you a sense of satisfaction, and even improve your chances for professional advancement.

Planning the Article

When you are thinking about writing an article for a trade journal, consider the following questions:

- Is your work or your knowledge of the subject original? If not, what is there about your approach that justifies publication?
- Will the significance of the article justify the time and effort needed to write it?
- What parts of your work, project, or study are most appropriate to include in the article?

To help you answer those questions, learn as much as possible about the periodical or periodicals to which you plan to submit an article and consult your colleagues for advice. Once you have decided on several journals, consider the following factors about each one:

- The professional interests and size of its readership
- The professional reputation of the journal
- The appropriateness of your article to the journal's goals, as stated on its masthead page or in a mission statement on its Web site
- The frequency with which its articles are cited in other journals

Next, read back issues of the journal or journals to find out information such as the amount and kind of details that the articles include, the length of the articles, and the typical writing **style**. (See also **context**.)

If your subject involves a particular project, begin work on your article when the project is in progress. That allows you to write the draft in manageable increments and record the details of the project while they are fresh in your mind. It also makes the writing integral to the project and may even reveal any weaknesses in the design or details, such as the need for more data.

As you plan your article, decide whether to invite one or more coauthors to join you. Doing so can add strength and substance to an article. However, as with all **collaborative writing**, you should establish a schedule, assign tasks, and designate a primary author to ensure that the finished article reads smoothly.

Gathering the Data

As you gather information, take notes from all the sources of primary and secondary **research** available to you. Begin your research with a careful review of the literature to establish what has been published about your topic. A review of the relevant information in your field can be insurance against writing an article that has already been published. (Some articles, in fact, begin with a **literature review**.) As you compile that information, record your references in full; include all the information you need to document the source. See also **documenting sources**.

Organizing the Draft

Some trade journals use a prescribed **organization** for the major sections. The following organization is common in scientific journals: introduction, materials and methods, results, discussion, and sources cited. Look closely at several issues of the journals that you target to determine how those or other sections are developed. If the major organization is not prescribed, choose and arrange the various sections of the draft in a way that shows your results to best advantage.

The best guarantee of a logically organized article is a good outline. (See **outlining**.) If you have coauthors, work from a common, well-developed outline to coordinate the various writers' work and make sure the parts fit together logically.

Preparing Sections of the Article

As you prepare an article, pay particular attention to a number of key sections and elements: the **abstract**, the **introduction**, the **conclusion**, **visuals** and **tables**, **headings**, and references.

Abstract. Although it will appear at the beginning of the article, write the abstract only after you have finished writing the body of the manuscript. Follow any instructions provided by the journal on writing abstracts and review those previously published in that journal. Prepare your abstract carefully—it will be the basis on which other researchers decide whether to read your article in full. Abstracts are often published independently and are a source of terms (called *keywords*) used to index, by subject, the original article for computerized information-retrieval systems.

Introduction. The introduction should discuss these aspects of the article:

- The **purpose** of the article
- A definition of the problem examined
- The **scope** of the article
- The rationale for your approach to the problem or project and the reasons you rejected alternative approaches
- Previous work in the field, including other approaches described in previously published articles

Above all, your introduction should emphasize what is new and different about your approach, especially if you are not dealing with a new concept. It should also demonstrate the overall significance of your project or approach by explaining how it fills a need, solves a current problem, or offers a useful application.

Conclusion. The conclusion section pulls together your results and interprets them in relation to the purpose of your study and the methods used to conduct it. Your conclusion must grow out of the evidence for the findings in the body of the article.

Visuals and Tables. Used effectively and appropriately, visuals can clarify information and reinforce the point you are making in the article. Design each visual for a specific purpose: to describe a function, to show an external appearance, to show internal construction, to display statistical data, or to indicate trends.

Headings. Headings are important for an article because they break the text into manageable portions. They also allow journal readers to understand the development of your topic and pinpoint sections of particular interest to them. Check the use of headings when you review back issues of the journal.

References. The references section (often titled "Works Cited") of the article lists the sources you used in the article. The specific format for listing sources varies from field to field. Usually the journal to which you submit an article will specify the form the editors require for citing sources. For a detailed discussion of using and citing sources, see **documenting sources** and **quotations**.

Preparing the Manuscript

Some trade journals recommend a particular style guide, such as *The Chicago Manual of Style*, or offer a style sheet with detailed guidelines on style and format. Such style sheets often include specific instructions about how to format the manuscript, how many copies to submit, and how to handle **abbreviations**, symbols, **mathematical equations**, and the like. The following guidelines are typical.

- Double-space the manuscript, leaving one-inch margins all around, and number each page.
- Provide specific, accurate, and self-explanatory captions for all figures and tables. Add *call-outs* (labels) to those illustrations that need them.
- Provide clear and accurately worded labels for drawings and other illustrations.

- Place any mathematical or chemical equations on separate lines in the text and number them consecutively.
- Check with the editor about the format, file type, or other special requirements for **photographs** or scanned images.

Obtaining Publication Clearance

After you have finalized your article, submit a copy to your employer for review before sending it to the journal. A review will ensure that you have not inadvertently revealed any proprietary information. Likewise, secure permission ahead of time to print information for which someone else holds the **copyright**. See also **plagiarism**.

transition

Transition is the means of achieving a smooth flow of ideas from sentence to sentence, **paragraph** to paragraph, and subject to subject. Transition is a two-way indicator of what has been said and what will be said; it provides **readers** with guideposts for linking ideas and clarifying the relationship between them.

Transition can be obvious.

▶ *Having considered* the technical problems outlined in this proposal, *we move next* to the question of adequate staffing.

Transition can be subtle.

▶ *Even if* technical problems could be solved, there *still remains* the problem of adequate staffing.

Either way, you now have your readers' attention fastened on the problem of adequate staffing, exactly what you set out to do.

Methods of Transition

Transition can be achieved in many ways: (1) using transitional words and phrases, (2) repeating keywords or key ideas, (3) using **pronouns** with clear antecedents, (4) using enumeration (1, 2, 3, or first, second, third), (5) summarizing a previous paragraph, (6) asking a question, and (7) using a transitional paragraph.

Certain words and phrases are inherently transitional. Consider the following terms and their functions:

FUNCTION	TERMS
Result	*therefore, as a result, consequently, thus, hence*
Example	*for example, for instance, specifically, as an illustration*
Comparison	*similarly, likewise, in comparison*
Contrast	*but, yet, still, however, nevertheless, on the other hand*
Addition	*moreover, furthermore, also, too, besides, in addition*
Time	*now, later, meanwhile, since then, after that, before that time*
Sequence	*first, second, third, initially, then, next, finally*

Within a paragraph, such transitional expressions clarify and smooth the movement from idea to idea. Conversely, the lack of transitional devices can make for disjointed reading. See also **telegraphic style**.

Transition Between Sentences

You can achieve effective transition between sentences by repeating keywords or key ideas from preceding sentences and by using pronouns that refer to antecedents in previous sentences. Consider the following short paragraph, which uses both of those means.

▶ Representative of many American university towns is Middletown. *This midwestern town*, formerly *a sleepy farming community*, is today the home of a large and vibrant *academic community*. Attracting students from all over the Midwest, *this university town* has grown very rapidly in the last ten years.

Enumeration is another device for achieving transition.

▶ The recommendation rests on *two conditions. First*, the department staff must be expanded to handle the increased workload. *Second*, sufficient time must be provided for training the new staff.

Transition Between Paragraphs

The means discussed so far for achieving transition between sentences can also be effective for achieving transition between paragraphs. For paragraphs, however, longer transitional elements are often required. One technique is to use an opening sentence that summarizes the preceding paragraph and then moves on to a new paragraph.

▶ One property of material considered for manufacturing processes is hardness. Hardness is the internal resistance of the material to the forcing apart or closing together of its molecules. Another property is ductility, the characteristic of material that permits it to be drawn into a wire. Material also may possess malleability, the property that makes it capable of being rolled or hammered into thin sheets of various shapes. Purchasing managers must consider these properties before selecting manufacturing materials for use in production.

The requirements of hardness, ductility, and malleability account for the high cost of such materials. . . .

Another technique is to ask a question at the end of one paragraph and answer it at the beginning of the next.

▶ New technology has always been feared because it has at times displaced some jobs. However, it invariably creates many more jobs than it eliminates. Almost always, the jobs eliminated by technological advances have been menial, unskilled jobs, and workers who have been displaced have been forced to increase their skills, which resulted in better and higher-paying jobs for them. *In view of these facts, is new technology really bad?*

Certainly technology has given us unparalleled access to information and created many new roles for employees. . . .

A purely transitional paragraph may be inserted to aid readability.

▶ The problem of poor management was a key factor that caused the weak performance of the company.

Two other setbacks to the company's fortunes also marked the company's decline: the loss of many skilled workers through the early retirement program and the intensification of the rate of employee turnover.

The early retirement program resulted in engineering staff . . .

If you provide logical **organization** and have prepared an outline, your transitional needs will easily be satisfied and your writing will have **unity** and **coherence**. During **revision**, look for places where transition is missing and add it. Look for places where it is weak and strengthen it.

trip reports

A trip report provides a permanent record of a business trip and its accomplishments. It provides managers with essential information about the results of the trip and can enable other staff members to benefit from the information. See also **reports**.

A trip report is normally written as a **memo** or an **e-mail** and addressed to an immediate superior, as shown in Figure T–5. The subject line identifies the destination and dates of the trip. The body of the report explains why you made the trip, whom you visited, and what you

> From: James D. Kerson <jdkerson@psys.com>
> To: Roberto Camacho <rcamacho@psys.com>
> Sent: Wed, 14 Jan 2009 12:16:30 EST
> Subject: Trip to Smith Electric Co., Huntington, West Virginia, January 5–6, 2009
>
> Attachments: 📄 Expense Report.xls (25 KB)
>
> ---
>
> I visited the Smith Electric Company in Huntington, West Virginia, to determine the cause of a recurring failure in a Model 247 printer and to fix it.
>
> **Problem**
> The printer stopped printing periodically for no apparent reason. Repeated efforts to bring it back online eventually succeeded, but the problem recurred at irregular intervals. Neither customer personnel operating the printer nor the local maintenance specialist was able to solve the problem.
>
> **Action**
> On January 5, I met with Ms. Ruth Bernardi, the Office Manager, who explained the problem. My troubleshooting did not reveal the cause of the problem then or on January 6.
>
> Only when I tested the logic cable did I find that it contained a broken wire. I replaced the logic cable and then ran all the normal printer test patterns to make sure no other problems existed. All patterns were positive, so I turned the printer over to the customer.
>
> **Conclusion**
> There are over 12,000 of these printers in the field, and to my knowledge this is the first occurrence of a bad cable. Therefore, I do not believe the logic cable problem found at Smith Electric Company warrants further investigation.
>
> =================================
> James D. Kerson, Maintenance Specialist
> Printer Systems, Inc.
> 1366 Federal St., Allentown, PA 18101
> (610) 747-9955 Fax: (610) 747-9956
> jdkerson@psys.com
> www.psys.com
> =================================

FIGURE T–5. Trip Report Sent as E-mail (with Attachment)

accomplished. The report should devote a brief section to each major activity and may include a **heading** for each section. You need not give equal space to each activity—instead, elaborate on the more important ones. Follow the body of the report with the appropriate **conclusions** and recommendations. Finally, if required, attach a record of expenses to the trip report.

trouble reports

The trouble report is used to analyze such events as accidents, equipment failures, or health emergencies. For example, the report shown in Figure T–6 describes an accident involving personal injury. The report assesses the causes of the problem and suggests changes necessary to prevent its recurrence. Because it is usually an internal document, the trouble report normally follows the **memo** format. See also **reports**.

In the subject line of the memo, state the precise problem you are reporting. Then, in the body of the report, provide a detailed, precise description of the problem. What happened? Where and when did the problem occur? Was anybody hurt? Was there any property damage? Was there a work stoppage?

Consolidated Energy, Inc.

To: Marvin Lundquist, Vice President
Administrative Services

From: Kalo Katarlan, Safety Officer *KK*
Field Service Operations

Date: August 19, 2009

Subject: Field Service Employee Accident on August 5, 2009

The following is an initial report of an accident that occurred on Wednesday, August 5, 2009, involving John Markley, and that resulted in two days of lost time.

Accident Summary
John Markley stopped by a rewiring job on German Road. Chico Ruiz was working there, stringing new wire, and John was checking with Chico about the materials he wanted for framing a pole. Some tree trimming had been done in the area, and John offered to help remove some of the debris by loading it into the pickup truck he was driving. While John was loading branches into the bed of the truck, a piece broke off in his right hand and struck his right eye.

Accident Details
1. John's right eye was struck by a piece of tree branch. John had just undergone laser surgery on his right eye on Monday, August 3, to reattach his retina.
2. John immediately covered his right eye with his hand, and Chico Ruiz gave him a paper towel with ice to cover his eye and help ease the pain.

FIGURE T–6. Trouble Report (Using Printed Memo)

> 7. On Thursday, August 6, John returned to his eye surgeon. Although bruised, his eye was not damaged, and the surgically reattached retina was still in place.
>
> **Recommendations**
> To prevent a recurrence of such an accident, the Safety Department will require the following actions in the future:
>
> - When working around and moving debris such as tree limbs or branches, all service crew employees must wear safety eyewear with side shields.
> - All service crew employees must always consider the possibility of shock for an injured employee. If crew members cannot leave the job site to care for the injured employee, someone on the crew must call for assistance from the Service Center. The Service Center phone number is printed in each service crew member's handbook.

FIGURE T–6. Trouble Report (Using Printed Memo) *(continued)*

▶ **ETHICS NOTE** Because insurance claims, workers'-compensation awards, and even lawsuits may hinge on the information contained in a trouble report, be sure to include precise times, dates, locations, treatment of injuries, names of any witnesses, and any other crucial information. (Notice the careful use of language and factual detail in Figure T–6.) Be thorough and accurate in your analysis of the problem and support any judgments or conclusions with facts. Be objective: Always use a neutral **tone** and avoid assigning blame. If you speculate about the cause of the problem, make it clear to your **readers** that you are speculating. See also **ethics in writing**. ◆

In your **conclusion**, state what has been or will be done to correct the conditions that led to the problem. That may include, for example, recommendations for training in safety practices, improved equipment, and protective clothing.

try to

The phrase *try and* is colloquial for *try to*. For technical writing, use *try to*.

▶ Please try ~~and~~ *to* finish the report by next week.

U

unity

Unity is singleness of **purpose** and focus; a unified **paragraph** or document has a central idea and does not digress into unrelated topics.

The logical sequence provided through **outlining** is essential to achieving unity. An outline enables you to lay out the most direct route from **introduction** to **conclusion**, and it enables you to build each paragraph around a topic sentence that expresses a single idea.

Effective **transition** helps build unity, as well as **coherence**, because transitional terms clarify the relationship of each part to what precedes it.

up

Adding the word *up* to **verbs** often creates a redundant phrase. See also **conciseness**.

▶ You must open ~~up~~ the exhaust valve.

usability testing

Usability refers to whether **readers** can use a document or Web site to easily fulfill their goals or accomplish tasks. For example, the usability of a technical **manual** might refer to the ability of readers to perform a procedure accurately and smoothly with the aid of the **instructions**.*
The usability of a retail Web site might refer to the ability of visitors to understand the contents quickly and navigate the site easily to order a product.

Usability testing helps produce a document that reduces the learning curve, allows more functionality with less effort, and increases productivity. At the same time, by focusing development on users, companies

---

*Usability engineering, or *user-centered design* (UCD), refers to the process of developing products and systems with the user's needs in mind so that ideally usability is built into both products and their documentation from the beginning.

also reap benefits in reduced costs and increased customer satisfaction in the following ways:

- Usable documents can enhance the organization's reputation.
- The documents are functional and more likely to be used.
- Product and document changes can be made before they become expensive.
- Efficient document development and training are facilitated.
- The need for updates and maintenance releases is minimized.

Usability testing is a common way to involve users in the development process—for example, by testing periodic drafts on users and then revising the document in response to the test results. The process typically begins with the establishment of specific, quantitative, and measurable goals for documents. Next, the documents are designed to fulfill those goals. Finally, tests must be conducted with representative users to detect problems and determine whether the established goals have been achieved.

Usability testing involves teams of skilled usability specialists, interface designers, and technical writers. Usability testing has three main goals:

- To create a document that is easy to use and allows users to accomplish the tasks outlined in that document
- To detect potential problems for users as well as to guide designers in resolving such issues, minimizing the risk of releasing ineffective documents
- To enable companies to avoid repeating mistakes when developing future documents

Test participants typically are members of the target **audience** for the document. If the participants encounter problems with the document, it is likely that actual users will experience similar problems. For example, if page 15 of a tax form is unclear to test participants, it is likely to be confusing to most taxpayers.

Usability tests can include one or more of the following methods.

- *User testing.* Testers observe and record the actions of test participants who perform real tasks using the document.
- *Protocols.* Test participants make comments aloud as they read documents and perform document tasks, to reveal their thought processes, attitudes, and reasons for decision-making.
- *Comprehension tests.* Users complete tests to determine whether they understand and can recall document features, such as **visuals** or **tables**.

- *Surveys and interviews.* Testers interview users both before and after they read documents to determine their comprehension and attitudes and the **clarity** of the documents.

Analyzing the results of those test methods can help document designers detect problems and determine whether the document's **purposes** have been met. If test participants have trouble navigating a document or quickly locating specific items, for example, the organization as well as **layout and design** are revised. See also **forms design**, **interviewing for information**, and **questionnaires**.

usage

Usage describes the choices we make among the various words and expressions available in our language. The lines between standard English and nonstandard English and between formal and informal English are determined by those choices. Your guideline in any situation requiring such choices should be appropriateness: Is the word or expression you use appropriate to your **audience** and your subject? When it is, you are practicing good usage.

This book contains many entries that focus on various usages. For a complete list of the usage entries, which appear in italics throughout the book, see "Commonly Misused Words and Phrases" on pages 627–28. An up-to-date **dictionary** is also an invaluable aid in your selection of the right word.

> **WEB LINK** | **Online Usage and Style Guides**
>
> Bartleby.com provides classic reference books online, including the *American Heritage Book of English Usage*. For this and additional useful links, see *bedfordstmartins.com/alredtech* and select *Links for Handbook Entries*.

utilize

Do not use *utilize* as a long variant of *use*, which is the general word for "employ for some purpose." *Use* will almost always be clearer and less pretentious. See **affectation**.

V

vague words

A vague word is one that is imprecise in the context in which it is used. Some words encompass such a broad range of meanings that there is no focus for their definition. Words such as *real*, *nice*, *important*, *good*, *bad*, *contact*, *thing*, and *fine* are often called "omnibus words" because they can have so many meanings and interpretations. In speech, our vocal inflections help make the meanings of such words clear. Because you cannot rely on vocal inflections when you are writing, avoid using vague words. Be concrete and specific. See also **abstract / concrete words** and **word choice**.

VAGUE	It was a *productive* meeting. [Why was it productive?]
SPECIFIC	The meeting resolved three questions: pay scales, fringe benefits, and workloads.

verbals

Verbals are derived from **verbs** but function as **nouns**, **adjectives**, and **adverbs**. The three types of verbals are gerunds, infinitives, and participles.

Gerunds

A gerund is a verbal ending in *-ing* that is used as a noun. A gerund can be used as a subject, a direct **object**, the object of a **preposition**, a subjective **complement**, or an **appositive**.

- *Drawing* is an important engineering skill. [subject]
- I find *drawing* interesting. [direct object]
- We were unprepared for their *coming*. [object of preposition]
- Seeing is *believing*. [subjective complement]
- My primary departmental function, *programming*, occupies about two-thirds of my time on the job. [appositive]

Only the possessive form of a noun or **pronoun** should precede a gerund.

▶ *John's* working has not affected his grades.

▶ *His* working has not affected his grades.

Infinitives

An infinitive is the bare, or uninflected, form of a verb (for example, *go, run, fall, talk, dress, shout*) without the restrictions imposed by **person** and **number**. Along with the gerund and the participle, it is one of the nonfinite verb forms. The infinitive is generally preceded by the word *to*, which, although not an inherent part of the infinitive, is considered to be the sign of an infinitive. An infinitive is a verbal and can function as a noun, an adjective, or an adverb.

▶ *To expand* is not the only objective. [noun]

▶ These are the instructions *to follow*. [adjective]

▶ The company struggled *to survive*. [adverb]

The infinitive can reflect two **tenses**: the present and (with a helping verb) the present perfect.

▶ to go [present tense]

▶ to have gone [present perfect tense]

The most common mistake made with infinitives is using the present perfect tense when the simple present tense is sufficient.

▶ I should not have tried to ~~have gone~~ *go* so early.

Infinitives formed with the root form of transitive verbs can express both active and (with a helping verb) passive **voice**.

▶ to hit [present tense, active voice]

▶ to have hit [present perfect tense, active voice]

▶ to be hit [present tense, passive voice]

▶ to have been hit [present perfect tense, passive voice]

A split infinitive is one in which an adverb is placed between the sign of the infinitive, *to*, and the infinitive itself. Because they make up a grammatical unit, the infinitive and its sign are better left intact than separated by an intervening adverb.

▶ To ~~initially~~ build a client base, experts recommend networking with friends and family.

However, it may occasionally be better to split an infinitive than to allow a sentence to become awkward, ambiguous, or incoherent.

AMBIGUOUS	She agreed immediately *to deliver* the specimen to the lab. [This sentence could be interpreted to mean that she agreed immediately.]
CLEAR	She agreed *to* immediately *deliver* the specimen to the lab. [This sentence is no longer ambiguous.]

Participles

A participle is a verb form that functions as an adjective. Present participles end in *-ing*.

▶ *Increasing* costs forced us to reduce our staff.

Past participles end in *-ed*, *-t*, *-en*, *-n*, or *-d*.

▶ What are the *estimated* costs?

▶ Repair the *bent* lever.

▶ Return the *broken* part.

▶ What are the metal's *known* properties?

▶ The story, *told* many times before, was still interesting.

The perfect participle is formed with the present participle of the helping verb *have* plus the past participle of the main verb.

▶ *Having gotten* [perfect participle] a large bonus, the *smiling* [present participle], *contented* [past participle] technician worked harder than ever.

A participle cannot be used as the verb of a sentence. Inexperienced writers sometimes make that mistake, and the result is a **sentence fragment**.

▶ The committee chairperson was responsible. His vote being the decisive one. [, his]

▶ The committee chairperson was responsible. His vote being the decisive one. [was]

For information on participial and infinitive phrases, see **phrases**.

verbs

DIRECTORY
Types of Verbs   549
Forms of Verbs   550
   Finite Verbs   550
   Nonfinite Verbs   550
Properties of Verbs   550

A verb is a word or group of words that describes an action ("The copier *jammed* at the beginning of the job"), states how something or someone is affected by an action ("He *was disappointed* that the proposal was rejected"), or affirms a state of existence ("She *is* a district manager now").

Types of Verbs

Verbs are either transitive or intransitive. A *transitive verb* requires a direct **object** to complete its meaning.

▶ They *laid* the foundation on October 24.
[*Foundation* is the direct object of the transitive verb *laid*.]

▶ Rosalie Anderson *wrote* the treasurer a letter.
[*Letter* is the direct object of the transitive verb *wrote*.]

An *intransitive verb* does not require an object to complete its meaning. It makes a full assertion about the subject without assistance (although it may have **modifiers**).

▶ The engine *ran*.

▶ The engine *ran* smoothly and quietly.

A *linking verb* is an intransitive verb that links a **complement** to the subject.

▶ The carpet *is* stained.
[*Is* is a linking verb; *stained* is a subjective complement.]

Some intransitive verbs, such as *be*, *become*, *seem*, and *appear*, are almost always linking verbs. A number of others, such as *look*, *sound*, *taste*, *smell*, and *feel*, can function as either linking verbs or simple intransitive verbs. If you are unsure about whether one of those verbs is a linking verb, try substituting *seem*; if the sentence still makes sense, the verb is probably a linking verb.

- Their antennae *feel* delicate.
  [*Seem* can be substituted for *feel*—thus *feel* is a linking verb.]
- Their antennae *feel* delicately for their prey.
  [*Seem* cannot be substituted for *feel*; in this case, *feel* is a simple intransitive verb.]

Forms of Verbs

Verbs are described as being either finite or nonfinite.

Finite Verbs. A finite verb is the main verb of a **clause** or sentence. It makes an assertion about its subject and often serves as the only verb in its clause or sentence. ("The telephone *rang*, and the receptionist *answered* it.") See also **sentence construction**.

A helping verb (sometimes called an *auxiliary verb*) is used in a verb **phrase** to help indicate **mood**, **tense**, and **voice**. ("The phone *had* rung.") Phrases that function as helping verbs are often made up of combinations with the sign of the infinitive, *to* (for example, *am going to*, *is about to*, *has to*, and *ought to*). The helping verb always precedes the main verb, although other words may intervene. ("Machines *will* never completely *replace* people.")

Nonfinite Verbs. Nonfinite verbs are **verbals**—verb forms that function as **nouns**, **adjectives**, or **adverbs**.

A *gerund* is a noun that is derived from the *-ing* form of a verb. ("*Seeing* is *believing*.") An *infinitive*, which uses the root form of a verb (usually preceded by *to*), can function as a noun, an adverb, or an adjective.

- He hates *to complain*. [noun, direct object of *hates*]
- The valve closes *to stop* the flow. [adverb, modifies *closes*]
- This is the proposal *to consider*. [adjective, modifies *proposal*]

A *participle* is a verb form that can function as an adjective.

- The *rejected* proposal may be resubmitted when the concerns are addressed.
  [*Rejected* is a verb form that is used as an adjective modifying *proposal*.]

Properties of Verbs

Verbs must (1) agree in **person** with personal pronouns functioning as subjects, (2) agree in tense and **number** with their subjects, and (3) be in the appropriate voice.

> **ESL TIPS** for Avoiding Shifts in Voice, Mood, or Tense
>
> To achieve clarity in your writing, you must maintain consistency and avoid shifts. A shift is an abrupt change in voice, mood, or tense. Pay special attention when you edit your writing to check for the following types of shifts.
>
> **VOICE**
>
> ▶ The captain permits his crew to go ashore, but ~~they are not~~ *he does not permit them* ~~permitted~~ to go downtown.
>
> [The entire sentence is now in the active voice.]
>
> **MOOD**
>
> ▶ Reboot your computer, and ~~you should~~ empty the cache.
>
> [The entire sentence is now in the imperative mood.]
>
> **TENSE**
>
> ▶ I was working quickly, and suddenly a box ~~falls~~ *fell* off the conveyor belt and ~~breaks~~ *broke* my foot.
>
> [The entire sentence is now in the past tense.]

Person is the term for the form of a personal pronoun that indicates whether the pronoun refers to the speaker, the person spoken to, or the person (or thing) spoken about. Verbs change their forms to agree in person with their subjects.

▶ I *see* [first person] a yellow tint, but she *sees* [third person] a yellow-green hue.

Tense refers to verb forms that indicate time distinctions. The six tenses are past, past perfect, present, present perfect, future, and future perfect.

Number refers to the two forms of a verb that indicate whether the subject of a verb is singular ("The copier *was* repaired") or plural ("The copiers *were* repaired").

Most verbs show the singular of the present tense by adding *-s* or *-es* (he *stands*, she *works*, it *goes*), and they show the plural without *-s* or *-es* (they *stand*, we *work*, they *go*). The verb *to be*, however, normally changes form to indicate the singular ("I *am* ready") or plural ("We *are* ready").

Voice refers to the two forms of a verb that indicate whether the subject of the verb acts or receives the action. The verb is in the *active voice* if the subject of the verb acts ("The bacteria *grow*"); the verb is in the *passive voice* if it receives the action ("The bacteria *are grown* in a petri dish").

WEB LINK	Conjugation of Verbs

The conjugation of a verb arranges all forms of the verb so that the differences caused by the changing of the tense, number, person, and voice are readily apparent. For a chart showing the full conjugation of the verb *drive*, see *bedfordstmartins.com/alredtech* and select *Links for Handbook Entries*.

very

The use of **intensifiers** like *very* is tempting, but the word can usually be deleted.

▶ The board was ~~very~~ worried about a possible product recall.

When you do use intensifiers, clarify their meaning.

▶ The patient's resting heart rate was *very* fast; it was measured at 180 beats per minute.

via

Via is Latin for "by way of." The term should be used only in routing instructions.

▶ The package was shipped *via* FedEx.

▶ Her project was funded ~~via~~ *as a result of* the recent legislation.

visuals

Visuals can express ideas or convey information in ways that words alone cannot by making abstract concepts and relationships concrete. Visuals can show how things look (drawings, photographs, maps), represent numbers and quantities (graphs, tables), depict processes or relationships (flowcharts, schematic diagrams), and show hierarchical relationships (organizational charts). They also highlight important information and emphasize key concepts succinctly and clearly.

Many of the qualities of good writing—simplicity, clarity, conciseness, directness—are equally important when creating and using visuals. Presented with clarity and consistency, visuals can help **readers** focus on key portions of your document, presentation, or Web site. Be aware, though, that even the best visual will not be effective without **context**—and most often context is provided by the text that introduces the visual and clarifies its purpose

The following entries in this book are related to specific visuals and their use in printed and online documents, as well as in **presentations** (see that entry for presentation graphics).

drawings 154	**maps** 319
flowcharts 192	**organizational charts** 362
formal reports 195	**photographs** 377
global graphics 230	**tables** 519
graphs 235	**Web design** 561
layout and design 295	**writing for the Web** 570

Selecting Visuals

Consider your **audience** and your **purpose** carefully in selecting visuals. You would need different illustrations for an automobile owner's manual or an auto dealer's Web site, for example, than you would for a mechanic's diagnostic guide. Figure V–1 can help you select the most appropriate visuals, based on their purposes and special features. Jot down visual options when you are considering your **scope** and **organization**.

▸ **ETHICS NOTE** Be aware that visuals have the potential of misleading readers when data are selectively omitted or distorted. For example, Figure G–6 shows a graph that gives a misleading impression of lab-test failures because the scale is compressed, with some of the years selectively omitted. Visuals that mislead readers call the credibility of you and your organization into question at the least—and they are unethical. The use of misleading visuals can even subject you and your organization to lawsuits. ✦

Integrating Visuals with Text

Consider the best locations for visuals before you begin **writing a draft**. Your goal should be to use visuals where they will best advance your purpose, aid your readers, and integrate smoothly within your text.

CHOOSING APPROPRIATE VISUALS

TO SHOW OBJECTS AND SPATIAL RELATIONSHIPS

DRAWINGS CAN...

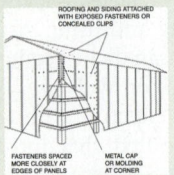

- Depict real objects difficult to photograph
- Depict imaginary objects
- Highlight only parts viewers need to see
- Show internal parts of equipment in cutaway views
- Show how equipment parts fit together in exploded views

PHOTOGRAPHS CAN...

- Show actual physical images of subjects
- Record an event in process
- Record the development of phenomena over time
- Record the as-found condition of a situation for an investigation

TO DISPLAY GEOGRAPHIC INFORMATION

MAPS CAN...

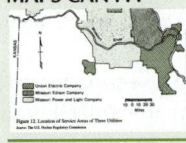

- Show specific geographic features of an area
- Show distance, routes, or locations of sites
- Show the geographic distribution of information (e.g., populations by region)

TO SHOW NUMERICAL AND OTHER RELATIONSHIPS

TABLES CAN...

Divisions	Employees
Research	1,052
Marketing	2,782
Automotive	13,251
Consumer Products	2,227

- Organize information systematically in rows and columns
- Present large numerical quantities concisely
- Facilitate item-to-item comparisons
- Clarify trends and other graphical information with precise data

BAR & COLUMN GRAPHS CAN...

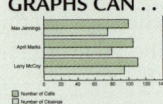

- Depict data in vertical or horizontal bars and columns for comparison
- Show quantities that make up a whole
- Visually represent data shown in tables

FIGURE V–1. Chart for Choosing Appropriate Visuals (*continued on next page*)

LINE GRAPHS CAN...

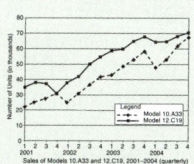

- Show trends over time in amounts, sizes, rates, and other measurements
- Give an at-a-glance impression of trends, forecasts, and extrapolations of data
- Compare more than one kind of data over the same time period
- Visually represent data shown in tables

PICTURE GRAPHS CAN...

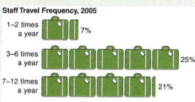

- Use recognizable images to represent specific quantities
- Help nonexpert readers grasp the information
- Visually represent data shown in tables

PIE GRAPHS CAN...

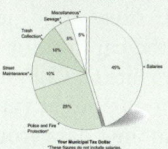

- Show quantities that make up a whole
- Give an immediate visual impression of the parts and their significance
- Visually represent data shown in tables or lists

TO SHOW STEPS IN A PROCESS OR RELATIONSHIPS IN A SYSTEM

FLOWCHARTS CAN...

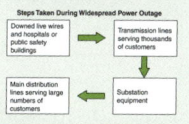

- Show how the parts or steps in a process or system interact
- Show the stages of an actual or a hypothetical process in the correct direction, including recursive steps

TO SHOW RELATIONSHIPS IN A HIERARCHY

ORGANIZATIONAL CHARTS CAN...

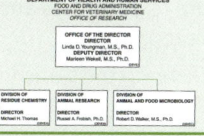

- Give an overview of an organization's departmental components
- Show how the components relate to one another
- Depict lines of authority within an organization

TO SUPPLEMENT OR REPLACE WORDS

SYMBOLS OR ICONS CAN...

- Convey ideas without words
- Save space and add visual appeal
- Transcend individual languages to communicate ideas effectively for international readers

FIGURE V–1. Chart for Choosing Appropriate Visuals (*continued*)

One way to use visuals wisely is to make their placement a part of your **outlining** process. At appropriate points in your outline, either make a rough sketch of the visual, if you can, or write "illustration of . . . ," noting the source of the visual and enclosing each suggestion in a text box. You may also include sketches of visuals in your thumbnail pages, as discussed in **layout and design**.

When you write the draft, place visuals as close as possible to the text where they are discussed—in fact, no visual should precede its first text mention. Refer to graphics (such as drawings and photographs) as "figures" and to tables as "tables." Clarify for readers why each visual is included in the text. The amount of description you should provide will vary, depending on your readers' backgrounds. For example, nonexperts may require lengthier explanations than experts need.

ETHICS NOTE Obtain written permission to use copyrighted visuals—including images and multimedia material from Web sites—and acknowledge borrowed material in a source line below the caption for a figure and in a footnote at the bottom of a table. Use a site's "Contact Us" page to request approval. Acknowledge your use of any public (uncopyrighted) information, such as demographic or economic data from government publications and Web sites, with a source line. See also **copyright**, **documenting sources**, and **plagiarism**. ✦

WRITER'S CHECKLIST Creating and Integrating Visuals

CREATING VISUALS

- Keep visuals simple: Include only information needed for discussion in the text and eliminate unneeded labels, arrows, boxes, and lines.
- Position the lettering of any explanatory text or labels horizontally; allow adequate white space within and around the visual.
- Specify the units of measurement used, make sure relative sizes are clear, and indicate distance with a scale when appropriate.
- Use consistent terminology; for example, do not refer to the same information as a "proportion" in the text and a "percentage" in the visual.
- Define **abbreviations** the first time they appear in the text and in figures and tables. If any symbols are not self-explanatory, include a key, as in Figure G–10.
- Give each visual a caption or concise **title** that clearly describes its content, and assign figure and table numbers if your document contains more than one illustration or table.
- Refer to visuals in the text of your document by their figure or table numbers.

Writer's Checklist: Creating and Integrating Visuals (continued)

INTEGRATING VISUALS

- Clarify for readers why each visual is included in the text and provide an appropriate description.
- Place visuals as close as possible to the text where they are discussed, but after their first text mention.
- Refer to visuals in the text of your document as "figures" or "tables" and by their figure or table numbers.
- Consider placing lengthy, detailed visuals in an **appendix** and refer to it in the text.
- In documents with more than five illustrations or tables, include a section following the **table of contents** titled "List of Figures" or "List of Tables" that identifies each by number, title, and page number.
- Check the editorial guidelines or recommended style manual when preparing visuals for a **trade journal article**.

voice

In grammar, *voice* indicates the relation of the subject to the action of the **verb**. When the verb is in the *active voice*, the subject acts; when it is in the *passive voice*, the subject is acted upon.

ACTIVE David Cohen *wrote* the newsletter article.
[The subject, *David Cohen*, performs the action; the verb, *wrote*, describes the action.]

PASSIVE The newsletter article *was written* by David Cohen.
[The subject, *the newsletter article*, is acted upon; the verb, *was written*, describes the action.]

The two sentences say the same thing, but each has a different emphasis: The first emphasizes *David Cohen*; the second emphasizes *the newsletter article*. In technical writing, it is often important to emphasize who or what performs an action. Further, the passive-voice version is indirect because it places the performer of the action behind the verb instead of in front of it. Because the active voice is generally more direct, more concise, and easier for **readers** to understand, use the active voice unless the passive voice is more appropriate, as described on page 559. Whether you use the active voice or the passive voice, be careful not to shift voices in a sentence.

▶ David Cohen corrected the inaccuracy as soon as ~~it was identified~~ ~~by~~ the editor. [insert: *identified it*]

Using the Active Voice

Improving Clarity. The active voice improves **clarity** and avoids confusion, especially in **instructions** and **manuals**.

> PASSIVE   Sections B and C *should be checked* for errors.
> [Are they already checked?]
>
> ACTIVE   *Check* sections B and C for errors.
> [The performer of the action, *you*, is understood: (You) *Check* the sections.]

Active voice can also help avoid **dangling modifiers**.

> PASSIVE   Hurrying to complete the work, the cables *were connected* improperly.
> [*Who* was hurrying? The implication is the cables were hurrying!]
>
> ACTIVE   Hurrying to complete the work, the technician *connected* the cables improperly.
> [Here, *hurrying to complete the work* properly modifies the performer of the action: *the technician*.]

Highlighting Subjects. One difficulty with passive sentences is that they can bury the performer of the action in **expletives** and prepositional **phrases**.

> PASSIVE   It *was reported by* the testing staff that the new model is defective.
>
> ACTIVE   The testing staff *reported* that the new model is defective.

Sometimes writers using the passive voice fail to name the performer—information that might be missed.

> PASSIVE   The problem *was discovered* yesterday.
>
> ACTIVE   The maintenance technician *discovered* the problem yesterday.

Achieving Conciseness. The active voice helps achieve **conciseness** because it eliminates the need for an additional helping verb as well as an extra **preposition** to identify the performer of the action.

PASSIVE   Arbitrary changes in policy *are resented by* employees.
ACTIVE   Employees *resent* arbitrary changes in policy.

The active-voice version takes one verb (*resent*); the passive-voice version takes two verbs (*are resented*) and an extra preposition (*by*).

Using the Passive Voice

The passive voice is sometimes effective or even necessary. Indeed, for reasons of tact and diplomacy, you might need to use the passive voice to avoid accusing others.

ACTIVE   Your staff *did not meet* the quota last month.
PASSIVE   The quota *was not met* last month.

◆ ETHICS NOTE Be careful, however, not to use the passive voice to evade responsibility or to obscure an issue or information that readers should know.

▶ Several mistakes *were made*. [*Who* made the mistakes?]
▶ It *has been decided*. [*Who* has decided?]

See also **ethics in writing**. ✦

When the performer of the action is either unknown or unimportant, of course, use the passive voice. ("The copper mine *was discovered* in 1929.") When the performer of the action is less important than the receiver of that action, the passive voice is sometimes more appropriate. ("Ann Bryant *was presented* with an award by the president.")

When you are explaining an operation in which the reader is not actively involved or when you are explaining a process or a procedure, the passive voice may be more appropriate. In the following example, anyone—it really does not matter who—could be the performer of the action.

▶ Area strip mining *is used* in regions of flat to gently rolling terrain, like that found in the Midwest. Depending on applicable reclamation laws, the topsoil *may be removed* from the area *to be mined*, *stored*, and later *reapplied* as surface material during reclamation of the mined land. After the removal of the topsoil, a trench *is cut* through the overburden to expose the upper surface of the coal to be mined. The overburden from the first cut *is placed* on the unmined land adjacent to the cut. After the first cut *has been completed*, the coal *is removed*.

Do not, however, simply assume that any such explanation should be in the passive voice; in fact, as in the following example, the active voice is often more effective.

▶ In the operation of an internal combustion engine, an explosion in the combustion chamber *forces* the pistons down in the cylinders. The movement of the pistons in the cylinders *turns* the crankshaft.

Ask yourself, "Would it be of any advantage to the reader to know the performer of the action?" If the answer is yes, use the active voice, as in the previous example.

> **ESL TIPS** for Choosing Voice
>
> Different languages place different values on active-voice and passive-voice constructions. In some languages, the passive is used frequently; in others, hardly at all. As a nonnative speaker of English, you may have a tendency to follow the pattern of your native language. But remember, even though technical writing may sometimes require the passive voice, active verbs are highly valued in English.

wait for / wait on

Wait on should be restricted in writing to the activities of hospitality and service employees. ("We need extra staff to *wait on* customers.") Otherwise, use *wait for*. ("Be sure to *wait for* Ms. Garcia's approval.") See also **idioms**.

Web design

Designing Web sites and pages requires that you stay current with changes in technology, so the best sources for guidance are on the Web itself.* To find many useful tutorials, use a search engine to search for *tutorial* along with other keywords such as *HTML*, *XHTML*, or *Web design*. For advice on developing your site's content for online **readers**, see **writing for the Web**.

> **WEB LINK**   **Web Design Resources**
>
> For Mike Markel's helpful overview of the process and principles of designing Web sites, see *bedfordstmartins.com/alredtech* and select *Tutorials*, "Designing for the Web." For links to more tutorials as well as sites that provide advice for improving accessibility for people with disabilities, select *Links for Technical Writing*.

Audience and Purpose

Many of the principles covered throughout this book apply to designing Web sites. For example, when planning a Web site and designing Web pages, carefully consider your **purpose** and **audience**. Most professional Web sites have well-defined goals, such as reference, marketing, education, publicity, or advocacy. One way to help you achieve your goals is to create a clear purpose statement, as shown in the following examples.

---

*Some useful sites include *www.w3.org*, *webmonkey.com*, and *trace.wisc.edu/world/web/*.

EXTERNAL SITE	The purpose of this site is to enable our customers to locate product information, place online orders, and contact our customer-service department.
INTERNAL SITE	The purpose of this site is to provide SNR Security Corporation employees, suppliers, and partner organizations with a single, consistent, and up-to-date resource for materials about SNR Security.

As these examples show, the two general kinds of sites are external and internal. External Web sites target an audience from the entire Internet. Internal Web sites are designed for audiences either on an intranet (a network within an organization) that is not accessible to audiences outside that organization or an extranet (a password-protected site) available exclusively to trusted partners and suppliers to provide them with common data and documents.

Access for People with Disabilities

Design elements such as colorful graphics, animation, and streaming video and audio can be barriers to people with impaired vision or hearing or those who are color-blind. Use the following strategies to meet the needs of such audiences.

- Avoid frames, complex tables, animation, JavaScript, and other design elements incompatible with text-only browsers and adaptive technologies, such as voice or large-print software.
- Provide HTML versions of pages and documents whenever possible because this format is most compatible with the current generation of screen readers.
- Include text-equivalent captions with graphic or audio elements.
- Design for the color-blind reader by making meaning independent of color. For example, rather than asking users to "Click on the green button for more information," label the button ("Click Here") or embed a link in a sentence: "See our catalog for more information."

You may want to consider offering different options for site visitors, such as full-graphics, light-graphics, and text-only versions.

DIGITAL TIP

Testing Your Web Site

You might take advantage of Web sites that will test a limited sample of your site's features for free. (They also offer greatly expanded testing for a fee.) For more on this topic, see *bedfordstmartins.com/alredtech* and select *Digital Tips*, "Testing Your Web Site."

WRITER'S CHECKLIST: Designing Web Pages

Many of the guidelines on typography in **layout and design** are applicable to Web pages; however, not all apply, so keep the following in mind:

- Work with the Webmaster or site administrator to optimize your site for speed of access and to maintain technical and design standards. On campus, consult your instructor or campus computer support staff about standards for posting content.
- Draft a navigation chart or map of your site early in the design process to make information logically accessible in the fewest possible steps (or clicks).
- Anchor links on relevant words in the sentence ("For more information about employment opportunities, visit Human Resources.")
- Avoid overusing complex graphics and animation that can clutter or slow access to your site. For high-resolution graphics, consider using thumbnails (images reduced to 10 to 15 percent of the original file size) that link to the original-size image.
- Use lighter colors for backgrounds and use typeface colors that contrast (but do not clash) with background colors.
- Limit the number of typeface colors as well as styles; use boldface type and uppercase letters sparingly. Use underlined text only for links.
- Use sans serif fonts for text passages; they are generally more legible on computer screens than are serif fonts.
- Separate text from graphics with generous blank space (the equivalent of white space).
- Limit your text lines to 50 to 70 characters (or 10 to 12 words) for readability.
- Use a consistent style for **headings** and subheadings.
- Block-indent text sections that you expect viewers to read in detail.
- Test your design and load time by viewing it on alternate browsers and slow modems.
- Incorporate the name of your organization and its logo in a banner at the top of each page.
- Include a link to your site's Privacy Statement (see page 573).
- Check that all your links work, particularly after site changes.

when / where / that

When and if (or *if and when*) is a colloquial expression that should not be used in writing.

▶ When ~~and if~~ funding is approved, you will get the position.

▶ ~~When and~~ *If* ~~if~~ funding is approved, you will get the position.

In phrases using the *where . . . at* construction, *at* is unnecessary and should be omitted.

▶ Where is his office ~~at~~?

Do not substitute *where* for *that* to anticipate an idea or fact to follow.

▶ I read in the newsletter ~~where~~ *that* research funding will increase.

whether

Whether communicates the notion of a choice. The use of *whether or not* to indicate a choice between alternatives is redundant.

▶ The client asked whether ~~or not~~ the proposal was finished.

The phrase *as to whether* is clumsy and redundant. Either use *whether* alone or omit it altogether.

▶ ~~As to whether we will~~ *We have decided to* commit to a long-term ~~contract, we have decided to do so.~~ *contract.*

while

While, meaning "during an interval of time," is sometimes substituted for connectives like *and*, *but*, *although*, and *whereas*. Used as a connective in that way, *while* often causes ambiguity.

▶ Ian Evans is a media director, ~~while~~ *and* Joan Thomas is a vice president for research.

Do not use *while* to mean *although* or *whereas*.

▶ ~~While~~ *Although* Ryan Sims wants the job of engineering manager, he has not yet applied for it.

Restrict *while* to its meaning of "during the time that."

▶ I'll have to catch up on my reading *while* I am on vacation.

white papers

A white paper is a document that announces a new product or a newly developed technical or business solution to a common problem. White papers function as (1) educational tools that provide credibility for an organization and (2) marketing tools to establish a company's market position as an industry leader. White papers are often posted on a company's Web site, distributed to potential customers at trade shows and conferences, and published at Web sites that feature white papers from various sources.

Using Persuasion

White papers are promotional documents, so they must be interesting to read, reflect strong empathy with the reader, and use effective **persuasion**. White papers require that writers shift from a technology-centered point of view to the **"you" viewpoint**, as suggested in the following example from Darren K. Barefoot's article "Ten Tips on Writing White Papers."*

TECHNOLOGY CENTERED	Buy a DVD player because it has a 128-bit oversampling and advanced virtual surround sound.
READER CENTERED	Buy a DVD player because you want better sound and a crisper picture.

If your white paper is targeted to nonspecialists, leave the details to the technical **manuals**, sales **brochures**, and press releases. Focus on your readers' concerns and explain clearly how your company or product can solve their specific problems.

Preparing to Write

Begin by checking with management to determine their goals for the white paper and how it fits into their overall strategy. Then analyze

---

*Darren K. Barefoot, "Ten Tips on Writing White Papers," *Intercom* February 2002, 12–13.

your **audience** to determine whether your readers are technical experts or nonexperts such as executives. Plan your language and approach appropriately. For example, experts understand and appreciate technical terminology, whereas executives tend to focus on the bottom line and the benefits of solving a problem.

Decide on the most appropriate **method of development** and then create an outline that offers your product or solution as the most effective way to solve the readers' problem. An outline helps provide **coherence** and **transition** to your draft so that readers can follow the development of a new technology or product as the white paper moves smoothly from the **introduction** to the **conclusion**. Define any new terms, avoid **affectation**, and follow the principles of effective **technical writing style**.

As you consider your **scope**, make certain that your white paper deals with only one product or solution; for multiple topics, use multiple white papers. The length of your white paper depends on the complexity of your product, but try to keep it as brief as possible. If your document runs more than ten pages, include a **table of contents**.

Planning Layout and Design

A good design can make the most complex information look accessible and enable your readers to retrieve information easily. A white paper should make good use of **visuals** to help readers grasp and retain your message. Use such visuals carefully, however. A complex visual that requires paragraphs of explanation is more likely to hinder your reader's understanding than enhance it. Make the best use of **headings**, **lists**, call-outs, text boxes, and other devices that are described in **layout and design**.

Developing Major Sections

The *introduction* to a white paper should attract your readers' interest by explaining the problem that needs to be solved. Your readers might be persuaded to accept your message more readily if they believe your company is developing innovative products, as in the following example:

> ▶ The rings of Saturn have puzzled astronomers ever since Galileo discovered them in 1610 using the first telescope. Recently, even more rings have been discovered. . . .
>
> Our company's Scientific Instruments Division designs and manufactures research-quality, computer-controlled telescopes that promise to solve the puzzles of Saturn's rings by enabling

scientists to use multicolor differential photometry to determine the rings' origins and composition.

Use the *body* of a white paper to provide supporting evidence that your product or solution is the best way to solve the readers' problem. Be careful not to be too biased toward your own product, however, especially early in your white paper. You might start by explaining what led to the development of the problem and how its solution impacts all customers in the industry. Describe the kind of technology that can solve the problem, and develop a high standard for the needed solution. Then explain precisely how your way of solving the customer's problem is the best way to meet that standard. Provide persuasive supporting evidence from outside sources. See **research** and **documenting sources**.

The *conclusion* of a white paper should summarize the problem and how your solution best solves it. As a marketing tool, your white paper should conclude by telling readers how to contact your company for more information.

Revising and Getting Clearances

Ask management, subject-matter experts, and perhaps your legal department to read the finished draft to make certain it meets management's objectives, is technically accurate, and stays within legal requirements. Be sure to cite any outside sources used and get permission from any source quoted. See **copyright** and **plagiarism**.

> **WEB LINK** **White Papers**
>
> For links to online collections of white papers, see *bedfordstmartins.com/alredtech* and select *Links for Handbook Entries*.

who / whom

Who is a subjective case pronoun, and *whom* is the objective case form of *who*. When in doubt about which form to use, substitute a personal pronoun to see which one fits. If *he*, *she*, or *they* fits, use *who*.

▶ *Who* is the training coordinator?
  [You would say, "*She* is the training coordinator."]

If *him*, *her*, or *them* fits, use *whom*.

▶ It depends on *whom*?
  [You would say, "It depends on *them*."]

who's / whose / of which

Who's is the contraction of *who is*. ("*Who's* scheduled today?") *Whose* is the possessive case of *who*. ("Consider *whose* budget should be cut.")

Normally, *whose* is used with persons, and *of which* is used with inanimate objects.

- ▶ The employee *whose* car had been towed away was angry.
- ▶ Completing a Ph.D. in engineering is an achievement *of which* to be proud.

If *of which* causes a sentence to sound awkward, *whose* may be used with inanimate objects. (Compare "The business the profits *of which* steadily declined" with "The business *whose* profits steadily declined.")

-wise

Although the **suffix** *-wise* often seems to provide a tempting shortcut in writing, it leads more often to **affectation** than to economical expression. It is better to rephrase the sentence.

- ▶ Our department ~~rates~~ high ~~efficiencywise~~.
  (has a) (efficiency rating.)

The *-wise* suffix is appropriate, however, in instructions that indicate certain space or directional requirements (*lengthwise*, *clockwise*).

word choice

Mark Twain once said, "The difference between the right word and almost the right word is the difference between 'lightning' and 'lightning bug.'" The most important goal in choosing the right word in technical writing is the preciseness implied by Twain's comment. Vague words and abstract words defeat preciseness because they do not convey the writer's meaning directly and clearly.

    VAGUE   It was a *productive* meeting.
    PRECISE   The meeting resulted in the approval of the health-care benefits package.

In the first sentence, *productive* sounds specific but conveys little information; the revised sentence says specifically what made the meeting "productive." Although **abstract words** may at times be appropriate to your topic, using them unnecessarily will make your writing difficult to understand.

Being aware of the **connotations** and denotations of words will help you anticipate your reader's reactions to the words you choose. (See also **audience**, **connotations / denotations**, and **readers**.) Understanding **antonyms** (*fresh/stale*) and **synonyms** (*notorious/infamous*) will increase your ability to choose the proper word. Make other **usage** decisions carefully, especially in technical contexts, such as **average / median / mean** and **biannual / biennial**.

Although many of the entries throughout this book will help you improve your word choices and avoid impreciseness, the following entries should be particularly helpful:

affectation	22	**euphemisms**	180
biased language	46	**idioms**	248
buzzwords	57	**jargon**	285
clichés	71	**logic errors**	312
conciseness	90	**vague words**	546

A key to choosing the correct and precise word is to keep current in your reading and to be aware of new words in your profession and in the language. In your quest for the right word, remember that there is no substitute for an up-to-date **dictionary**. See also **English as a second language**.

> **WEB LINK** | **Wise Word Choices**
>
> For online exercises on word choice, see *bedfordstmartins.com/alredtech* and select *Exercise Central*.

writing a draft

You are well prepared to write a rough draft when you have established your **purpose** and reader's needs, considered the **context**, defined your **scope**, completed adequate **research**, and prepared an outline (whether rough or developed). (See also **audience**, **outlining**, and **readers**.) Writing a draft is simply transcribing and expanding the notes from your outline into **paragraphs**, without worrying about **grammar**, refinements of language, or **spelling**. Refinement will come with **revision** and **proofreading**. See also "Five Steps to Successful Writing."

Writing and revising are different activities. Do not let worrying about a good opening slow you down. Instead, concentrate on getting your ideas on paper—now is not the time to polish or revise. Do not wait for inspiration—treat writing a draft as you would any other on-the-job task.

> **WRITER'S CHECKLIST** Writing a Rough Draft
>
> - Set up your writing area with whatever supplies you need (laptop, notepads, reference material, and so on). Avoid distractions.
> - Resist the temptation of writing first drafts on the computer without planning.
> - Use a good outline as a springboard to start and to write quickly.
> - Give yourself a set time in which you write continuously, regardless of how good or bad your writing seems to be. But don't stop if you are rolling along easily — keep your momentum.
> - Start with the section that seems easiest. Your readers will neither know nor care that the middle section of the document was the first section you wrote.
> - Keep in mind your readers' needs, expectations, and knowledge of the subject. Doing so will help you write directly to your readers and suggest which ideas need further development.
> - When you come to something difficult to explain, try to relate the new concept to something with which the readers are familiar, as discussed in **figures of speech**.
> - Routinely save to your hard drive and create a backup copy of your documents on separate disks or on the company network.
> - Give yourself a small reward — a short walk, a soft drink, a brief chat with a friend, an easy task — after you have finished a section.
> - Reread what you have written when you return to your writing. Seeing what you have already written can return you to a productive frame of mind.

writing for the Web

In the workplace, those who write content for Web sites are not always the same people who design the sites. This entry provides guidelines only for those writing for an online **audience**; for more information on Web-site design, see **Web design**.

⬥ **ETHICS NOTE** For questions about the appropriateness of content you plan to post on the Web, check with your Webmaster or manager to determine if your content complies with your organization's Web policy. On campus, consult your instructor or campus computer support staff about standards for posting Web content. See **ethics in writing**. ✦

Crafting Content for Web Pages

Because most **readers** scan Web pages for specific information, the way you write and organize your information will greatly affect your readers' understanding. State your important points first, before any detailed supporting information. (This method is called *inverted pyramid organization*.) Keep your writing **style** simple and straightforward, and avoid such directional cues as "as shown in the example below" that make sense on the printed page but not on a Web page. Use the following techniques to make your content more accessible to readers.

Headings. Break up dense blocks of text with short **paragraphs** that stand out and can be quickly scanned and absorbed. Use informative **headings** to help readers identify topics and decide at a glance whether to read a passage. Headings also clarify text by highlighting structure and organization as they also signal **transitions** from one topic to the next.

Lists. Use bulleted and numbered **lists** to break up dense paragraphs, reduce text length, and highlight important content.

Keywords. To help search engines and your audience find your site, use keywords in the first fifty or so words of your text.

WITHOUT KEYWORDS	We are proud to introduce a new commemorative coin honoring our bank's founder and president. The item will be available on this Web site after December 3, 2009, which is the hundredth anniversary of our first deposit.
WITH KEYWORDS	The new *Reynolds* commemorative coin features a portrait of *George G. Reynolds*, the founder and president of *Reynolds Bank*. The coin can be purchased after December 3, 2009, in honor of the hundredth anniversary of the first *Reynolds* deposit.

Graphics. Graphics provide visual relief from text and make your site attractive and appealing. Use only **visuals** that are appropriate for your audience and **purpose**, however, and use graphics that load quickly.

Hyperlinks. Use internal hyperlinks to help readers navigate the information in your site. If a passage of text is longer than two or three screens, create a table of contents of hyperlinks for it at the top of the Web page and link each item to the relevant content further down the page. Use external hyperlinks to enrich coverage of your topic with information outside your site and to help reduce content on your page. When you do, consider placing an icon or a text label next to the hyperlink to inform users that they are leaving the host site. In general, avoid too many hyperlinks within text paragraphs because they can distract readers, make scanning the text difficult, and tempt readers to leave your site before reaching the end of your page.

Fonts. Font sizes and styles affect screen legibility. Because computer screens display fonts at lower resolutions than does printed text, sans serif fonts work better for online text passages. Consult your Webmaster (or your manager) about your organization's font preferences. Do not use all capital letters or boldface type for blocks of text because they slow the reader. For content that contains special characters (such as for mathematical or chemical content), consult your Webmaster about the best way to submit the files for HTML (hypertext markup language) coding or post them as PDF files.

Line Length. Line length also affects readability because short line lengths reduce the amount of eye movement necessary to scan text. Optimal line length is approximately half the width of the screen. To achieve this optimal length, draft text that is between 50 and 70 characters (or between 10 and 12 words) to a line.

Writing for a Global Audience

When you write for public access sites, eliminate expressions and references that make sense only to someone very familiar with American English. Express **dates**, clock times, and measurements consistent with international practices. For visuals, choose symbols and icons, colors, representations of human beings, and captions that can be easily understood, as described in **global communication** and **global graphics**. See also **biased language** and **English as a second language**.

Linking to Reputable Sites

Links to outside sites can expand your content. However, review such sites carefully before linking to them. Is the site's author or sponsoring organization reputable? Is its content accurate, current, and unbiased? Does the site date-stamp its content with notices such as "This page was last updated on January 1, 2009"? (For more advice on evaluating Web sites, see **research**.) Link directly to the page or specific area of an outside site that is relevant to your users, and be sure that you provide a clear **context** for why you are sending your readers there.

Posting an Existing Document

If you post electronic files of existing paper documents to a Web site, try to retain the document's original sequence and page layout. If you shorten or revise the original document for posting to the Web, add a notice informing readers that it differs from the printed original.

Regardless of format or version used, be sure to follow these practices:

- Obtain permission from the **copyright** holder for the use of copyrighted text, **tables**, or images. See also **plagiarism**.
- Ask the Webmaster or site administrator about the preferred file format to submit for coding and posting. On campus, consult your instructor or campus computer support staff.
- Request that the Webmaster optimize any slow-loading graphics files for quick access.
- Review the document internally before it is posted to the public site: Is it the correct version? Is any information missing? Do all links work and go to the right places? See **proofreading**.
- Ask the Webmaster to create a single-file version of the document (a version formatted as a single, long Web page) for readers who will print it to read offline.

◼ **ETHICS NOTE** Document sources of information—text, images, streaming video, and other multimedia material—or of help received. Seek prior approval before using any copyrighted information. Documenting your sources not only is required but also bolsters the credibility of your site. To document your sources, either provide links to your source or use a citation, as described in **documenting sources**. ◆

Protecting the Privacy of Users

Put a link on your Web page to the site's privacy statement, particularly if you solicit comments about your content or have an **e-mail** link for unsolicited comments for site users. A privacy statement informs site visitors about how the site sponsor handles solicited and unsolicited information from individuals, its policy on the use of cookies,* and legal action it takes against hackers. Inform users if you intend to use their information for marketing or to share their information with third parties, and give visitors the option of refusing you permission to use or share their information.

---

*"Cookies" are small files that are downloaded to your computer when you browse certain Web pages. Cookies hold information, such as your user name and password, so you do not need to reenter it each time you visit the site.

> **DIGITAL TIP**
>
> **Using PDF Files**
>
> Converting documents such as reports, articles, and brochures to PDF files allows you to retain the identical look of the printed documents. The PDF pages will display on-screen exactly as they appear on the printed page. Readers can read the document online, download and save it, or print it in whole or in part. For more on this topic, see *bedfordstmartins.com/alredtech* and select *Digital Tips*, "Using PDF Files."

"you" viewpoint

The "you" viewpoint places the reader's interest and perspective foremost. It is based on the principle that most readers are naturally more concerned about their own needs than they are about those of a writer or a writer's organization. See **audience**.

The "you" viewpoint often, but not always, means using the words *you* and *your* rather than *we*, *our*, *I*, and *mine*. Consider the following sentence that focuses on the needs of the writer and organization (*we*) rather than on those of the reader.

▶ *We must receive* your signed invoice before *we can process* your payment.

Even though the sentence uses *your* twice, the words in italics suggest that the **point of view** centers on the writer's need to receive the invoice in order to process the payment. Consider the following revision, written with the "you" viewpoint.

▶ *So you can receive* your payment promptly, please send your signed invoice.

Because the benefit to the reader is stressed, the writer is more likely to motivate the reader to act. See also **persuasion**.

In some instances, as suggested earlier, you may need to avoid using the **pronouns** *you* and *your* to achieve a positive **tone** and maintain goodwill. Notice how the first of the following examples (with *your*) seems to accuse the reader. But the second (without *your*) uses **positive writing** to emphasize a goal that reader and writer share—meeting a client's needs.

ACCUSATORY	*Your* budget makes no allowance for setup costs.
POSITIVE	The budget should include an allowance for setup costs to meet all the concerns of our client.

As this example illustrates, the "you" viewpoint means more than using particular pronouns or adopting a particular writing **style**. By placing the readers' interests at the center, you can achieve your **purpose** not only in **correspondence** but also in **proposals**, **brochures**, many **reports**, and **presentations**.

your / you're

Your is a possessive **pronoun** ("*your* wallet"); *you're* is the contraction of *you are* ("*You're* late for the meeting"). If you tend to confuse *your* with *you're*, use the search function of your word processor to review both terms during **proofreading**.

Acknowledgments (continued)

Figure D–12. Exploded-View Drawing. From Xerox Corporation, *WorkCentre XD Series User Guide*, page 6. Copyright © Xerox Corporation. Used with permission.

Figure D–13. Cutaway Drawing (Hard Drive). From *Hard Disk Quick Reference* by P. D. Moulton and Timothy S. Stanley. Copyright © 1989 by Que Publishing. Reprinted with the permission of Pearson Computer Publishing, a division of Pearson Education.

Figure F–2. Flowchart Using Labeled Blocks. Information from PEPCO, "How We Restore Power," Lines 33, no. 7 (July 2004). Reprinted with permission.

Figure F–3. Flowchart Using Pictorial Symbols. Infographic: FDA *Consumer Magazine* November–December 2004/Renée Gordon. U.S. Food and Drug Administration. 5600 Fishers Lane. Rockville, MD 20857-0001. 1-888-INFO-FDA. *www.fda.gov.*

Figure F–5. Formal Report (Cover Memo). Reprinted with the permission of Susan Litzinger, a student at Pennsylvania State University, Altoona.

Figure F–7. Online Form with Typical Components. Copyright © 2008 Facebook, Inc. Reprinted with permission.

Figure G–4. International Organization for Standardization Symbols. From "Graphical Symbols to Address Consumer Needs" by John Perry, in the *ISO Bulletin,* March 2003, Volume 34, No. 3, Copyright © 2003 by ISO. Reprinted with permission.

Figure G–7. Bar Graph (Quantities of Different Items During a Fixed Period.) Copyright © 2008 by ASAPS. Reprinted with permission.

Figure I–6. "CharBroil Use and Care Manual (Warning in a Set of Instructions)." Reprinted with the permission of W. C. Bradley Company, 1997, Columbus, Georgia.

Figure N–2. Newsletter Article. "Online Solution Speeds Video Delivery." From *Connection* (January 2007): 7. Reprinted with the permission of Ken Cook Company.

Figure N–3. Company Newsletter (Front Page). "Volvo EPA Solution a Model for 2007." From *Connection* (October–November 2006): 1. Reprinted with the permission of Ken Cook Company.

Figure P–2. Photo (of Aircraft Door). Copyright © 2003 by Ken Cook Company. Reprinted with the permission of Ken Cook Company.

Figure R–6. Résumé (Highlighting Professional Credentials). Prepared by Kim Isaacs, Advanced Career Systems, Inc. Reprinted with permission.

Figure R–10. Advanced Résumé (Combining Functional and Chronological Elements). Prepared by Kim Isaacs, Advanced Career Systems, Inc. Reprinted with permissions.

Index

a / an, 1, 40
   as adjectives, 13, 14
a few / few, 189
a lot / alot, 2
a while / awhile, 44
abbreviations, 2–5
   in abstracts, 9
   acronyms and initialisms, 2, 11–12
   capitalization of, 62
   commas with, 82
   common scholarly, 4–5
   defining, 556
   in e-mail, 163
   in instant messaging, 256–57
   in international correspondence, 269
   of Latin words, 160, 177
   list of, in reports, 198
   for measurements, 4
   for names and titles, 4
   for organization names, 3–4
   periods with, 375
   plurals of, 3, 34
   spelling out uncommon, 181
   in tables, 519
   in titles of works, 531
   Web resources for, 5
   writer's checklist for, 3
above, 6
abridged dictionaries, 124
absolute phrases, 79
absolute words, 14, 176–77
absolutely, 6
abstract nouns, 26, 349
abstract words, 6–7
   affectation and, 22
   ethics in using, 178
   figures of speech for clarifying, 190
   as gobbledygook, 233
   impreciseness of, 568–69
abstracts, 7–9
   vs. executive summaries, 181

   in formal reports, 197, 203
   for trade journal articles, 535
   writing strategies for, 8–9
academic titles, 3, 4
accept / except, 9
acceptance / refusal letters, for
     employment, 9–11, 275–76
   samples of, 10, 276
accident reports, 541
   sample of, 340
accuracy
   in application letters, 39
   in correspondence, 110–11, 162
   of instructions, 263, 316
   in note-taking, 348
   in résumés, 472
   revising for, 488
accuracy / precision, 11
accusative case, 64
acknowledgment letters, 11
   sample of, 11
acronyms and initialisms, 2, 3, 11–12
action verbs, 64, 484
actions, imperative mood for, 317
activate / actuate, 12
active listening, 307–9
active voice, 44, 515, 552, 557, 558–59
   for conciseness, 92, 558–59
   for emphasis, 168
   in formal writing, 513
   for highlighting subjects, 558
   for improving clarity, 558
   infinitives and, 547
   for instructions, 259, 263
   for newsletters, 345
   revising for, 489
   in sentence construction, 499
activities section, in résumés, 485
activity reports, 402
   sample of, 404
actually, 448

actuate / activate, 12
ad hoc, 12
adapt / adept / adopt, 12
addresses, street
  commas with, 81
  inside, in letters, 304
  numbers in, 354–55
  in résumés, 482
  return, in letters, 301–4
adept / adopt / adapt, 12
adjective clauses, 15, 171
adjectives, 12–16, 335. *See also* modifiers
  as absolute words, 14
  adverbs modifying, 19–20
  articles, 1, 13, 39–41
  commas with, 16, 80–81
  comparison of, 14
  converting verbs into, 505
  with count nouns, 170
  demonstrative, 13–14, 170
  descriptive, 13
  ESL tips for, 15
  indefinite, 14
  irregular, 14
  limiting, 13–14
  nouns as, 16, 350, 388, 517
  numeral, 14
  phrases as, 380, 381
  placement of, 15–16
  possessive, 14, 170
  with prepositions, 390
  relative, 91
  verbals as, 546, 547, 548, 550
adjustment letters, 16–19
  context in, 99
  full adjustments, 18–19
  partial adjustments, 19
  samples of, 17, 18
Adobe Acrobat, 74
Adobe PDF files, 463, 488, 574
adopt / adapt / adept, 12
advanced searches, 464
adverbs, 19–21, 335. *See also* modifiers
  commas with, 80
  comparison of, 20–21
  conjunctive, 20
  intensifiers, 264, 335
  irregular, 21
  nouns as, 350
  ordinal numbers as, 14
  phrases as, 380, 381
  placement of, 21
  with prepositions, 390
  prepositions confused with, 390
  questions answered by, 20
  split infinitives and, 547–48
  types of, 20
  verbals as, 546, 547, 550
affect / effect, 21–22
affectation, 22–23
  allusions and, 31
  buzzwords, 58
  ethics in writing and, 22
  euphemisms, 180
  foreign words as, 195
  formal style as, 103, 514
  as gobbledygook, 233
  nominalizations as, 346
  overusing abbreviations, 2
  Web resources for, 22
  wordiness as, 92
affinity, 23
aforementioned, 6
aforesaid, 6, 22
age bias, in writing, 48
agenda, for meetings, 324–25
agreement, 23–29. *See also* pronoun-antecedent agreement; subject-verb agreement
  with collective nouns, 26, 28, 115, 350
  with demonstrative adjectives, 13, 28–29
  with *each*, 24, 27, 160
  with *everybody / everyone*, 28, 180
  gender, 24, 27–28, 226, 409
  with *none*, 346
  with *nor / or*, 347
  number, 24–26, 28–29, 351–52, 409
  with *one of those . . . who*, 357–58
  of person, 24, 550, 551
  with possessive pronouns, 407
  pronoun-antecedent, 24, 27–29, 409
  subject-verb, 24–27, 499, 551

ain't, 234
aircraft names, 284
all-capital letters
  in e-mail, 163–64
  for emphasis, 169, 297
  in Web design, 572
all ready / already, 29
all right / alright, 30
all together / altogether, 30
allude / elude / refer, 30
allusion / illusion, 30
allusions, 30–31, 269
almost / most, 31
also, 31
ambiguity, 32–33
  commas for preventing, 77, 81
  incomplete comparisons, 32
  missing or misplaced modifiers, 32, 335–36
  with prepositional phrases, 381
  in pronoun references, 32, 405
  in telegraphic writing, 523
  using *that* for, 527
  word choice and, 32–33, 70
American Heritage College Dictionary, The, 124
American Medical Association (AMA), 153
American National Standards Institute, 263
American Psychological Association. *See* APA documentation
American Society of Indexers, 251
America's Job Bank, 289
among / between, 45
amount / number, 33
ampersands, 4, 33, 132
an. *See* a / an
analogy, 190
  definition by, 118
  in description, 120
and
  both . . . and usage, 52, 96
  commas with, 83
  in compound subjects, 26–27, 29
and / or, 33
anecdote, opening with, 279
annotated bibliography, 48–49, 312

antecedents, 406. *See also* pronoun-antecedent agreement
  agreement with pronouns, 24, 27–29, 409
  compound, 29
  implied, 405
  personal pronouns and, 409
  references to, 32, 405, 453
antonyms, 33, 529, 569
anxiety, presentation, 400
APA documentation, 130, 132–38. *See also* documenting sources
  for articles in periodicals, 133–34, 135
  for books, 132–33
  compared with other styles, 130–31
  documentation models, 132–36
  for electronic sources, 134–35
  in-text citations, 132, 137
  for multimedia sources, 135–36
  other sources, 136
  for quotations, 447
  sample list of references, 138
apostrophes, 34
  in contractions, 34, 101
  forming plurals with, 34
  indicating omissions, 34
  showing possession, 34, 65, 387–88
appendixes, 34–35
  in formal reports, 199
  in proposals, 413, 419
  in requests for proposals, 458
  visuals in, 557
application letters, 35–39
  résumés and, 471
  samples of, 36–38
  structure and parts of, 35–39
  writing style for, 513
appositives, 39
  case of, 39, 65, 408
  commas with, 78–79
  following proper names, 62
  gerunds as, 546
  nouns as, 350
arguments, loaded, 314
articles (*a / an / the*), 39–41
  a / an usage, 1, 40
  count nouns and, 170

articles (*continued*)
  definite, 39–40, 41
  eliminating in telegraphic style, 41, 522
  ESL tips for, 40, 170–71
  indefinite, 1, 39–40
  as limiting adjectives, 13, 14
  the usage, 41
  in titles of works, 41, 61, 530
  when not to use, 40
Articles by Topic (Inc.com), 465
articles in periodicals. *See also*
      newsletter articles; trade journal articles
  APA documentation of, 133–34, 135
  electronic, documenting, 135, 141, 147
  evaluating, 466–67
  IEEE documentation of, 140, 141
  MLA documentation of, 146, 147
  online databases and indexes for, 461–62
  titles for, 61, 95, 444, 529–30, 531
  with unknown author, 134, 140, 146
  writing style for, 513
as / *because* / *since*, 41
as / *like*, 306–7
as / *than*, and pronoun case, 64, 65
as much as / *more than*, 41–42
as such, 42
as to whether, 564
as well as / *both*, 42
ASCII résumés, 480, 487
assure / *insure* / *ensure*, 263
asterisks, 163–64
astronomical terms, 61
atlases, 463
attachments, e-mail, 100, 163, 164
attention-getting statement, opening with, 393
attributive adjectives, 16
audience, 42–43
  analyzing needs of, xvi–xviii, 42–43, 389, 392
  building goodwill for, 104–5
  content management for, 97
  defining terms for, 43, 117
  diverse, writing for, 43
  draft writing and, 569

  for environmental impact statements, 175
  layout and design and, 296, 297
  for manuals, 316
  for newsletters, 343
  pace and, 367
  for proposals, 413
  purpose and, 435
  selecting the medium for, 494
  tone and, 532
  usability testing and, 543–45
  usage and, 545
  for visuals, 553
  for Web sites, 561–62, 570
  for white papers, 565
  writing style and, 103, 513
  "you" viewpoint for, 575–76
augment / *supplement*, 43
author-date method of documentation, 130. *See also* APA documentation
authorization request, in proposals, 421
authors, documenting
  in bibliographies, 48
  corporate, 133, 134, 139, 141, 146, 147
  of electronic sources, 134–35, 141, 147
  multiple, 133, 139, 145
  multiple, in citations, 132
  unknown, 48, 134, 140, 141, 146, 147
auxiliary verbs, 550
average / *median* / *mean*, 44
awhile / *a while*, 44
awkwardness, 44
  ambiguity and, 33
  expletives and, 182
  inverted word order, 506
  revising, 489
  word choice and, 70
  writer's checklist for eliminating, 44

back matter
  for formal reports, 196, 199–200
  for proposals, 419
background
  in environmental impact statements, 174–75

opening with, 278
in proposals, 420, 433
bad / badly, 45
bad-news patterns. *See also* negative messages
  in correspondence, 106–8
  cultural differences in, 265
  positive writing and, 386–87
  refusal letters, 449–52
bar graphs, 236–38, 240, 554
barely, 153
bcc: notations, 163, 306
be sure to / be sure and, 45
because, reason is, 448
because / since / as, 41
because of / due to, 159
behavioral interviews, 274
beside / besides, 45
between / among, 45
between you and me / between you and I, 46, 390
bi- / semi-, 46
biannual / biennial, 46
biased evidence, 313
biased language, 46–48. *See also* sexist language
  ethics in writing and, 46, 178
  revising for, 489
  sexist language, 46–47, 226, 241
bibliographies, 48–49. *See also* documenting sources
  annotated, 48–49, 312
  in formal reports, 199
  IEEE style, 131, 139–42, 144
  organization names in, 3
  in proposals, 419
  for research, 462
biennial / biannual, 46
bilingual dictionaries, 125
blind-copy notations (bcc:), 163, 306
block indents, in Web design, 563
block quotations, APA style for, 132, 447
blogs, 49–52, 497
  documenting, 135, 148
  ethics in writing for, 51–52
  external, 49–50
  internal, 50–51
  organizational, 49–52
  writing strategies for, 51–52
body
  for feasibility reports, 188
  for formal reports, 196, 198–99, 210–14
  of presentations, 394
  for proposals, 418, 419
  for reports, 454
  of tables, 520
  for white papers, 567
body depictions, for global graphics, 231, 232
body movement, in presentations, 399, 400
boilerplate material, copyright and, 102
boldface type. *See also* typography
  for emphasis, 169, 298
  for headings, 298
  in Web design, 572
books
  APA documentation of, 132–33, 134
  capitalization in titles, 61, 95
  evaluating, 466
  IEEE documentation of, 139–40
  italics for titles, 283, 444, 531
  MLA documentation of, 145–46, 147
  online catalogs for, 461
Boolean operators, 464
both / as well as, 42
both . . . and, 52, 96
box heads, 519
boxes
  for organizational charts, 362
  in page design, 300
brackets, 52
  in equations, 322
  for parenthetical items, 52, 373
  for *sic*, 52, 446
brainstorming, 53–54
  in job searches, 286
  during presentations, 396
  for research, 459
  for résumé writing, 481
brochures, 54–57
  designing, 54–57
  documenting, 136, 142, 149
  italics for titles, 283, 531

brochures (*continued*)
  using repurposed content for, 455
  visuals for, 155
  Web resources for, 57
  writer's checklist for, 57
budget, in proposals, 433
bulleted lists, 56, 110, 259, 309, 310, 456. *See also* lists
business names. *See* organization names
but, commas with, 83
buzzwords, 57–58, 264
  as affectation, 22
  as gobbledygook, 233
  Web resources on, 58

call for action, in conclusions, 94–95
callouts. *See* labels
campus career services, 288–89
can / may, 59
cannot, 59
capital appropriations proposals, 414
capitalization, 59–62. *See also* all-capital letters
  of abbreviations, 62
  for adjectives of origin, 15
  of events and concepts, 61
  of first words, 60, 76, 114, 305, 445, 530
  of letters of alphabet, 62
  in lists, 310
  of organizations, 61
  of professional and personal titles, 62
  of proper nouns, 59–60, 349
  for quotations, 60, 445
  of specific groups, 60
  of specific places, 60–61
  in subject lines, 61, 110
  of titles of works, 41, 61, 95, 391, 530
captions, with visuals, 300, 379, 536
cardinal adjectives, 14
career counselors, 288–89
career summary, in résumés, 483
case, 62–66, 407–9
  agreement of, 24
  of appositives, 39, 65, 408
  objective case, 64, 407, 408
  possessive case, 64–65, 387–89, 407, 408
  subjective case, 63–64, 407–8
  tests for determining, 65–66, 408–9
cause
  definition by, 118
  false, 313
  words expressing, 41
cause-and-effect method of development, 66–67, 331, 507
  evaluating evidence, 66–67
  linking causes to effects, 67
cc: notations, 163, 166, 306
CD-ROM publications
  dictionaries, 124, 125
  documenting, 135
cell phones, 495
center on / center around, 67
centuries, format for writing, 115–16
chalkboards, 396
channel, 494n. *See also* selecting the medium
chapters and sections
  in feasibility reports, 188
  in manuals, 317
  numbers for, 355
  titles of, 62, 284, 444, 531
charts. *See also* visuals
  documenting, 141
  flowcharts, 192–94, 555
  organizational, 362, 555
  for presentations, 396, 398
Chemical Engineering, 533
Chicago Manual of Style (*CMS*), 140, 141, 515, 536. *See also* documenting sources
choppy writing, 505
chronological method of development, 68, 331, 507
  for literature reviews, 312
  in narratives, 339
  for résumés, 483–84
  sample of, 69
CIO's Resource Centers, 465
circular definitions, 117
circumlocution, 91
citations. *See also* documenting sources
  APA style, 130, 132
  IEEE style, 130, 139
  MLA style, 131, 145
  sample pages, 137, 143, 151

cite / *sight* / *site*, 68
claims
  adjustment letters for, 16–19
  complaint letters, 87–88
clarity, 68–70
  active voice for, 558
  commas for, 81
  conciseness and, 90, 259
  in correspondence, 109–10
  defining terms for, 116–17
  grammar and usage for, 234–35
  hyphens for, 246
  punctuation for, 434
  semicolons for, 498
  in technical writing, 521–22
  transitions for, 537–39
  word choice and, 568–69
classification, 127–29. *See also* division-and-classification method of development
clauses, 70–71. *See also* dependent clauses; independent clauses
  adjective, 171
  commas with, 78–79
  missing subjects in, 503
  restrictive and nonrestrictive, 470–71
clichés, 70, 71
  as figures of speech, 190
  as gobbledygook, 233
  revising for, 489
climactic order, for emphasis, 167
clip-art images, 155, 158
closed-ended questions, 221
closings. *See also* conclusions
  for application letters, 39
  for correspondence, 108–9, 305
  cultural differences in, 265–69
  for e-mail, 164–65
  goodwill, 106
  for meetings, 328
  for presentations, 394–95
  for refusal letters, 452
clustering, 53–54
coherence, 71–72
  clarity and, 68
  outlining and, 72, 363, 566
  in paragraphs, 370
  pronoun references and, 405

revising for, 488
unity and, 543
collaborative writing, 72–75
  active listening and, 309
  coherence in, 72, 73
  of environmental impact statements, 176
  for journal articles, 534
  managing conflict, 74
  outlining in, 363
  of proposals, 414
  purpose in, 435
  team tasks in, 73–74
  Web sites for, 497
  writer's checklist for, 74–75
collective nouns, 349, 350
  agreement with, 26, 28, 115, 350
colloquialisms, 41
  fine as, 191
  in informal writing, 514
  as nonstandard English, 173
  quotation marks for, 444
colons, 75–76
  capitalization of word following, 60, 76
  for enumeration, 498
  with numbers, 76
  with quotation marks, 76, 444, 445
  in salutations, 75, 304
  in sentences, 75
  in titles and citations, 75
  unnecessary use of, 76
color
  in brochures, 57
  in global graphics, 230, 232
  in graphs, 238
  in maps, 319
  in page design, 299
  in warnings, 261, 263
  in Web design, 563
color-blind readers, 562
column graphs, 236–38, 554
columns
  in newsletters, 345
  page design for, 298, 299
  in tables, 519–20
comma splices, 77, 490
commands, 337, 499
  in instructions, 259

commas, 77–83
  with adjectives, 16, 80–81
  with appositives, 39
  avoiding unnecessary, 83
  for clarifying and contrasting, 81
  for dates, 81–82, 115
  for enclosing elements, 78–79, 114, 470
  with interjections, 80, 264–65
  for introducing elements, 79–80
  for items in series, 80–81, 83
  for linking independent clauses, 20, 78, 82, 500
  with names, 81–82
  with numbers, 81–82, 354
  for omissions, 81
  with other punctuation, 82–83
  with quotation marks, 80, 83, 375, 444, 445
common / mutual, 338
common knowledge, 384
common nouns, 60, 349
communication
  global, 228–30
  listening, 307–9
  selecting the medium, 494–97
company information, in requests for proposals, 457
comparative form
  absolute words and, 176–77
  of adjectives, 14
  of adverbs, 20–21
compare / contrast, 84
comparison, 84–85
  determining basis of, 85
  incomplete, 32
  like and *as* for, 307
comparison method of development, 85–87, 331
complaint letters, 87–88
  context of, 99
  responding to, 16–19, 452
  sample of, 88
complement / compliment, 88
complements, 88–89, 500. *See also specific types*
  linking verbs and, 549
  types of, 89
  verbals as, 171

completeness, revising for, 488
complex sentences, 168, 501
compliment / complement, 88
complimentary closings, 60, 305
component content-management systems, 98
components, definition by, 118–19
compose / constitute / comprise, 89–90
compound antecedents, 29
compound-complex sentences, 501
compound predicates, 83, 500
compound sentences, 70, 168, 500
compound subjects, 500
  commas with, 83
  possessive case of, 388
  pronoun case and, 65–66, 408
  subject-verb agreement with, 26–27
compound words, 90
  hyphens for, 245
  plurals of, 351
comprehension tests, 544
comprise / compose / constitute, 89–90
computer technology. *See also* Digital Tips
  backing up files, 570
  clip-art images, 155
  for collaborative writing, 74
  content-management systems, 97–98
  for designing forms, 222
  desktop publishing software, 300, 345
  incorporating tracked changes, 489
  for indexing, 252
  for layout and design, 300
  for letters, 301
  for mathematical equations, 320–21
  for newsletters, 345
  for note-taking, 348
  for presentations, 396–97
  search-and-replace command, 92, 489
  style and format templates, 200
concepts, capitalization of, 61
conciseness, 90–92. *See also* wordiness
  in abstracts, 9
  active voice for, 558–59
  avoiding bluntness, 104
  in blog writing, 51
  causes of wordiness and, 90–91

clarity and, 70, 259
compound predicates for, 500
in international correspondence, 265
overdoing, 91
revising for, 489
subordination for, 516
telegraphic style, 522–23
using abbreviations, 2–3
writer's checklist for, 91–92
conclusions, 93–95. *See also* closings
　in abstracts, 7, 8
　for feasibility reports, 188
　for formal reports, 199, 215–16
　lists for, 299
　for newsletter articles, 343
　opening with summary of, 278
　for progress reports, 402
　for proposals, 418, 419, 421, 432, 433
　for reports, 455
　revising, 488
　for test reports, 527
　for trade journal articles, 536
　for trip reports, 540
　for trouble reports, 542
　for white papers, 567
concrete nouns, 349
concrete words, 6–7
conditional statements, subordination for, 516–17
confidentiality
　blogs and, 52
　e-mail and, 162
　forms and, 218
　instant messaging and, 258
conflict
　in collaborative writing, 74
　in meetings, 327
conjugation of verbs, 552
conjunctions, 95–96. *See also specific types*
　capitalization and, 61, 95
　in compound subjects, 26–27
　eliminating in telegraphic style, 522
　types of, 71, 95–96
conjunctive adverbs, 20
　for comma splices, 77
　commas with, 20, 82
　in compound sentences, 500

for connecting clauses, 71, 96
list of common, 20, 82, 96
for run-on sentences, 376
semicolons with, 497
connotation / denotation, 96
connotations
　of synonyms, 518, 529
　of visuals, 231–32
　word choice and, 569
consensus, 96
consistency
　proofreading for, 412
　in repetition, 453–54
　in résumé format, 472
　revising for, 489
　for visuals, 398, 556
　in voice, mood, tense, 551
constitute / comprise / compose, 89–90
contact lists, 256, 258
content management, 97–98, 455
　systems, 97–98
　Web resources for, 98
　writing for, 97
context, 98–100
　assessing, 99–100
　in bad-news patterns, 106
　design and, 296
　global communication and, 229
　global graphics and, 230
　listening and, 307
　openings and, 35, 100
　in preparation stage, 389
　purpose and, 435
　in refusal letters, 451
　repurposing for, 455
　of requests for proposals, 457
　scope and, 493
　selecting the medium and, 494
　signaling, 100
　style and, 513
　tone and, 532
　for visuals, 553
　word choice and, 6
　in writing process, xvi–xviii
continual / continuous, 100
continuing pages
　for letters, 306
　for memos, 329
continuing tables, 521

contractions, 101
  apostrophes in, 34
  in global communication, 265–69
contrast
  commas for, 81
  sentence variety for, 506
contrast / compare, 84
conversational style
  in correspondence, 103
  in informal writing, 514
  for newsletters, 345
  nonstandard English, 173–74
"cookies," 573
coordinate nouns, possessive case of, 388
coordinate series, 80
coordinating conjunctions
  for comma splices, 77
  commas with, 78, 83
  in compound sentences, 500
  for connecting clauses, 71, 78, 95
  list of, 78, 95
  for run-on sentences, 376, 490
  semicolons with, 498
  in subordination, 516
  in titles of works, 61, 95, 530
copy notations (cc:), 163, 166, 306
"copyleft" Web material, 102, 384
copyright, 101–2. *See also* documenting sources
  citing permissions, 198
  common knowledge and, 384
  ethics in writing and, 52, 101, 102, 179
  exceptions, 101–2
  permissions, 101
  plagiarism and, 384
  repurposing and, 456
  for visuals, 153, 155, 380
  for Web documents, 573
  Web resources for, 101
Corel Presentations, 396
corporate names. *See* organization names
correlative conjunctions, 96
  double negatives and, 154
  in parallel structure, 371
correspondence, 102–11. *See also* e-mail; letters; memos
  accuracy of, 110–11, 162

acknowledgment letters, 11
adjustment letters, 16–19
application letters, 35–39
clarity and emphasis in, 109–10
complaint letters, 87–88
cover letters, 111
documenting, 136, 142, 149
e-mail, 162–67
employment acceptance / refusal letters, 9–11, 275–76
good-news / bad-news patterns, 106–8
goodwill tone for, 104–5
headings in, 110
informal reports as, 454
inquiries and responses, 253–55
international, 265–70
letters, 301–6
lists in, 110
memos, 328–29
methods of development for, 361
names and titles in, 4, 304–5
openings and closings, 108–9, 276, 277
outlining for, 363
point of view in, 386
reference letters, 449
refusal letters, 449–52
resignation letters, 467–69
test reports as, 527
writer's checklist for, 110–11
writing style for, 103, 513, 514
cost analysis, in proposals, 421, 427
Council on Environmental Quality, 174
count nouns, 14, 25, 33, 349
  articles and, 170
  ESL tips for, 169–70
  fewer / less usage and, 189
cover letters, 111
  for e-mail attachments, 100, 163
  for formal reports, 196, 201
  for meeting agendas, 325, 326
  for proposals, 419, 422–23
  providing context in, 100
  for questionnaires, 439, 442
  samples of, 111, 201, 326, 422–23
Craigslist, 466
Creative Commons, 456
credible / creditable, 112
criteria / criterion, 112

critique, 112
cross-references, in index, 252
cultural differences. *See also* global communication; global graphics; international correspondence
    in communication styles, 228–30, 327
    in directness of messages, 106, 229, 265
    in listening, 307
currency, of secondary sources, 460
curriculum vitae, 471n. *See also* résumés
cutaway drawings, 155, 157

-*d* endings, 524, 548
dangling modifiers, 113
    active voice for, 558
    ambiguity and, 32
    infinitive phrases as, 382–83
    participial phrases as, 382
dashes, 114, 245
    for emphasis, 114, 169
    for enclosing elements, 114, 498
data / datum, 115
databases, online
    APA documentation of, 135
    IEEE (*CMS*) documentation of, 141
    MLA documentation of, 147
    for periodicals, 461–62
    plain-text résumés for, 487
    search strategies, 463–65
dates, 115–16
    apostrophes in, 34
    capitalization of, 61
    commas with, 81–82, 115
    on forms, 222
    in global communication, 270
    in headers or footers, 241
    for international correspondence, 115, 269, 354, 509
    in letter heading, 301–4
    numbers in, 353, 354
    plurals of, 34
    slashes in, 509
    in titles, 197, 419, 530
days. *See also* dates
    capitalization of, 61
    format for writing, 81–82, 115
deadlines, in proposals, 421

deceptive language, avoiding, 178, 377, 386
decimal numbering system
    for headings, 243–44
    for outlining, 364–65
decimals
    punctuation with, 82, 376
    writing as numbers, 353
declarative sentences, 501
decreasing-order-of-importance method of development, 331, 361
    for FAQs, 184
    for reference letters, 449
    sample of, 359
defective / deficient, 116
defining terms, 116–17. *See also* definitions
    for audience, 43, 117, 233
    in executive summaries, 181
    technical terminology, 175, 281
definite / definitive, 117
definite articles, 39–40, 41
definition method of development, 117–20, 331
definitions. *See also* word meanings
    by analogy, 118
    by cause, 118
    circular, 117
    by components, 118–19
    dictionaries for, 123–25
    by exploration of origin, 119
    extended, 117–18
    formal and informal, 116
    in glossaries, 233
    "is when" and "is where," 117
    negative, 116–17, 119–20
    opening with, 279, 281
deletions. *See* omissions
delivery schedules
    in proposals, 421, 427
    in requests for proposals, 457
delivery techniques, for presentations, 398–400
demonstrative adjectives, 13–14
    count nouns and, 170
    number agreement with, 13, 28–29
demonstrative pronouns, 406
denotation / connotation, 96
denotations, and word choice, 569

dependent clauses, 70
  in sentence construction, 500–501
  sentence faults and, 503
  subordinating conjunctions with, 96
  subordination of, 516–17
description, 120–22
  samples of, 120, 121
  schematics for, 155
  spatial method of development for, 510, 511
descriptive abstracts, 7–9
descriptive adjectives, 13, 15–16
descriptors, in electronic résumés, 487
design. *See* layout and design
desktop publishing software, 300, 345
despite / in spite of, 122
detailed solutions, in proposals, 420, 426
details
  in description, 120
  interesting, opening with, 278
diagnosis / prognosis, 122
dialectal English, 173
dictionaries, 123–25
  abridged, 124
  documenting entries, 133, 140, 146
  ESL, 124–25
  for research, 462
  sample entry, 123
  subject, 125
  unabridged, 124
  for usage, 545
  Web resources for, 124
  for word choice, 569
differ from / differ with, 125
different from / different than, 125
Digital Tips. *See also* computer technology
  conducting meetings from remote locations, 324
  creating an index, 252
  creating an outline, 365
  creating styles and templates, 200
  incorporating tracked changes, 489
  leaving an away-from-desk message, 166
  preparing an ASCII résumé, 487
  proofreading for format consistency, 412
  reviewing collaborative documents, 74
  sending e-mail attachments, 164
  testing your Web site, 562
  using PDF files, 574
  wikis for collaborative documents, 72
direct address, 126
  commas with, 78, 80, 126
  for emphasis, 168
direct objects, 89, 356
  gerunds as, 546
  nouns as, 349
  pronoun case and, 64
  transitive verbs and, 293, 356, 549
  verbals as, 171
direct quotations, 445. *See also* quotations
  commas with, 80, 445
  quotation marks for, 443
direct statements, for emphasis, 168
disabilities, people with
  bias toward, 47–48
  Web design for, 562
discreet / discrete, 126
discussion groups, for research, 459
disinterested / uninterested, 126
division-and-classification method of development, 126–29, 331
document design. *See* layout and design
document elements. *See* report elements
document-management systems, 97–98
document titles. *See* titles of works and documents
documenting sources, 129–53. *See also* APA documentation; authors, documenting; IEEE documentation; MLA documentation
  APA style, 132–38
  for avoiding plagiarism, 383–84
  bibliographies, 48–49
  in blog writing, 51, 52
  colon use in, 75
  common abbreviations for, 4–5
  common knowledge, 384
  comparison of major styles, 130–31
  copyrighted materials, 102
  in formal reports, 199

Index **591**

IEEE style, 139–44
in-text citations, 132, 139, 145
journal articles, 535
in literature reviews, 312
MLA style, 145–52
in note-taking, 347, 348
paraphrasing and, 372, 383–84
purposes of, 129–30
quotations, 445, 447
samples of, 137–38, 143–44, 151–52
style manuals for, 131, 153
for trade journal articles, 536
for visuals, 136, 149, 153, 155, 380, 395, 520, 556
for Web sites, 573
for white papers, 567
Dogpile, 464
double negatives, 153–54, 386, 453
draft writing
integrating visuals in, 553–56
outlining for, 363, 365, 569
revising, 488–89
for trade journal articles, 535
writer's checklist for, 570
in writing process, xxi, xxiii, 569–70
drawings, 154–58. *See also* visuals
functions of, 554
vs. photographs, 154, 379
types of, 154–55
writer's checklist for, 157
due to / because of, 159
dummy, in page design, 300

each, 160
gender agreement with, 27
subject-verb agreement with, 24, 27, 160, 409
each other, 407
EBSCOhost's Academic Search Premier, 461
economic / economical, 160
-ed endings, 524, 548
edited collections, documenting, 133, 140, 146
editing. *See* proofreading; revision
editions, book, documenting, 133, 139, 146
education section, of résumé, 483

educational material, copyright and, 101
effect. *See* cause-and-effect method of development
effect / affect, 21–22
e.g. / i.e., 5, 160
either . . . or, 96, 346–47
agreement with, 26–27, 29
in parallel structure, 371
electronic résumés, 289, 480, 486–88
electronic sources. *See also* Internet research strategies
APA documentation of, 134–35
evaluating, 465–67
IEEE (*CMS*) documentation of, 140–41
MLA documentation of, 147–48
online catalogs, 461
online databases and indexes, 461–62
with unknown author, 134, 141
elegant variation, 22
ellipses, 161
periods as, 375
in quoted material, 446
elliptical constructions, commas with, 81
elude / refer / allude, 30
e-mail, 162–67. *See also* correspondence
abbreviations in, 163
addresses, 163, 166, 482
away-from-desk message, 166
bad-news patterns for, 107–8
confidentiality and, 162
design considerations, 163–64
direct address in, 126
documenting, 135, 136, 141, 148
ethics in writing and, 162
functions of, 162, 164–65, 495
interviewing for information by, 272
length of, 493
netiquette rules, 163
openings for, 108, 277
point of view in, 386
research using, 459
responding to, 163, 164
résumés sent via, 487–88
reviewing, 162

e-mail (*continued*)
  salutations and closings, 164–65
  sending attachments, 100, 163, 164
  signature blocks for, 165
  subject line of, 61, 110, 163, 530
  titles for, 530
  tone for, 513, 532
  trip reports as, 539–40
  writer's checklist for, 163, 166–67
  writing style for, 163, 173
emoticons, 163
emphasis, 167–69
  active voice for, 168
  clarity and, 68, 109–10
  climactic order for, 167
  colons for, 75
  conjunctions for, 95
  dashes for, 114, 169
  direct statements for, 168
  figures of speech for, 190
  highlighting devices for, 163–64, 169, 283, 297–98
  intensifiers for, 168, 264
  pace and, 367
  parallel structure for, 371
  position for, 167
  really and *actually* for, 448
  repetition for, 168, 453
  sentence construction and, 167–68
  sentence variety and, 506
  subordination for, 516
  word order and, 498–99, 502
employment agencies, 290–91
employment experience, in résumé, 483–84
employment process
  acceptance / refusal letters, 9–11
  application letters, 35–39
  job descriptions, 285–86
  job interviews, 272–76
  job searches, 286–91
  reference letters, 449
  resignation letters, 467–69
  résumés, 471–88
  returning job seekers, 485–86
  salary negotiations, 275, 491–93
enclosure notations, 163, 166, 305–6
Encyclopedia of Careers and Vocational Guidance, The (Hopke), 462

encyclopedias
  documenting entries, 133, 140, 146
  for research, 462
end notations, in correspondence, 305–6
end punctuation. *See* sentence endings
endnotes. *See* footnotes / endnotes
English, varieties of
  colloquial, 173
  dialectal, 173
  localisms, 173
  nonstandard, 173–74, 545
  slang, 173–74
  standard, 173, 513, 545
English as a second language (ESL), 169–72
  adjective clauses, 171
  adjectives, 15
  articles, 40, 170–71
  assigning gender, 226
  avoiding shifts in voice, mood, tense, 551
  choosing voice, 560
  considering audiences, xvii
  count and mass nouns, 169–70
  determining mood, 337
  dictionaries, 124–25
  gerunds and infinitives, 171
  idioms, 248
  possessive pronouns, 407
  present perfect verb tense, 172
  progressive verb tenses, 172, 526
  punctuating numbers, 354
  sentence construction, 502
  stating an opinion, 386
  subject of sentence, 499
  Web resources for, 172
ensure / assure / insure, 263
enterprise content-management systems, 98
entry lines and fields, in forms, 222–23
enumeration
  punctuation for, 498
  for transitions, 370, 537, 538
environmental impact statements (EISs), 174–76
  audiences for, 175

background and scope of, 174–75
collaborative guidelines for, 176
ethics in writing for, 176
guidelines for writing, 175–76
Web resources for, 176
equal / unique / perfect, 176–77
-er / -est endings, 14, 20–21
-es endings
  for noun plurals, 350, 387
  for plural numbers, 353
  for verbs, 351, 551
-ese endings, 22
essay titles, 444, 531
etc., 5, 177
ethics in writing, 177–79
  affectation and, 22
  biased language, 46
  copyright and, 101, 102, 153, 155
  e-mail and confidentiality, 162
  for environmental impact statements, 176
  euphemisms and, 180
  for forms, 218
  I / we usage, 386
  for instant messaging, 258
  logic errors and, 312, 313
  for online social communities, 289
  for organizational blogs, 51–52
  paraphrasing, 372
  passive voice usage, 178, 559
  persuasive writing, 377
  plagiarism and, 347, 383–84
  positive writing, 386
  for quotations, 443, 445
  for reference letters, 449
  repurposing and, 456
  for requests for proposals, 458
  for research, 460
  for résumés, 472
  revision process and, 489
  for sales proposals, 420
  for technical writing, 521
  for trouble reports, 542
  using graphs, 236, 553
  for visuals, 153, 155, 236, 380, 395, 553, 556
  Web resources for, 179
  for Web writing, 570–71, 573
  writer's checklist for, 179

ethnic groups
  bias toward, 47
  capitalization of, 60
etymologies, in dictionaries, 123
euphemisms, 180
  as affectation, 22
  ethics in using, 178
  as gobbledygook, 233
evaluating sources, 465–67
  Web resources for, 467
  writer's checklist for, 466–67
events, capitalization of, 61
every
  gender agreement with, 27
  subject-verb agreement with, 27
everybody / everyone, 28, 180
evidence
  biased or suppressed, 313
  guidelines for evaluating, 66–67
  in proposals, 412
examples, in extended definitions, 118
except / accept, 9
exclamation marks, 83, 180–81
  with interjections, 180, 264
exclamatory sentences, 501
executive summaries, 181
  in environmental impact statements, 175
  in formal reports, 198, 205–6
  in proposals, 413, 419, 424
  sample of, 424
  writer's checklist for, 181
experience section, of résumé, 483–84
experiments, for research, 459
explanations, in bad-news patterns, 106
explanatory notes, in formal reports, 199
explanatory phrases, as sentence fragments, 505
expletives, 182, 529
  awkwardness and, 44
  passive voice and, 558
  subject position and, 499
  wordiness and, 91
explicit / implicit, 182
exploded-view drawings, 155, 156
exploration of origin, definition by, 119
exposition, 182

extended definitions, 117–18
external proposals, 414
extranet, 562
eye contact, in presentations, 399, 400

face-to-face meetings, 495, 496. *See also* meetings
fact, 183
facts, vs. opinions, 183, 313
fair use criteria, 101
false cause, 313
false conclusions, 66
false impressions, and ethics in writing, 178
FAQs (Frequently Asked Questions), 183–85
   organization of, 184
   placement of, 185
   questions to include in, 184
   writer's checklist for, 185
faulty logic, in sentences, 504
faulty parallelism, 371–72
faulty subordination, 503
faxes, 496
feasibility reports, 185–88
   sample of, 186–87
   sections in, 188
   titles for, 530
   Web resources for, 188
Federal Business Opportunities, 414
FedStats, 465
female / male, 188
few / a few, 189
fewer / less, 189
figuratively / literally, 189
figures. *See also* visuals
   list of, 197, 419, 557
   numbering, 197, 355, 379
figures of speech, 189–91
   as slang, 173
   in technical writing, 521
films
   documenting, 135, 141, 148
   titles of, 61, 95, 283, 531
fine, 191
finite verbs, 550
first / firstly, 191
first person, 376
   antecedents and, 409

   in narratives, 339
   point of view, 385
first words, capitalization of, 60, 76, 114, 305, 445, 530
FirstSearch, 461
Five Steps to Successful Writing, xv–xxii
flames, e-mail, 163
flammable / inflammable / nonflammable, 192
flip charts, 396
flowcharts, 192–94, 555
fonts, 296, 572. *See also* typography
footers. *See* headers and footers
footnotes / endnotes
   in formal reports, 199
   for graphs, 239
   for tables, 520
for example
   vs. *e.g. / i.e.*, 160
   in sentence fragments, 505
forceful / forcible, 194
forecast, opening with, 279
foreign words in English, 194–95
   affectation and, 22, 195
   italics for, 194, 283
foreword / forward, 195
forewords, in formal reports, 198
formal definitions, 116
formal proposals, 418–19
formal reports, 195–217, 454. *See also* report elements; reports
   back matter, 199–200
   body of, 198–99
   format for, 196–200, 218
   front matter, 196–98
   sample of, 200–217
formal style, 513–14. *See also* writing style
   affectation and, 22
   in correspondence, 103
   usage and, 545
format, 218. *See also* layout and design
   for activity reports, 402
   for formal reports, 196–200
   for letters, 301–6
   for memos, 329
   for newsletters, 345
   for organization, 361

for progress reports, 402
proofreading for consistency in, 412
for requests for proposals, 413–14, 418
for résumés, 481–85, 487
for titles of works, 530–32
using templates, 200, 301
former / latter, 218
forms, 218–23
  choosing response types, 221
  ethics in writing for, 218
  layout and design for, 221, 222–23
  online vs. paper, 220–21
  samples of, 219–20
  sequencing entries, 222
  wording questions, 222
  writing instructions for, 221
forms of discourse, 339. *See also* description; exposition; narration; persuasion
fortuitous / fortunate, 224
fractions
  numerals vs. words, 353
  slashes in, 508
front matter, 519
  for formal reports, 196–98
  for proposals, 419
full-block-style letter, 301, 302
full justification, 298
functional résumé, 483–84
functional shift, 14, 224, 374
fused sentences, 376, 490
future perfect tense, 524, 525
future progressive tense, 524, 526
future tense, 525

garbled sentences, 44, 225, 489
gathering information. *See also* research
  interviews, 270–72
  for presentations, 392–93
  questionnaires, 437–43
  for trade journal articles, 534–35
gender, 226. *See also* sexist language
  agreement, 24, 27–28, 226, 409
  avoiding, in global graphics, 231
  ESL tips for assigning, 226
  female / male usage, 188
  Ms. / Miss / Mrs. usage, 304, 338

  of nouns and pronouns, 226
  unknown, in salutations, 304
general and specific methods of development, 227–28, 331
generalizations, sweeping, 312
geographic features
  capitalization of, 60–61
  commas with, 82
  visuals for, 319–20, 554
gerund phrases, 383
gerunds, 383, 546–47, 550
  for actions, 317
  ESL tips for, 171
  object of, and pronoun case, 64
  possessive pronouns with, 388
gestures
  cultural differences in, 230
  in presentations, 399, 400
global communication, 228–30. *See also* cultural differences; international correspondence
  context and, 100, 229
  idioms and, 248
  international correspondence, 265–70
  visuals and graphics for, 230–32
  Web resources for, 230, 269
  Web writing and, 572
  writer's checklist for, 229–30
global graphics, 230–32
  for instructions, 260
  for manuals, 318
glossaries, 233
  in environmental impact studies, 175
  in formal reports, 200
  in instructions, 259
  in proposals, 413, 419
goals
  in manuals, 317
  purpose and, 435
gobbledygook, 22, 233
good-news patterns, in correspondence, 106, 108
good / well, 234
goodwill, establishing
  with acknowledgment letters, 11
  with adjustment letters, 16–19
  brochures for, 54
  in closings, 106

goodwill, establishing (*continued*)
  in correspondence, 104–5
  in refusal letters, 449–51
  writer's checklist for, 105
  "you" viewpoint for, 575–76
Google, 464, 465
Google Earth, 463
government agencies
  invitation for bids from, 413
  job searches and, 291
government publications
  documenting, 136, 142, 149
  as public domain, 102
government specifications, 512
grammar, 234–35
  proofreading for, 411
  revising for, 489
  Web resources for, 234
grant proposals, 421–33
"granularized" systems, 98n
graphics. *See* visuals
graphs, 235–40. *See also* visuals
  avoiding distortion in, 236, 553
  bar, 236–38
  functions of, 554–55
  line, 235–36
  picture, 239
  pie, 238–39
  for presentations, 398
  vs. tables, 235, 239, 519
  writer's checklist for, 239–40
groupware, 324
Guide to Reference Books (Balay), 462

had / have, 524, 525, 548
Hall, Edward T., 229
handbooks, for research, 462
hardly, 153
he / him / his, 376, 406
he / she, 241
  gender agreement with, 27, 226
  sexist language and, 27, 47, 226, 241, 409
headers and footers, 241–42, 298
  on continuing pages, 306, 329, 331
headings, 242–44
  for brochures, 55, 56
  in correspondence, 110, 301–4
  decimal numbering system, 243–44
  in document design, 298
  on forms, 221
  general style for, 242–43
  in instructions, 260
  italics for, 284
  levels of, 242, 244, 364–65
  in manuals, 316, 317
  for meeting minutes, 332
  in newsletters, 345
  in outlining, 364–65
  parallel structure for, 371
  for résumés, 482
  in table of contents, 197, 519
  in tables, 519
  for trade journal articles, 536
  in trip reports, 540
  typography for, 297
  in Web design, 563, 571
  for white papers, 566
  writer's checklist for, 244
helping verbs, 383, 500, 550
hierarchical relationships, visuals for, 362, 555
highlighting devices
  for e-mail, 163–64, 165
  for emphasis, 164, 169, 283, 297–98
  ethics in writing and, 178
  for warnings, 260–61
historical terms, 61
honors and activities section, of résumé, 485
HotJobs, 486
humor
  in global communication, 269
  malapropisms, 315
  quotation marks for, 444
hyperbole, 190
hyperlinks. *See* links
hyphens, 245–47
  clarity and, 246
  for compound words, 90, 245
  as highlighting cue, 165
  with modifiers, 245
  with prefixes, 246, 389
  with spelled-out numbers, 116, 245
  with suffixes, 246
  for word division, 246–47
  writer's checklist for, 246–47

I (personal pronoun)
  between you and . . ., 46, 390

Index 597

me / my / mine, 376, 406
  in technical writing, 385–86
  usage in résumés, 484
icons
  functions of, 555
  in Web design, 299–300
ideas for consideration, in conclusions, 94
idioms, 248
  clarity and, 70
  in dictionaries, 123
  in global communication, 248, 265, 269
i.e. / e.g., 160
IEEE documentation, 130–31, 139–44
  for articles in periodicals, 140, 141
  for books, 139–40
  compared with other styles, 130–31
  documentation models, 139–42
  for electronic sources, 140–41
  in-text citations, 139, 143
  other sources, 141–42
  sample bibliography, 144
  sample page, 143
IEEE Transactions on Communications, 533
-ies endings
  for noun plurals, 350
  for plural numbers, 353
if and when, 564
illegal / illicit, 249
illusion / allusion, 30
illustrations. *See* visuals
images, in description, 120
imitation, and affectation, 23
imperative mood, 337
  for actions, 317
  for conciseness, 92
  for instructions, 259, 263
  for nonsexist language, 241
imperative sentences, 501
impersonal point of view, 385–86
implications, in conclusions, 94
implicit / explicit, 182
imply / infer, 249
imprecision, and affectation, 23
impression, and affectation, 23
in / into, 249
in order to, 249–50
in re, 448

in regard to, 453
in spite of / despite, 122
in terms of, 250
in-text citations. *See* citations
increasing-order-of-importance method of development, 331–32, 361
  sample of, 360
indefinite adjectives, 14
indefinite articles, 1, 39–40
indefinite pronouns, 406
  one usage, 357
  possessive case of, 388–89
  subject-verb agreement with, 25, 409
independent clauses, 70
  in comma splices, 77
  connecting, 20, 71, 78, 82, 96, 497
  in run-on sentences, 376, 490
  in sentence construction, 500–501
indexes
  computer technology for, 252
  in formal reports, 200
  for manuals, 318
  periodical, 461–62
  subject directories, 465
indexing, 250–53
  compiling, 251
  cross-referencing, 252
  wording entries, 251
  writer's checklist for, 252–53
indicative mood, 337
  vs. subjunctive mood, 337–38
  wordiness of, 92
indirect objects, 89, 356
  nouns as, 350
  pronoun case and, 64
indirect quotations, 445. *See also* quotations
  punctuating, 80, 443
indiscreet / indiscrete, 253
individual, 253
industrial specifications, 510–12
infer / imply, 249
infinitive phrases, 382–83
infinitives, 382, 547–48, 550
  for actions, 317
  ESL tips for, 171
  split, 547–48
  subject of, and pronoun case, 64
  to as sign of, 382, 547–48, 550

inflammable / nonflammable / flammable, 192
Infomine, 465
informal definitions, 116
informal proposals, 418
informal reports, 454
informal style, 514–15. *See also* writing style
   in correspondence, 103
   usage and, 545
informal tables, 521
information models, for content management, 97
information sources. *See* Internet research strategies; library research strategies; research
informational brochures, 54–57
informational interviews, 288
informative abstracts, 7–9
InfoTrac, 461
-*ing* endings
   gerunds, 546, 550
   for present participles, 548
   in progressive form, 526
initialisms. *See* acronyms and initialisms
initials
   in correspondence, 305
   periods with, 376
initiation, and affectation, 23
inquiries and responses, 253–55
   for job searches, 291
   responding to inquiries, 255
   samples of, 254, 255
   for unsolicited proposals, 414
   writing inquiries, 254–55
insecurity, and affectation, 23
inside / inside of, 256
inside address, 304
insoluble / insolvable, 256
instant messaging (IM), 256–58, 495
   abbreviations in, 3, 257
   ethics in writing and, 258
   privacy and security with, 258
   sample of, 257
   writer's checklist for, 258
Institute of Electrical and Electronics Engineers, Inc. Standards Style Manual. See IEEE documentation

instructions, 258–63. *See also* manuals
   active voice for, 558
   for forms, 221
   methods of development for, 332, 507, 510
   point of view in, 385
   testing, 263
   usability of, 543
   visuals for, 154, 155, 258, 260
   warnings in, 181, 260–63
   writer's checklist for, 263
   writing style for, 259–60
insure / ensure / assure, 263
intensifiers, 264
   adverbs as, 335
   for emphasis, 168
   overuse of, 92
   really and *actually* as, 448
   redundant, 6
   very as, 552
intensive pronouns, 407
intention, in sentence construction, 501
interface, 264
interjections, 264–65
   commas with, 80
   exclamation marks with, 180
internal correspondence, 494–95
internal proposals, 414
international correspondence, 265–70. *See also* global communication
   avoiding euphemisms, 180
   avoiding figures of speech, 191
   context and, 100, 229
   cultural differences and, 228–30, 265
   dates in, 115, 269, 270, 354, 509
   idioms and, 248, 265
   indirect style for, 106, 229, 265
   punctuation for, 232
   salutations for, 165
   samples of, 266–68
   visuals and graphics for, 230–32
   Web resources for, 230, 269
   writer's checklist for, 269–70
   writing style for, 265–70
International Organization for Standardization (ISO) symbols
   for flowcharts, 192, 193
   for global communication, 231
   for warnings, 263

International System of Units (SI), 4
Internet. *See also* e-mail; Internet research strategies; Web design; Web resources; Web sites; writing for the Web
   addresses, 246, 509
   domain names, 460, 465–66
   for mailing newsletters, 345
   online forms, 218–23
   online postings, 135, 142
   primary research on, 459
Internet Public Library, The, 465
Internet research strategies, 463–65. *See also* electronic sources
   documenting sources, 134–35, 140–41, 147–48
   evaluating sources, 465–67, 572
   job searches, 289, 486
   library resources, 461–62
   online databases and indexes, 461–62
   search engines, 464
   subject directories, 465
   writer's checklist for, 464
interrogative adverbs, 20
interrogative pronouns, 406
interrogative sentences, 436, 501
interrupting elements, commas with, 79
interruptive person, in meetings, 327
interviewing for information, 270–72
   active listening for, 307–9
   choosing interviewees, 270
   compared with questionnaires, 437–38
   conducting the interview, 271
   expanding interview notes, 272
   for job searches, 288
   for newsletter articles, 341
   by phone or e-mail, 272
   preparation, 270–71
   for research, 459, 460
   for usability testing, 545
   writer's checklist for, 271
interviewing for a job, 272–76
   application letters, 35–39
   behavior and responses during, 274–75
   behavioral interviews, 274
   follow-up procedures, 275–76
   preparation for, 272–74

   résumés and, 471
   salary negotiations, 491–92
interviews, documenting, 136, 141–42, 148, 149
intimidation, and affectation, 23
intranet, 562
intransitive verbs, 293, 549–50
introductions, 276–81. *See also* openings
   conclusions consistent with, 95
   in correspondence, 109
   for feasibility reports, 188
   for formal reports, 198, 207–9
   full-scale, 280
   for lists, 309
   literature reviews in, 311
   for manuals, 280–81
   for newsletter articles, 343
   opening strategies, 277–80
   for presentations, 393–94
   for process explanations, 401
   for progress reports, 402
   for proposals, 418, 419, 420, 433
   providing context in, 100
   for reports, 454
   revising, 488
   routine openings, 277
   setting tone with, 533
   for specifications, 280–81
   for trade journal articles, 535–36
   when to write, 280
   for white papers, 566–67
introductory elements. *See also* sentence openings
   commas with, 79–80
   dangling modifiers as, 113
   eliminating wordy, 92
inverted pyramid organization, 571
inverted sentences
   for emphasis, 502
   subject-verb agreement in, 25
   for variety, 506
investigative reports, 281–83
   sample of, 282
invitation for bids (IFB), 413–14, 418, 420, 456n
irregardless, 453
"is when" and "is where" definitions, 117

ISO symbols. *See* International Organization for Standardization symbols
it
  as expletive, 182, 499
  as subject, 499
it / its, 376, 406
italics, 283–84
  for emphasis, 169, 283, 297–98
  for foreign words, 194, 283
  for proper names, 284
  for subheadings, 284
  for titles of works, 283–84, 531
  for words, letters, and numbers, 284
its / it's, 284

jammed modifiers. *See* stacked (jammed) modifiers
jargon, 285. *See also* technical terminology
  as affectation, 22
  ethics in using, 178
  as functional shift, 224
  in global communication, 265, 269
  as gobbledygook, 233
  in manuals, 317
  revising for, 489
  stacked modifiers and, 335
job descriptions, 285–86
  sample of, 287
  writer's checklist for, 286
job interviews. *See* interviewing for a job
job objective, in résumés, 482
job search, 286–91. *See also* employment process
  advertisements, 289–90
  campus career services, 288–89
  employment agencies, 290–91
  informational interviews, 288
  networking, 288
  résumés, 471–88
  trade and professional listings, 290
  Web resources, 289, 291, 486
  writing letters of inquiry, 291
job titles, 62
journals. *See* articles in periodicals; periodicals; trade journal articles
judgments, in conclusions, 93, 94
justification, in page design, 298

key terms
  defining, 116–17
  for index, 250–51
  repetition of, 168, 453, 537, 538
  for Web writing, 571
keys (legends)
  for graphs, 239
  for maps, 319
  for symbols, 194, 556
keyword searches
  for online databases, 461, 462
  résumé databases and, 487
  using abstracts, 535
  using search engines, 464
kind / type / sort, 13–14, 29
kind of / sort of, 28–29, 292
know-how, 292

labels
  for forms, 223
  for visuals, 122, 157, 192, 194, 238, 240, 319, 379, 536
laboratory reports, 293, 459
  chronological order for, 68
  sample of, 294–95
  vs. test reports, 527
lack of reason, 312
Latin words. *See also* foreign words in English
  common abbreviations for, 160, 177
  via, 552
latter / former, 218
lay / lie, 293–95
layout and design, 295–300. *See also* format; report elements; Web design
  analyzing audience needs, 43
  for brochures, 54–57
  captions, 300
  color in, 299
  columns, 299
  consistency in, 489
  context of, 99
  for e-mail, 163–64, 165
  ethics in writing and, 178

for forms, 221, 222–23
headers and footers, 298
headings, 242–44, 298
for instructions, 260
justification, 298
for letters, 301–6
lists, 299
for manuals, 316
for newsletters, 345
for organization, 361
for organizational blogs, 51
page-design elements, 298–99
for persuasion, 377
for presentation visuals, 396–97
proofreading for, 412
for repurposed content, 456
for résumés, 472
rules, 300
for technical writing, 522
typography, 296–98
usability testing for, 545
using thumbnails, 300
of visuals, 299–300, 556
Web resources for, 300
for white papers, 566
white space, 299
lectures, documenting, 150. *See also* presentations
left-justified (ragged-right) margins, 298
legal case titles, 283, 531
legal terms
 as affectation, 22
 ethics in using, 178
 illegal / illicit, 249
 party, 374
 quid pro quo, 443
 re, 448
legalese, 22, 233
legends. *See* keys (legends)
lend / loan, 300
length
 of paragraphs, 368–69
 of sentence, 505–6
less / fewer, 189
less / least, 14, 21
letterhead
 for correspondence, 301
 for formal contacts, 494

letters, 301–6. *See also* correspondence
 body of, 305
 common styles for, 218, 301, 302–3
 complimentary closings, 305
 continuing pages, 306
 end notations, 305–6
 for formal contacts, 494
 headings in, 301–4
 inside address, 304
 names and titles in, 304–5
 openings for, 108, 277
 salutations, 304–5
 samples of, 302–3
letters (alphabet)
 capitalization of, 62
 italics for, 284
 plurals of, 34
Lexis/Nexis Academic, 461
library research strategies, 460–63. *See also* Internet research strategies; research
 library homepage, 461
 online catalogs, 461
 online databases and indexes, 461–62
 reference works, 462–63
lie / lay, 293–95
-like, 517
like / as, 306–7
limiting adjectives, 13–14, 15–16
line drawings, 155
line graphs, 235–36, 240, 555
linking verbs, 45, 63, 549–50
links
 in blog writing, 51
 in Web design, 571–72
list-hosting service, 345
listening, 307–9
 active, 307–9
 fallacies about, 307
 in interviews, 274
 in meetings, 327
lists, 309–10
 of abbreviations and symbols, 198
 in brochures, 56
 climactic order for, 167
 colons for introducing, 75, 76, 309
 consistency for, 310
 in correspondence, 110

lists (*continued*)
  in document design, 299
  of figures and tables, 197, 419, 557
  formatting for, 310
  in instructions, 259
  numbered and bulleted, 259, 309, 375, 398
  organization names in, 3
  parallel structure for, 371
  in presentations, 310, 398
  stacked, 76
  in Web writing, 571
  in white papers, 566
  writer's checklist for, 310
literally / figuratively, 189
literature reviews, 310–12
  sample of, 311
  for trade journal articles, 534
litotes, 190
loaded arguments, 314
loaded sentences, 502
loan / lend, 300
localisms, 173
logic errors, 312–14
  awkwardness and, 44
  biased or suppressed evidence, 313
  correcting, 363
  double negatives and, 154
  ethical considerations for, 312, 313
  fact vs. opinion, 183, 313
  false cause, 313
  lack of reason, 312
  loaded arguments, 314
  mixed constructions, 334
  non sequiturs, 313
  in sentence construction, 504
  sweeping generalizations, 312
long variants, as affectation, 22
Longman Advanced American Dictionary, 125
Longman American Idioms Dictionary, 125
Longman Dictionary of American English, 125
loose / lose, 314
loose sentences, 501, 506–7
LSU Libraries Federal Agencies Directory, 465
-ly endings, 21, 245

magazines. *See also* periodicals
  documenting articles, 133, 140, 146
  visuals in, 299
mailing lists, 345
malapropisms, 315
male / female, 188
management
  activity reports and, 402
  job description writing, 285–86
  progress reports and, 402
manuals, 315–19. *See also* instructions
  actions and responses in, 317–18
  audience for, 316
  chapters and sections in, 317
  context of, 99
  decimal number heads for, 365
  introductions for, 280–81
  overview in, 317
  for research, 462
  types of, 315–16
  usability of, 543
  verb usage in, 317–18
  warnings in, 181, 319
  writer's checklist for, 318–19
MapQuest, 463
maps, 319–20
  documenting, 141
  functions of, 554
  Web resources for, 320
  writer's checklist for, 319
margins
  for brochures, 57
  for letters, 301
  in page design, 298
mass nouns, 25, 33, 349
  articles and, 171
  ESL tips for, 169–70
  fewer / less usage and, 189
  list of common, 169
mathematical equations, 320–22
  computer technology for, 320–21
  numbering, 321
  as part of text, 321
  positioning displayed, 321–22, 537
  slashes in, 508
may / can, 59
maybe / may be, 322

me
  between you and . . . , 46, 390
  my / mine, 376, 406
mean / average / median, 44
measurements
  abbreviations for, 4
  average / median / mean, 44
  in global communication, 270
  numbers for, 353
  precision of, 11
  subject-verb agreement with, 25
media / medium, 322–23
median / mean / average, 44
medical terms, defining, 119
medium. *See* selecting the medium
meetings, 323–28
  closing, 328
  conducting, 326–28
  conflict in, 327
  determining attendees, 323
  establishing agenda for, 324–25
  functions of, 496
  location for, 324
  minutes of, 325, 332–34
  planning for, 323–26
  purpose of, 323
  from remote locations, 323, 324
  for research, 459
  time and duration of, 323–24
  writer's checklist for, 328
memos, 328–29. *See also* correspondence
  additional pages for, 329, 331
  bad-news patterns for, 107–8
  as cover letters, 196, 201, 439
  format for, 329
  functions of, 328, 494–95
  length of, 493
  methods of development for, 359, 361
  openings for, 108, 277
  protocols for sending, 329
  samples of, 330–31
  subject line of, 61, 110, 530
  test reports as, 527
  titles for, 530
  tone for, 532
  trip reports as, 539
  trouble reports as, 541
  Web resources for, 329

Metacrawler, 464
metaphors, 191
metasearch engines, 464
methods of development, 329–32. *See also specific methods*
  cause-and-effect, 66–67
  chronological, 68
  clarity and, 68
  comparison, 85–87
  definition, 117–20
  division-and-classification, 126–29
  general and specific, 227–28
  order-of-importance, 358–61
  for organization, 361
  sequential, 507
  spatial, 510
  in writing process, xix–xx
metonyms, 191
Microsoft Encarta World Atlas, 463
Microsoft Encarta World English Dictionary, 124
Microsoft PowerPoint, 396
mind mapping, 53–54
minor sentences, 501
minutes of meetings, 332–34
  assigning responsibility for, 325
  chronological order for, 68
  sample of, 333
  writer's checklist for, 333–34
misplaced modifiers, 32, 335–36
misplaced participial phrases, 382
mixed constructions, 334
MLA documentation, 131, 145–52. *See also* documenting sources
  for articles in periodicals, 146, 147
  for books, 145–46
  compared with other styles, 130–31
  documentation models, 145–50
  for electronic sources, 147–48
  in-text citations, 145, 151
  for multimedia sources, 148
  other sources, 149–50
  sample pages, 151–52
  sample works cited, 152, 217
Modern Language Association. *See* MLA documentation
modest tone, in correspondence, 105
modified-block-style letter, 301, 303

modifiers, 334–36. *See also* adjectives; adverbs
  ambiguous use of, 32
  awkwardness and, 44
  dangling, 113
  hyphens with, 245
  misplaced, 32, 335–36
  redundant, 90
  restrictive and nonrestrictive, 470–71
  sentence construction and, 500
  squinting, 336
  stacked (jammed), 15, 16, 335
money, numbers for, 353–54
Monster.com, 289, 486
months. *See also* dates
  capitalization of, 61
  format for writing, 81–82, 115
mood, 336–38
  avoiding shifts in, 551
  ESL tips for, 337
  helping verbs and, 550
  imperative, 317, 337
  indicative, 337
  subjunctive, 337–38
more / most
  for comparison, 14, 21, 176–77
  number agreement with, 25
more than / as much as, 41–42
most / almost, 31
Ms. / Miss / Mrs., 304, 338
MSN, 464
multimedia sources
  APA documentation of, 135–36
  IEEE documentation of, 141
  MLA documentation of, 148
multivolume works, documenting, 133, 139, 146
musical works titles, 284, 444, 531
mutual / common, 338

names. *See* names and titles; organization names; proper nouns
names and titles
  abbreviations for, 4
  capitalization of, 62
  commas with, 81–82
  first name usage, 4, 165, 265, 304
  initials, 305, 376
  in letters, 304–5

Ms. / Miss / Mrs. usage, 304, 338
  in organizational blogs, 51
  on title page, 197
narration, 68, 339–40
  sample of, 340
National Environmental Policy Act (NEPA), 174
National Society of Professional Engineers (NSPE), 179
nationalities
  bias toward, 47
  capitalization of, 60
nature, 341
needless to say, 341
negative definitions, 116–17, 119–20
negative messages. *See also* bad-news patterns
  in correspondence, 105
  international correspondence and, 265
  positive writing and, 386–87
  refusal letters, 449–52
  resignation letters, 467–69
negative person, in meetings, 327
negatives, double, 153–54, 386, 453
neither . . . nor, 96, 346–47
  agreement with, 26–27, 29
  double negatives and, 154
  in parallel structure, 371
netiquette rules, for e-mail, 163
networking, for job search, 288
New York Times Index, 462
newsgroups, for research, 459, 463
newsletter articles, 341–43. *See also* articles in periodicals
  sample of, 342
  visuals in, 299, 343
  writer's checklist for, 343
newsletters, 343–45
  format and design for, 345
  internal blogs as, 50–51
  sample of, 344
  types of, 343–45
  typography for, 297
  visuals for, 155, 240, 345
newspapers
  for current information sources, 460
  documenting articles, 134, 140, 146
  italics for titles, 283, 531

online databases and indexes for, 462
no, commas with, 78
no one / not one, 346
nominalizations, 346
nominative case, 63
non sequiturs, 313
none, 346
nonfinite verbs, 550
nonflammable / flammable / inflammable, 192
nonrestrictive elements, 470–71. *See also* parenthetical elements
   commas with, 78–79, 470
   using *which* with, 471, 527
nonstandard English, 173–74, 545
nor / or, 346–47
   with compound subjects, 26–27, 29
not
   double negatives and, 153
   in verb phrases, 383
not only . . . but also, 96, 371
note-taking, 347–48
   in active listening, 308–9
   draft writing and, 569
   for interviews, 271, 272
   meeting minutes, 325, 332–34
   methods for, 348
   outlining and, 363–64
   paraphrasing, 372
   plagiarism and, 347
   for presentations, 400
   quotations and, 445
   for research, 459, 460
   writer's checklist for, 348
notes section, in formal reports, 199
noun phrases, 39
nouns, 349–51
   abstract, 349
   as adjectives, 16, 350, 388, 517
   collective, 26, 349, 350
   common, 60, 349
   concrete, 349
   coordinate, 388
   count, 14, 25, 33, 169–70, 349
   for electronic résumés, 487
   ESL tips for, 169–70
   functions of, 349–50
   gender and, 226
   mass, 25, 33, 169–70, 349
   as objects, 356
   phrases as, 380, 383
   plurals of, 65, 350–51, 387–88
   possessive case of, 62–63, 64–65, 387–88, 547
   pronouns modifying, 408–9
   proper, 59–60, 349
   as subject, 499–500
   types of, 349
   verbals as, 546–47, 550
Nucleonics Week, 533
number (grammar), 351–52. *See also* agreement
   indefinite pronouns and, 25, 409
   pronoun-antecedent agreement, 24, 28–29
   subject-verb agreement, 24–26, 409, 499, 550, 551
number (word)
   number / amount usage, 33
   subject-verb agreement with, 25–26
number-style method of documentation, 130. *See also* IEEE documentation
numbered lists, 110, 309, 310. *See also* lists
   for inquiries, 254
   in instructions, 259
   periods in, 375
numbering
   decimal headings, 243–44, 364–65
   in documents, 355
   of equations, 321
   for figures and tables, 197, 355, 379, 519
   in outlines, 364–65
numbers, 352–55. *See also* measurements
   in addresses, 354–55
   adjectives as, 14
   colons with, 76
   commas with, 81–82, 354
   dates, 354
   in documents, 355
   ESL tips for, 354
   fractions, 353
   hyphens with spelled-out, 116, 245
   italics for, 284

numbers (*continued*)
  mathematical equations, 320–22
  measurements, 353
  money, 353–54
  numerals vs. words, 14, 352–53
  ordinal, 14, 191
  in parentheses, 355, 373
  percentages, 353, 374
  periods with, 354, 375, 376
  plurals of, 34, 353
  slashes with, 508
  time, 354
numerical data
  in appendixes, 35
  in graphs, 235–40, 554–55
  sources for, 463
  in tables, 519, 554

object of preposition, 356, 390
  gerunds as, 546
  nouns as, 350
  pronoun case and, 64
  punctuation of, 76
objective case, 63, 64, 407, 408
objective complements, 89, 350
objectives
  opening with, 277
  in résumés, 482
objects, 356. *See also* direct objects; indirect objects; object of preposition
observation, direct, in research, 459, 460
Occupational Outlook Handbook, 291
of
  with demonstrative adjectives, 13, 28–29
  possessive case using, 387, 388
  redundant use of, 256, 365, 391
of which / who's / whose, 568
OK / okay, 356
omissions
  apostrophes for, 34
  commas for, 81
  dashes for, 114
  ellipses for, 161, 375, 446
  ethics in writing and, 178
  slashes for, 508
  in telegraphic writing, 522

omnibus words, 546
on / onto / upon, 357
one, 357
  subject-verb agreement with, 24
one another, 407
one of those . . . who, 357–58
online databases. *See* databases, online
online forms, 220–21
online postings
  documenting, 135, 142, 148
  subject lines of, 530
online resources. *See* electronic sources; Internet research strategies
online social communities, 289
online surveys, 438
only, 21, 358
open-ended questions, 221
openings. *See also* introductions
  for application letters, 35–38
  for bad-news letters, 106
  for correspondence, 108–9
  good news in, 108
  for international correspondence, 265–69
  for investigative reports, 283
  for presentations, 393–94
  providing context in, 35, 100
  for refusal letters, 451
  routine, 277
  for test reports, 527
OpenOffice.org Impress, 396
operators' manuals, 316
opinions
  ESL tips for stating, 386
  vs. facts, 183, 313
  point of view for, 386
or / nor, 346–47
  for compound subjects, 26–27, 29
oral presentations. *See* presentations
order-of-importance method of development, 66, 184, 331–32, 358–61
ordinal numbers, 191
  as modifiers, 14
  spelling out, 352–53
organization, 361–62. *See also* methods of development
  awkwardness and, 44
  coherence and, 71–72

outlining, 362–65
paragraph length and, 368
for presentations, 393–95
for résumés, 471–72, 481–85
for trade journal articles, 535
in writing process, xix–xx, xxiii
organization names
  abbreviations of, 3–4
  ampersands in, 4, 33
  capitalization of, 61
  in salutations, 304
  as singular nouns, 350
  on title page, 197, 419
  on Web sites, 563
organizational blogs, 49–52
organizational charts, 362, 555
organizational newsletters, 343–44
organizational sales pitch, in proposals, 421, 431
origins, definitions exploring, 119
orphans, 299
our / ours, 376, 406
outlining, 362–65
  abstracts and, 8–9
  advantages of, 363
  brainstorming and, 53
  clarity and, 68
  for coherence, 72, 363, 566
  decimal headings in, 244
  digital tips for, 365
  draft writing and, 363, 365, 569
  including visuals, 154, 361–62, 364, 556
  method of development for, 332
  for organization, 361–62, 363
  parallel structure for, 364
  steps for creating, 363–65
  for trade journal articles, 535
  transitions and, 539
  types of, 363
  unity and, 543
  writing paragraphs and, 363, 369
  in writing process, xx
outside [of], 365
over [with], 366
overhead transparencies, 396
overview
  in manuals, 317
  in presentations, 393

ownership. *See* possessive case
Oxford American Wordpower Dictionary for Learners of English, 125
Oxford English Dictionary, The, 124

pace, 367
  clarity and, 68–70
  inserting clauses or phrases for, 250, 506
  nominalizations for slowing, 346
  in presentations, 400
Pace University Library Internet Resources, 465
page design. *See* layout and design
page numbers
  in formal reports, 197
  in headers or footers, 241, 298
  number usage for, 354, 355
painting titles, 284, 531
pamphlets
  documenting, 136, 142, 149
  italics for titles, 283, 531
panels, brochure, 55–56
paragraphs, 367–70
  in e-mail, 164
  functions of, 367
  length of, 368–69
  outlining, 363, 369
  topic sentence of, 38, 368
  transitional, 537, 539
  transitions between, 72, 538–39
  unity and coherence in, 370, 488, 543
  writing, 369–70
parallel structure, 370–72, 515
  in abstracts, 9
  for *both . . . and* usage, 52
  faulty parallelism, 371–72
  for garbled sentences, 225
  for headings, 244
  in lists, 310
  for nonsexist language, 47
  in outlining, 364
  in sentence construction, 502
paraphrasing, 372
  in active listening, 308
  documenting sources and, 130, 383–84

paraphrasing (*continued*)
  ethical considerations for, 372
  indirect quotations as, 443
  in meeting minutes, 332
  note-taking and, 347
parentheses, 373
  brackets within, 52
  commas with, 82
  for enclosing elements, 114, 498
  in equations, 322
  numbers in, 355, 373
  periods with, 373, 375
parenthetical citations. *See* citations
parenthetical elements
  brackets for, 52, 373
  capitalization of, 62
  commas with, 78–79, 470
  dashes for, 114, 498
  nonrestrictive elements as, 470–71
  parentheses for, 373, 498
  for sentence variety, 506
part-by-part method of comparison, 85, 86
participial phrases, 381–82
  dangling and misplaced, 382
participles, 548, 550
  ESL tips for, 15
parts of speech, 374. *See also specific parts of speech*
  in dictionaries, 123
party, 374
passive voice, 552, 557, 559–60
  avoiding responsibility with, 178, 559
  awkwardness and, 44
  ethics in writing for, 559
  infinitives and, 547
  in laboratory reports, 293
  in technical writing, 522
  wordiness of, 92, 558–59
past participles, 15, 548
past perfect tense, 524
past progressive tense, 524, 526
past tense, 172, 524
PDF files. *See* Adobe PDF files
per, 374
percent / percentage, 374
percentages, numbers in, 353
perfect / equal / unique, 176–77

perfect participles, 548
perfect tense, 524
periodic sentences, 501, 507
periodical indexes, 7, 461–62
periodicals. *See also* articles in periodicals; trade journal articles
  abstracts in, 7
  for current information sources, 460
  documenting, 133–34, 140, 146
  italics for titles, 283, 444, 531
periods, 83, 375–76
  for abbreviations, 3, 4
  with ellipses, 161
  with numbers, 354, 375, 376
  with parentheses, 373, 375
  period faults, 376
  with quotation marks, 375, 444
  for run-on sentences, 376
permissions, for copyrighted material, 101
person, 376, 406. *See also* point of view
  agreement of, 24, 550, 551
  in narratives, 339
  point of view and, 385–86
  pronouns and, 409
person / party, 374
personal communications, documenting, 136, 142, 149
personal experience, opening with, 393
personal pronouns, 406
  antecedents and, 409
  forms of, 376
  point of view and, 385–86
personification, 191
persuasion, 377
  in application letters, 35
  ethics in writing for, 377
  in inquiry letters, 253
  in openings, 109, 280
  in presentations, 394
  in proposals, 412, 417
  in résumés, 472
  sample memo, 378
  in white papers, 565
  "you" viewpoint and, 575
phenomenon / phenomena, 377
photographs, 377–80. *See also* visuals
  vs. drawings, 154, 377–79

functions of, 554
in trade journal articles, 537
phrases
commas with, 78–79
functions of, 380
gerund, 383
infinitive, 382–83
noun, 383
participial, 381–82
prepositional, 380–81
restrictive and nonrestrictive, 470–71
verb, 383
pictorial symbols, in flowcharts, 192, 193
picture graphs, 239, 240, 555
pie graphs, 238–39, 240, 555
place names, 60–61
plagiarism, 383–84
documenting sources and, 102, 130
ethics in writing and, 179
note-taking and, 347
quotations and, 443
repurposing and, 456
Web resources for, 384
plain-language laws, 176
plain-text résumés, 487
planning, in collaborative writing, 73. *See also* outlining; preparation
play titles, 61, 95, 284, 531
pluperfect, 524
plurals. *See also* number (grammar)
for abbreviations, 3, 34
apostrophes for, 34
compound words, 90, 351
for foreign words, 195
nouns, 65, 350–51
of numbers, 34, 353
possessive case of, 387–88
podcasts, documenting, 135, 148
poem titles, 284, 531
point of view, 385–86. *See also* person; "you" viewpoint
consistent use of, 70, 370
for correspondence, 104–5
in narratives, 339
one usage, 357
person and, 376
"you" viewpoint, 575–76

polite tone
in correspondence, 105
in global communication, 265–69
political divisions (capitalizing), 60
portfolios, in résumés, 485
position, and emphasis, 167
positive writing, 386–87, 515. *See also* goodwill, establishing
for defining terms, 116–17
for double negatives, 154
ethics in writing for, 386
"you" viewpoint, 575
possessive adjectives, 14
count nouns and, 170
possessive case, 387–89
apostrophes for, 34, 387–88
for compound words, 90
ESL tips for, 407
of nouns, 62–63, 64–65, 387–88
of pronouns, 63, 64–65, 387, 388–89, 407, 408
post hoc, ergo propter hoc, 313
postal service abbreviations, 5, 301
practicing, for presentations, 398–99, 400
precision / accuracy, 11
predicate adjectives, 16
predicate nominative, 500
predicates, 500
complements in, 88–89
predictions, in conclusions, 94
preface, in formal reports, 198
prefixes, 246, 389
preparation, 389–90
analyzing audience needs, 42–43
determining purpose, 435
determining scope, 493
for research, 459
for specifications, 510
writer's checklist for, 389–90
in writing process, xvi–xviii, xxiii
prepositional idioms, 248
prepositional phrases, 356, 380–81, 390
commas with, 83
overuse of, 381
passive voice and, 558
showing possession, 387

prepositions, 380, 390–91. *See also* object of preposition
confused with adverbs, 390
at end of sentence, 390–91
errors with, 391
number agreement with, 13, 28–29
in titles of works, 61, 391, 530
present participle, 15, 548
present perfect tense, 524, 525
ESL tips for, 172
infinitives and, 547
present progressive tense, 524
ESL tips for, 172, 526
present tense, 172, 525
infinitives and, 547
for instructions, 259
presentations, 391–401
analyzing audience for, 392
body of, 394
closings, 394–95
delivery techniques for, 398–400
direct address in, 126
eye contact during, 399, 400
gathering information for, 392–93
introductions, 279, 393–94
methods of development for, 332, 358, 361
practicing, 398–99, 400
presentation anxiety, 400
purpose of, 392
software for, 396–97
structuring, 393–95
transitions in, 395
typography for, 297, 398
using lists in, 310, 398
visuals for, 155, 240, 395–98
writer's checklist for, 398, 400–401
pretentious language. *See* affectation
primary research, 459–60
principal / principle, 401
privacy statements, in Web design, 563, 573
problem statement
opening with, 277, 393, 535, 566–67
in proposals, 413, 420
process explanations, 401
flowcharts for, 192–93, 555
instructions, 258–63

organization for, 507
sample of, 522
product description, in proposals, 420, 425
professional journals, job listings in, 290. *See also* trade journal articles
prognosis / diagnosis, 122
progress and activity reports, 402–4
samples of, 403–4
titles for, 530
progressive tense, 524
ESL tips for, 172, 526
project description, in requests for proposals, 457
project management, for proposals, 414
projection, in presentations, 400
promotional writing
newsletters, 344–45
white papers, 565–67
pronoun-antecedent agreement, 24, 27–29. *See also* agreement
with compound antecedents, 29
in gender, 27–28, 226, 409
in number, 28–29
pronoun references, 32, 405, 453. *See also* antecedents
pronouns, 405–9. *See also* case; indefinite pronouns; personal pronouns; pronoun-antecedent agreement; relative pronouns
as adjectives, 14
case of, 62–66, 407–9
in compound subjects, 65–66, 408
demonstrative, 406
eliminating in telegraphic style, 522
gender and, 27–28, 226, 409
indefinite, 388–89, 406
intensive, 407
interrogative, 406
number and, 28–29, 409
objective case of, 64
person and, 376, 406, 409, 551
plurals of, 351
point of view and, 385–86
possessive, 64–65, 387, 388–89, 547
reciprocal, 407
reflexive, 406–7
relative, 406

sexist language and, 27–28, 47, 226, 241, 409
  as subject, 499
  subjective case of, 63–64
  for transitions, 537, 538
pronunciations, in dictionaries, 123
proofreader's marks, 410
proofreading, 411–12
  application letters, 39
  blog entries, 52
  e-mail, 162
  for format consistency, 412
  for spelling, 411, 512
  for typographical errors, 489
  writer's checklist for, 411
proper nouns, 349
  capitalization of, 59–60
  italics for, 284
proposal description, in requests for proposals, 457
proposal-evaluation criteria, in requests for proposals, 458
proposals, 412–33
  appendixes for, 34–35
  audience and purpose for, 413
  collaborative writing of, 72
  conclusions, 93–95
  context in, 98–99
  cover letters for, 100, 111
  environmental impact statements, 174–76
  executive summaries in, 181
  external, 414
  formal structure for, 418–19
  grant and research, 421–33
  informal structure for, 418
  internal, 414
  length of, 493
  as memos, 329
  methods of development for, 331–32, 510
  persuasive writing for, 412, 417
  project management for, 414
  requests for proposals, 456–58
  routine, 414
  sales, 420–21
  samples of, 415–17, 422–32
  solicited and unsolicited, 413–14
  special-purpose, 414, 415–17
  titles for, 529–30
  writer's checklist for, 417
ProQuest, 461
protocols, 544
pseudo- / quasi-, 434
public domain works, 102, 456
publication dates. *See also* dates
  in headers or footers, 241, 298
  for reports, 197
punctuation, 434
  apostrophes, 34
  brackets, 52
  colons, 75–76
  commas, 77–83
  dashes, 114
  ellipses, 161
  exclamation marks, 180–81
  for global communication, 232
  hyphens, 245–47
  parentheses, 373
  periods, 375–76
  question marks, 436–37
  quotation marks, 443–45
  semicolons, 497–98
  slashes, 508–9
  Web resources for, 434
purpose, 435
  in abstracts, 7, 8
  active listening and, 308
  of brochures, 54–55
  clarity and, 68
  design and, 296
  for feasibility reports, 187
  front matter for defining, 196
  in introduction, 280, 535
  of manuals, 317
  of meetings, 323
  needs of audience and, 42, 448
  in preparation stage, 389
  for presentations, 392
  for proposals, 413
  repurposing, 455–56
  for requests for proposals, 457
  scope and, 493
  selecting the medium and, 494
  style and, 513
  in titles, 529
  tone and, 532
  unity and, 543

purpose (*continued*)
　usability testing and, 545
　for visuals, 553
　in Web design, 561–62
　word choice and, 6
　in writing process, xvi

qualifications summary
　in proposals, 433
　in résumés, 483
quasi- / pseudo-, 434
question marks, 83, 436–37
questionnaires, 437–43
　advantages and disadvantages of, 437–38
　biased evidence and, 313
　form design for, 218–23
　preparing questions, 438–42
　for research, 459
　sample of, 439–41
　selecting recipients for, 438
　Web resources for, 438
　writer's checklist for, 442–43
questions
　adverbs asking, 20
　for forms, 221
　frequently asked (FAQs), 183–85
　in inquiry letters, 254
　for interviewing for information, 271
　job interview, 272–73
　pronouns asking, 406
　punctuation for indirect, 436
　punctuation for polite requests, 436
　for research, 460
　rhetorical, 393, 490
　as transitions, 537, 539
　verb phrases and, 383
　word order for, 502
　for writing trade journal article, 534
quid pro quo, 443
quiet person, in meetings, 327
quotation marks, 443–45
　for direct quotations, 443, 445
　punctuation with, 444–45, 498
　for titles of works, 284, 444, 531
　for words and phrases, 444
quotations, 445–47
　block, 132, 447
　in blog writing, 51
　capitalization of first word of, 60, 445
　colons with, 76, 444, 445
　commas with, 80, 83, 375, 444, 445
　direct, 443, 445
　documenting, 130, 445
　ellipses for omissions, 161, 446
　ethics in writing for, 443, 445
　with exclamation marks, 181
　incorporating into text, 446–47
　indirect, 445
　inserting words into, 52, 446
　in note-taking, 347, 348, 459
　opening with, 279, 393
　paraphrasing, 372, 383–84
　parenthetical citations following, 132, 139, 145
　periods in, 375, 444
　plagiarism and, 383–84, 445
　question marks with, 437
　within quotations, 443
　semicolons with, 444, 498
　using *sic*, 52, 446

racial groups
　bias toward, 47
　capitalization of, 60
radio programs
　documenting, 136, 148
　titles of, 444, 531
raise / rise, 448
rambling person, in meetings, 327
rambling sentences, 503–4
Random House Webster's College Dictionary, 124
re, 448
reactor, in listening, 308
readers, 448. *See also* audience
Readers' Guide to Periodical Literature, 461–62
really, 448
reason is [because], 448
reasoning errors. *See* logic errors
reciprocal pronouns, 407
recommendation letters, 449
recommendation reports, 186, 283
recommendations
　in abstracts, 7, 8
　in conclusions, 93, 94

in cover letters, 419
in feasibility reports, 188
lists for, 299
opening with summary of, 278
in reports, 199, 215–16, 455
in test reports, 527
redundancy, 90, 92. *See also* repetition
   eliminating, 448
   one usage as, 357
   with prepositions, 256, 365–66, 391, 543
refer / allude / elude, 30
reference letters, 449
   ethics in writing for, 449
   sample of, 450
reference works, 462–63
references
   avoiding *above*, 6
   in résumés, 485
references list (APA), 48, 130. *See also* APA documentation; documenting sources
   documentation models, 132–36
   in formal reports, 199
   sample page, 138
reflexive pronouns, 406–7
refusal letters, 449–52
   bad-news patterns for, 106–7
   employment, 9–11
   samples of, 10, 451
regarding / with regard to, 453
regardless, 453
relative adjectives, 91
relative pronouns, 406
   in adjective clauses, 171
   for combining ideas, 517
   for connecting clauses, 71
   in sentence fragments, 504
   subject-verb agreement with, 25, 26
   wordiness and, 91
religious groups
   bias toward, 47
   capitalization of, 60
repetition, 453–54
   awkwardness and, 44
   dashes before, 114
   for emphasis, 168, 453
   for transitions, 453, 537
   wordiness from, 90, 92

report elements. *See also* layout and design
   abstracts, 7–9, 197, 203
   appendixes, 34–35, 199
   back matter, 199–200
   bibliographies, 48–49, 199
   body, 198–99, 210–14, 454
   conclusions, 93–95, 199, 215–16, 455
   cover letters, 111, 196, 201
   executive summaries, 181, 198, 205–6
   explanatory notes, 199
   foreword, 198
   front matter, 196–98
   glossaries, 200, 233
   headers and footers, 241–42, 298
   headings, 242–44, 298
   index, 200
   introductions, 198, 207–9, 276–81, 454
   list of abbreviations and symbols, 198
   list of figures and tables, 197
   page numbers, 197
   preface, 198
   problem statement, 277
   recommendations, 199, 215–16, 455
   references / works-cited list, 199, 217
   routine openings, 276, 277
   sample report, 200–217
   table of contents, 197, 204, 218, 519
   text, 198
   title page, 197, 202
   titles, 529–30, 531
   visuals, 299–300, 379
reports, 454–55. *See also* report elements
   documenting, 136, 142, 149
   feasibility reports, 185–88
   formal reports, 195–217
   investigative reports, 281–83
   laboratory reports, 293
   as memos, 329
   methods of development for, 361
   point of view in, 386
   progress and activity reports, 402–4
   test reports, 527

reports (*continued*)
  tone for, 533
  trip reports, 539–40
  trouble reports, 541–42
  types of, 454
repurposing, 455–56
  boilerplate material, 102
  for context, 455
  ethics in writing for, 456
  for medium, 456
requests for proposals (RFP), 413–14, 420, 456–58
  ethics in writing for, 458
  format for, 413–14, 418
  government specifications in, 512
  Web resources for, 458
research, 459–67. *See also* Internet research strategies; library research strategies
  for collaborative writing, 73
  direct observation, 459, 460
  ethical considerations for, 460
  evaluating sources, 465–67
  Internet research strategies, 463–65
  interviews, 270–72
  for job interviews, 273
  laboratory reports, 293
  library research strategies, 460–63
  literature reviews, 310–12, 534
  for newsletters, 341–42, 345
  note-taking for, 347–48
  for presentations, 392–93
  primary, 459–60
  questionnaires for, 437–43
  reference works for, 462–63
  secondary, 460
  for specifications, 510
  for trade journal articles, 534–35
  in writing process, xviii–xix, xxiii
research methods, in abstracts, 7, 8
research plan, in proposals, 433
research proposals, 421–33
resignation letters, 467–69
  samples of, 468, 469
respectful tone, in correspondence, 105
respective / respectively, 469–70
responder, in listening, 308
response letters. *See* inquiries and responses

responses
  to e-mail, 163, 164
  in form design, 221
  indicating, in manuals, 318
  to inquiries, 255
  in job interviews, 275
responsibility
  ethics in writing and, 178, 179, 559
  statement of, in proposals, 421, 430
restrictive and nonrestrictive elements, 470–71. *See also* parenthetical elements
résumés, 471–88
  analyzing background for, 481
  in appendixes, 35
  application letters and, 35, 471
  electronic, 480, 486–88
  e-mail attached, 487–88
  ethics in writing for, 472
  organization for, 481–85
  posting online, 289
  for returning job seekers, 485–86
  salary negotiations and, 485, 491
  samples of, 472–81
  scannable and plain-text, 487
  Web resources for, 481
reviewing. *See also* revision
  for biased language, 46
  in collaborative writing, 73–74
  e-mail, 162
  literature reviews, 310–12
  manuals, 319
revision, 488–89. *See also* reviewing
  achieving conciseness, 91–92
  for coherence, 72
  in collaborative writing, 74
  proofreading and, 411
  for sentence construction, 489, 507
  for transitions, 488, 539
  writer's checklist for, 488–89
  in writing process, xxi–xxii, xxiii–xxiv
RFP. *See* requests for proposals (RFP)
rhetorical questions, 490
  opening with, 393, 490
  as titles, 490, 530
rise / raise, 448

routine proposals, 414
rules
  in page design, 300
  in tables, 520
run-on sentences, 376, 490

-*s* endings
  for noun plurals, 350, 387–88
  for plural numbers, 353
  for verbs, 351, 551
salary negotiations, 491–93
  benefits and job perks, 492–93
  job interviews and, 275, 491–92
  résumés and, 485, 491
  Web resources for, 492
sales brochures, 54–57
sales proposals, 420–21
  conclusions for, 94
  ethics in writing for, 420
  sample of, 422–32
  structure for, 420–21
  using repurposed content for, 455
  Web resources for, 421
salutations
  capitalization in, 60
  colons in, 75
  commas in, 82
  for e-mail, 164–65
  first name usage, 165, 265, 304
  for international correspondence, 165, 265
  for letters, 304–5
  for unknown gender, 304
sans serif typeface, 297, 563, 572
scarcely, 153
schedules
  chronological order for, 68
  for collaborative writing, 73
  delivery, in proposals, 421, 427
  delivery, in requests for proposals, 457
  for research proposals, 433
schematic diagrams, 155, 158
scholarly abbreviations, 4–5
scientific names, 61
scope, 493
  in abstracts, 7
  in environmental impact statements, 174–75
  in introduction, 188, 277–78, 280, 535
  in literature reviews, 311–12
  in preparation stage, 389–90
  of test reports, 527
  in titles, 529
  for white papers, 566
  in writing process, xviii
Scope of Work section, in requests for proposals, 457
screen names, 256
screening (shading)
  in graphs, 236, 238
  in maps, 319
  in page design, 299
sculpture titles, 284, 531
search engines, 463–65
second pages. *See* continuing pages
second person, 376
  antecedents and, 409
  point of view, 385
secondary research, 460
sections. *See* chapters and sections
selecting the medium, 494–97
  context and, 99
  factors involved in, 494
  for repurposed content, 456
  scope and, 493
  in writing process, xviii
-self / *-selves*, 406
semi- / *bi-*, 46
semicolons, 497–98
  for comma splices, 77
  in compound sentences, 500
  with conjunctive adverbs, 20, 82
  for items in series, 80, 82
  for linking independent clauses, 497
  in long sentences, 498
  with quotation marks, 444, 498
  for run-on sentences, 376, 490
  with strong connectives, 497
sentence construction, 498–502. *See also* sentence types
  for abstracts, 9
  avoiding loaded sentences, 502
  awkwardness and, 44
  capitalization of first word, 60
  clauses, 70–71
  comma splices, 77

sentence construction (*continued*)
  emphasis and, 167–68
  ESL tips for, 499, 502
  in formal style, 513
  garbled sentences, 225
  for global communication, 269
  guidelines for effective, 501–2
  intention in, 501
  mixed constructions, 334
  period faults, 376
  predicates, 500
  revising, 489, 507
  run-on sentences, 376, 490
  sentence faults, 503–4
  sentence fragments, 504–5
  sentence structures, 500–501
  sentence variety, 505–7
  stylistic use in, 501
  subject-verb-object pattern, 499, 501–2, 518
  subjects, 499–500
sentence endings
  ESL tips for, 502
  with lists, 310
  within parentheses, 373, 375
  prepositions, 390–91
  punctuation for, 180–81, 375, 436
sentence faults, 503–4
  faulty logic, 504
  faulty subordination, 503
  missing subjects, 503
  missing verbs, 504
  rambling sentences, 503–4
  run-on sentences, 490
sentence fragments, 70, 503, 504–5
  with participles, 548
  period faults, 376
  rewriting, 505
sentence length, and emphasis, 167
sentence openings. *See also*
    introductory elements
  avoiding *also*, 31
  commas with, 79–80
  conjunctions as, 95
  dangling modifiers, 113
  ESL tips for, 502
  with numbers, 352
sentence outlines, 363
sentence types
  complex, 501
  compound, 70, 500
  compound-complex, 501
  declarative, 501
  emphasis and, 168
  exclamatory, 501
  imperative, 501
  interrogative, 501
  loose, 501, 506–7
  minor, 501
  periodic, 501, 507
  simple, 70, 500
sentence variety, 505–7. *See also*
    sentence types
  for formal writing, 513–14
  for garbled sentences, 225
  length of sentence, 505–6
  loose and periodic sentences for, 506–7
  subordination for, 505, 506, 507, 516
  using parenthetical elements, 506
  word order, 506
sentence word order. *See* inverted
    sentences; word order
sentences, transitions between, 538
sequential method of development, 68, 332, 507
  in flowcharts, 192
  in instructions, 259–60
  lists for, 299
  in narratives, 339
series of items
  commas with, 80–81, 83
  etc. with, 177
  parallel structure for, 372
  question marks with, 436
  semicolons with, 80, 82, 498
serif typeface, 297
service, 507
service manuals, 316
set / sit, 508
sexist language, 46–47. *See also* biased
    language
  female / male usage, 188
  imperative mood for, 241
  Ms. / Miss / Mrs. usage, 304, 338
  occupational descriptions, 47
  pronoun usage, 27–28, 47, 226, 241, 409
  unparallel terms, 47

shall / will, 508, 525
she / he, 241
   gender agreement with, 27, 226
   sexist language and, 27, 47, 226, 241, 409
she / her / hers, 376, 406
ship names, 284
short story titles, 444
sic, 5, 52, 446
sight / site / cite, 68
signature blocks, e-mail, 165
signatures
   for letters, 305
   for memos, 329
similes, 191
simple sentences, 70, 168, 500
since / as / because, 41
single-source publishing, 455n
sit / set, 508
site / cite / sight, 68
site preparation, in proposals, 421, 428
skills and abilities section, of résumé, 484–85
slang
   as nonstandard English, 173–74
   quotation marks for, 444
slashes, 508–9
slides, for presentations, 396–97
so / so that / such, 509
Society for Technical Communication, 533
software, for presentations. *See also* computer technology; Digital Tips
solicited proposals, 413–14
some / somewhat, 509
some time / sometime / sometimes, 510
sort / kind / type, 13–14, 29
sort of, 292
source lines, for visuals, 239, 520, 556
sources, citing. *See* documenting sources
sources of information. *See also* research
   evaluating, 465–67
   in proposals, 412
   for research, 459–60
   in writing process, xix
spacing
   for brochures, 57
   for forms, 223
   for letters, 301
   in Web design, 563
   white space, 299
spam, e-mail, 163
spatial method of development, 332, 510
   sample of, 511
spatial relationships, visuals depicting, 554
special-purpose manuals, 316
special-purpose proposals, 414, 415–17
specific-to-general method of development, 228, 331
specifications, 510–12
   decimal number heads for, 243–44
   government, 512
   industrial, 510–12
   introductions for, 280–81
speeches, documenting, 150. *See also* presentations
spelled-out words, hyphens for, 246
spelling, 512
   dictionaries for, 123
   proofreading for, 411, 512
   spell checkers, 489, 512
split infinitives, 547–48
squinting modifiers, 336
stacked (jammed) modifiers, 335
   as gobbledygook, 233
   order for, 15
   series of nouns as, 16
staffing, in proposals, 421
standard English, 173, 513, 545
standardized symbols, 192–94
statistics. *See* numerical data
status reports, 402
Stedman's Medical Dictionary, 125
steps. *See* sequential method of development
stereotypes. *See* biased language; sexist language
strata / stratum, 513
stubs, in tables, 520
style, 513–16. *See also* technical writing style; writing style
   formal writing, 513–14
   informal writing, 514–15
   for trade journal articles, 536–37
   Web resources for, 515
   writer's checklist for, 515

style manuals, 3, 131, 153, 515, 545
Stylebook and Briefing on Media Law (Associated Press), 515
stylistic use, in sentence construction, 501
subheadings, 284. *See also* headings
subject (of document), introducing, 280
subject (of sentence), 499–500. *See also* subject-verb agreement
  active voice for highlighting, 558
  compound, 500
  ESL tips for, 499, 502
  gerunds as, 546
  nouns as, 349
  position of, 499
  problems with missing, 503, 504
subject dictionaries, 125
subject directories, 465
subject lines
  capitalization in, 61, 110
  for e-mail and memos, 110, 163, 530
  for test reports, 527
  for trip reports, 539
  for trouble reports, 541
subject-verb agreement, 24–27, 499, 551. *See also* agreement
  abstract nouns and, 26
  with collective nouns, 26, 350
  compound subjects and, 26–27
  indefinite pronouns and, 25, 409
  with intervening words and phrases, 24, 25
  inverted word order and, 25
  with measurement subjects, 25
  relative pronouns and, 25, 26
  special problem words, 26
  subjective complements and, 25
  titles of works and, 26
  with word *number*, 25–26
subject-verb-object pattern, 499, 501–2, 518
subjective case, 63–64, 407–8
subjective complements, 89
  adjectives as, 15
  gerunds as, 546
  nouns as, 350
  pronouns as, 63–64
  subject-verb agreement with, 25
subjunctive mood, 337–38

subordinate clauses, 70. *See also* dependent clauses
subordinating conjunctions
  for connecting clauses, 71, 96
  list of common, 96
  in sentence fragments, 504
  for subordination, 516
subordination, 516–17
  in abstracts, 9
  avoiding overlapping, 517
  awkwardness and, 44
  clarity and, 68
  for comma splices, 77
  for conciseness, 91
  effective use of, 70
  faulty, 503
  in formal writing, 514
  for garbled sentences, 225
  for loaded sentences, 502
  for run-on sentences, 490
  for sentence variety, 505, 506, 507, 516
  vs. using parentheses, 373
subscription newsletters, 344–45
substantives, 500
such / so / so that, 509
such as, in sentence fragments, 505
suffixes, 246, 517
summaries
  career, 483
  in conclusions, 93, 94
  executive, 181
  meeting minutes, 332
  for note-taking, 445
  opening with, 278
  for transitions, 537
summarizing statements, dashes before, 114
summary list, in manuals, 316
superlative form
  of adjectives, 14
  of adverbs, 20–21
supplement / augment, 43
surveys. *See also* questionnaires
  in investigative reports, 283
  online, 438
  questionnaires, 437–43
  for research, 459
  for usability testing, 545

sweeping generalizations, 312
symbols
    in flowcharts, 192–94
    functions of, 555
    for global communication, 230, 231
    for graphs, 239, 240
    in instructions, 263
    keys for, 556
    list of, in reports, 198
    for manuals, 318, 319
    proofreader's marks, 410
    in tables, 519
synonyms, 518
    coordinated, 91
    for defining terms, 116
    in dictionaries, 123, 518
    in thesaurus, 529
    word choice and, 569
    vs. word repetition, 453–54
syntax, 518

tables, 519–21. *See also* visuals
    for comparisons, 85, 87
    elements of, 519–21
    functions of, 554
    vs. graphs, 235, 239, 519
    informal, 521
    list of, 197, 557
    numbering, 197, 355, 519
    screening (shading) in, 299
    for trade journal articles, 536
tables of contents, 519
    decimal headings in, 244
    for FAQs, 184
    in formal reports, 196, 197, 204, 218
    parallel structure for, 371
    placement of, 519
    in proposals, 419
    for white papers, 566
team writing. *See* collaborative writing
Technical Communication, 533
technical manuals. *See* manuals
technical terminology
    affectation and, 22
    audience and, 522
    defining for readers, 175, 281
    in glossaries, 233
    in instructions, 259, 263
    jargon, 285
    in manuals, 316, 317
    in proposals, 413
    in white papers, 566
    word choice and, 569
technical writing style, 513, 521–22.
    See also IEEE documentation;
    style; writing style
    active voice for, 557–58
    ethics in writing and, 178–79, 521
    exposition for, 182
teleconferences, 495–96
telegraphic style, 522–23
    caution with, 91, 104
    eliminating articles, 41
    in instructions, 259
telephone calls, 495–96
    interviewing for information by, 272
television programs
    documenting, 136, 141, 148
    titles of, 283, 444, 531
templates
    for letters, 301
    for style and format, 200
"Ten Tips on Writing White Papers"
    (Barefoot), 565
tenant / tenet, 523
tense, 523–26, 551
    avoiding shifts in, 526, 551
    ESL tips for, 172, 526
    helping verbs and, 550
    for infinitives, 547
    in narratives, 339
    present perfect, 172
    present progressive, 172
territorial person, in meetings, 327
test reports, 293, 459, 527
    sample of, 528
testing. *See* usability testing
text, in formal reports, 198
than / as, and pronoun case, 64, 65
that
    dual function of, 406
    with indirect quotations, 443, 445
    one of those . . . , 357–58
    for restrictive clauses, 471
that / those
    as demonstrative adjectives, 13, 406
    number agreement with, 13, 25, 26, 28

that / where, 564
that / which / who, 26, 527–29
the, 13, 41
there, as expletive, 182, 499, 529
there / their / they're, 529
thesaurus, 518, 529
they / them / their / theirs, 376, 406
third person, 376
    antecedents and, 409
    in narratives, 339
    point of view, 385
this / these
    as demonstrative adjectives, 13, 406
    number agreement with, 13, 28–29
    pronoun references with, 405
thought-provoking statement, in conclusions, 95
thumbnails, in page design, 300
time
    in global communication, 269, 270
    numbers for, 354
time periods, capitalization of, 61
title page
    in formal reports, 196–97, 202
    in proposals, 419
titles, 529–32. *See also* names and titles; titles of works and documents
titles of works and documents, 529–32
    articles in, 41, 61, 530
    capitalization of, 41, 61, 95, 391, 530
    colons in, 75
    coordinating conjunctions in, 61, 95, 530
    formatting, 530–32
    in headers or footers, 241, 298
    importance of, 529
    italics for, 283–84, 531
    memos, e-mails, and Internet postings, 530
    newsletter articles, 343
    prepositions in, 61, 391, 530
    question marks in, 436
    quotation marks for, 284, 444, 531
    reports and long documents, 197, 202, 529–30
    as rhetorical question, 490, 530
    setting tone with, 529, 533

    special cases, 532
    subject-verb agreement with, 26
to
    in prepositional phrases, 382
    as sign of infinitive, 382, 547–48, 550
to / too / two, 532
tone, 532–33
    in abstracts, 7
    for adjustment letters, 16
    building goodwill with, 104–5
    for complaint letters, 87
    context and, 99
    for correspondence, 104–5
    for newsletters, 341, 345
    for refusal letters, 451
    style and, 513
    in technical writing, 521
    of titles, 529, 533
    for trouble reports, 542
    "you" viewpoint, 575
topic outlines, 363, 369
topic sentences, 38, 368
    in abstracts, 8
    paragraph unity and, 370
    placement of, 368
topics
    in headers or footers, 241, 298
    headings for highlighting, 242, 244
trade journal articles, 533–37. *See also* articles in periodicals
    abstracts for, 7
    gathering data, 534–35
    literature reviews in, 311
    obtaining publication clearance for, 537
    online databases and indexes for, 461–62
    organizing draft, 535
    planning for, 534
    preparing manuscript, 536–37
    sections and elements of, 535–36
    style format for, 536–37
trade journals, job listings in, 290
train names, 284
training manuals, 315–16
training requirements, in proposals, 421, 429

transitions, 537–39
   between paragraphs, 538–39
   between sentences, 538
   clarity and, 68
   coherence and, 72
   commas with, 79
   conjunctions as, 95
   eliminating in telegraphic style, 522
   enumeration for, 370, 537, 538
   for garbled sentences, 225
   methods of, 537–38
   in narratives, 339
   outlining and, 363
   for paragraph coherence, 370
   in presentations, 395
   in process explanations, 401
   for quotations, 447
   repeating keywords, 453, 537, 538
   revising for, 488, 539
   semicolons with, 497
   unity and, 543
   for Web writing, 571
   words and phrases for, 537–38
transitive verbs, 293, 356, 549
transmittals, 111. *See also* cover letters
transparencies, overhead, 396
trip reports, 539–40
   chronological order for, 68
   sample of, 540
   scope of, 493
trouble reports, 541–42
   chronological order for, 68, 331
   ethics in writing for, 542
   samples of, 69, 541–42
try to, 542
tutorials, 315
Twain, Mark, 568
two / to / too, 532
type / sort / kind, 13–14, 29
typographical errors, checking for, 489
typography, 296–98. *See also* highlighting devices
   for brochures, 57
   for electronic résumés, 487
   for e-mail, 163–64
   for emphasis, 297–98
   for presentations, 297, 398
   typeface and type size, 296–97
   in Web design, 563, 572

unabridged dictionaries, 124
underlining
   for emphasis, 169
   for italics, 283
uninterested / disinterested, 126
unique / perfect / equal, 176–77
unity, 543
   clarity and, 68
   in paragraphs, 370
   revising for, 488
   transitions and, 539, 543
unpublished data, documenting, 136
unsolicited proposals, 414
up, 391, 543
upon / on / onto, 357
uppercase letters. *See* all-capital letters
us / our / ours, 376, 406
usability engineering, 543n
usability testing, 543–45
   for forms, 220
   for global graphics, 232
   for instructions, 263
   main goals of, 544
   for manuals, 318
   methods for, 544–45
   for Web design, 562, 563
usage, 545
   clarity and, 70
   dictionaries on, 123
   grammar and, 234–35
   Web resources for, 545
user-centered design (UCD), 543n
user manuals, 315
user testing, 544
utilize, 545

vague words, 70, 546
   clichés as, 71
   as gobbledygook, 233
   impreciseness of, 568
vendor description, in proposals, 431, 457
vendor qualifications, in requests for proposals, 457–58
verb tense. *See* tense
verbals, 546–48
   ESL tips for, 171
   gerunds, 546–47
   infinitives, 547–48

verbals (*continued*)
  as nonfinite verbs, 550
  participles, 548
  in sentence fragments, 504
verbs, 549–52. *See also* mood; subject-verb agreement; tense; voice
  adding *up* to, 543
  adjectival forms, 15
  agreement with subject, 24–27, 550, 551
  conjugation of, 552
  converting into adjectives, 505
  ESL tips for, 502
  finite and nonfinite, 550
  forms of, 550
  intransitive, 293, 549–50
  linking, 549–50
  for manuals, 317–18
  mood of, 336–38
  phrases as, 380, 383
  plurals of, 351–52
  as predicates, 500
  with prepositions, 390
  problems with missing, 504
  properties of, 550–52
  tense of, 523–26
  transitive, 293, 356, 549
  types of, 549–50
  using simple and direct, 317–18
  voice of, 557–60
  weak, 346
very, 552
via, 552
videoconferencing, 496
videos, documenting, 135, 141, 148
viruses
  in e-mail, 164
  in instant messaging, 258
visual logic, in page design, 298
visuals, 552–57
  in appendixes, 35
  boxes for, 300
  in brochures, 56
  for description, 122
  in dictionaries, 123
  documenting and copyright of, 136, 149, 153, 155, 380, 395, 520, 556
  drawings, 154–58, 554
  ethical considerations, 153, 155, 236, 380, 395, 553, 556
  in executive summaries, 181
  flowcharts, 192–94, 555
  for global communication, 230–32, 260
  graphs, 235–40, 554–55
  guidelines for selecting, 554–55
  icons, 299–300
  indexing, 251
  for instructions, 258, 260
  integrating with text, 553–56
  labels and captions for, 122, 157, 300, 379
  layout and design for, 299–300
  list of figures and tables, 197, 419, 557
  for manuals, 316, 318
  maps, 319–20, 554
  for newsletters, 155, 240, 299, 343, 345
  numbering, 197, 355, 379, 519
  organizational charts, 362, 555
  outlining and, 154, 361–62, 364
  photographs, 377–80, 554
  for presentations, 395–98
  references to, 6
  in spatial descriptions, 510
  symbols and icons, 555
  tables, 519–21, 554
  in technical writing, 522
  in trade journal articles, 536, 537
  types of, 553
  for Web design, 563, 571
  for white papers, 566
  writer's checklist for, 556–57
voice, 552, 557–60. *See also* active voice; passive voice
  active voice, 558–59
  avoiding shifts in, 551
  ESL tips for, 560
  helping verbs and, 550
  passive voice, 559–60
  in presentations, 400, 401
voice mail, 496

wait for / wait on, 561
warnings
  exclamation marks for, 181

in instructions, 260–63
italics for, 298
in manuals, 319
sample of, 262
we
  in technical writing, 385–86
  us / our / ours, 376, 406
weak verbs, 346. *See also*
  nominalizations; voice
Web addresses, 246, 509
Web content-management systems, 97
Web design, 561–63. *See also* writing
  for the Web
  audience and purpose in, 561–62
  forms, 218–23
  hyperlinks in, 571–72
  icons, 299–300
  for people with disabilities, 562
  posting documents, 573
  privacy statements, 573
  for résumés, 486
  testing, 562, 563
  typography for, 297, 572
  using repurposed content for, 456
  visuals for, 379
  Web resources for, 561
  writer's checklist for, 563
  writing for the Web, 570–74
Web resources
  for abbreviations and acronyms, 5
  for affectation, 22
  for argument, 314
  for avoiding plagiarism, 384
  blogs, 49–52
  for brochures, 57
  for buzzwords, 58
  for conjugation of verbs, 552
  for content management, 98
  dictionaries, 124
  for environmental impact statements, 176
  for ESL skills, 172
  for ethical codes, 179
  for evaluating sources, 467
  for feasibility reports, 188
  for global communication, 230, 269
  for grammar, 234
  for job searches, 289
  for layout and design, 300

for maps, 320
for memos, 329
occupational resources, 291
online surveys, 438
for prepositional idioms, 248
for presentation slides, 396
for punctuation, 434
for requests for proposals, 458
for résumés, 481
for salary information, 492
for sales proposals, 421
for sending e-mail attachments, 164
for style manuals, 153, 515, 545
for Web design, 561
for white papers, 567
for word choice, 569
Web sites. *See also* Internet
  APA documentation of, 134
  copyrighted materials from, 101, 102
  for current information sources, 460
  evaluating, 467
  FAQs on, 183–85
  functions of, 496–97
  IEEE documentation of, 140–41
  indexes for, 251
  internal and external, 562
  italics for titles, 531
  MLA documentation of, 147
  posting résumés on, 289, 486–87
  white papers on, 565
Web subject directories, 465
Webster's Third New International Dictionary, Unabridged, 124
well / good, 234
when / where / that, 564
when and if, 564
where / that, 564
where . . . at, 564
whether, 564
whether . . . or, 96
which
  dual function of, 406
  for nonrestrictive clauses, 471, 527
  number agreement with, 26
which / who / that, 26, 527–29
while, 564–65
white papers, 565–67
  getting clearances for, 567

white papers (*continued*)
  interesting-detail openings for, 278
  layout and design for, 566
  major sections for, 566–67
  persuasion in, 565
  preparation for, 565–66
  Web resources for, 567
white space, in page design, 299, 301.
  See also spacing
whiteboard software, 324
whiteboards, 396
who, one of those . . . , 357–58
who / that / which, 26, 527–29
who / whom, 25, 26, 567
who / whom / what / which, 406
who / whom / which / that, 406
whole-by-whole method of
  comparison, 85, 86
who's / whose / of which, 568
widows, 299
Wikipedia, 102, 384, 462, 466
wikis, 497
  for collaborative writing, 72
  documenting, 135, 148
will / shall, 508, 525
WinZip, 164
-wise, 568
with regard to / regarding, 453
word choice, 568–69
  abstract / concrete words, 6–7
  in adjustment letters, 18
  affectation and, 22
  ambiguity and, 32–33, 70
  buzzwords, 57–58
  clarity and, 70
  clichés, 71
  colloquialisms, 173
  connotation / denotation and, 96
  deceptive language, 178, 377, 386
  dictionaries for, 123–25, 569
  ethics in writing and, 178
  figures of speech, 189–91
  for global communication, 265–69
  gobbledygook, 233
  idioms, 248
  jargon, 285
  localisms and slang, 173–74
  malapropisms, 315

outdated words, 22
  revising for, 489
  standard vs. nonstandard English,
    173–74
  thesaurus for, 529
  tone and, 533
  usage conventions and, 234–35,
    545
  vague words, 546
  Web resources for, 569
word divisions
  in dictionaries, 123
  hyphens in, 246–47
word meanings. *See also* definitions
  capitalization and, 59
  connotation / denotation in, 96
  dictionaries for, 123–25
word order. *See also* sentence
  construction
  emphasis and, 498–99, 502
  for sentence variety, 506
  subject-verb-object pattern, 499,
    501–2, 518
  syntax, 518
word-processing software. *See*
  computer technology; Digital Tips
wordiness. *See also* conciseness
  affectation and, 22
  causes of, 90–91
  of passive voice, 92, 558–59
  rambling sentences, 503–4
  revising for, 489
  using *that*, 527
  wordy phrases, 41, 92, 183, 249–50,
    341, 381
words used as words
  italics for, 284
  plurals of, 34
work plan, in proposals, 421
work schedules, in proposals, 433
works-cited list (MLA), 48, 131. *See
  also* documenting sources; MLA
  documentation
  documentation models for, 145–50
  in formal reports, 199
  sample page, 152, 217
Workscope Description, in requests for
  proposals, 457

Writer's Checklist
  achieving conciseness, 91–92
  communicating globally, 229–30
  correspondence and accuracy, 110–11
  creating and integrating visuals, 556–57
  creating and using drawings, 157
  creating and using maps, 319
  creating graphs, 239–40
  designing a brochure, 57
  designing a questionnaire, 442–43
  designing Web pages, 563
  developing an effective style, 515
  developing an FAQ, 185
  eliminating awkwardness, 44
  evaluating print and online sources, 466–67
  indexing, 252–53
  instant messaging privacy and security, 258
  interviewing successfully, 271
  managing your e-mail and reducing overload, 166–67
  observing workplace netiquette, 163
  planning and conducting meetings, 328
  preparing for and delivering a presentation, 400–401
  preparing manuals, 318–19
  preparing minutes of meetings, 333–34
  preparing to write, 389–90
  proofreading in stages, 411
  revising your draft, 488–89
  taking notes, 348
  using abbreviations, 3
  using headings, 244
  using hyphens to divide words, 246–47
  using lists, 310
  using search engines and keywords, 464
  using tone to build goodwill, 105
  using visuals in presentations, 398
  writing a rough draft, 570
  writing collaboratively, 74–75
  writing ethically, 179
  writing executive summaries, 181
  writing instructions, 263
  writing international correspondence, 269–70
  writing job descriptions, 286
  writing newsletter articles, 343
  writing persuasive proposals, 417
writing a draft, 569–70. *See also* draft writing
writing block, in forms, 223
writing for the Web, 570–74. *See also* Web design
  documenting sources, 573
  ethics in writing for, 570–71, 573
  fonts, 572
  for global audiences, 572
  graphics, 571
  headings, 571
  hyperlinks, 571–72
  keywords, 571
  line length, 572
  lists, 571
  organizational blogs, 51–52
  posting documents, 573
  privacy statements, 573
writing line, in forms, 223
writing process
  brainstorming, 53–54
  checklist for, xxiii–xxiv
  for content management, 97
  five steps for success in, xv–xxii
  writing a draft, 569–70
writing style
  for abstracts, 9
  for correspondence, 103
  for e-mail, 163, 173
  for environmental impact statements, 175–76
  formal, 513–14
  for global communication, 265–70
  informal, 514–15
  for instructions, 259–60
  for newsletters, 341
  plain-language laws and, 176
  standard vs. nonstandard English, 173–74
  technical, 521–22

writing style (*continued*)
   using contractions, 101
   for Web writing, 571
WWW Virtual Library, 465

Yahoo!, 464
years. *See also* dates
   format for writing, 81–82, 115
   plurals of, 34

yes, commas with, 78
"you" viewpoint, 42, 575–76
   in correspondence, 104–5
   in FAQs, 185
   in persuasive writing, 377
   in white papers, 565
you / your / yours, 376, 406
your / you're, 576

Commonly Misused Words and Phrases

A
a/an   1
a lot   2
above   6
absolutely   6
accept/except   9
accuracy/precision   11
activate/actuate   12
ad hoc   13
adapt/adept/adopt   13
affect/effect   23
affinity   26
all ready/already   32
all right   32
all together/altogether   33
allude/elude/refer   33
allusion/illusion   34
almost/most   34
also   34
amount/number   36
and/or   36
as/because/since   44
as much as/more than   45
as such   45
as well as   45
augment/supplement   46
average/median/mean   46
awhile/a while   46

B
bad/badly   48
be sure to   48
beside/besides   48
between/among   48
between you and me   49
bi-/semi-   49
biannual/biennial   49
both . . . and   52

C
can/may   59
cannot   59
center on   68
cite/sight/site   68
compare/contrast   84
complement/compliment   87

compose/constitute/comprise   90
consensus   97
continual/continuous   99
credible/creditable   114
criteria/criterion   116
critique   116

D
data   119
defective/deficient   120
definite/definitive   121
despite/in spite of   125
diagnosis/prognosis   127
differ from/differ with   130
different from/different than   130
discreet/discrete   130
disinterested/uninterested   130
due to/because of   162

E
each   163
economic/economical   163
e.g./i.e.   163
equal/unique/perfect   177
etc.   178
everybody/everyone   181
explicit/implicit   185

F
fact   187
female/male   190
few/a few   191
fewer/less   191
figuratively/literally   191
fine   193
first/firstly   193
flammable/inflammable/
   nonflammable   194
forceful/forcible   196
foreword/forward   197
former/latter   220
fortuitous/fortunate   226

G
good/well   238

H
he/she 246

I
illegal/illicit 253
imply/infer 253
in/into 253
in order to 253
in terms of 254
indiscreet/indiscrete 257
individual 257
inside/inside of 259
insoluble/insolvable 260
insure/ensure/assure 266
interface 267
its/it's 287

K
kind of/sort of 294
know-how 294

L
lay/lie 295
lend/loan 303
like/as 304
loose/lose 311

M
maybe/may be 319
media/medium 319
Ms./Miss/Mrs. 336
mutual/common 337

N
nature 340
needless to say 340
none 345
nor/or 345

O
OK/okay 356
on/onto/upon 356
one 356
one of those . . . who 356
only 357
outside [of] 364
over [with] 364

P
party 372
per 372
percent/percentage 373
phenomenon/phenomena 376
principal/principle 401
pseudo-/quasi- 433

Q
quid pro quo 442

R
raise/rise 447
re 447
really 449
reason is [because] 449
regarding/with regard to 453
regardless 453
respective/respectively 472

S
service 511
set/sit 511
shall/will 512
so/so that/such 513
some/somewhat 513
some time/sometime/sometimes 514
strata/stratum 517

T
tenant/tenet 527
that/which/who 531
there/their/they're 533
to/too/two 536
try to 546

U
up 547
utilize 549

V
very 558
via 558

W
wait for/wait on 567
when/where/that 573
whether 574
while 574
who/whom 577
who's/whose/of which 577
-wise 578

Y
your/you're 585

Model Documents and Figures by Topic

Use the following list as a quick reference for finding selected samples of technical writing and visuals by topic. See also the complete Contents by Topic on the inside front cover of this book. For additional model documents, see the companion Web site at *bedfordstmartins.com/alredtech*.

Technical Writing Documents and Elements

- A–2 Informative Abstract (for an Article) 8
- B–3 Brochure (Cover Panel) 55
- B–4 Brochure (Inside Panels) 56
- F–1 Feasibility Report 186
- F–5 Formal Report 201
- I–5 Illustrated Instructions 261
- I–6 Warning in a Set of Instructions 262
- I–10 Investigative Report 282
- J–1 Job Description 287
- L–1 Laboratory Report 294
- L–8 Literature Review 311
- N–2 Newsletter Article 342
- N–3 Company Newsletter (Front Page) 344
- P–4 Progress Report 403
- P–5 Activity Report 404
- P–7 Special Purpose Internal Proposal 415
- P–8 Sales Proposal 422
- Q–1 Questionnaire 439
- T–4 Test Report 528
- T–5 Trip Report Sent as E-mail (with Attachment) 540
- T–6 Trouble Report (Using Printed Memo) 541

Design and Visuals

- D–14 Schematic Diagram 158
- F–2 Flowchart Using Labeled Blocks 192
- F–3 Flowchart Using Pictorial Symbols 193
- F–4 Common ISO Flowchart Symbols (with Annotations) 193
- F–6 Form (for a Medical Claim) 219
- F–7 Online Form with Typical Components 220
- G–3 Graphics for U.S. and Global Audiences 231
- G–4 International Organization for Standardization Symbols 231
- G–5 Double-line Graph (with Shading) 235
- G–6 Distorted and Distortion-Free Expressions of Data 236
- G–7 Bar Graph (Quantities of Different Items During a Fixed Period) 237
- G–8 Bar (Column) Graph (Showing the Parts That Make Up the Whole) 237
- G–9 Pie Graph (Showing Percentages of the Whole) 238
- G–10 Picture Graph 239
- H–2 Headings Used in a Document 243
- M–1 Map 320
- T–1 Elements of a Table 520
- V–1 Chart for Choosing Appropriate Visuals 554

Correspondence

- A–5 Acknowledgment 11
- A–6 Adjustment Letter (When Company Is at Fault) 17
- A–7 Partial Adjustment (Accompanying a Product) 18
- C–6 Complaint Letter 88
- C–8 A Poor Bad-News Message 106
- C–9 A Courteous Bad-News Message 107
- C–10 A Good-News Message 108
- C–11 Cover Message 111
- I–2 Inquiry 254
- I–3 Response to an Inquiry 255